Embryology of Flowering Plants

Terminology and Concepts

Volume 1

Generative Organs of Flower

EDITED BY T.B. BATYGINA

Science Publishers, Inc.

Enfield (NH), USA Plymouth, UK

CIP data will be provided on request.

SCIENCE PUBLISHERS, INC.
Post Office Box 699
Enfield, New Hampshire 03748
United States of America

Internet site: *http://www.scipub.net*

sales@scipub.net (marketing department)
editor@scipub.net (editorial department)
info@scipub.net (for all other enquiries)

ISBN 1-57808-188-2

Published by Science Publishers Inc., Enfield, NH, USA
Printed in India.

V.L. Komarov Botanical Institute

Russian Academy of Sciences

DEPARTMENT OF EMBRYOLOGY

AND REPRODUCTIVE BIOLOGY

RUSSIAN FOUNDATION

FOR BASIC RESEARCH

Contributors:

G.K. Alimova, T.B. Batygina, J. Bednara, N.N. Bhandari, F. Bouman, A.M. Bugara, B. Chaudhry, A.A. Chebotaru, P. Chitralekha, M.F. Danilova, G.El-Ghazaly, O. Erdelská, N.I. Gabarayeva, H.N. Gerassimova-Navashina, I.N. Golubovskaya, G.E. Gvaladze, D.A. Hamilton, O.P. Kamelina, G.A. Komar, V.A. Konnova, I.A. Korchagina, E.L. Kordyum, N.N. Kruglova, J.P. Mascarenhas, A.P. Melikyan, H.L. Mogensen, Z.I. Nikiticheva, V.F. Ogorodnikova, I.N. Orlova, V.P. Pechenitsyn, V.A. Poddubnaya-Arnoldi, O.B. Proskurina, B. Rodkiewicz, I.D. Romanov, S.D. Russell, R. Shah, I.I. Shamrov, S.V. Shevchenko, V.K. Simonenko, E.S. Teryokhin, V.N. Tikhomirov, G.E. Titova, V.E. Vasilyeva, M.R. Vijayaraghavan, M.T.M. Willemse, N.A. Zhinkina, G.Ya. Zhukova

Explain the meanings of words and you will deliver mankind from half its errors.

R. Descartes

This differentiation in terms is not a mere play on words, but is necessary for correct insight into the nature of things.

L. Van der Pijl

Preface

The principal aim of the present publication (in three volumes) is to summarize classical and current concepts of the generative organs of flowers, their structure and development, and seed formation processes. In preparing this publication, extensive use was made of the five-volume book *Comparative Embryology of Flowering Plants* (1981, 1983, 1985, 1987, 1990). The book contains ample material that can be utilized in theoretical generalizations, analyzing the distribution of features (or their uniqueness) and evolutionary transformations of structures. This offers vast possibilities for revising existing classifications and concepts and developing new ones.

While preparing the present publication, the contributors have drawn on the works of prominent botanists and plant embryologists, such as *Anatomie der Angiospermen-Samen* (1926) by F. Netolitzky, *Embryologie der Angiospermen. Archegoniaten* (1929) and *Vergleichende Embryologie der Angiospermen* (1931) by K. Schnarf, *An Introduction to the Embryology of Angiosperms* (1950) by P. Maheshwari, *Foundations of the Evolutionary Morphology of Angiosperms* (1964) by A. Takhtajan, *General Embryology of Angiosperms* (1964) and *Cytoembryology of Angiosperms: Principles and Perspectives* (1976) by V.A. Poddubnaya-Arnoldi, *Systematic Embryology of Angiosperms* (1966) by G.L. Davis, *The Seeds of Dicotyledons* (1976) by E. Corner, and *Embryology of Angiosperms* (1984) by B.M. Johri (ed.). P.A. Baranov's original work *The History of Plant Embryology* (1955) also proved particularly helpful. The author has summarized information on the main stages of development of embryological science and emphasized the outstanding contribution of Russian plant embryologists to world science. Formulations of individual terms and concepts were made with reference to the *Latin-Russian Dictionary for Botanists* (1957) by N.N. Zabinkova and M.E. Kirpichnikov, *Russian-Latin Dictionary of Botanical Terms* (1977) by M.E. Kirpichnikov and N.N. Zabinkova, and other dictionaries and manuals.

The present publication is a unique and unconventional endeavour to combine the terminological dictionary principle with a monographic description of individual structures and processes. Concomitantly, the traditional scope of embryology is expanded. The problems of seed dormancy and germination, ecological aspects of plant reproduction and other issues are also discussed.

The publication comprises three volumes. Volume 1 is divided into three parts: **Flower**, **Anther**, and **Ovule**. Volume 2, titled *The Seed*, comprises the following principal parts: **Double Fertilization, Endosperm, Perisperm, Embryo, Seed, Seed Dormancy** and **Germination**. Volume 3 is devoted to reproduction systems.

In contrast to the traditional dictionary structure, terminological entries are arranged on according to subject matter rather than in alphabetical order, to give a comprehensive idea of the processes occurring in the generative organs of the flower.

The text consists of a number of general, theoretical and terminological articles dealing with individual structures and processes. In each entry, the definition of the term is given, its semantics and historical background are clarified, and information on the genesis and functions of various embryonal structures and generally recognized classifications of generative structures and their development provided. For example, microsporangium wall development is treated in conformity with the accepted classification of G. Davis (1966); the principles and names in the description of types of embryo sac development follow the classification of I.D. Romanov (1971, 1981).

At the same time, a number of new or modified (refined) classifications and new theoretical concepts are profiled in the book, all of them based on up-to-date information; modern notions about evolutionary transformations of generative structures are also dealt with.

The occurrence and distribution of the structures described are illustrated by examples in flowering plants drawn from the above-mentioned 5-volume book *Comparative Embryology of Flowering Plants*.

Articles contributed by foreign authors appear without much editing. Also included in the book (albeit somewhat summarized) are papers by classical researchers which retain their topicality even today. These are articles by H.N. Gerassimova-Navashina (1951, 1954, 1958) on the evolution of male and female gametophytes, I.D. Romanov (1981) on the principles of embryo sac classification and by V.A. Poddubnaya-Arnoldi (1976) on megaspore tetrads.

The number of references had to be kept to a minimum because of space limitations, the references cited bearing, for the most part, upon the priority of discoveries and terms and the uniqueness of the feature discovered. No reference is made to the author(s) where a popular opinion is expressed or the feature described is widespread. Each article or entry is provided with illustrative material, such as drawings, micrographs (LM, TEM, SEM), or diagrams, collected in Plates at the back of the book.

Z.I. Nikiticheva

Acknowledgements

As editor of this volume, I express my deep gratitude to the team of contributing authors. I am especially thankful to colleagues from various parts of the world who kindly accepted the invitation to participate in the preparation of the manuscript for publication, namely N.N. Bhandari, B. Chaudhry, P. Chitralekha, R. Shah and M.R. Vijayaraghavan (India), F. Bouman and M.T.M. Willemse (The Netherlands), B. Rodkiewicz and J. Bednara (Poland), O. Erdelská (Slovakia), G. El-Ghazaly (Sweden), D.A. Hamilton, J. Mascarenhas, H.L. Mogensen and Scott D. Russell (USA); and also our colleagues from CIS countries: E.L. Kordyum, A.M. Bugara, and V.K. Simonenko (The Ukraine), and G.E. Gvaladze (Georgia).

My special thanks are due to fellow workers in the Department of Embryology and Reproductive Biology (V.L. Komarov Botanical Institute), especially Z.I. Nikiticheva, O.P. Kamelina, and G.K. Alimova for their active participation in the preparation of the present publication. I am also grateful to Publication Secretaries G.Ya. Zhukova and I.I. Shamrov; to N.A. Zhinkina and L.B. Pankova for beautiful illustrations; to L.M. Rotenfeld for excellent photography; and to G.E. Titova, E.A. Bragina, G.M. Anisimova, O.N. Voronova, N.P. Storozhuk, O.G. Butuzova, E.E. Evdokimova, A.A. Torshilova and A.A. Zakharova for their valuable technical assistance.

My deep gratitude to professor M.E. Kirpichnikov for his invaluable advice on terminology and to V.E. Vasilyeva, A.M. Zakharov and I.I. Shamrov for editing the book.

I am deeply grateful to professors E.S. Teryokhin and I.P. Ermakov, for their assistance in formatting this volume.

Articles written in Russian were translated into English by G.K. Lakhman, I.V. Putro, N.L. Kurchanov.

Financial support from the **Russian Foundation for Basic Research** in publishing the Russian version of Volume 1 is gratefully acknowledged.

Tatyana Batygina

Introduction

Plant embryology, as a science studying the mechanisms of origin and the pattern of early development of a plant organism, is now in its heyday due to the general progress in natural sciences. Plant embryologists are currently investigating not only the development of embryo and seed as a whole, but also complicated morphogenetic processes involved in asexual reproduction, as well as various abnormalities occurring during sexual reproduction. Instead of remaining a purely morphological branch of science, embryology is nowadays acquiring the features of morphophysiological and ecomorphological science. Morphogenetic and morphophysiological correlations determining normal development of the embryo and seedling cannot be established without investigating plant–environment interactions which change during ontogenesis.

The objective of modern embryology is control of the development of organisms. This aim can only be achieved provided there is a close collaboration between plant embryologists and experts in cytology, physiology, genetics, ecology and plant breeders. **Embryological evidence is becoming increasingly essential for theoretical and experimental studies on plant reproduction**. Exciting discoveries in the twentieth century, such as experimental haploidy, parasexual hybridization, propagation of plants via somatic cell culture etc., are all of great general biological significance. They cannot be used to full advantage, however, without further elaboration of the embryological aspects of these phenomena. Moreover, today plant embryology per se offers a number of fundamental discoveries pertaining to reproduction which are of great practical importance, e.g. cytoplasmic male sterility (CMS), apomixis, etc.

At the junction of embryology, anthecology, carpology, genetics, physiology and plant breeding, a new and rapidly evolving discipline, **reproductive biology**, has emerged. **Elaboration of theoretical foundations of sexual and asexual reproduction** is one of the objectives of reproductive biology. Reproductive biology involves studies of various interrelated stages of ontogenesis: flower organogenesis, anthesis, pollination, fertilization, embryogenesis, seed maturation, dispersal and germination, and propagation by seeds. All of these are in one way or another connected with embryology **in the broad sense**.

The **primary objective** is to tackle the problems of amphimixis and apomixis, to elucidate the mechanisms of embryogenesis, incompatibility and self-incompatibility, as well as the patterns of variation in potential and actual seed productivity and, consequently, in crop yield.

The development of reproductive biology is especially important for the introduction and repatriation of rare and vanishing plants, as well as of plants of great agricultural importance. Cytoembryological studies concerned with the effect of pollutants on reproductive structures (anthecology). constitute part of the environmental protection programme. Investigations of this kind have become increasingly important in recent years (the age of ecological stresses), because it is the reproductive structures that suffer most from adverse environmental factors.

Of special interest are apomixis phenomena requiring full-scale studies. Parthenogenesis, apospory, diplospory and polyembryony call for further intensive investigations to meet practical needs. Such types of agamospermy as vivipary and adventive embryony, i.e., possible intermediate forms between vegetative reproduction and propagation by seeds, merit special attention from plant embryologists.

Elucidation of evolutionary patterns of ontogenesis in conjunction with the problems of phylogeny and systematics remains a major line of inquiry in modern embryology. Relatively conservative, embryological features can be employed to supplement and refine our present notion of the systematic position of some orders, families, genera and even species. One urgent task is to investigate poorly studied taxa. This is essential for refining the criteria of evolutionary and taxonomic significance of embryological features.

New data accumulated by descriptive and experimental biology, morphology, reproductive and developmental biology should stimulate studies in evolutionary embryology. An important contribution is also forthcoming from some recently published monographs, particularly those concerned with various aspects of **descriptive** and **comparative embryology** (Poddubnaya-Arnoldi, 1976, 1982; Johri (ed.), 1984; Johri *et al.*, 1992; *Comparative Embryology of Flowering Plants*, 1981-1990; and others), as well as from a more comprehensive analysis of works of the founders of plant embryology, namely W. Hofmeister, A.S. Famintzyn, A.T. Bolotov, R. Souèges, K. Schnarf, J. Goroschankin, N. Zheleznov, W.I. Belyaev, S.G. Nawaschin, D.A. Johansen, P. Maheshwari, H.N. Gerassimova-Navashina, V.A. Poddubnaya-Arnoldi, I.D. Romanov, M.S. Yakovlev and others.

In the 5-volume work *Comparative Embryology of Flowering Plants*, families are characterized according to the A. Takhtajan (1980) system. It was compiled by embryologists of the Komarov Botanical Institute (Russian Academy of Sciences, St. Petersburg) and summarizes the results of research in descriptive embryology, from the beginning of the nineteenth century to the present day. The publication represents a wide panorama of development and differentiation of embryonal structures in dicotyledons (Volumes 1 to 4) and monocotyledons (Volume 5) covering a full range of forms found in these plants. The compendium is intended as a manual and guide for continuing work in reproductive biology and contains summarized evidence on sexual and asexual reproduction (including apomixis) in economic, rare and vanishing plants.

While constituting a basis for in-depth studies in various fields of embryology of seed plants, the aforesaid work concomitantly demonstrates the need for refining existing classifications of embryonal structures and processes in flowering plants and for their typification. In recent years, a number of new classifications have been worked out by Russian scientists: that of the modes and types of reproduction in angiosperms; types of embryo sac and embryo development; and anther, tapetum, and apomixis. A unique concept of the reproductive system of flowering plants has been developed and a new type of asexual reproduction—embryoidogenic—recognized.

As new evidence accumulates, classifications exhibit increasingly higher objectivity and better approximation to actual evolutionary processes. This encourages deeper involvement of embryologists in research on systematics and phylogeny of flowering plants. However, theoretical foundations of classification of embryonal structures have not been sufficiently elaborated. The importance of these classifications for systematics and phylogeny can be increased through progress in our understanding of the evolutionary processes occurring in plants, the forces (ecological, genetic, etc.) behind this evolution, and the patterns of evolutionary changes in plant ontogenesis. This suggests that embryonal structures, as entities most comprehensively embodying the peculiarities of the evolution of ontogeny, deserve special attention, and that all these problems and new classifications need to be discussed at scientific conferences and symposia, with the participation of both embryologists and taxonomists.

Within the comparatively short period of its development, plant embryology has become a ramification of biology.

In evolutionary embryology, a new approach to the study of evolutionary adaptation of plant organisms—**an ecomorphological approach**—has emerged. Studies with parasitic plants have revealed that underlying the evolutionary transformations of their reproductive structures are such general biological phenomena as metamorphosis, neoteny and reduction. It is from this point of view that hypotheses on the possible mechanisms of evolution of the monocotyledonous condition of the embryo have been considered in the context of existing heterochrony and syncotyly hypotheses.

Ecological embryology is the most important line of investigation of early ontogenesis directed towards identifying critical periods in it. Plasticity and tolerance of reproductive systems are also the subject matter of ecological embryology. In this connection it is important to study embryonal structures (ovules, embryos, etc.) with a view to assessing the potentialities and sustainability of biological systems.

This line of inquiry merges with **population embryology** concerned with morphogenetic and phenotypic variability within a population (life cycle variation and reproductive system diversity).

The immunological direction in embryology is expected to play an important role in overcoming the barriers of cross-incompatibility.

Recent trends incorporate **synthesis of embryological and genetic evidence**, i.e., **genetic embryology**. N.K. Koltsov (1935) had already pointed out that 'it is only the integration of the two disciplines [experimental embryology and genetics—T.B.] with one another, as well as with cytology and biochemistry, that will create a single science capable of solving problems of general biology'. All the problems related to morphogenesis—differentiation, specialization, evaluation of characters, definition of such concepts as 'gene and character', 'genotype and phenotype'—are to a greater or lesser extent the concern of embryology and genetics.

Recent years have witnessed a rapid development of **experimental embryology** which, unlike descriptive embryology, provides answers not only to the question of how complex morphological processes of embryonal development evolve, but also why specific processes show this or that particular trend.

As far back as 1950, the outstanding plant embryologist P. Maheshwari predicted that the future of embryology would undoubtedly lie with experimental embryology. The needs of practical breeding (e.g. uncovering the causes of sterility and partially filled ear, with subsequent correction; use of adventive embryos; etc.) have spurred the development of experimental plant embryology.

Plants regenerated from cultured cells, tissues, organs and embryos (embryoculture) have become, especially in recent years, a model for studying mechanisms of differentiation and molecular genetic laws of morphogenesis of embryonal structures. Some interesting results have been obtained on the effects of various abiotic factors on plant morphogenesis and possible means of formation of embryoids and other structures suggested. Success of an embryoculture depends on the correct choice of developmental stage of the generative structure to be used in the in-vitro culture as well as on the state of donor plants. Of vital importance is the **elaboration of a theoretical basis for the in-vitro cultivation** of vegetative and generative structures (ovary, anther, ovule, embryo, endosperm, embryo sac, sperm, etc.). The experimental approach, combined with thorough knowledge of the genesis of reproductive structures, allows one to predict the morphogenetic pathways of regenerants, to determine the optimal developmental stage of the embryo to be used for embryoculture, and so forth. This direction merges with cell engineering and biotechnology.

Embryological evidence becomes increasingly important in designing genetic and breeding programmes and developing effective biotechnology methods since it provides an insight into the processes leading to normal seed formation and uncovers the causes of abnormalities arising during embryogenesis.

A systems approach to generative structures used in the in-vitro culture elaborated by embryologists of the Komarov Botanical Institute has enabled some theoretical assumptions to be made, in particular concerning the autonomy of the embryo, critical periods in the development of generative structures, and so on. These principles are applied in breeding work and have resulted in the production of new plant forms.

It has become increasingly apparent that **applied embryology** holds much promise. Among the problems to be addressed, the following should be mentioned.

Ecological problems: development of special methods for producing normal seeds of rare and economic plants; conservation of plant genetic resources, i.e., establishing a genebank (seeds, embryos, gametes); estimating the degree of plasticity and tolerance of reproductive systems, and the proportions of different modes of reproduction within various taxa of plants exposed to adverse factors.

Biotechnology problems: development of effective methods for mass production (propagation) of new forms and varieties as well as of rare and vanishing plant species. This can be done on the basis of embryoidogenesis (somatic embryogenesis) and gemmorhizogenesis (organogenesis).

Agricultural problems: increasing actual production capacity; development of new forms and varieties via distant and parasexual hybridization; overcoming the barriers of cross-incompatibility; etc.

Progress in embryology is largely dependent on application of modern methods of electron microscopy, autoradiography, tracer labelling, modelling, microsurgery, etc. These methods open up new vistas for research at different levels of organization: molecular, cellular, tissue, organ, organismal, and population.

Unique data on fine structure and functions of generative organs and tissues have been obtained with the aid of such methods as fluorescence (including immunofluorescence), phase-contrast, electron (TEM, SEM), Nomarski-interference microscopy, as well as application of methods of modern cell biology: immunocytochemistry, time-lapse photography, video-image processing, etc. Some other findings are also worthy of mention: sperm dimorphism, feasibility of enzymatic isolation of such structures as generative cell and sperms, embryo sac, as well as of individual cells of the female gametophyte. Utilization of endoenzyme markers has enabled elucidation of specific features of meiosis.

Quantitative mRNA studies of pollen grains and sperms have provided new evidence on their fine structure, mRNA as well as specific genes controlling microsporogenesis were identified in mature anthers.

Major advances in embryology made in recent decades call for a new comprehensive survey, such as that offered by the present book. The system approach was adopted for treating the genesis of particular embryonal structures; reproduction of flowering plants is viewed from a modern perspective.

It is worth mentioning that recent international congresses and symposia dedicated to plant embryology (Italy, 1988; USSR, 1990; USA, 1992; Austria, 1994) have reaffirmed the need for a publication such as this. Not only plant embryologists, but also geneticists, physiologists, cytologists, and biotechnology specialists participated in the symposia. Researchers in various fields of biology are very much in need of generalized embryological data and in particular unified terminology.

All the foregoing prompted preparation of the present 3-volume publication, *Embryology of Flowering Plants: Terminology and Concepts*. Participating in the preparation of volume 1 have been plant embryologists and morphologists of the Komarov Botanical Institute (Russian Academy of Sciences), their colleagues from CIS (the Ukraine and Georgia) and from other countries of the world: the USA, the Netherlands, Poland, Slovakia, India, and Sweden. In this publication, the results of studies, the hypotheses and conceptions advanced by Russian embryologists are presented more comprehensively than in any of the latest monographs of plant embryology (Johri [ed.], 1984; Johri et al., 1992).

It is hardly possible for a large team of authors to maintain uniformity of style. It was the Editor's task, therefore, to impose uniformity and to avoid unnecessary repetition of material in different sections. Individual articles, along with an objective presentation of factual evidence, unavoidably reflect the author's own view of the structures and processes described. This makes some of the articles open to discussion. But disputes are essential for the progress of science. In this respect, the present publication will be useful to research workers, experts in plant embryology, anatomy, taxonomy, genetics, physiology, cytology and ecology; to specialists in molecular biology, biotechnology, plant growing and to plant breeders; as well as to college and university students and lecturers of biology and to all those interested in biology. This 3-volume work on plant embryology should hopefully demonstrate the great role played by modern embryology in solving problems of general biology in this age of ecological stresses. It ought also to further theoretical and experimental studies in plant embryology. Any criticisms and useful suggestions are welcome and will be taken into account in future work.

Contents

MICROSPOROGENESIS

MICROSPORE

POLLEN GRAIN

PART THREE—OVULE

Overview

MEGASPORANGIUM

INTEGUMENTS

OTHER SPECIAL STRUCTURES OF THE OVULE

EMBRYO SAC

TYPES OF EMBRYO SAC DEVELOPMENT

TRANSPORT OF SUBSTANCES IN THE OVULE

PART ONE—FLOWER

OVERVIEW

FLOWER—the most important reproductive structure of angiosperms (Plate 1). Processes such as micro- and megasporogenesis, pollination and fertilization, embryo development and lastly seed and fruit formation take place in it. The flower is most frequently regarded as a metamorphosed, monaxonic, shortened shoot exhibiting limited growth and performing a reproductive function.

Flowers vary widely in structure, colour and size, ranging from a few millimetres to one metre and more in diameter, as in *Rafflesia*.

The flower crowns the stem, the main or a lateral one. The portion of the stem below the flower, normally leafless, is termed the **pedicel**. Pedicels vary in size in different plants and play an important role in the formation of various types of inflorescences (umbels, corymbs, racemes, etc.). Sometimes, the pedicel is shortened, hardly distinguishable or even lacking; in the latter case, the flowers are said to be sessile.

The pedicel gradually turns into the floral axis, or **floral receptacle**. The receptacle is usually broader than the pedicel and shows poorly distinguishable internodes. The receptacle nodes carry all flower organs, both sterile (**sepals** and **petals**) and fertile (**stamens** and **carpels**). The receptacle may vary in shape—oblong, flattened, convex, etc. In some angiosperms, the receptacle varies greatly in size and coloration during fruit maturation and serves as an attractant for biotic agents assisting in fruit dispersal (as in strawberry, brier, etc.).

Sepals and petals together form the **perianth**. Sepals normally serve to protect the inner flower parts, especially prior to flower opening (i.e., within the flower bud). When green, they also act as accessory photosynthetic organs. Occasionally, sepals become petaliform and function as petals, thus attracting vector organisms effecting cross-pollination (as in *Delphinium*, wolfsbane and trollflower belonging to the Ranunculaceae). Sepals may also serve to protect the developing fruits, be involved in formation of fruit structure and assist in fruit dispersal. As biologically polyfunctional structures, sepals are subject to various changes. They are of leaf origin and appear to have originated from top vegetative leaves. Transitions from top leaves to sepals are prominent in members of some families (e.g. Paeoniaceae, Calycanthaceae, Dilleniaceae, etc.).

Petals forming the **corolla** usually perform a function over and above those of sepals. At early stages of flower development, they also serve to protect the inner floral organs but later, in an open flower, their primary function is to attract pollinators and to ensure successful cross-pollination. The origin of petals is not the same in different groups of flowering plants. In most cases, petals are of staminal origin (Ranunculaceae, Papaveraceae, Nymphaeaceae, Rosaceae, and many others). However, in a number of families, mostly archaic ones, such as Magnoliaceae, Paeoniaceae, Winteraceae and Schisandraceae, petals, like sepals, are of leaf origin. Therefore, it is common practice now to distinguish andropetals (petals originating from stamens) and bracteopetals (those originating from bracts).

The corolla size, shape and coloration vary widely and are associated with the biology of pollination. Petals often fuse at the margins to form a sympetalous corolla. Such a corolla evolves from a choripetalous corolla independently in different evolutionary lines of angiosperms. When the perianth is differentiated into calyx and corolla, it is called a double one. In cases where petals are either absent or there is no marked difference between sepals and petals, the perianth is termed simple. A simple perianth is said to be corolla-like if its lobes are showy, or calyciform if its coloration is unattractive, exhibiting various shades of green.

From the perianth inwards, stamens are arranged in cycles, and the carpels are located in the centre of the flower. The collection of stamens carried by a flower is referred to as the **androecium** and that of carpels as the **gynoecium**.

Flower organs are initiated on the floral receptacle in the form of meristematic primordia, normally in a strictly specified sequence: from the periphery to the centre (centripetally), i.e., sepal primordia are formed first, then petal, stamen and, lastly carpel primordia. But not infrequently, the sequence is partially altered. Moreover, members of some taxa (including large ones) exhibit a centrifugal sequence of initiation of the androecium parts (when polyandrous), i.e., from the centre to the periphery.

The spiral or acyclic arrangement of flower parts on the receptacle appears to have been the initial one in the evolution of angiosperms. In living angiosperms, circular or cyclic flowers predominate, with flower parts forming distinct cycles, whorls or circles (e.g. 5 petals in one circle, 5 petals in the next one, 10 stamens in two circles, and a pistil composed of 5 carpels). Among the more common types are pentacyclic, tetracyclic (especially where one cycle of stamens is lacking) and tricyclic flowers. However, a reduction or an increase in the cycle number (oligotaxy and pleiotaxy, respectively) may occur. Finally, there are hemicyclic flowers whose outer phyllomes (the entire perianth or the calyx alone) are arranged in cycles, and the inner ones (stamens and pistils) in a spiral (i.e., acyclically).

In cyclic flowers, the rule of cycle alternation is normally observed, i.e., the members of each subsequent cycle alternate with those of the adjacent cycles rather than lying opposite them. Occasionally, this rule is violated. Thus, if a flower has two cycles of stamens (i.e., is diplostemonous), then the outer stamens may lie opposite the petals rather than sepals, a condition termed obdiplostemony.

Flowers may be **unisexual** (with androecium and gynoecium) or **bisexual** (with either androecium or gynoecium alone). Moreover, bisexual flowers may occur on a single plant, as in oak, birch and maize, and the plant then termed monoecious, or occur on different plants, as in poplar, sea buckthorn and hemp, the plant then termed dioecious. When both unisexual and bisexual flowers occur on a single plant, as in many members of Compositae and Cucurbitaceae, the plant is said to be polygamous. It is currently believed that bisexual flowers have evolved from unisexual ones and that dioecy in them is a more recent phenomenon than monoecy.

A very important feature of the flower is its symmetry, primarily determined by the arrangement of perianth segments. If a flower possesses radial symmetry, i.e., can be bisected vertically by any number of planes to form two identical halves, it is termed actinomorphic (regular). Most angiosperms have actinomorphic flowers.

In the course of specialization, associated primarily with the biology of pollination, actinomorphic flowers evolved into zygomorphic (irregular) ones in many angiosperms. Unlike actinomorphic flowers, zygomorphic ones possess only one plane of symmetry, i.e., division into two identical parts (one-half being the mirror-image of the other) can only be made along one longitudinal plane, as in violet or pea. In addition to regular (i.e., actinomorphic) and irregular (zygomorphic) flowers, there also exist asymmetrical flowers which cannot be bisected along the plane of symmetry to produce two identical (symmetrical) halves, e.g. canna.

To facilitate studies of the flower structure, starting in the middle of the XIXth century, scientists began using special diagrams and formulas. The most precise method of constructing a floral diagram is, for all practical purposes, performing a section of a mature flower together with the cataphyllary leaf and floral bracts, if any, and accurate representation of the resultant picture. The diagram offers a means of determining the number of flower parts, their interrelationships, relative positions (arrangement), type of flower symmetry, etc.

A more concise, but no less illustrative way of providing a general view of the flower structure is by means of a formula (flower formula). Using the accepted conventional symbols and numbers, the flower formula may also incorporate such features as type of symmetry, acyclic or cyclic arrangement of flower parts, number of components, their fusion, increase in number due to separation, position of the ovary, etc.

Emergence during the evolutionary history of plants of a structure such as the flower was an aromorphosis event of great biological significance. In angiosperms the flower provided an opportunity for involving numerous biotic and abiotic agents (vectors) in the cross-pollination process, thereby offering them considerable advantages over other groups of higher plants.

With respect to the origin of the flower, there are several prevailing views. Most commonly, the flower is regarded as a metamorphosed, shortened shoot exhibiting limited growth and adapted to perform specific functions (the euanthous or strobilar theory). In doing so, the flower is assumed to have originated from the strobilus (cone) of a specific group of gymnosperms, in particular the Bennettitaceae and Cycadaceae. However, it is currently believed that both the flower of angiosperms and strobili of gymnosperms share a common origin in the reproductive shoots of the then strobilus-free seed ferns (Pteridospermae). According to other views, the flower evolved from a set of shoots rather than from a simple shoot and hence is more likely to be homologous to the inflorescence ('pseudanthium'), although specific morphological solutions here may vary widely (pseudanthous theory). There is also the telome theory treating the flower as a modification of telomes of the earliest higher plants. Finally, some authors believe that the enormous polymorphism of angiospermous flowers makes it impossible to trace them back to a single, original prototype and, consequently, flowers in different taxa are not necessarily strictly homologous.

Be that as it may, in practice the **flower** is, in most cases, most conveniently likened to a **simple, short shoot** having an **axial part** (floral receptacle or torus) and **leaf-shaped phyllomes**, both **sterile** (perianth) and **fertile** (stamens and pistils).

Of course, estimates of the degree to which morphological features of flowers of angiosperms are primitive or, conversely, evolutionarily advanced depend on the existing views of the origin of this group of plants. Based on prevailing concepts, most workers regard the following features as primitive: a convex or conical receptacle, acyclic arrangement of flower parts, multiplicity and indeterminacy (non-fixiform condition) of its parts, actinomorphic condition, absence of fused segments, phylloid stamens without a well-differentiated filament and usually with a conspicuous supraconnective, apocarpous gynoecium, semiclosed carpels, peripterous stigmas, etc. On the other hand, alternative features—flat or point receptacle, fixed number of cyclically arranged segments, sympetalous condition, coenocarpy, etc.—are considered evolutionarily advanced features.

ANDROECIUM (Greek: *aner*, gen. *andros-man* and *oikos* = home)—a collection of stamens. The term was proposed by De Candolle in 1827.

Stamens in a flower are arranged in a strictly specified fashion. Moreover, in members of comparatively more archaic families, they are spirally arranged and usually numerous, their number indefinite. However, in most angiosperms stamens are arranged in cycles or whorls, their number strictly definite and usually not large.

Stamens may fuse, such that either filaments (as in Leguminosae) or anthers (as in Compositae) become united. Fusion rarely proceeds so far as to obliterate the outlines of anthers (as in some members of Cucurbitaceae). Stamens may also fuse, over a greater or smaller part of their length, to the perianth, especially to the corolla. They also tend to fuse to the style (e.g. gynostemium in Orchidaceae). All types of stamen fusion have evolved due to various modes of pollination and are generally, extremely permanent.

As mentioned earlier, in the course of evolution of angiosperms, the number of stamens in flowers reduced, becoming strictly definite. However, in some evolutionary lines the number of stamens again increased, which is typical for example of Cactaceae and some members of Rosaceae. This increase in number of stamens is associated with the biology of pollination and is normally characteristic of flowers lacking nectar and in which pollen grains serve as an attractant for biotic agents effecting cross-pollination.

STAMEN—a generative organ of the flower. It is a microsporophyll that has been transformed in the course of evolution (Plates 2 and 3).

In most flowering plants, the stamen consists of a proximal sterile part, the **filament**, and a distal fertile part, the **anther**. Filaments vary in shape, size and longevity. Typically, this is a provisional organ but occasionally, having become ligneous, stamens attached to the fruit persist (as in *Kingia*). Appendages of various origin performing diverse functions are sometimes formed on the filament. Most commonly, the occurrence of these adaptations is associated with a particular mode of pollination. The stamen of most species of flowering plants has a vascular bundle. Normally, the bundle extends along the whole length of the stamen, terminating in the connective tissue or at the base of the anther. In some plants, the staminal vascular bundle is partially or completely reduced.

The anther is composed of one or several **microsporangia**, in whose locules **pollen** is formed. The **anther connective** is a sterile tissue linking the microsporangia in a structural unity, the anther. The connective may project above the microsporangia, forming appendages varying in shape and size. In most species of flowering plants (95%), the connective has a single vascular bundle; in cereals, bundle sheath cells differentiate round the vascular bundle (Batygina, 1974b). The anther shape is largely determined by the connective shape, size and degree of differentiation.

The filament, the connective and its distal appendages are microsporophyll parts. Therefore, it is not quite correct morphologically to say that the anther is 'attached' to the stamen. Based on the nature of the connection between the anther and the filament, several types of stamens (ingrown, versatile, etc.) have been distinguished which are viewed as taxonomic features.

In some primitive members of Magnoliaceae, Degeneriaceae and others, the microsporangia differentiate and develop inside a lamellar, trinervate microsporophyll. In these species (e.g. *Degeneria vitiensis*), the stamen is not differentiated into filament and anther. In some species (e.g. members of genus *Juglans*), the filament has undergone in the course of evolution, a reduction resulting in sessile stamens. However, sessile stamens may be a result of the effects of a number of other factors.

The number of microsporangia and their position within the microsporophyll are not the same in different taxa. Changes in location of microsporangia are partially traceable to their genesis, but generally these changes are difficult to identify since they are a result of long-term evolutionary transformations (adaptation to different modes of pollen dispersal). Two types of disposition of microsporangia have been distinguished: **abaxial** and **adaxial**. In addition, the disposition of microsporangia within an anther is also determined by the mode of anther dehiscence: **extrorse, introrse, lartrorse** and variations of these. All these features are taken into consideration by taxonomists and are essential for understanding the evolution of microsporangia and microsporophyll in general.

In most flowering plants, the anther is **tetrasporangiate**, i.e., carries four microsporangia arranged in pairs. However, many members of Adoxaceae, Malvaceae, Phyllidraceae and others have **bisporangiate** anthers. There are families (Araceae, Asclepiadaceae, Asteraceae, Moringaceae, etc.) in which both bisporangiate and tetrasporangiate anthers have been found in a single flower. Anthers carrying eight and more microsporangia also exist.

The **anther** or half of it is referred to as the **synangium** (Eames, 1961), and stamens or anthers that have fused to form a single structure, as the **synandrium** (Takhtajan, 1980b).

The chamber in which pollen is formed is described by various terms: microsporangium, anther locule, pollen sac, pollen chamber, and theca. Morphologically, most of these terms are inaccurate, because they fail to describe the structure adequately. In some species, in the course of maturation of both tetra- and bisporangiate anthers, the number of microsporangia 'decreases', as it were, to a single one because they 'fuse' as a result of degeneration of walls (septa) between the adjacent microsporangia. Such a reduction in the microsporangium number may be due to other factors, including abortion. Regrettably, Davis (1966) attaches diagnostic value to these secondary modifications and erroneously regards tetrasporangiate anthers with disrupted septa as bilocular, and bisporangiate ones as unilocular. Therefore, the term 'unilocular anther', for example, may have quite different morphological meanings. In view of this, the term 'anther locule' loosely equated by many workers to the term 'microsporangium' should not be used. The same applies to the term 'theca' which, Eames (1961) believes, should be avoided for the same reason. Suffice it to say that in some members of Cucurbitaceae and in genus *Callitriche*, all four microsporangia are united in a single 'theca'.

In mature lartrorse anthers of some plant species, microsporangia fail to unite since they occur in isolated positions in the sporophyll corners, such that the 'theca' represents a single sporangium and the anther is a 'tetrathecal' one (Eames, 1961). The term 'theca' is common in Russian literature and is used to denote two lateral microsporangia (dorsal and ventral) of a pair within a tetrasporangiate anther.

In some members of Nymphaeaceae, Calycanthaceae and Magnoliaceae, a proportion of stamens become sterile and are then termed **staminodia** (Latin *stamen* and Greek *oidos* = kind). Staminodia vary in shape and size, often representing intermediate forms between the stamen and the petal.

Reduced stamens, occasionally found in pistillate flowers, are also referred to as staminodia. In some plant species they function as nectaries (as in *Coptis*). Staminodia have a vascular supply characteristic of stamens, but in the case of reduced stamens (as in some members of Scrophulariaceae), the vascular bundle is either completely lacking in the stamens or no more than its vestiges are present in the floral receptacle.

In the course of evolution of certain groups of plants, some parts of the stamen—filament, connective—as well as the number of microsporangia, number of microsporangium wall layers and number of pollen grains have undergone varying degrees of reduction. Thus, for instance, in lamellar sporophylls the connective tissue is usually massive whereas in such a specialized family as Poaceae it is represented by no more than a group of cells. A simultaneous reduction may occur in both the number of microsporangia and the distal appendage of the connective; in the parasitic plant *Arceuthobium* (Viscaceae), a single kidney-shaped microsporangium persists in the anther.

Gynoecium (Latin *gyne* = woman and Greek *oikos* = house, abode)—a combination of carpels carried by an angiospermous flower (Plate 4). The flower can possess one or more carpels which may be free, i.e., not fused (**apocarpous gynoecium**; Greek *apo* = one and *karpos* = fruit), or united in a structural unity (**syncarpous gynoecium**; Greek *syn* = together).

With respect to the gynoecium, the term '**pistil**', strongly engrained in the literature, has also been applied. It refers either to each carpel in an apocarpous gynoecium (simple pistil) or to the entire syncarpous gynoecium (compound pistil). Some botanists (e.g. Eames, 1961) suggest that the term 'pistil' be abandoned; however, the term, as defined above, continues to be useful even today.

Carpel—a generative organ of the flower (Plate 4). It is considered by many authors to be homologous to the megasporophyll. The carpel is a phylloid structure which, in an **apocarpous gynoecium,** is conduplicate, i.e., folded along the midrib such that its adaxial surface occurs inside the chamber and its margins come more or less closer to each other, draw together or fuse. The mode of carpel folding is regarded as a taxonomic feature (Eames, 1961). In many taxa, during pistil formation the straight, non-involute margins of the carpels are folded to form a marginal zone. This zone seems to have undergone reduction in width. In another type of apocarpous gynoecium, the carpel margins first become involute and only later do they fold and fuse by their abaxial surfaces, such that the marginal vascular bundles of carpels are inverted. At maturity, the carpel with a reduced marginal zone is often hard to distinguish from a carpel with involute margins.

The fusion of carpels in a **syncarpous gynoecium** occurs in one of three ways: adjacent carpels, closed or united after the first pattern (with non-involute margins) or the second one (with involute margins) fuse by their flattened, abaxial sides with one another. Occasionally they additionally fuse to the styliform carpophore, resulting in a **coenocarpous** (Greek *koinos* = common) bilocular or multilocular gynoecium. Where the partitioning septa disintegrate, a unilocular gynoecium composed of two to several carpels, the so-called **lysicarpous gynoecium** (Latin *lysis* = disintegration, dissolution; Greek *lysos* = dissolution, degradation) is formed. In a syncarpous gynoecium of the third type, initially or subsequently, non-conduplicate carpels, completely or partially unfolded, fuse with one another at their margins, resulting in a unilocular, so-called **paracarpous gynoecium** (Latin *parallelicus* = parallel). These principal methods of fusion of carpels may vary due to secondary modifications. An example of such a modification is to be found in members of Brassicaceae, Boraginaceae and Lamiaceae where the locules of the gynoecium composed of two fused carpels are divided in two by a pseudoseptum formed by placental outgrowths at the margins of carpels.

In evolutionally more advanced taxa, the carpel of an apocarpous gynoecium differentiates into a lower, broadened fertile portion, the **ovary**, in whose chamber (locule) ovules are formed, and an upper, tapered, sterile portion, the **stylodium**. In the most primitive taxa, stylodia are lacking. Formation of a syncarpous gynoecium primarily involves fusion of fertile parts of carpels (ovaries) as well as of stylodia to form a **style**.

Along the carpel margins in the more primitive taxa and on stylodia or the style in more advanced ones, a kind of secretory tissue differentiates which secretes mucilage and sugary fluid that serve to entrap pollen and aid in its subsequent germination. These carpel parts are termed **stigmata**. Depending on shape and position, the following types of stigma are distinguished: sutural ('stigmatic crest'), peripterous, apical (capitate, globular, oval, spherical, concave, cylindrical, plumose, lobate, penicillate, etc.). In some primitive taxa, a so-called 'inner stigma' is formed which sometimes completely lines the ovary chamber (Eames, 1961). When the style or stylodium beneath an apical stigma is reduced, the stigma is termed **sessile**.

PLACENTA (Greek *placus* = cake, scone)—the **site of ovule initiation on the carpel**. It may be a rather large outgrowth which occasionally almost completely fills the ovary lumen (Plate 4).

According to a number of authors, the most primitive type of placentation is the laminar (diffuse or surface) one which results in the ovules being scattered over the entire inner surface of the carpel, as in Butomaceae and Limnocharitaceae. Most flowering plants have submarginal or sutural placentae. In cases like these, each carpel carries two placentae running close to its margins. Sometimes a placenta is said to be sutural, marginal, but there is no general agreement among authors regarding this (Eames, 1961; Takhtajan, 1964).

The type and position of the placenta depend on the method of carpel fusion and type of gynoecium (Puri, 1951). Placentation is most commonly **submarginal** and **sutural** (subsutural) in an apocarpous gynoecium, **central-angular** (or central-sutural) in a coenocarpous, **columnar** (or central) in a lysicarpous, and **parietal** in a paracarpous gynoecium. In a unilocular ovary, the placenta may be located at the very base of the ovary, in which case placentation is termed **basal**. In other taxa, placental and chalazal tissues are linked in a structural unity, the so-called **placentochalaza** (for details, see Part III: Funiculus).

PART TWO—ANTHER

OVERVIEW

ANTHER (Latin *anthera,* from Greek *anthos* for flower)—the fertile part of the stamen where, in microsporangia, microsporogenesis occurs and pollen grain formation and maturation proceeds (Plate 5). The term was introduced by Linnaeus. The first morphological description of anther and pollen dates back to the late seventeenth century (Grew, 1672; Camerarius, 1694). In the early eighteenth century, Linnaeus defined all the parts of the stamen and anther. The terminology proposed by him has been universally accepted.

In anther development, three periods are distinguished: premeiotic, meiotic, and postmeiotic (Kamelina, 1981a). The premeiotic period is marked by an enhanced mitotic activity resulting in the formation of an anther wall and sporogenous tissue, that is, microsporangia are formed. It is completed with cessation of divisions in the sporogenous tissue and by transformation of sporogenous cells into microsporocytes. During the meiotic period, microsporangium wall layers differentiate and meiosis proceeds in microsporocytes, resulting in tetrads of haploid microspores. In the postmeiotic period, pollen grains develop and mature concurrent with the associated changes in the microsporangium wall structure.

In a mature anther, locular fluid is lacking, with only those wall layers persisting which are involved in anther dehiscence. Anther dehiscence is most commonly longitudinal, occasionally valvular (Berberidaceae), by pores (Ericaceae), or by other means. In many species, anther dehiscence is preceded by disruption of septum separating microsporangia within a theca, such that locules of two microsporangia unite and then open by a single pore. At the site of the thecal opening, some plant species (e.g. members of Liliaceae, Bignoniaceae and others) develop ridges of enlarged epidermal and endothecial cells (stomium) which are closed by a few small cells. These latter cells are ruptured to effect dehiscence. However, anthers without a stomium carrying small closing epidermal and endothecial cells are more common, with the cells forming no fibrous thickenings. In some species, microsporangia remain separated within the anther right up to the time of dehiscence (Dipsacaceae, Morinaceae, Valerianaceae and others).

Evolution of the anther of flowering plants was towards reduction in size of the anther connective and a lower number of vascular bundles, number of layers in the microsporangium wall as well as altered positions of microsporangia and changes in other features.

MICROSPORANGIUM (Greek *micros* = small, *sporá* = seed, spore, and *angeion* = conceptacle, reservoir)—the sporogenous portion of the anther where the development of microspores and subsequently of pollen grains occurs. Alternative terms: anther locule, pollen sac (Plate 5).

In a typical case, at early stages of anther development, the anther primordium is represented by a meristem enclosed in the epidermis. Later, four protuberances become conspicuous. In each of these, one or several subepidermal cells develop into archesporial cells. Successive periclinal divisions of an archesporial cell and its derivatives result in the formation of a microsporangium wall and a sporogenous tissue. On the anther connective side, the inner portion of the tapetum and, in some species, also the other layers identical to the microsporangium wall layers develop from the active meristem adjoining the sporogenous tissue.

Tetrasporangial anthers are characteristic of most flowering plants. In some species, however, the number of microsporangia per anther may be lower, namely one (Diapensiaceae), two (Moringaceae, Marantaceae and some members of Monimiaceae), and three (Daphniphyllaceae, along with four), or higher, i.e., eight (Zannichelliaceae, Cymodoceaceae, along with four) and even as many as fifty (Mimosaceae). Variation in microsporangium number may be due to various factors: reduction in part of the anther or its transformation into a sterile appendage (Cannaceae, Marantaceae), fusion of a pair of microsporangia at an early stage of anther development—a sporadic occurrence (Daphniphyllaceae, Nitrariaceae), stamen splitting (Adoxaceae), or anther primordium fusion at early stages, it may also be due to formation of additional septa within a microsporangium. In some families (Myristicaceae, Aristolochiaceae), the anthers unite to form synandria carrying numerous microsporangia.

The microsporangia carried by a single anther may be equal or different in size, with microsporangia located on the abaxial side of the anther being smaller than those on the adaxial side (Eupomatiaceae, Ceratophyllaceae, Berberidaceae, Nandinaceae). Microsporangia vary in shape from oval, to rounded to horseshoe-shaped (in species with the anther connective outgrowths projecting into the locule, as in Lamiaceae, Bignoniaceae, and some members of Campanulaceae). Microsporangia also vary in size. At early developmental stages, they may exhibit differential numbers of sporogenous tissue layers and different microsporocyte sizes, with the pollen grain number and size varying at later stages.

In microsporangia of a single anther, the processes of microsporogenesis and pollen grain development normally occur synchronously. But occasionally they are asynchronous. Thus in Peganaceae, all meiotic stages—from the interphase of meiosis I (at the base of the anther) to telophase (towards the anther tip)—can be observed to occur successively in one and the same microsporangium (Kapil and Ahluwalia, 1963). Asynchronous development is also observable in different microsporangia of a single anther, as in Bignoniaceae and Zygophyllaceae (Kamelina, 1985e; Kamelina *et al.*, 1990). In many plants, postmeiotic microsporangia are filled with locular fluid surrounding the microspores which disappears in the course of anther maturation and dehydration.

Epidermis (Greek *epi* = on, above, atop and *derma* = skin)—the ground tissue completely enclosing the anther. Synonym: **exothecium** (Plate 6).

At early stages of anther development, the epidermal cells of the microsporangium wall and anther connective are not dissimilar. They look like a meristem, their characteristic feature being that they only divide anticlinally. But as the wall layers become differentiated, these differences are reflected in altered sizes and shapes of cells and nuclei, presence of storage reserves, cytoplasm density and viability of the cells proper.

In the course of development, epidermal cells of the **microsporangium wall** undergo changes in size, shape, condition of organelles, and in other parameters. In most cases, the epidermis persists in a mature anther, but occasionally shrinks or peels off (Nitrariaceae, Tetradiclidaceae). In some plants, most commonly where the anther dehisces by pores or in some other way, rather than longitudinally, the wall of a mature anther is represented by the epidermis alone (Bignoniaceae). In specific cases, the epidermis develops fine fibrous thickenings (Empetraceae). The microsporangium wall epidermis is covered by a cuticle which varies in thickness (very thick in *Peganum harmala*) and shape (e.g. stellate, spinulate in Coriariaceae, comb-shaped in Morinaceae). In some species, the cuticle on the outer tangential wall arises very early, even prior to meiosis (Elaeagnaceae, Brexiaceae). In early developmental stages, epidermal cells often contain starch (Valerianaceae, Dipsacaceae), certain species accumulate tannins (Brexiaceae, Cyrillaceae, Myristicaceae), with calcium oxalate crystals (genera *Ixerba* and *Malacocarpus*) and chloroplasts (Zannichelliaceae, Ruppiaceae) also being present. In some species, the epidermis develops hairs (Calycanthaceae) and specific structures in the form of acute spicules aiding in the mechanics of anther dehiscence (*Incarvillea*), and trichomes (peltate glandules in *Stachys trinervis*, Lamiaceae). Occasionally, stomata occur in the epidermis (Geraniaceae).

Connective—the sterile part of the anther joining all the microsporangia together. It is essentially an extension of the filament.

Connectives may vary in structure, shape and in the number of tissue layers present in them. In some species, the connective is very massive (Magnoliaceae, Eupomatiaceae, Degeneriaceae, Trapaceae, Brexiaceae), in others (which are many) it is not bulky, being sometimes represented by no more than a few tissue layers (Tetradiclidaceae). In some aquatic plants, air pockets occur between the cells of the connective (Zannichelliaceae). Occasionally, the connective develops multicellular outgrowths—placentoids—which grow into each microsporangium (Eames, 1961), thus giving a horseshoe-like appearance to the connective (Lamiaceae, Verbenaceae, Bignoniaceae, some members of Campanulaceae). These outgrowths may also originate at the septum between the microsporangia (Lamiaceae, Verbenaceae). The anther connective cells accumulate starch (Nandinaceae, Bignoniaceae), tannins (Escalloniaceae, Brexiaceae, Marantaceae), and calcium oxalate crystals (Peganaceae). Within the anther connective of some plants, specific storage cells occur which accumulate various substances: fats (Pittosporaceae), mucilage of polysaccharide nature (Nitrariaceae), etc.

In some species, at later stages of anther development, the layer of the connective cells adjoining the locule undergoes differentiation which follows the pattern seen in the endothecial cells with characteristic fibrous thickenings in the cells (Datiscaceae, Parnassiaceae, Tamaricaceae, Escalloniaceae and others). Fibrous thickenings also occur in subepidermal cells of the connective (Daphniphyllaceae, Magnoliaceae, Cneoraceae, Peganaceae and others), in its placentoids (some members of Campanulaceae) and in the subjacent layers (Brexiaceae, Morinaceae). A vascular bundle(s) runs along the central part of the connective (they are numerous in *Ostrowskia magnifica*, Campanulaceae).

The anther connective is a polyfunctional structure which performs mechanical (joins together the two thecae of an anther) and storage functions and is involved in transport of nutrients from the filament and maternal plant to the microsporangia.

MICROSPORANGIUM WALL

Microsporangium Wall—structure comprising layers of microsporangium cells enclosing the sporogenous tissue (Plate 7). The origin and development of the

microsporangium wall is one of the most knotty questions in plant embryology. Warming (1873) was the first to note the **centrifugal** mode of the microsporangium wall development in angiosperms. Under this mode of development, the **tapetum** (if it originates from the parietal layer) is the **initial product** of the **parietal layer** development. By contrast, under the **centripetal** mode of the microsporangium wall development, the **tapetum** is an end product of the **parietal layer** differentiation (Carniel, 1961). In speaking of the centrifugal and centripetal modes of the parietal layer development, we mean a vector running from the centre of a microsporangium to its margin or, conversely, from the margin to its centre. Due to complex morphogenetic transformations in the anther, the microsporangium wall of most taxa exhibits a variety of modes of development (Teryokhin *et al.*, 1993). A distinction needs to be drawn between the **formed wall** of a microsporangium which shows a finite, taxon-specific number of layers, and the **wall of a mature microsporangium** at the moment of dehiscence. In the course of anther maturation, further differentiation and specialization of cells of all the microsporangium wall layers occurs, usually concurrent with meiosis and pollen formation. In most angiosperms, the formed microsporangium wall is composed of the **epidermis, endothecium, middle layer** and **tapetum**. It is thicker in primitive taxa than in more advanced ones. An increase in the microsporangium wall massiveness normally occurs due either to an increased number of middle layers (Magnoliaceae, Degeneriaceae, Ranunculaceae) or to a higher number of the tapetum layers (Magnoliaceae, Trochodendraceae, Schisandraceae and others). In some members of Rubiaceae, Oleaceae and other families, the tapetum is multilayered only on the connective side of the anther. In members of Magnoliaceae and Schisandraceae, the endothecium is non-uniformly bilayered one while in Scrophulariaceae (*Chelone glabra*) it is represented by two to six layers. However, a multilayered endothecium has also been reported in some primitive families (Bhandari, 1971; see also *Comparative Embryology of Flowering Plants*, 1981-1990). During the evolutionary history of many plant species, reductions have occurred both in number of the microsporangia per se (*Arceuthobium pusilum*) and in number of their different layers. In spite of the impressive progress in our understanding of the genesis of angiosperm microsporangia, there remain some disputable points related to the earlier stages of development and crucial for elucidation of the mechanism of microsporangium formation. As rightly pointed out by Bhandari (1984), the topicality of the problem stems from the scarcity of studies on early stages of microsporangium development. However, evidence summarized in *Comparative Embryology of Flowering Plants* (1981-1990) and findings of some Russian and other (FSU) plant embryologists (Batygina *et al.*, 1963; Sladkov and Grevtsova, 1988; Teryokhin *et al.*, 1993; and others) make it possible to raise and discuss a number of issues concerning the morphogenetic regulation of development of the anther as a plant (flower) organ. One such important issue is the origin of some layers of the microsporangium wall, especially on the anther connective side. The microsporangium wall was long thought to be formed only on the outside of the microsporangium (Warming, 1873; Schnarf, 1929, 1931; and others). That is why in subsequent studies of the microsporangium wall the primary emphasis was also on its formation on the outside of the microsporangium only, i.e., on the side opposite to the connective. Since the early 1960s, attention has been given to the formation of the microsporangium wall both on the outside of the microsporangium and on the connective side. Carniel (1961) was the first to notice that in *Zea mays* the microsporangium wall completely encloses the sporogenous tissue, **the tapetum is derived from the parietal layer**, but the origin of the latter did not seem to be the principal concern of the author. However, Batygina and co-workers (1963) and Batygina (1974b, 1987a) came to the conclusion that in *Triticum* the tapetum is also a derivative of the parietal layer but has a **dual origin**: **a portion of it** (on the connective side) is differentiated from cells of the **meristem**, and on the outside, the parietal layer cells **are derived from archesporial cells**. There are at least two ways in which the tapetum can be formed on the connective side of the microsporangium: either **from the anther meristem adjoining the connective**, or **from the meristem which is a component of the microsporangium itself.**

Periasamy and Swamy (1966) suggested a '**dual origin of the tapetum**' in microsporangia of flowering plants. In doing so, they implied that **on the connective side the tapetum** is **invariably** formed from the **meristem of the connective, and not from the parietal layer**. Some authors even now hold this view (Kamelina, 1981a,b,c,d, 1991). Later, this assumption was corroborated for a number of taxa of flowering plants by studies on the origin of the tapetum (Vijayaraghavan and Ratnaparkhi, 1973; Nikiticheva, 1987; *Comparative Embryology of Flowering Plants*, 1981-1990). Indeed, at least some of the taxa of flowering plants can be noted to exhibit dual origin of certain layers of the microsporangium wall or its initials, although morphogenetic pathways leading to its formation are not the same in members of different taxa. However, an **alternative pathway of tapetum formation**, i.e., **from parietal tissue**, as previously demonstrated for *Zea mays* and *Triticum* (Carniel, 1961; Batygina *et al.*, 1963) and confirmed for *Triticale* (Bhandari and Khosla, 1982) must not be ruled out. The authors have clearly demonstrated the tapetum in microsporangia of the aforesaid cereals to be formed **exclusively from the parietal tissue, such that it envelops the sporogenous tissue in a concentric layer**. For lack of detailed studies on the earliest stages of microsporangium development, the origin of the parietal layer and other microsporangium wall layers is still an open question. It suffices to mention that in specific cases with a multicelled archesporium, the possibility was demonstrated of the parietal layer (in all its parts) being formed from sporogenous tissue (Shreve, 1906; Holmgren, 1913; Batygina *et al.*, 1963). However, the evidence for the tapetum being differentiated from the sporogenous tissue (Maheshwari, 1950; Bhandari, 1971, 1984; Lakshmanan and Poornima, 1988) needs to be verified.

Thus, the problem of microsporangium wall genesis, important as it is for understanding the level, nature and mechanisms of regulation of microsporangium formation, cannot yet be considered completely solved.

Parietal Layer (Latin *paries* = wall)—a microsporangium wall layer originating by periclinal division of archesporial cells and giving rise to other layers. The term was coined by Nägeli (1842) while describing the tissues of a young anther; all the wall layers were termed parietal. Primary and secondary parietal layers are distinguished. The former arises from an archesporial cell(s) and its cells divide periclinally. Depending on the mode of microsporangium wall development, the following may be derived from the primary parietal layer: a tapetum and a secondary parietal layer (in the case of centrifugal type of development), an endothecium and a secondary parietal layer (centripetal type), or two secondary parietal layers (basic type) (Davis, 1966). Periclinal division(s) of cells of the secondary parietal layer(s) results in the formation of an endothecium and a middle layer (centrifugal type), a tapetum and a middle layer (centripetal type), or all the layers of the wall; of an endothecium, two middle layers and a tapetum (basic type). The cells of parietal layers (parietal cells) exhibit features of the formative tissue (meristem).

ENDOTHECIUM (Greek *endon* = inside and *theke* = conceptacle, reservoir)—an outer layer of the microsporangium wall located below the epidermis (Plate 8). The term was coined by Purkinje (1830) to refer to all the layers of the wall of a dehiscent anther lying below the epidermis. The endothecium is a derivative of either the primary or secondary parietal layer, resulting in differential timing of its differentiation: earlier in the former case and later in the latter. In a young anther, endothecial cells differ little from epidermal cells or from middle layer cells. Endothecial cells become significantly enlarged during the meiotic period in some plants and, more frequently, during the postmeiotic period in others. Starting at the first mitosis of pollen grains, the so-called **fibrous thickenings** (bands) are formed on anticlinal and inner tangential cell walls of the endothecium. These are mainly oriented at right angles to the epidermal layer. In some instances, no fibrous thickenings are formed, whereas radial and inner tangential walls of the envelope become greatly thickened (Nandinaceae), or all the envelopes with sieve pores become thickened (some members of Magnoliaceae). In individual cases, the endothecial cell envelopes do not become thickened, nor are fibrous bands formed in them (Empetraceae, Epacridaceae). This is also the case in some aquatic plants (Zannichelliaceae, Cabombaceae, Ceratophyllaceae) and cleistogamous ones.

At early developmental stages, endothecial cells show the presence of starch; occasionally they accumulate tannins (Elaeagnaceae, Ebenaceae, Brexiaceae). In most plants, the endothecium persists in the wall of a mature anther and is involved in anther dehiscence. In some cases (*Incarvillea*), an endothecium with fibrous thickenings persists only in the vicinity of the stomium.

Typically, the anther wall endothecium may be monolayered, occasionally irregularly bilayered (*Maranta*), or multilayered.

MIDDLE LAYER (takes its name from the fact that it is located between the tapetum and the endothecium; Coulter, 1898)—a derivative of the primary and/or secondary parietal layer. The number of middle layers within a microsporangium wall varies between one (in rare instances, entirely absent) and 5-7 (Nandinaceae, Marantaceae) (Plate 9). Middle layers increase in number due to additional periclinal divisions (often irregular ones) occurring in one or two middle layers already formed. At early stages, middle layer cells are not dissimilar to epidermal and endothecial cells and often show the presence of starch. During the meiotic period, flattening of middle layer cells and nuclei occurs followed by their degradation and lysis in the postmeiotic period or by partial enlargement and expansion of the cells after tapetum disintegration. In many cases wherein the middle layer is represented by a single layer, it is transitory and, with commencement of meiosis, gradually degenerates and disintegrates prior to the onset of lysis of the tapetum (Dipsacaceae, Valerianaceae). In other instances, it remains functional following tapetum lysis, up to the time of pollen grain maturity (Poaceae). Where numerous, middle layers persist for a long time, gradually disintegrating centripetally (Daphniphyllaceae, Nandinaceae). Where there are two to three middle layers in an anther, they occasionally show differential developmental patterns. The middle layer(s) adjoining the tapetum disintegrates, and that adjacent to the endothecium persists in the mature anther. Moreover, the surviving cells of the middle layer develop fibrous bands (thickenings) which, more often, differ in shape from those formed in the endothecium (Daphniphyllaceae, Datiscaceae, Morinaceae and others) but occasionally are similar to them (Escalloniaceae, Winteraceae). In some instances, the middle layer is retained in the wall of a mature anther but its envelopes show no fibrous thickenings (Actinidiaceae, Davidiaceae, Peganaceae). The two middle layers in anthers of *Beta vulgaris* differ in cell ultrastructure: the outer layer cells exhibit a more developed plastid apparatus (plastidome) and those of the inner layer a more developed endoplasmic reticulum (ER) as well as a higher ribosome density and a lower degree of vacuolation (Golubeva, 1989). The cells of the persisting middle layers sometimes accumulate tannins (Brexiaceae, Myrsinaceae). In Araceae, the middle layer adjoining the tapetum is involved in the formation of a periplasmodium (Pacini and Juniper, 1983). The numerous middle layers found in anthers of Marantaceae secrete a mucilage of polysaccharide nature into the locule (Kamelina, 1990a,b). Carotene and lipid bodies have been reported to occur within middle layers (Liliaceae) (Reznickova, 1984).

Thus the middle layer has multiple functions: storage (deposition of starch and other nutrients), transport of assimilates, secretion (including involvement in pollenkitt formation) and mechanical, concerned with opening of anther wall.

FIBROUS THICKENINGS—specific bulges of cell walls of various anther tissues (Plate 8). They occur in the endothecium (in most plants), in persisting middle layers of a mature anther (Datiscaceae, Morinaceae, Zygophyllaceae), in the anther connective subepidermis (Daphniphyllaceae, Magnoliaceae), in the connective outgrowths (placentoids) (*Ostrowskia*), and, very rarely, in the anther epidermis (Empetraceae). Fibrous thickenings are formed in differentiated cells, generally prior to or during mitosis I in pollen grains. Upon their formation, the inner tangential wall of the cell becomes considerably thickened (which is particularly the case with the endothecium), with fibrous bands extending outwards to radial walls and terminating near the outer tangential wall or, occasionally, spanning it. Fibrous thickenings exhibit the following chemical composition: alpha-cellulose (De Fossard, 1969), pectin and lignin. They have been observed to produce intense phosphorescence in polarized light.

Fibrous thickenings vary in band thickness, disposition and shape. Although Eames (1961) has recorded a multitude of structural variants, there is no comprehensive classification of fibrous thickening types. For some families, such as Araceae (French, 1985a,b, 1986), Asteraceae (Dormer, 1962; Vincent and Getliffe, 1988), Poaceae and Restionaceae (Manning and Linder, 1990) and Orchidaceae (Feudenstein, 1991), more detailed classifications have been proposed of the endothecial fibrous thickenings as they are thought to be an important taxonomic feature.

The primary function of fibrous thickenings is that of aiding in the mechanics of anther dehiscence. Upon cell dehydration, they (endothecial and middle layer bands) act as a spring forcing the walls of the anther thecae apart. Obviously, the fibrous bands of the anther connective and placentoid tissues function during anther dehiscence as a mechanism assisting in the expulsion of pollen grains from the anther.

TAPETUM (Greek *tapes* = lining, carpet)—a polyfunctional tissue immediately adjacent to the sporogenous tissue of the anther and ensuring effective meiosis, normal development of microspores and pollen grain maturation. Alternative term: lining (Plate 10).

The tapetum may be homomorphic or heteromorphic when its inner portion is multilayered or is represented by large papillate cells (Lamiaceae, Bignoniaceae and others). In some instances, the tapetum develops transient auxiliary partitions (septa or trabecules) between sporogenous cells (Gentianaceae, Balsaminaceae). Tapetal cells may be mono-, bi- and multinucleate.

Tapetum and Orbicules (Ubisch Bodies)

The tapetal cells form a well-defined layer which surrounds the microsporogenous tissue in the anther. They show the following characteristics: (1) they are distinctly enlarged and always ephemeral; (2) the cytoplasm is rich in ribosomes and active organelles; (3) tapetal cells may be multinucleate or polyploid; (4) various irregular mitotic divisions and nuclear fusions take place in the tapetal cells; (5) their rapid and intense activity ends in degeneration of the cytoplasm (Plate 11).

Two main types of tapetum tissue are distinguished:

1. **Glandular (secretory) type**, in which the cells remain in their original position and disintegrate later.
2. **Amoeboid type (invasive** or **periplasmodial)**, in which the protoplasts of the cells penetrate between the pollen mother cells and the developing pollen grains where they fuse with one another to form a tapetal periplasmodium.

In some cases, the tapetal cells repeatedly enter into periods of hypersecretory activity until near maturity of pollen grains. For example, in *Nymphaea* (Rowley *et al.*, 1990) there are two distinct intervals during which tapetal cells protrude into anther locules. Following each interval of locular invasion, then retract and become arranged as a palisade around the loculus.

As information accumulates, the border between types of tapetum tissue becomes less clear. Variations in tapetum morphology may be an adaptation to pollen grain size, shape, mode of dispersal, pollenkitt and probably other factors (Pacini, 1990).

Functions of the tapetum. The main function of tapetum tissue is that of supplying nutrients to the developing microspores. In addition, the tapetal cells may take part in the following activities:

1. Secretion of the enzyme callase to dissolve the callosic wall of the tetrad (Stieglitz, 1977) and release the microspores.
2. Secretion of polysaccharides into the loculus during free microspore stage and their subsequent absorption by the developing microspores (Pacini and Franchi, 1983).
3. Formation of exine precursors which, according to recent results of degradation experiments (Schulze Osthoff and Wiermann, 1987), could be phenols or p-coumaric acid (Wehling *et al.*, 1989). The male gametophyte, along with the tapetal cells, contributes to formation of the exine.
4. Tapetal cells may have other activities resulting in the formation of various structures characteristic of some families. For example, in Onagraceae tapetal cells play a role in the formation of fine flexible threads known as viscin threads, which are an extension of the outer layer of the exine (Hesse, 1984). In family Compositae (Asteraceae), the tapetum forms an acetolysis-resistant membrane outside the sporogenous tissue (Heslop-Harrison, 1969). This membrane is known as the culture sac or peritapetal membrane (Shivanna and Johri, 1985).
5. Formation of orbicules (Ubisch bodies).

6. Formation of pollenkitt, a hydrophobic layer composed mainly of lipids and carotenoids (Pacini and Casadoro, 1981; Rowley and El-Ghazaly, 1992). Pollenkitt is generally deposited on the exine surface and helps to bind pollen grains together, probably for efficient insect pollination.
7. Formation of tryphine, a mixture of hydrophobic and hydrophilic substances. Tryphine is usually deposited on the pollen surface and possibly helps in protection of pollen grains and in insect pollination (Dickinson and Lewis, 1973).

Orbicules. Orbicules were first observed by Rosanoff (1865) to occur in close association with the tapetum and to have a composition similar to the exine of pollen grains. Useful light microscopical observations were published later by Schnarf (1923) and Ubisch (1927). Schnarf (1923) was apparently the first to describe such exine-like bodies in *Lilium martagon*. Kosmath (1927) suggested that these bodies should be called 'Ubisch granules', because of the contribution Ubisch made towards their characterization. Heslop-Harrison (1962) used the term 'plaques' to describe such bodies, and Erdtman *et al.* (1961) proposed the term 'orbicules'. These structures were too small to be investigated solely by light microscopy; with the advent of the electron microscope they were easily reidentified and it became possible to consider their morphology and ontogeny.

Orbicules are generally spheroidal structures found in the anthers of many genera of angiosperms, both monocotyledons and dicotyledons, and many gymnosperms. Orbicular shape varies in different species. It may be rod-like, doughnut-shaped, oval or rounded triangular or plate-like with undulating or jagged margin, etc. Orbicules frequently fuse into large compound aggregates, as in *Oxalis pes-caprae*. The morphological variation of orbicules might be of taxonomic importance (Raj and El-Ghazaly, 1987; El-Ghazaly, 1989).

The wall of orbicules apparently consists of sporopollenin, a material of conspicuous durability. The surface pattern of orbicules and exine can be remarkably similar (El-Ghazaly and Jensen, 1986a).

Concerning the function of orbicules, many definitive functions have been attributed by various investigators. These include, for example, a transport mechanism for conveyance of sporopollenin, or providing a non-wettable layer lining the anther locule from which pollen grains are readily detached, or they may be associated with pollen dispersal (for a detailed review, see Echlin, 1971a, b; Pacini *et al.*, 1985).

Development of orbicules. This begins in the cytoplasm of tapetal cells on small globular bodies (pro-orbicules) with a diameter of about 0.1 μm. The origin of the pro-orbicules seems to be associated with the endoplasmic reticulum of the tapetal cells. At the sporogenous stage, numerous membrane-bound pro-orbicules appear in the tapetal cytoplasm. The membrane coat of the pro-orbicules fuses with the plasma membrane of the tapetal cell and young orbicules pass through the plasma membrane and arrange themselves on its surface; sporopollenin starts to accumulate on the surface of the pro-orbicules until their wall is developed. The orbicules apparently remain attached to the plasma membrane of the tapetal cells and are located in the outer and inner tangential layers and the radial sides of these cells (El-Ghazaly and Jensen, 1986a; Hesse, 1986; Rowley and Walles, 1987).

Tapetal Membrane—structure completely enclosing each cell of a secretory tapetum (Plate 12). The term was coined by Banerjee (1967) to describe an acetolysis-resistant membrane which is formed on the surface of tapetal cell walls on the anther locule side. Synonyms: **tapetal pellicle** (Romanov, 1970c), **orbicular wall** (Christensen *et al.*, 1972), **pellicle** (Cousin, 1979).

In addition to the tapetal membrane enveloping each tapetal cell (Ogorodnikova, 1976a, b, 1986, 1990; Golubeva, 1983), an acetolysis-resistant membrane was discovered which binds all the tapetal cells on the middle layer side. It is variously called the extratapetal membrane (Heslop-Harrison, 1969), peritapetal membrane (Dickinson, 1970) or peritapetal pellicle (Reznickova, 1984). Tapetal and extratapetal membranes are formed during post-tetrad stages, concurrently with the formation of the pollen grain exine. The structure of these membranes, as viewed from the surface side and in cross-sections of anthers acetolysed in concentrated sulphuric acid, was examined under an electron microscope. The extratapetal membrane structure, as viewed from the surface, appears as an electron-light layer with sporopollenin globules, and in cross-sections, as an electron-dense line which extends up to and onto the radial walls of the two neighbouring cells and forms a V-shaped lamella (Walles and Rowley, 1979). The tapetal membrane structure has been found to vary according to the membrane topography. This prompted the introduction of specific terms for particular tapetal cell walls: the inner tangential wall, the outer tangential wall and radial walls. The inner tangential wall is a cell wall facing the cavity of the anther locule, the outer tangential wall is a cell wall presented to the middle layer cells, while radial walls face the neighbouring tapetal cells. In plants with secretory tapetum, four types of the tapetal membrane structure can arbitrarily be recognized.

I. **Pinaceae**-type. The membrane structure of all cell walls is fairly uniform. It is represented by fibrillar material with orbicules embedded in it on the inner tangential wall. This type of membrane structure occurs in members of Pinaceae, Schisandraceae, Simmondsiaceae, Moraceae, Chenopodiaceae, Alliaceae and Ruscaceae.

II. **Rosaceae**-type. This type of tapetal membrane differs from the previous one by lack of orbicules in fibrillar material and is found in members of such families as Rosaceae, Fabaceae, Linaceae and Polygonaceae.

III. **Solanaceae**-type. The tapetal membrane structure is represented by a network of strands completely enclosing each cell. Orbicules are attached to the membrane on the inner tangential wall. A structure like this occurs in members of the Solanaceae family.

IV. **Poaceae**-type. The membrane structure varies with membrane position. The tapetal membrane of the inner tangential wall comprises three components: a perforated layer, a network of strands and orbicules. Structurally, the tapetal membrane of the radial walls generally resembles that on the inner tangential wall, differing in that orbicules occur over approximately one-third the surface of the radial walls, closer to the anther locule. The tapetal membrane of radial walls partially overlaps the outer tangential wall while rounding the exterior corners of the cell. The membrane of the outer tangential wall is made up of fibrillar material. This composite and unique structure of the tapetal membrane is characteristic of the Poaceae family.

To date, the tapetal membrane structure has been studied by light- and electron microscopic techniques in members of 30 families with the secretory tapetum.

In plants with the **amoeboid type of tapetum** (Asteraceae, Caprifoliaceae and others), no membrane is formed other than the extratapetal one, which appears as an electron-light layer with sporopollenin globules (Heslop-Harrison, 1969).

The chemical composition analysis of the extratapetal and tapetal membranes, like that of the pollen grain exine, shows them to be made up of an acetolysis-resistant substance belonging to the sporopollenins. The functional significance of the extratapetal membrane is still unclear. It is assumed to form a *culture sac* for retention of pollen grains (Heslop-Harrison and Dickinson, 1969). In grasses, the tapetal membrane on inner tangential walls presumably ensures a single-layer arrangement of pollen grains in the anther locule owing to the exine bacula 'adhering' to the spinules of orbicules (Romanov, 1970c). It is believed that orbicules aid in the shedding of mature pollen from the anther, since ripe pollen grains are readily detached from their non-wettable surface (Heslop-Harrison and Dickinson, 1969). The structure of the tapetal membrane components, like that of the pollen grain exine, exhibits stability and constitutes an important taxonomic feature. Orbicules vary in shape, size and number according to the genus.

Sporopollenin—a biopolymer contained in the exine of pollen grains, in the exosporium of spores, as well as in the cells of some fungi and algae at certain stages of their life cycle. The term was coined by Zetzsche (Zetzsche *et al.*, 1937).

Sporopollenin is a highly resistant (both physically and chemically) polymer and is only slightly soluble in 2-ethanolamine and chromic acid. The chemical composition of sporopollenin has not been elucidated, nor have its precursors and their sites of synthesis been clearly identified (Southworth, 1990a, b). The first report suggested sporopollenin to be an oxidized hydrocarbon with a chemical composition between $C_{90}H_{134}O_{20}$ and $C_{90}H_{150}O_{33}$ (Zetzsche *et al.*, 1937). Brooks and Shaw (1968, 1978) believed sporopollenin to be formed via oxidation copolymerization of carotenoids and carotenoid esters. The authors observed a hundredfold increase in carotenoid content during zygospore formation in the fungus *Mucor mucedo*, as well as incorporation of ^{3}H beta-carotene in sporopollenin of some fungi and of ^{14}C carotenoids in the pollen of *Lilium* and *Cucurbita*. More recently, however, formation of the sporopollenin wall of *Cucurbita pepo* pollen was found to be only slightly affected by carotenoid synthesis inhibition (Prahl *et al.*, 1985). Spectroscopic analysis of the chemically resistant material of the outer cell wall of the alga *Botryococcus braunii* showed this substance to be distinctly of non-carotenoid nature (Berkaloff *et al.*, 1983; Largeau *et al.*, 1984). Moreover, in the sporopollenin fraction of the developing microspores of tulip, phenylalanine was detected (Prahl *et al.*, 1986) which, the authors presume, is involved in sporopollenin biosynthesis as a direct precursor of phenols. It is not inconceivable that terpenoid metabolism is implicated in sporopollenin biosynthesis. Sporopollenin samples isolated from pollen of plants belonging to different taxonomic categories were also analyzed by ^{13}C nuclear magnetic resonance spectroscopy (Guilford *et al.*, 1988). Sporopollenin has been found to be not a unique substance, but rather a combination of related biopolymers (Hemsley *et al.*, 1992) derived from saturated precursors such as fatty acids. Electron cytochemical studies of developing microspores of *Liriodendron chinensis* employing enzymes prior and subsequent to fixation provided evidence supporting the assumption of lipoproteins conceivably being precursors of sporopollenin (Gabarayeva, 1990; 1991a). Following isolation of sporopollenin from pollen of *Pinus mugo* and its examination by mass spectrometry, the resultant spectrum was found to be dominated by peaks characteristic of p-coumaric acid (Wehling *et al.*, 1989). Chemical degradation of sporopollenin using AlJ_3 also showed p-coumaric acid to be a major product of its degradation, suggesting that p-coumaric acid is a genuine structural unit of the sporopollenin skeleton (Fig. 1) (Wehling *et al.*, 1989; Wiermann and Gubatz, 1992).

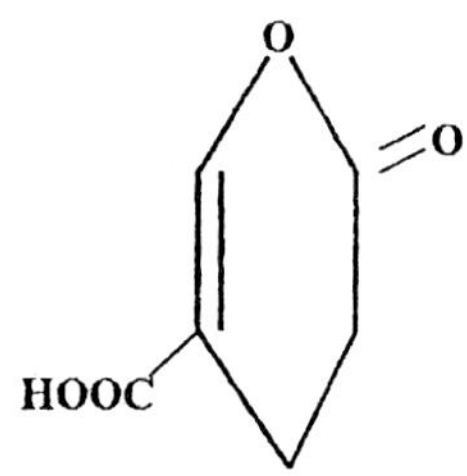

p-Coumaric acid

Fig. 1: Sporopollenin

Southworth *et al.* (1988) developed a technique for exine and cytoplasm purification with a view to producing sporopollenin antibodies. Preliminary results indicated that any chemical extractions removing non-sporopollenin components of pollen grains reduced the frequency of binding of antibodies to pollen grains, suggesting that a proportion of antibodies had previously bound to protein or other components of exines (Southworth *et al.*, 1988). On the other hand, the findings of Rowley and coworkers (Rowley and Prijanto, 1977; Rowley *et al.*, 1981b; Rowley, 1990) on exine structure showing the exine to be composed of glycocalyx components 'embedded' in sporopollenin, inspire little hope for the possibility of producing pure (uncontaminated) sporopollenin. As for the intention of identifying the site of sporopollenin synthesis through the use of antibodies, this task appears quite hopeless since its precursors in the cytoplasm exhibit a distinctly different chemical composition, and the antibodies are not expected to interact with them. It is known that no traces of intracellular sporopollenin have been found, the cell secreting nothing more than a substance which is a precursor of sporopollenin and which is not polymerized until it lands on the surface of microspores. It is considered to be a fairly well-established fact that in most plants sporopollenin precursors are synthesized by the microspore itself during the tetrad period and predominantly by the tapetum in the post-tetrad one. Opinions differ as to their sites of synthesis in various cellular organelles, with preference increasingly given in recent years to endoplasmic reticulum (ER) cisternae. It has been suggested that synthesis of sporopollenin precursors is accomplished by short tubular cisternae of the smooth endoplasmic reticulum (SER) of microspores (Dunbar, 1973; Gabara, 1974; Dickinson, 1976; Gabarayeva, 1991b). Interestingly, in terpenoido-genic cells of many plant (and animal) species, SER was demonstrated to be involved in the synthesis and transport of lipids terpenes and steroids (Vassilyev, 1977), i.e., substances belonging in the same groups in which presumably sporopollenin precursors belong. Other unusual forms of ER, namely aggregated ones—the so-called chain-mail ER and 'zebra' ER—are most likely to be implicated in the synthesis of sporopollenin precursors in members of the Magnoliaceae family (Gabarayeva, 1990; 1991a). It is essential to highlight possible functions of the exine sporopollenin. In addition to some widely known functions (mechanical protection of pollen after shedding, protection from desiccation, UV rays and radiation; storage of proteins, recognition and self-incompatibility), other functions have also been proposed. The point is that carotenoids contain free radicals and conjugated double bonds which are responsible for their antioxidant activity. This has led some authors to suggest that a thin carotenoid-bearing layer or a layer of some other chemical composition showing numerous conjugated double bonds may inactivate various chemical compounds and inhibit certain reactions, thereby preventing protoplasts from being oxidized (Kronestedt-Robards and Rowley, 1989). Thus the protective role of exine manifested during pollen grain development is likely to stem from the fact that the antioxidant contained in it is active under low partial pressure of oxygen within the anther (prior to anthesis). By contrast, when the oxygen content is high (after pollen shedding), it exhibits a high level of autooxidation (Burton and Ingold, 1984).

Tapetum: Ultrastructural Aspects

The tapetum ultrastructure has been studied in various representatives of flowering plants: *Paeonia tenuifolia* (Marquardt *et al.*, 1968), *Zea mays* (Chebotaru, 1969a, 1972), *Lilium longiflorum* (Heslop-Harrison and Dickinson, 1969), *Mahonia aquifolium* (Roland-Heydacker, 1979), *Disporum smilacinum* (Takahashi and Sohma, 1979), *Primula obconica* (Stevens and Murray, 1981), *Ziziphora bibersteiniana* and *Thymus caramarjamicus* (Kurbanov, 1981), *Lilium henryi* (Hrich, 1983), *Austrobaileya maculata* (Zavada, 1984), *Secale cereale*, *Triticum aestivum* and *Avena sativa* (Ogorodnikova, 1976b; El-Ghazaly and Jensen, 1985), *Michelia fuscata* (Gabarayeva, 1986), *Tillandsia pallidoflavens*, *Tilia platyphyllos* (Hesse and Hess, 1993). The cited works have been comprehensively reviewed by Bhandari (1984).

The amoeboid (periplasmodial) tapetum has been studied in various plants: *Tradescantia bracteata* (Mepham and Lane, 1969), *Helianthus annuus* (Horner, 1977), *Mahonia aquifolium* (Roland-Heydacker, 1979) and other species. Tapetal cells of *Tradescantia* are rich in plastids, mitochondria, ribosomes, large polysaccharidic formations and electron-dense bodies with granular contents. Raphides not associated with membrane structures have been found within the periplasmodium (Bhandari, 1984). In the tapetum of *Helianthus annuus*, styloid crystals have been detected. Numerous Golgi vesicles have been shown to be involved in the formation of multivesicular bodies. Similar formations have also been found in the secretory tapetum of *Zea mays* (Chebotaru, 1972), *Ipomea purpurea* and *Helleborus foetidus* (Bhandari, 1984).

During microsporogenesis, the tapetum undergoes definitive degradation: by the time the tetrads of microspores are formed, its cell walls are dissolved, protoplasts with all kinds of organelles penetrating between them. In the bulk of periplasmodium outflow, plastids, like other cellular organelles, undergo redifferentiation, as evidenced by their structural reorganization. The cytoplasm again shows the presence of dictyosomes, rough ER, microtubules and the formation of electron-light vesicles carrying hydrolytic enzymes. The latter coalesce with the callosic wall of microspore tetrads, thereby speeding up its disintegration.

During resorption of cells, their vacuolization is intensified, the cytoplasm stays electron-dense and the number of plastoglobules in plastids increases. Concurrent with degradation processes occurring in the tapetum, tryphine constituents—polysaccharide bodies, flattened lipid globules and other cytoplasmic fragments—are deposited as a surface layer on the exine (Bhandari, 1984).

The secretory (glandular) tapetum is most commonly found in flowering plants and has been studied by electron microscopic techniques in a wide range of species: *Poa annua, Allium cepa, Beta vulgaris, Citrus limon, Sorghum bicolor, Capsicum annuum, Antirrhinum majus, Gentiana acaulis, Pelargonium zonale, Kalanchoe obtusa, Lycopersicum peruvianum, Lens culinare* (all the species are cited in Bhandari, 1984), *Zea mays* (Chebotaru, 1969a, 1972), *Primula obconica* (Stevens and Murray, 1981), *Austrobaileya maculata* (Zavada, 1984), *Secale cereale, Triticum aestivum, Avena sativa* (Ogorodnikova, 1986) and others. The secretory tapetum cells are uni-, bi-, tetra- and multinucleate. They are connected with one another by numerous plasmodesmata which can also be seen between the tapetum and the intermediate layer cells and microsporocytes (at early stages of development). In binucleate cells of *Zea mays* tapetum (Chebotaru, 1972), large bundles of microtubules have been found along the radial wall of the cell. Similar structures have been discovered in *Avena, Olea* and *Helleborus* (Bhandari, 1984). The function and origin of these structures remain unknown. The reports of Cooper (1952) and Takats (1962) continue to be of some interest; they have demonstrated synthesis in the tapetal cell cytoplasm of nucleic acids (DNA, RNA) and proteins, most of the synthesis occurring during meiosis. However, this problem has not been solved completely. Nor has the manner in which nutrients and building blocks are supplied to microsporocytes by tapetal cells been conclusively established. Takats (1962) believes that nucleic acids produced within the tapetal layer undergo depolymerization and resynthesis on reaching the microsporocytes. Cytophotometric studies (Taylor, 1959; Moss and Heslop-Harrison, 1967; Reznickova, 1984) have demonstrated that biosynthesis of major high molecular weight compounds occurs independently in the developing microsporocytes, microspores, pollen grains and tapetum of maize. It has been suggested that accumulation of RNA and proteins within the cells of the tapetum is associated with formation of this tissue and its specific functions. However, this appears to be a highly implausible assumption.

Studies on tapetogenesis[1] in *Zea mays* allow the following stages (S) to be distinguished: S1—tapetum differentiation; S2—division of cells and their polyploidization; S3—enhanced physiological and biochemical activity, increased levels of nucleoproteins, amino acids, vitamins, activation of organelles (plastids, mitochondria, the Golgi apparatus, ER, etc.), synchronous interaction with metabolic processes occurring in microspore mother cells and in cells of other layers of the microsporangium wall; S4—onset of wave of lytic activity, commencement of the so-called tapetum degeneration and occurrence of morphophysiological changes in microspore mother cells, secretion of sporopollenin precursors and sporopollenin formation; S5—'periplasmodization' of tapetal cells, structural reorganization of organelles. We shall attempt to outline what takes place during tapetogenesis at the submicroscopic level.

---

[1]Tapetogenesis or ontogenesis of the tapetum is a series of morphological and physiological changes which occur in the time interval between initiation of a tapetum and its histolysis (Chebotaru, 1972).

It has been established that the onset of the dissolution of callose enveloping the tetrads of microspores coincides with that of pectocellulosic tapetal cell walls. Plasmodesmal connections (cytomictic channels) between cells of *Zea mays* (Chebotaru, 1972) expand to the extent that they can readily be penetrated by mitochondria. By the uninucleate microspore stage the tapetal cells lack walls and their protoplasts are only bounded by the plasmalemma. Subsequent to dissolution of cell walls, 'periplasmodium' migrates into the space between pollen grains, occasionally reaching the microsporangium centre. Chloroplasts, amyloplasts as well as intense activity of the Golgi apparatus, mitochondria and ER can be seen in such 'periplasmodial' structures (Chebotaru, 1969b, c). These phenomena occur in a short period of time, however. What happens to the tapetal cell cytoplasm and its organelles during the fourth stage (S4) of tapetogenesis?

The matrix of tapetal cells is electron-dense and rich in organelles. A major part of the protoplast is represented by prominent nuclei tightly appressed against one another. Unlike the cells of other anther wall layers (endothecium, epidermis), the tapetal cells of *Zea mays* show no presence of large vacuoles during S4–S5. A handful of small vacuoles occurring around the time of completion of the (tapetal cell) nucleus division are mostly scattered around the periphery of the cell.

Resorption of organelles and degeneration processes generally occur in the presence of vacuoles. The opposite situation occurs in the endothecial and epidermal cells. A noteworthy feature of the tapetal cell matrix is its comparatively high density. The predominant cytoplasm constituents are ribosomes, free (in polysomes) and membrane-bound, and rough ER. By the time intense degeneration sets in, numerous poorly differentiated mitochondria and plastids have undergone characteristic structural changes. Myeline structures arise within the protoplast. The plasmalemma, clearly discernible throughout the cell circumference, shows abundant pinocytic invaginations which sometimes turn into peculiar intracytoplasmic passages (cytomictic channels) on the outer cell surface facing the microspores.

The nuclei of tapetal cells are fairly large, harbour one or several nucleoli and numerous chromatin structures exhibiting high activity (complete disappearance of these structures does not occur). This condition is clearly observable throughout tapetogenesis. Upon Palade fixation and contrasting with uranyl acetate and lead tetroxide, chromatin structures within the tapetum of *Zea mays* become readily detectable (Chebotaru, 1969a, 1972).

The karyolymph of tapetal nuclei is electron-transparent, contains electron-dense fibrillar structures and abundant heterochromatic segments ultrastructurally resembling the nucleolus, but more coarse-grained. The literature abounds in opinions on the role of polyploidization of tapetal cells and hence on increased content of nucleic acids in them, although no accurate description of fusion of such nuclei has been provided. Ultrastructural studies suggest only a close juxtapositioning.

Plastids (leuco-, amylo-, chloro- and chromoplasts) are poorly differentiated. Well-developed chloroplasts, like those occurring in the endothecial cells, have not been found in the tapetum. At the uninucleate pollen grain stage, plastids differ in structure of intracellular membrane systems.

A distinctive feature of plastids entering the phase of intense disintegration is the matrix delamination resulting in complete lysis of organelles. Membrane structures disappear, whereas a single or numerous osmiophilic globules arise which are arranged in clusters. In general, the tapetal cell plastids undergo obliteration identical to that in other cellular organelles. Plastids of the released 'periplasmodium' undergo, along with reactivation of other organelles, a peculiar reorganization to form the intracellular membrane system.

Mitochondria are numerous, uniformly distributed throughout the protoplast, mostly spherical, less often oblong, variable in thickness, large-size (megamitochondria), with dense matrix and abundant long cristae. In places, mitochondria are surrounded by ER channels. Within the periplasmodium, mitochondria, as the structures retaining morphophysiological activity over the longest period of time, become more electron-dense and even show signs of fragmentation around the organelle periphery. With increasing lytic activity, they become less electron-dense, however, and the cristae in the central section of the matrix disappear, with just a bare handful of them persisting close to the organelle walls.

During early tapetogenesis, the Golgi apparatus releases numerous vesicles fusing to form multivesicular bodies which migrate towards the plasmalemma. The Golgi apparatus is represented by numerous elongated dictyosomes; the terminal cisternae which assume a sac-like shape carry electron-dense material. During S1 (tapetal cell differentiation), the Golgi apparatus is highly active; during S3–S4, dictyosomes reach maximum length, are less active and persist until the 'periplasmodium' is completely lysed.

The endoplasmic reticulum (ER) is of the rough type and very well developed. Numerous ER channels form areas of ergastoplasm, surround nuclei, plastids and mitochondria, and in places anastomose. The ER and dictyosomes are concerned with transport of sporopollenin precursors as well as supply of nutrients to the developing microspores. There are reports of a rough ER transforming into a smooth one during tapetal cell histolysis.

The ribosomes are abundant throughout the tapetogenesis; they are either attached to the ER membranes, forming continuous polysome chains (4 to 8 components), or free.

Retrospective analysis of tapetogenesis shows that the tapetum possesses the largest metabolic potential compared to the other microsporangium wall layers and is adapted to utilizing intermediate and end products of cell metabolism of other wall layers (Chebotaru, 1970).

New Approach to Classifying Tapetum Types

The tapetum can be classified into a number of types according to its structural features. Most classifications, starting with the typology proposed by Goebel (1905), recognize two types of tapetum: secretory (synonyms: glandular, cellular, parietal) and amoeboid (synonyms: periplasmodial, invasive). Classified under the latter type are cases of periplasmodium formation. Following Tischler (1915), two types of periplasmodium have been distinguished: true and false (synonyms: typical and atypical respectively). In doing so, various authors use different classification criteria (Tischler, 1915; Wunderlich, 1954; Carniel, 1963; Davis, 1966; Poddubnaya-Arnoldi, 1976). Claussen (1927) distinguished four subtypes of periplasmodium.

Present-day classifications incorporate a wide variety of forms and structural/ functional features of the tapetum (Pacini *et al.*, 1985; Pacini, 1990; Pacini and Franchi, 1991) and take account of all its developmental phases in conjunction with anther genesis (Kamelina, 1978, 1981a, 1988a, 1991).

Using as the basis of tapetum classification not only structural but also temporal differences in tapetum development outlined by Carniel (1963), Kamelina (1978) advanced the concept of 'tapetum **reorganization**'. This occurs in a number of taxa during the postmeiotic period. It was also proposed that **amoeboid** and **periplasmodial** tapeta as well as 'false' and 'true' periplasmodia be recognized, with the status of **basic type** reserved for the latter. To avoid confusion in terminology, it was suggested that 'false' periplasmodium be referred to as the **syncytium** and viewed as a form of cellular tapetum reorganization. Electron microscopic studies confirmed these assumptions (Pacini and Juniper, 1983; Pacini and Keijzer, 1989).

Thus, the **basic types of tapetum** in flowering plants are the **cellular** (secretory) and **periplasmodial tapetum** (Fig. 2).The first type includes the cellular tapetum without reorganization and that with reorganization (four forms) which takes place after the tetrad period (during the postmeiotic period of anther development), resulting in the formation of a **syncytium, amoeboid tapetum, bordering symplast** and **tapetal tissue**. More than one variation (mode) of development has been reported also in the periplasmodial type of tapetum (Kamelina, 1991).

Cellular tapetum: characterized by cellular structure and parietal location over the entire period of its existence (except when tapetal reorganization occurs). The primary wall is retained in tapetal cells until the tetrad stage. Then, following its dissolution, a secondary cell wall is formed which contains sporopollenin and, in many cases, carries orbicules on the inner tangential and radial sides of each cell. The cell contents are gradually lysed, with no more than a sporopollenin tapetal membrane studded with orbicules persisting in a mature anther.

This type of tapetum (cellular, without reorganization) has been recorded in 229 families of dicotyledons and 38 families of monocotyledons.

Periplasmodial tapetum (Greek *peri* = around, *plasma* = liquid or gel-like part of the cell, and *eidos* = shape), or true periplasmodium according to Carniel (1963): **characterized by formation of a coenocytic structure at early (premeiotic and meiotic) periods of anther development**. Tapetal cell wall dissolution occurs immediately after tapetal layer formation, such that in the periplasmodial tapetum, unlike the cellular one, no plasmodesmata arise between the tapetal cells and microsporocytes, the cell walls showing no cellulose (only pectin has been reported; Albertini *et al.*, 1987). In some cases, fusion of cytoplasm and formation of periplasmodium occur at the same time as the periplasmodium fills the microsporangium at the microsporocyte stage (Ruppia-variation). In other cases, the cytoplasm of individual protoplasts fuses to form a homogeneous mass in the parietal layer, enveloping the microsporocyte complex with subsequent penetration into the microsporangium locule at the tetrad stage (Potamogeton-variation) or during release of individual microspores from the tetrads (Zannichellia-variation). The tapetal nuclei divide within the parietal layer of individual protoplasts or in the total mass of cytoplasm, spreading subsequently throughout the periplasmodium. The microsporocytes, microspore tetrads and individual microspores develop in direct contact with the plasmalemma of the common protoplast of the periplasmodium. The sporopollenin-containing membrane is only present on the outer periplasmodium protoplast face adjoining the anther tissues.

This type of tapetum has been reported in 21 families of monocotyledons.

Tapetum reorganization: structural and functional rearrangement of a cellular tapetum taking place during the early postmeiotic period of anther development. It is a

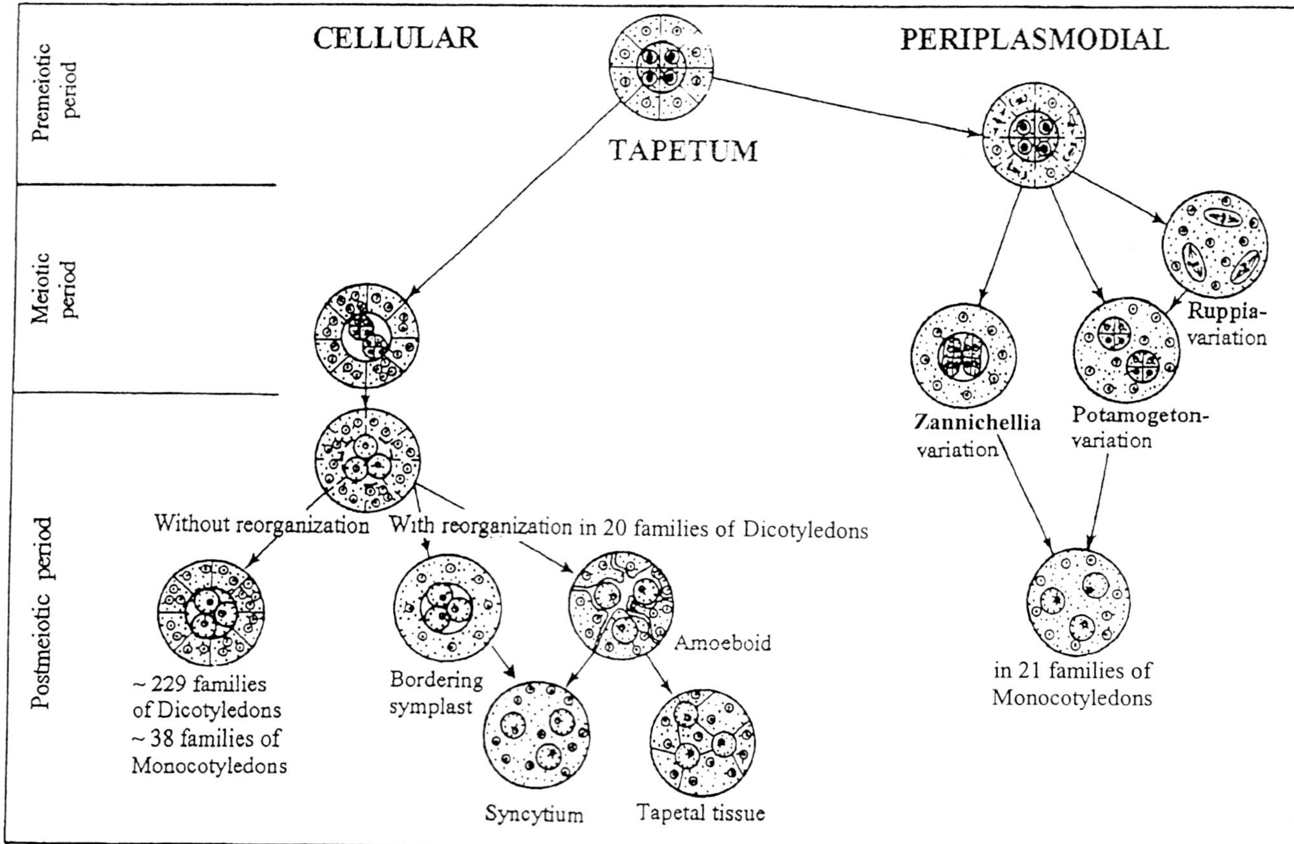

Fig. 2: Tapetum types in flowering plants.

genetically determined feature common to many plant species. Reorganization disturbances result in pollen sterility. Tapetum reorganization has been recorded in 20 families of dicotyledons (Plate 13).

Syncytium (Greek *syn* = together and *kytos* = cell) (= false periplasmodium according to Carniel (1963): form of cellular tapetum reorganization which is characterized by tapetal cell wall dissolution during early postmeiotic stages of anther development, invasion of the microsporangium locule by tapetal protoplasts and subsequent fusion of their cytoplasm. The homogeneous mass of the periplasmodium fills up the space between microspores during their vacuolation stage.

This form of tapetum reorganization has been recorded in Winteraceae, Malvaceae, Rubiaceae, Caprifoliaceae, Valerianaceae, Asteraceae and other plant families.

Amoeboid tapetum (= non-syncytial invasive tapetum according to Pacini and Franchi, 1991): form of cellular tapetum reorganization which is characterized by tapetal cell wall dissolution during early postmeiotic stages of anther development and invasion of the microsporangium locule by tapetal protoplasts which fill the space between microspores. Each protoplast retains its plasmalemma in the process, resulting in no fusion of their cytoplasm, the boundaries between individual protoplasts being invariably visible (i.e., protoplasts retain their individuality).

This form of tapetum reorganization occurs in some members of Lauraceae, Papaveraceae, Rubiaceae and Asteraceae.

Bordering symplast (Greek *syn* = together, and *plastos* = formed, moulded):—form of cellular tapetum reorganization characterized by tapetal cell wall dissolution during early postmeiotic stages of anther development with subsequent fusion of the cytoplasm of individual protoplasts to form a multinuclear protoplast within the parietal layer without invading the microsporangium locule.

This form of tapetum reorganization has been described in Berberidaceae (Budell, 1964).

Tapetal tissue: form of cellular tapetum reorganization characterized by tapetal cell wall dissolution during early postmeiotic stages of anther development with subsequent invasion of the microsporangium locule by tapetal protoplasts to fill the space between microspores. Between the protoplasts, secondary cellulosic walls are formed, with the tapetum again becoming cellular at the stage of vacuolated microspores. Neither sporopollenin nor orbicules have been found in tapetal cell walls. Tapetal cell lysis occurs by degrees, with the cell walls the last to be lysed. No tapetal tissue is preserved in a mature anther.

This form of cellular tapetum reorganization has been documented in Dipsacaceae (Kamelina and Yakovlev, 1976; Kamelina, 1980, 1987a).

It has been proposed that the tapetum whose walls persist over the entire course of development, as is typical of some species with pollen aggregated in polyads and massulae (Mimosaceae, Asclepiadaceae, Orchidaceae), and the tapetum of grasses (due to its peculiar tapetum-microspore interactions) be viewed as separate forms of the cellular tapetum (referred to as the parietal tapetum by the authors, Pacini and Franchi, 1991).

In identifying the type of tapetum, the latter should be considered a developing structure, i.e., the presence or absence of reorganization and the following criteria should be taken into account: the tapetum structure before and during meiosis (to identify the basic type) and at the stage of vacuolated microspores (to identify the form of reorganization), the degree and site of the protoplast cytoplasm fusion (to identify the form of reorganization). In characterizing a tapetum, it is also necessary to describe such features as the number of nuclei in its cells (uninucleate or multinucleate), presence of tapetal membrane with orbicules or of the extratapetal membrane alone, uniformity and heteromorphic condition, and other morphological features.

Tapetum cytology, ultrastructure and functions have been comprehensively covered in excellent reviews (Bhandari, 1984; Pacini *et al.*, 1985; Albertini *et al.*, 1987; Pacini, 1990; Pacini and Franchi, 1991).

New Approach to Classifying Modes of Microsporangium Wall Formation

A classification of modes of the microsporangium wall formation was first proposed nearly 40 years ago (Batygina *et al.*, 1963). That classification was based on original evidence from microsporangium wall development studies in Poaceae and Ericaceae, as well as on the centrifugal microsporangium wall development sequence discovered by Warming (1873) and on the centripetal one described by Carniel (1961). Descriptions of the centripetal sequence of development of microsporangium wall layers were found in older literature as well, but it was Carniel who recorded this mode of development in *Zea mays* as a centripetal morphogenetic process. We use the term 'microsporangium wall' to describe the distal (lying opposite to the connective) part of the microsporangium tegmen, where the aforementioned mechanisms are most conspicuous.

The classification proposed by Batygina *et al.* (1963) incorporated evidence for the microsporangium wall being formed from the parietal layer and for the tapetum arising from the marginal cells of the secondary archesporium. Davis (1966) advanced her version of the classification based on the behaviour of the secondary parietal layers and distinguished four modes of microsporangium wall development. To the already known two types—'centrifugal' ('dicotyledonous' according to Davis, 1966; 'Solanaceae' according to Batygina *et al.*, 1963) and 'centripetal' ('monocotyledonous' according to Davis, 1966; 'Umbelliferae' according to Batygina *et al.*, 1963)—she added two new types: 'basic' and 'reduced'. In view of the fact that in most of the flowering plant taxa studied the microsporangium wall arises from the primary parietal layer while possible formation of the tapetum from secondary archesporium cells is still an open question, we shall restrict our consideration here to the modes of microsporangium wall formation from the primary parietal layer.

According to the evidence presented in *Comparative Embryology of Flowering Plants* (1981-1990), the microsporangium wall development has been studied in 150 families of the 400 described in the classification of Takhtajan (1980a). The centrifugal (dicotyledonous) mode of development has been described in 67 families of dicotyledons and 2 families of monocotyledons and the centripetal (monocotyledonous) mode in 23 families of dicotyledons and 26 families of monocotyledons. The basic type (mode) has been described in 27 families of dicots and the reduced one in 2 families of monocots.

The factual evidence summarized in *Comparative Embryology of Flowering Plants* enables a more comprehensive treatment of the data available with a view to refining the criteria underlying classification of modes of the microsporangium wall development in angiosperms.

It may be mentioned that attempts at refining the data on microsporangium wall development have already been made, mainly for the purpose of refining the nomenclature (Batygina and Shamrov, 1981a; Grevtsova and Sladkov, 1983; Sladkov and Grevtsova, 1988). As discussion of the full range of relevant questions is beyond the scope of this article, we shall consider only the principal ones of fundamental importance.

In the course of development of the microsporangium wall, three types of tissue are formed in it (in addition to the epidermis which is common to the anther as a whole).

Endothecium—a mechanical (strengthening) monolayer or multilayer tissue assisting in the mechanics of anther dehiscence. Not all taxa possess of this tissue in mature anthers (i.e., the tissue is facultative). The cellulosic bands of the cell wall thickenings (fibrous thickenings) common to this type of tissue are typically formed in subepidermal cells of the microsporangium wall. For this reason, the subepidermal layer has generally been referred to as the endothecium already at the time of its formation, which does not always adequately describe its subsequent functions. It is worth noting that formation of the cell wall thickenings in the subepidermal layer occurs much later (normally during microspore formation).

In many species, no fibrous thickenings whatsoever are formed in the microsporangium wall cells[2]. They are not formed in anthers dehiscing by pores, as in members of Melastomataceae, Ericaceae, Alangiaceae, Asclepiadaceae, Solanaceae and Scrophulariaceae (Solntseva, 1983, 1985; Yakobson and Teryokhin, 1983; Alimova, 1987; Naumova, 1987a, b; Nikiticheva, 1987; Zhukova and Poddubnaya-Arnoldi, 1987). Nor are they formed in a number of aquatic plants (Batygina and Shamrov, 1981b), namely in those exhibiting hydrophilous pollination (hypohydrophily): members of Ceratophyllaceae, Elatinaceae, Cymodoceaceae, Najadaceae and others (Batygina and Shamrov, 1981a,b; Sokolovskaya, 1983; Kamelina, 1990a, b; Teryokhin, 1990a, b). It follows that the endothecial tissue has adaptive value with respect to pollination modes.

Fibrous thickenings in subepidermal layer cell walls are also lacking when the function of anther dehiscence is taken over by the epidermis (exothecium). Examples are provided by Ranunculaceae (*Anemone, Aquilegia*), Polygonaceae (*Oxyris*) (Sokolovskaya, 1981; Solntseva, 1983) and some other families.

Where fibrous thickenings are formed not only in the subepidermal cells but also in those of the subjacent layers, it would be more appropriate to speak of a multilayered endothecial tissue rather than fibrous thickenings in the middle layer cells. The multilayered endothecium has been described in members of Cannaceae, Celastraceae, Rosaceae (Mandrik and Petrus, 1985; Naumova, 1987b; Shevchenko, 1990) and others.

Occasionally, uniform thickenings (fibrous or of a different shape) are formed in both subepidermal and epidermal cells (*Arceuthobium*, Viscaceae), or in cells of the epidermis, subepidermal layer or subjacent layers. This has been reported, for example, in *Aeginetia* belonging to family Orobanchaceae, as well as in Rafflesiaceae,

[2]Evidence on the properties of the microsporangium wall in the taxa included in *Comparative Embryology of Flowering Plants* is presented from this point onwards, in accordance with this publication. Information on the original studies can be found in the references cited.

Zygophyllaceae, and Saxifragaceae (Teryokhin and Anisimova, 1981; Korobova, 1985; Kamelina, 1985e; Teryokhin, 1987b, c).

It is evident from the above examples that the endothecial tissue varies in origin and its formation is not directly associated with differentiation of one microsporangium wall layer or another. Therefore, we propose that all cell layers lying between the epidermis and tapetum of the formed microsporangium wall be called **middle layers** comprising the **middle layer tissue**. In the course of further specialization, the outer middle layer(s) is transformed into an endothecium, whereas the inner middle layer(s) turns into a nutritive tissue.

Nutritive tissue—a monolayer or multilayer, provisional[3], facultative tissue supplying nutrients to the sporogenous tissue. Its facultative nature is evidenced by its absence from the microsporangium wall in some members of Lemnaceae, Najadaceae and Zannichelliaceae (Ilyina, 1990; Kamelina and Teryokhin, 1990; Teryokhin, 1990a, b). In the case of multilayer nutritive tissue and multilayer endothecium, it is not possible to distinguish between these two tissues until the appearance in endothecial cells of specific thickenings or prior to the onset of degeneration of the nutritive tissue cells.

Tapetum—a monolayer or multilayer provisional tissue which is obligatory in the microsporangium wall. It performs three major functions in the microsporangium: supplying nutrients to the microspores, exine formation, as well as synthesis and release of material utilized in the course of tryphine and pollenkitt deposition (Bhandari, 1984). A distinctive feature of tapetal tissue is early specialization of its cells taking place shortly after their differentiation. The increase in number of tapetal layers in anthers of members of certain families presumably occurs following specialization of the tapetum as a tissue (Kamelina, 1985a-e; Teryokhin, 1987a).

Specific features of development of the main microsporangium wall tissues allow three types of morphogenetic processes involved in its formation to be distinguished.

Differentiating divisions—periclinal divisions giving rise to functionally distinct tissues. Differentiating divisions are exemplified by archesporial cell divisions resulting in the formation of sporogenous tissue and primary parietal tissue; and divisions of cells of the primary or secondary parietal layer giving rise to a tapetum. Divisions resulting in multiplication of cells or in higher numbers of middle layers are not differentiating in the strict sense of the word, since the endothecium and nutritive tissue are formed at final stages of microsporangium development. The false effect of differentiation is only produced in the case of the standard model of microsporangium, viz. 'three layers—three tissues'.

Forming (formative, according to the definition of Ivanovskaya, 1983) divisions—periclinal or anticlinal divisions resulting in multiplication of cells or layers of cells in a tissue. Forming divisions are exemplified by divisions resulting in multiplication of tapetal cells and in higher numbers of tapetal cell layers and middle layers as precursors of endothecium and nutritive tissue.

Tissue specialization. Unlike the above morphogenetic processes, this one is not associated with cell division. Uniform changes occurring in individual cells and extending to neighbouring cells lead to the formation of a certain type of tissue, as

[3]Provisional structures are transient structures which, unlike definitive ones, disappear in the course of development.

suggested by the appearance of fibrous thickenings during differentiation of the endothecium.

The proposed **classifications of types (modes) of microsporangium wall development** are based on the **criterion of directed sequence of differentiating divisions leading to the initiation of tapetum formation**, a principle already described by Warming (1873) and Carniel (1961) and employed by Batygina and coworkers (1963) in their classification and more recently adopted by Davis (1966) for the dicotyledonous and monocotyledonous types. The directionality of differentiating divisions is most pronounced in standard models of microsporangium wall: 'three cell layers—three tissues' (tapetum, middle layer and subepidermal layer termed endothecium). The directionality of differentiating divisions clearly manifests itself in the sequence of formation of layers: tapetum—a derivative of the primary parietal layer, under the centrifugal developmental sequence; tapetum—a derivative of the secondary (inner) parietal layer, under the centripetal developmental sequence. Thus, the timing of tapetum initiation during ontogeny is a major criterion in identifying the type of microsporangium wall development.

In view of this, it is essential that the criterion used to distinguish the types of development be incorporated in their names, viz. centrifugal and centripetal. This is all the more appropriate as correspondence between types of microsporangium wall development and taxa of different standing (systematic position) is often lacking.

The criterion (principle) of the secondary parietal layer behaviour proposed by Davis (1966) for determining the type of development is, in our view, inapplicable, if for no other reason than that a tapetum can be assigned to the centrifugal type of development even on the basis of periclinal divisions of the primary parietal layer cells. In view of this as well as the fact that the classification under consideration belongs among the modes (types) of microsporangium wall formation from a parietal layer, the names of developmental types proposed by us earlier (Batygina *et al.*, 1963) and based on their original description in the respective taxa (Solanaceae and Umbelliferae types) should also be abandoned.

The diversity of modes (types) of angiosperm microsporangium wall development from a parietal layer is, in our view (Teryokhin *et al.*, 1993), restricted to the above two types (see the schematic representation). Davis (1966) appears to have had no reasons for distinguishing the basic type as an original one comparable, by the criterion described above, with the centrifugal and centripetal types. Under both the centripetal and the basic types of development, the tapetum is derived from the secondary (inner) parietal layer. All this allows only the status of a variation of the centripetal type of development to be reserved for the basic type. This variation can be termed **complicate**. In the **typical** (three-layered) variation of the centripetal type of development, the subepidermal middle layer formation is initiated as a result of periclinal divisions of the primary parietal layer cells, whereas in the complicated variation it does so as a result of repeated periclinal divisions of the secondary (outer) parietal layer cells. In multilayer models, the final subepidermal middle layer can, therefore, originate by any cycle of periclinal divisions of the cells of preceding subepidermal layers. All these developmental variations conform to the criterion adopted for the centripetal type of development, with the outer parietal layer essentially not belonging to the category of parietal layers since the divisions of its cells are not differentiating ones.

From the above reasoning it is clear that the reduced type cannot be recognized as a basic type of development either, being essentially a variation of the centrifugal type reduced to two layers (**reduced** variation). The tapetum in this case also originates from the primary parietal layer. That the reduced variation belongs to the centrifugal type of development is suggested by evidence for the formation in some species belonging to families Lemnaceae and Zannichelliaceae of a middle layer in absolute conformity with the centrifugal type of microsporangium wall development (Caldwell, 1899; Lodkina, 1971; Ilyina, 1990; Kamelina and Teryokhin, 1990).

It follows that the **hierarchic classification** based on recognition of **two types (centrifugal and centripetal) with several variations within each type adequately describes the diversity of evolutionary transformations of the microsporangium wall.**

Among the variations of the centripetal type of multilayered microsporangium wall development, the one described in the Nymphaeaceae (Batygina and Shamrov, 1981a) occupies a special place. Under this type (mode) of development, the tapetum is derived from the tertiary (inner) parietal layer. This variation can be termed **prolonged**, since the centripetal sequence of divisions resulting in formation of a tapetum is the most pronounced in it. Formally, this mode of development falls as it were outside the scope of the criterion underlying the centripetal type of development. Therefore, the criterion of the centripetal type should probably be expanded to include the following: tapetum derived from the secondary parietal layer or from its subsequent derivatives (in contrast to the centrifugal type in which the tapetum adjoins the inner surface of the microsporangium wall already following divisions of cells of the primary parietal layer).

Thus, the hierarchic classification of types (modes) of microsporangium wall development includes two types: centrifugal (the complicate, typical and reduced variations) and centripetal (the prolonged, complicate and typical variations) and can be schematically represented as shown in Figs. 3 and 4.

The foregoing suggests the need for some modifications in the terminology describing microsporangium wall development. In principle, we should seek to retain the established terms. Therefore, the suggestion of Sladkov and Grevtsova (1988) that, for example, the term 'initial (subepidermal) cells' be substituted for 'archesporium' appears unacceptable. The history of plant embryology attests that in describing the microsporangium wall the term 'archesporium' is invariably used to denote cells with dual fate (Schnarf, 1929; Goebel, 1932; Maheshwari, 1950; and others). To narrow the concept of archesporium to denote early stages of sporogenous tissue development, as suggested by Sladkov and Grevtsova, would not be quite correct. It seems worthwhile to retain the terms 'primary parietal layer' and 'secondary parietal layer' and to coin such terms as 'tertiary parietal layer', 'nutritive tissue' and 'middle layer tissue'

In conclusion, let us take a look at the possible aspects of evolutionary (phylogenetic) interpretation of modes (types) and variations of angiosperm microsporangium wall development and the feasibility of applying systematized relevant data to taxonomy. First and foremost, it should be borne in mind that **microsporangium wall development has been studied to date in members of less than half of the families comprising the division of flowering plants**. Like most other embryological features, modes of microsporangium wall development, contrary to the contentions of Davis (1966), are not always taxon-specific at the level of family, order, etc.

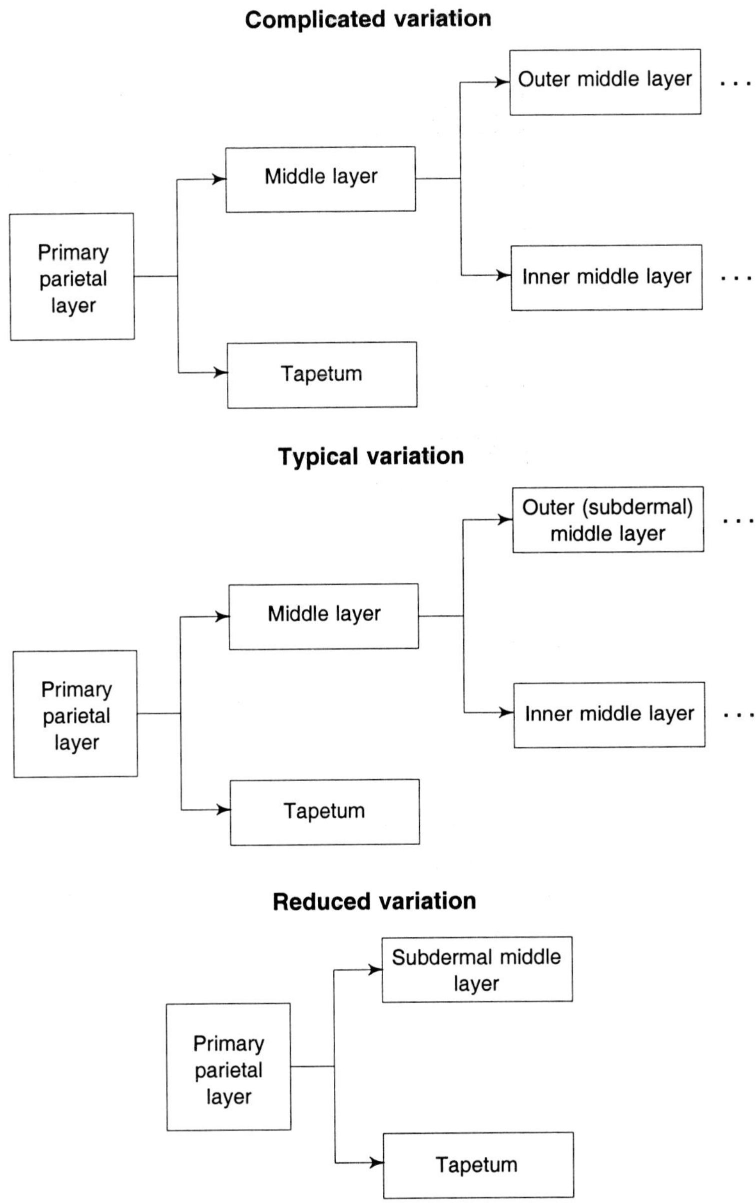

Fig. 3: Centrifugal type of microsporangium wall development. *Note*: Triple dots indicate the possibility of forming divisions and formation of new layers.

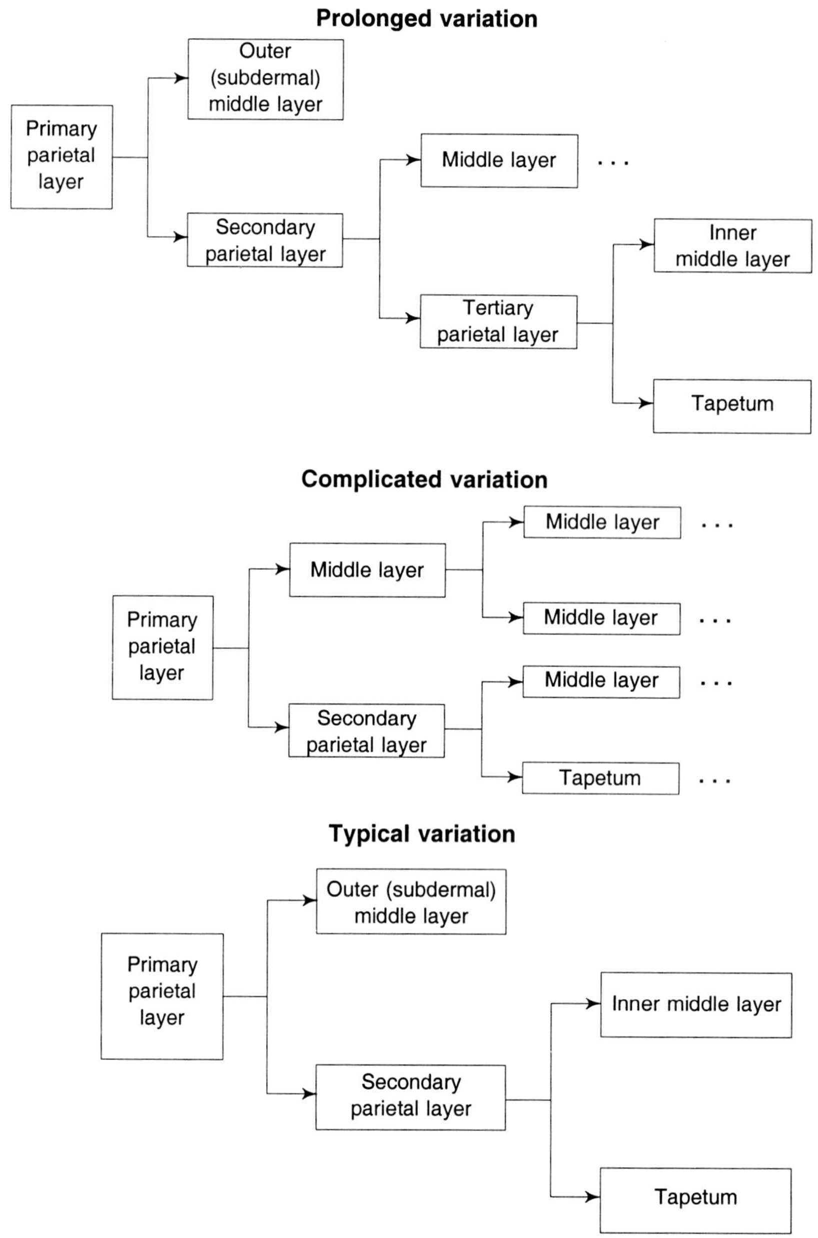

Fig. 4: Centripetal type of microsporangium wall development. *Note:* Triple dots indicate the possibility of forming divisions and formation of new layers.

In some higher taxonomic categories, the taxa studied exhibit similar modes (types) of microsporangium wall formation. Thus, for example, the centrifugal type of development has been described in orders Ebenales, Gentianales, Dipsacales, Polemoniales, and the centripetal in orders Commelinales and Liliales. On the other hand, both the centrifugal and centripetal types (with the basic type included in the latter) have been described in a number of families, such as Capparaceae, Brassicaceae, Sterculiaceae, Euphorbiaceae, Rosaceae, Zygophyllaceae, Apiaceae, Verbenaceae and Solanaceae (Batygina and Kolesova, 1983; Belyaeva and Rodionova, 1983; Rodionova, 1983; Savina, 1983; Kamelina, 1985b; Mandrik and Petrus, 1985; Freiberg, 1987a, b; Grevtsova, 1987; Zhukova and Poddubnaya-Arnoldi, 1987).

In families Lemnaceae and Zannichelliaceae, the standard and reduced variations of the centrifugal type have been described (Ilyina, 1990; Kamelina and Teryokhin, 1990).

The general distribution of modes (types) of microsporangium wall development among the angiosperms allows one to infer the non-concurrence of the course of evolutionary transformations of the microsporangium wall with the phylogenetic trends of flowering plants, suggesting the relative importance of the evidence for taxonomists. The occurrence of different types of microsporangium wall development in a single taxon (e.g. family) suggests possible parallel evolutionary transformations of microsporangium wall components.

Specific directions of evolutionary transformations of the microsporangium wall have not been conclusively established. As indicated earlier, the reduced variation of the 'centrifugal' type of development constitutes in some cases the final stage of the microsporangium wall adaptation for the purposes of hydrophilous pollination. At the same time, a similar direction of changes resulting in a reduced variation in Lemnaceae whose members are pollinated in the atmosphere has been something of a mystery.

The question of modes of evolutionary transformations of the microsporangium wall is also a challenge. An increase or decrease in the overall number of microsporangium wall layers evidently occurs due to variation in the number of final periclinal divisions. Concomitantly, reduced variation is strongly dependent on the ontogenetic timing of tapetum differentiation. Microsporangium wall development in angiosperms is as yet imperfectly understood, calling for more intensive comparative embryological studies of microsporangium wall components.

MICROSPOROGENESIS

Microsporogenesis—a process in which microspores are formed from meiotically dividing pollen mother cells developing within microsporangia (Plate 14). Pollen mother cells (microsporocytes) derive from the cells of sporogenous tissue. These cells, with dense cytoplasm and prominent nuclei, enter meiosis losing their plasmodesmal connections with the tapetum. Then pectocellulosic walls of microsporocytes undergo hydrolysis and are replaced by new callose walls. During these processes plasmodesmata between microsporocytes are disrupted, but they are linked by wide (1-2 μm) strands of cytoplasm running through the cytomictic channels of the callose walls. The channels are the most conspicuous at the zygotene-pachytene stages. They facilitate intercellular transport and make it possible to exchange macromolecules and even cytoplasmic organelles. As a consequence, physiologically active substances may spread very quickly in all microsporocytes of the anther loculus. This apparently synchronizes to some extent the development and phases of meiotic divisions (Bhandari, 1984; Bell, 1989).

During early prophase I, microsporocytes become isolated by thickened callose walls of restricted permeability. It is suggested that in effect the microsporocytes are not influenced by the correlating systems of the parent plant and therefore may pursue meiotic development pathways of their own (Dickinson, 1982).

The microsporocyte mitochondria and plastids dedifferentiate to a very simple form. ER cisternae whorl and insulate portions of cytoplasm in multilayered inclusions similar to those in megasporocytes. Concurrent with reshaping of ER there is a fall in cytoplasmic RNA, both ribosomal and messenger. These changes have been interpreted as leading to at least partial destruction of the specific information molecules and preparation of the ground for the transcripts (mRNAs) specific to the gametophytic phase to be translated; but the sporophytic incompatibility gene information survives and is expressed in germinating pollen grains. However, alternative interpretations of the effect of cytoplasmic changes have been proposed (Souvre *et al.*, 1987; Dickinson and Li, 1988; Bell, 1989).

Nuclear envelopes and nucleoli are very active during prophase I. The nuclear envelope produces intranuclear vesicles which seem to influence the position of chromosomes. The nucleolus discharges numerous micronucleoli which, following anaphase I, are apparently left in the cytoplasm as nucleoloids (Souvre *et al.*, 1987).

The first meiotic division leads to the formation of a dyad. After meiosis I, a tetrad of microspores is set up, usually in tetrahedral arrangement. The tetrad and the microspores within it are enclosed by continuous, often thick, callose walls (Kapil and Tiwari, 1978b). In very few plants (e.g. Cyperaceae), three postmeiotic nuclei degenerate leaving one nucleus in a single microspore.

There are two basic patterns of cytokinesis in microsporocytes: simultaneous (mostly in dicotyledons) and successive (mostly in monocotyledons).

Successive cytokinesis. Nuclear divisions of meiosis I and II are immediately followed by phragmoplast and callose cell plate formation.

Simultaneous cytokinesis. After meiosis I, the cell plate is not formed and two dyad nuclei lie in the common cytoplasm. Meiosis II is followed by simultaneous cytokinesis between four postmeiotic nuclei and formation of the internal callose walls of a tetrad.

In many plants with simultaneous cytokinesis, cytoplasmic organelles aggregate in a regular fashion during microsporogenesis (*Nymphaea, Impatiens, Orchis*). Plastids and mitochondria invade the equatorial plane of the meiosis I phragmoplast, forming there an aggregation which separates a dyad into two mononucleate domains. In some plants, a partially active cytokinetic apparatus builds an incomplete callose cell plate inside the organelle aggregation (in *Nymphaea*).

During interkinesis, the phragmoplast and incomplete cell plate disappear, but the organelle aggregation, in the shape of a disc, persists until conclusion of meiosis II. It then changes its shape and separates the four postmeiotic nuclei. The aggregation disperses while the walls between microspores are being formed.

Some plants (e.g. *Malva*) show a different pattern of organelle aggregation. Late prophase I plastids and mitochondria aggregate in the layer coating the nucleus. After meiosis I and meiosis II, each nucleus is again coated by organelles. Subsequently the walls between these coated nuclei are formed simultaneously. Organelle aggregations presumably play the role of the cell plates in preventing coalescence of nuclei and meiosis II spindles in the coenocytic microsporocytes (Rodkiewicz and Duda, 1988).

Microspores within tetrads become active, their callose walls undergo rapid dissolution, but before being released each microspore sets up a new wall, a primexine, thus entering a period of intensive gene transcription (Mascarenhas, 1989).

Archesporium (Greek *arche* = beginning and *sporá* = seed, spore)—a group of subepidermal cells of the microsporangium giving rise to the sporogenous tissue and microsporangium wall layers (Plate 15).

It was first described by Warming (1873) in his classical work *Untersuchungen über pollenbildende Phyllome und Kaulome* as 'initial pollen (mother) cells'. Subepidermal cells or their derivatives are transformed into initial cells.

Goebel (1880) proposed that the 'initial cells of the sporogenous tissue' described by Warming (1873) (both for micro- and megasporangia) be given a more concise name, viz., 'archesporium'.

Later, Schnarf (1929), having summarized evidence in the literature, pointed out that archesporial cells differ from the surrounding cells by a higher cytoplasm density and a prominent nucleus with a nucleolus.

Currently there is no consensus of opinion as to archesporium formation. Most authors hold that archesporial cells are differentiated within the subepidermal layer of the anther lobe (Maheshwari, 1950; Fagerlind, 1961; Batygina *et al.*, 1963; Korobova, 1971b; Batygina, 1974b, 1987a; Poddubnaya-Arnoldi, 1976; Kordyum, 1978; Shamrov, 1981b; Lakshmanan and Poornima, 1988). Other authors (Grevtsova and Jalilova, 1992) argue that these cells originate by divisions of initial subepidermal cells; still others (Lodkina, 1971) are of the opinion that the 'concept of "archesporium" is a highly arbitrary one, pointing only to the site in the organ primordium where seats of floral meristem have been preserved'.

Sporogenous Tissue—a tissue originating by differentiating periclinal divisions of archesporial cells, resulting in the formation of both sporogenous cells and parietal layer cells (Plate 16). The term was coined by Goebel (1880). The first detailed description of the tissue was provided by Schnarf (1929) who found its cells to differ significantly from the surrounding cells in their size and in the size of their nuclei, as well as by a higher cytoplasm density.

The structure of sporogenous tissue is not the same in different species. In some cases, the cells are arranged in a single longitudinal tier, in others in several tiers. Sporogenous cells divide actively and, upon completion of mitoses, transform into microsporocytes. During divisions, sporogenous cell nuclei become greatly enlarged in volume, with the volume of the cells themselves changing only slightly. The consequences of increased nuclear volume are drastic changes in the nucleocytoplasmic ratio and altered physical and chemical conditions within the nucleus, which cannot but affect cell metabolism (Reznickova, 1975, 1979, 1984). Sporogenous cells show high RNA levels and very low carbohydrate content (Vijayalakshmi and Lakshmanan, 1988).

Microsporocyte (Greek *cytos* = a hollow vessel + *micros* + *sporá*)—a sporogenous cell entering meiosis. Synonyms: microspore mother cell, meiocyte (Plate 17).

Microsporocytes differ from cells of the surrounding tissues by denser cytoplasm, larger size and prominent nuclei (Maheshwari, 1950; Batygina *et al.*, 1963; Chebotaru, 1972; Poddubnaya-Arnoldi, 1976). Prior to meiosis, microsporocytes increase significantly in size. During the premeiotic interphase, DNA replication (Taylor, 1953, 1959; Plaut, 1953; Moss and Heslop-Harrison, 1967; Reznickova and Ostroverkhov, 1970) and histone synthesis inhibition (Bogdanov *et al.*, 1973; Bogdanov and Strokov, 1976; Strokov, 1977) occur within the microsporocyte nuclei. The condensed chromatin of microsporocyte nuclei is arranged peripherally, close to the nuclear membrane. The cytoplasm shows the presence of ribosomes, well-differentiated plastids, mitochondria and short ER cisternae (Bhandari and Sharma, 1988).

During initiation of meiosis, callose begins to deposit around the microsporocytes (Sato Joshoniko, 1988; Blackmore and Burnes, 1988; Rodriguez-Garcia and Alche, 1992), with callose synthesis starting prior to leptotene and reaching maximum during pachytene. The callosic wall varies in thickness (Echlin and Godwin, 1968; Biddle, 1979) and does so according to the stage of meiosis (Waterkeyn, 1981; Rodriguez-Garcia and Alche, 1992). Formation of a multilayered callosic wall is possible (Skvarla and Larson, 1966; Waterkeyn, 1981). Callose is deposited between the plasma membrane and the primary cellulosic wall to form, in the long run, a continuous layer. Involved in callose formation are dictyosomes (Heslop-Harrison, 1966a, b), endoplasmic reticulum (Skvarla and Larson, 1966) and lipids (Willemse, 1971). Upon completion of meiosis, the callosic wall is dissolved. This is due to changes in pH and callase activity (Izhar and Frankel, 1971). Some authors (Vijayaraghavan and Shukla, 1977) believe no callosic wall is formed in the case of compound pollen or pollinia. There is evidence (Zee and Siu, 1990) suggesting that the callosic wall may be absent between microspores within a tetrad. Grasses show a lack of callose on the microsporocyte side facing the tapetum and the presence of a massive callosic wall in the form of a crest projecting into the microsporangium locule on the opposite side of the cells (Romanov, 1966, 1970a–c; Batygina, 1974b).

Callose is a polysaccharide composed of β-1,3-glucans (Rowley, 1959; Heslop-Harrison, 1968a, b; Waterkeyn, 1981). One function of callose is to provide physical and chemical isolation of microspores from one another during their formation and from the surrounding somatic tissues (Heslop-Harrison, 1964; Heslop-Harrison and Mackenzie, 1967; Waterkeyn, 1981). The callosic wall is assumed to play an important role in determining the primexine surface structure and pattern (Larson and Lewis, 1963; Waterkeyn and Bietait, 1969; Ford, 1971).

Many angiosperms possess plasmodesmata which link microsporocytes and cytoplasmic connections that penetrate the callosic walls of adjacent microsporocytes. Plasmodesmata may turn into broad cytomictic channels through which various cytoplasmic constituents and even plastids can pass (Pacini and Cresti, 1978). Exchange via plasmodesmata of dissolved substances present in the cytoplasm ensures synchronous early development of microsporocytes (Heslop-Harrison, 1966a, b).

Meiosis (Greek *meiosis* = diminution, reduction) was discovered by van Beneden in 1884 during cytological studies of development of the worm *Parascaris equorum*

(Plate 18). Meiosis occupies a central position in the sexual reproduction cycle of eukaryotes with alternation of diploid and haploid phases. **It is a special type of cell division whereby the chromosome number is exactly halved.** At meiosis, each cell carrying a diploid set of chromosomes gives rise to four haploid cells. **The halving (reduction) of the chromosome number constitutes the biological essence of meiosis and maintains the diploid chromosome number in successive generations.**

Distinctive Cytological Features of Meiosis and Mitosis

In an organism with a diploid chromosome number, each chromosome (except for sex chromosomes in the heterogametic sex) is represented by two copies (one copy from each parent). These pairs of chromosomes are referred to as homologous chromosomes, or homologues. Pairs of homologous chromosomes enter both mitosis and meiosis in a reduplicated state. Each homologue is represented by two sister chromatids joined by a centromere.

The principal differences between meiosis and mitosis concern their end products. In mitosis, a mother cell divides to form two daughter cells which are genetically identical to each other and to the parent cell from which they arose. Each daughter cell inherits one copy of each homologue, so that the diploid chromosome number and DNA content are maintained. For this to occur, each mitotic division of a nucleus is accompanied by separation of the reduplicated sister chromatids. Sister chromatids undergo congressional movement and finally come to lie at the spindle equator (prometaphase), reach their metaphase arrangement on the metaphase plate such that their kinetochores (centromeres) are aligned on the equatorial plane and orient to opposite spindle poles (metaphase). This allows regular anaphase separation of homologous chromosomes and their movement to opposite poles.

The principal result of meiosis is halving of the chromosome number. This occurs due to two events specific to meiosis: homologous chromosome pairing and one separation of centromere per two nuclear divisions during meiosis. The pairing of homologous chromosomes takes place during the first prophase of meiosis with involvement of a specific structure termed the synaptonemal complex (SC). The pairing partners (bivalents) are arranged in the equatorial plane of the spindle such that the centromeres of homologues orient to opposite poles, the sister chromatids of each of the homologues being connected with one of the poles by their kinetochore fibres. This pattern of centromere orientation results in each daughter cell inheriting both copies of either homologue during meiotic anaphase I. Being diploid in DNA content, daughter cells show halving of the centromere number. These cells differ from the mother cell and from each other in the origin of parental chromosomes due to reshuffling of paternal and maternal chromosomes, the number of different patterns of inheritance (assortment) being 2n (n = haploid chromosome number), and to crossing over within pairs of homologues.

Formation of haploid cells is completed during the second meiotic division which proceeds without replication. The chromosomes become arranged at the equator of the newly forming spindle, with sister chromatids moving to opposite poles, as in mitosis.

Normal Meiotic Sequence and Some Features of Nucleic Acid and Protein Metabolism

Two meiotic divisions are distinguished, each of which consists of several successive stages. The first meiotic division includes the interphase, prophase I, metaphase I, anaphase I and telophase I. The second meiotic division starts with a short stage called interkinesis followed by prophase II, metaphase II, anaphase II, and telophase II.

Interphase. During the S phase of premeiotic interphase, DNA synthesis occurs. The duration of this stage is much longer than that of the corresponding mitotic stage. The bulk (99.7%) of the DNA is synthesized in this period, with the rest (0.3%) of the DNA being replicated during prophase I. The type of DNA synthesis occurring during the premeiotic interphase is termed delayed DNA replication. Cells with reduplicated chromosomes enter prophase I, each chromosome being represented by two chromatids.

Prophase I. This is the most important stage of meiosis. It accounts for 90% of the total meiotic cycle duration and lasts in different organisms from a few days to tens of years. Prophase I is divided into five stages: leptotene, zygotene, pachytene, diplotene and diakinesis.

Leptotene. At this stage, chromosomes appear as a tangle of fine threads. Each chromosome binds, by its ends, to the nuclear membrane by means of attachment discs. Along each chromosome, a lateral axial element is organized which will become one of the lateral elements of the synaptonemal complex during synapsis of homologous chromosomes. The number of chromosomes observable at leptotene corresponds to the diploid set and while the total DNA content and number of chromatids to the tetraploid set.

A leptotene protein (L protein) has been isolated which is only synthesized within meiotic cells (meiocytes). It is not bound to the nuclear membrane and its molecular weight is 73,000 dalton. In the absence of Mg^{2+}, the protein is represented by a monomeric form. The L protein inhibits zygotene DNA synthesis in normal and achiasmate (lacking chromosome pairing and crossing over) genotypes, suggesting that its function is associated with neither pairing nor crossing over. Nor does it affect the synthesis of total nuclear DNA in meiotic and somatic cells. The L protein is believed to be responsible for the irreversible determination of the cell entry into meiosis following the S phase.

Zygotene. This stage of meiosis differs from the preceding one by initiation of homologous chromosome pairing. Homologous pairing initiation sites ('contact points') lie on the nuclear membrane. The ends of homologous chromosomes somehow recognize these sites on the nuclear membrane. Meiotic chromosome pairing is initiated at 'contact points' from which it proceeds along the chromosomes (pairing partners) in a zipper-like fashion. A tripartite protein structure consisting of two lateral and one central element and termed the synaptonemal complex (see above) is formed in regions of homologous chromosome pairing.

Pachytene. At this stage, the homologues are completely paired. The pairing configuration resulting from synapsis of two homologous chromosomes during prophase I is called a bivalent. Formation of the SC is completed over the entire length of the chromosome, the central element of SC showing at this stage the presence of recombination nodules which are thought to be responsible for crossing over.

During pachytene, reciprocal exchanges of segments between homologous non-sister chromatids (crossing over) occurs.

At these two stages of prophase I, maximum levels of synthesis of certain DNA fractions and of activity of specific enzymes involved in the more important cytogenetic processes of homologous chromosome pairing and crossing over are observable.

In zygotene, zygotene DNA (zygDNA) is synthesized, via a semiconservative replication mechanism. zygDNA is rich in G-C base pairs, which explains its higher buoyant density (1.712 g/cm^3). For comparison, total nuclear DNA contains 40% of the G-C base pairs, with a buoyant density of 1.700 g/cm^3. zygDNA is largely composed of unique sequences, each of which is 5 to 10 kb long and is uniformly distributed along the chromosome length. Meiotic behaviour of zygDNA is somewhat unusual. This refers not only to the fact that its replication is delayed to the zygotene. Nor is its synthesis completed at this stage, small gaps persisting at the ends of newly synthesized DNA strands up to anaphase I. Replication of zygDNA is inhibited by the L protein. The same protein protects zygDNA gaps against cleavage by the S-I endonuclease. zygDNA is assumed to have a role in pairing and recombination.

Pachytene DNA (pachDNA) has been identified in the pachytene of meiosis. Its replication is of the repair type (repair replication). Two fractions have been recognized in pachDNA: 30% of this DNA is made up of repetitive DNA sequences 5 to 7 kb long while the rest are short moderately repetitive DNA sequences about 200 base pairs long, reiterated 700 to 2000 times within the genome. Gaps in pachDNA are immediately removed via repair synthesis.

Pachytene DNA was found to contain sites subject to single-strand breaks on exposure to a specific endonuclease with an optimal pH value of 5.2. pachDNA accessibility to endonuclease activity generating single-strand breaks is controlled by a group of small (125 base pairs) molecules of nuclear RNA (pachytene small nuclear RNA, psnRNA) in transcription of which RNA polymerase III is involved. These RNA molecules are only synthesized provided true homologous chromosome pairing occurs. Identification of the psnRNA fraction enabled elucidation of the molecular organization of the pachDNA regions undergoing repair synthesis. DNA sequences complementary to psnRNA proved to be excision repair sites. pachDNA is assumed to be involved in crossing over which, in turn, relies on repair DNA synthesis for completion.

A specific endonuclease cleaves double-stranded DNA generating single-strand breaks, which is suggested by the occurrence of single-stranded 3′P and 5′OH extensions.

At these stages of prophase I of meiosis, proteins have been identified promoting reassociation of single-stranded DNA regions. These belong to DNA binding proteins. Two of the proteins are assumed to play an important role in crossing over. The U protein (DNA unwinding protein) is active in the presence of ATP, its activity reaching maximum during zygotene-pachytene. It exhibits two important properties: binding or maintaining of double stranded DNA and unwinding of double-stranded DNA.

The R protein (DNA reannealing protein) ensures DNA reannealing. It is present within the cell only in complexes with lipids. In molecular properties, the R protein is similar to the protein encoded by the 32 gene of phage T4 and plays a key role in genetic recombination. The R protein requires true homologous pairing for its activation.

In meiotic cells of lily, mouse and man, other enzymes have also been detected, such as exonucleases, polynucleotide kinases and lygases.

Thus, during the meiotic prophase a host of enzyme systems are activated which control (1) readiness of a cell for meiosis, (2) homologous chromosome pairing, (3) crossing over, etc.

Diplotene. At this stage, the intimate contact between paired homologues begins to disappear, since the SC is disassembled and dispersed. The paired chromosomes begin to repel one another and move apart, with each homologous chromosome of the bivalent regaining its individuality. But the homologues continue to be held together by one or several chiasmata at sites where crossing over has occurred. In organisms convenient for visualization of diplotene under the microscope (e.g. Acrididae), it is clearly observable that only two of the four chromatids of a bivalent are involved in chiasma formation.

In oocytes (developing egg cells), the diplotene may extend in time over months and years, since at this stage of oogenesis chromosomes undergo coiling and synthesize RNA supplying the egg cell with reserve substances. In some cases, diplotene chromosomes exhibit a particularly intense RNA synthesis. Chromosomes like these (lamp-brush chromosomes) occur, for example, in insects, amphibians and man.

Diakinesis. Chromosomes at this stage undergo further contraction by coiling, detach their ends from the nuclear membrane, the SC completely disintegrates and RNA synthesis terminates. Diakinesis completes prophase I of meiosis with all its complex cytogenetic events and sets the stage for separation of homologous chromosomes.

Metaphase I. At this stage of meiosis, the nucleoli disappear, the nuclear envelope dissolves and formation of a spindle is completed. Bivalents line up at the equatorial plate in such a way that homologous centromeres are equidistant from the equator and orient to opposite poles (see above). Chiasmata hold the homologues together until their separation in anaphase. In metaphase I, the stage is set for an orderly assortment of homologous chromosomes into daughter cells.

Anaphase I. Homologous chromosomes start to move toward opposite cell poles (one chromatid from each pair goes to the respective pole of the cell, led by its centromere). Molecular and physical mechanisms of anaphase movement are still unclear. One thing is clear: in anaphase I, separation of homologues (at chiasmata) occurs and, having parted, the homologous chromosomes acquire the ability to move to opposite poles. Each chromosome in anaphase I is represented by a chromatid pair joined by a centromere (dyad). Centromeres are the first to arrive at the poles. Half the number of chromosomes move to each pole, i.e., the complete haploid set which contains two copies of one homologue from each bivalent.

Telophase I. Outlines of chromosomes at this stage are no longer clearly observable under the microscope, the nuclear membrane is restored and, following cell division (cytokinesis), two daughter cells are formed. This stage completes the first meiotic division.

Thus, as a result of the first meiotic division each daughter cell inherits two copies of either homologue and therefore exhibits a diploid DNA content. However, these cells differ from normal diploid cells in two respects: (1) both DNA copies of each chromosome originate from one of the two homologous chromosomes that were present in the original mother cell, and (2) the cell inherits these two copies in the form of two tightly associated (by a centromere) sister chromatids comprising a single chromosome. The DNA content reduces from a tetraploid (4C), exhibited by the cell entering meiosis, to a diploid (2C) one (C = DNA content per haploid chromosome set). This is the condition of daughter cells prior to the second meiotic division.

The **second meiotic division** sets in after a short resting period (interkinesis) and follows the pattern (i.e., sequence of events) seen in mitosis. Since no DNA synthesis occurs during the resting period, one might get the impression that upon completion of the first meiotic division, chromosomes enter the second one straight away. Prophase II is short in all organisms. In prophase II and metaphase II, as in anaphase I, chromosomes appear as two chromatids joined by a centromere. The nuclear envelope dissolves as soon as a new spindle is formed. Then the second meiotic division rapidly proceeds through a succession of phases: metaphase II, anaphase II, and telophase II. As in mitosis, kinetochore fibres of sister chromatids orient to opposite cell poles. Sister chromatids are held together (by the centromere) at the metaphase plate until their separation in anaphase II. In telophase II, a nuclear envelope is formed around each of the four haploid nuclei, cytokinesis takes place, and this brings meiosis to a close.

Thus a single event of DNA synthesis during the premeiotic interphase, homologous chromosome pairing, crossing over and one division of centromeres per two successive meiotic divisions, all represent a sequence of events resulting in the transition from the 2n to the 1n chromosome set, the formation of four haploid cells from a single diploid cell, and a vast genetic diversity of gametes.

The above is the normal meiotic sequence typical of the majority of plant and animal species.

Types of Meiosis

In addition to the classical, two-step meiosis described above, other types of meiosis occur in nature. Of particular interest is the one-step meiosis first described by Cleveland (1947) in flagellates. In essence, it amounts to the omission of the first meiotic division associated with chromosome pairing, crossing over and chiasma formation. One-step meiosis has been described in Sporozoa, some species of rhizopods, and other protozoa (for details, see Raikov, 1975).

Meiosis in organisms with **non-localized centromeres** (the Coccid- and Luzula-types) merits separate consideration. It differs from the classical type in that the first meiotic division is a completely equational one. This is due to homologous chromosomes being arranged tandemly in diplotene-metaphase I of meiosis, resulting in auto-orientation of bivalents instead of co-orientation. In prophase II, homologous chromosome pairing occurs again (see Khvostova and Bogdanov, 1975).

Asymmetrical cytokinesis. Egg cell maturation during oogenesis of different organisms is a significant feature. (Of the entire complex process of egg-cell formation, we only consider two meiotic divisions.) The main distinctive feature of meiosis is that during oogenesis a single haploid egg cell, large in size and rich in nutrient substances, is formed from one diploid cell with a reduplicated (4C) DNA content. This is achieved by two features of meiosis exhibited during oogenesis: (1) a reversible arrest of meiosis at prophase I and (2) asymmetrical division of cytoplasm following each meiotic division.

The arrest of meiosis at diplotene extending in time from a few days to several tens of years is necessary for formation of the egg-cell envelope, which plays an important part in fertilization. Paired chromosomes in this case assume the form of 'lamp-brushes' characteristic of chromosomes actively synthesizing RNA.

As a result of asymmetrical cytoplasm division, the products of the first meiotic division—secondary oocytes—differ markedly in size while exhibiting equal DNA contents and chromosome numbers. One of these is a small polar body, the other a large cell which continues its development.

The second meiotic division again ends in asymmetrical cytokinesis which results in the formation of a large, mature egg cell and a second polar body. Both cells are haploid in DNA content and chromosome number. The first and second polar bodies undergo degradation.

Haploid males of honey-bee (drones) can only form sperms with a haploid chromosome number due to the fact that they have an evolutionally established abnormal meiosis associated with blocking of distribution of chromosomes to the poles. This results in all univalents remaining at the cell centre during the first meiotic division. Asymmetrical cytokinesis insulates a small portion of the cytoplasm without chromosomes, giving rise to a large cell with the original haploid chromosome number and a small cell showing no chromosomes and a small amount of cytoplasm. The second meiotic division follows the same route as normal mitosis, differing only in the formation of two haploid cells of different size as a result of asymmetrical cytokinesis. It is only the larger cell which develops into a viable sperm. Wasp male meiosis follows the same pattern.

Achiasmate meiosis proceeds without crossing over and chiasma formation. It is common in the heterogametic sex of butterflies, dipterans and praying mantis. One group of individuals (e.g. females of the silkworm *Bombyx mori* L.), exhibit a normal, but not disappearing during diplotene, SC which, having become modified, persists within the cell right up to metaphase I; crossing over is absent and no chiasmata are formed. In another group of individuals (exemplified by *Drosophila melanogaster* males), the SC formation is completely blocked and crossing over and chiasmata are absent. Thus, achiasmate meiosis in these two groups of organisms produces similar cytogenetic effects—no crossing over and chiasma formation—but genetic mechanisms blocking these processes appear to differ: in the former group, blocking can be due to genes directly concerned with recombination (e.g. the *mei9* gene in *Drosophila*), in the latter it results from the activity of asynaptic genes blocking the formation of SC.

Apomictic plant species vary in mode of transition from regular meiosis to mitosis. Variation concerns the first meiotic division and is manifested in varying degrees of disruption of homologous chromosome pairing and in blocking of crossing over. In some parthenogenetic animal species (*Carassius auratus*, *Daphnia pulex*), haploid males of Hymenoptera and Protozoa, the first meiotic division may be omitted (one-step meiosis), in others it is partially disrupted, such that crossing over is blocked and regular assortment of homologous chromosomes is disturbed (as in *Poeciliopsis Cx*).

Cytochemistry of Meiosis (Plate 19)

Meiosis is one of the most integrated biological processes, representing a special type of division of differentiating cells. Cytochemical techniques, due to their specific nature, are still incapable of solving the problems encountered; these are amenable to solution by biochemical and molecular biological techniques. However, far from all resources of the cytochemical approach to the study of meiosis have been used. The cytochemical approach can, in some cases, provide useful information on the metabolic status of the meiosing cells, thus appreciably supplementing the results from studies employing molecular biological techniques.

Application of cytochemical methods enables early metabolic changes occurring in cells prior to their entering the meiotic path of development to be detected. Plate 19 shows the results from a cytophotometric study of DNA accessibility to acid hydrolysis (Feulgen staining) in lily (*Lilium candidum*) sporogenous tissue cells at the stages of initiation and completion of mitoses. In both cases, hydrolysis curves exhibit a single peak. However, for the stage of completion of mitoses, this peak is shifted towards shorter hydrolysis. Thus, nuclear DNA at the stage of completion of mitoses is less resistant to acid hydrolysis, suggesting an increase in degree of chromatin decondensation in sporogenous cells coinciding in time with completion of proliferative activity.

The results of a cytochemical assay for RNA and total protein detectable through gallocyanine chromic alum and naphthol yellow staining also suggest perceptible metabolic changes in sporogenous cells occurring in the course of differentiation. At the stage of completion of mitoses, a substantial increase is observed in content of the indicated high-molecular compounds within both the nucleus and cytoplasm.

As mentioned earlier, the bulk of the DNA synthesis occurring during the meiotic cycle is completed in the S phase of premeiotic interphase. Cytochemical analysis of lily microsporocyte nuclei in late premeiotic interphase (G phase) showed these to have a DNA content corresponding to 4C. The DNA content values for individual stages of prophase I and metaphase I remain unchanged, becoming 1C in the nuclei of the forming tetrads.

Cytochemical determinations of histones showed that in contrast to DNA, their content during premeiotic interphase is not doubled. In late premeiotic interphase, the content of histones detectable through staining with fast green corresponds to approximately 3C. The concentration corresponding to 4C is only observable in diplotene-diakinesis. So, while the DNA and histone content doubling occurring during the mitotic cycle proceed in parallel and are completed in the S phase, in premeiotic interphase the histone synthesis lags behind the DNA synthesis, which is also the case during most of prophase I.

The currently available biochemical evidence supports the data obtained using cytochemical methods. Furthermore, not only the suppression of histone synthesis in premeiotic interphase has been corroborated, but also the dynamics of its individual fractions in meiotic prophase I has been elucidated (Bogdanov and Strokov, 1976). The fractions H2a, H2, H3 and H4 have been classed as early-synthesis ones and H1 as a late-synthesis fraction. Along with these fractions, a specific fraction—HM (histone of meiosis)—falling into the category of early-synthesis histones has been detected in lily microsporocytes (Scheridan and Stern, 1967).

Of particular interest are studies of chromatin structural patterns in premeiotic interphase and in prophase I. This problem can also be solved by applying cytochemical methods. Plate 19 shows the curves of acid hydrolysis resistance of nuclear DNA from lily microsporocytes. Without dwelling in detail on hydrolysis curves, two important corollaries of the results obtained are worth noting. First, a higher intensity of chromatin decondensation and its transformation into a more heterogenous state are observed in the late premeiotic interphase. Second, in passing through individual stages of meiotic prophase I, chromosome decondensation reaches maximum during zygotene and pachytene. Towards the end of prophase, an increase in chromosome condensation occurs, reaching maximum in metaphase I.

Based on the results obtained, a logical conclusion suggests itself of a relationship between the increased chromosome decondensation in zygotene-pachytene and the delayed histone synthesis. This finding is consistent with an earlier assumption that the delayed histone synthesis results in chromosome decondensation being maintained at a particular level providing favourable conditions for chromosome pairing and crossing over (Bogdanov and Strokov, 1976). At the same time, this specific decondensation level sustains a high metabolic activity of microsporocytes during prophase I, ensuring synthesis of RNA and proteins required both for meiosis per se and for young microspores (Reznickova, 1984).

The currently available cytochemical evidence suggests that a considerable increase in the RNA and total protein content occurs during premeiotic interphase (Moss and Heslop-Harrison, 1967; Reznickova, 1984). Synthesis of the indicated high-molecular compounds proceeds throughout prophase I. However, temporal variation in the intensity of this synthesis is not the same in different plant species.

The distinctive feature of RNA synthesis in meiotic prophase I is that chromosomal RNA accounts for the bulk of the RNA synthesized. Thus, a cytochemical study of lily microsporocytes failed to reveal a noticeable reduction in nuclear RNA content (gallocyanine chromic alum staining) even where the nucleolus became greatly reduced in volume, resulting in a fairly sharp decrease in its RNA content. Evidence from electron cytochemical studies indicated that RNA synthesized during prophase I occurred in peri- and interchromosomal regions (Porter *et al.*, 1982).

The foregoing suggests that underlying meiotic cell differentiation are events occurring at the molecular level. The first visible changes in cell metabolism are presumably observable already in the sporogenous tissue. Accumulation of total intracellular proteins can probably be regarded as a factor reducing the rate of cell division and subsequently completely suppressing mitoses. It is well known, for example, that cells whose total protein content has reached a particular level fail to enter mitosis even on induced proliferation (Magakyan *et al.*, 1987).

The passage of cells through premeiotic interphase and prophase I is marked by changes in chromatin and chromosome structure as well as by increased RNA and protein contents. The duration of meiotic prophase I with its characteristic chromosome structure is predetermined, presumably as early as in premeiotic interphase, and associated with delayed histone synthesis.

Microspore Tetrad (Greek *tetra* = four + *micros* + *sporá*)—a structure composed of four haploid cells arising in the course of two meiotic divisions of a microsporocyte (Plate 20).

Formation of tetrads and release of individual microspores in flowering plants were first described by Nageli (1842). Based on the timing and mode of cell plate formation, Hofmeister (1861, 1867) distinguished **two types of tetrad formation: successive** or succedane and **simultaneous**. Subsequently a more detailed classification of tetrad formation types was proposed (Sampson, 1969).

Under the successive type of tetrad formation, each meiotic division results in the formation of cell plates. The phragmoplast is laid down during the telophase of heterotypic division (meiosis I) in the centre of the microsporocyte and grows centrifugally, dividing the cell into two equal cells. In a similar fashion, the cells of a dyad are divided following homeotypic division (meiosis II), which results in the formation of four cells. Under the simultaneous mode of tetrad formation, cytokinesis is completely absent following meiosis I. Four microspore cells arise simultaneously after meiosis II via initiation of furrows in the form of microsporocyte wall invaginations from the cell circumference to the centre and division of the microsporocyte protoplast into four cells. In individual cases, an incomplete cell plate (*Zygoginum*) or its evanescent precursor (*Pseudowintera, Magnolia*) is formed. In some plants, centrifugal cell plate formation occurs at the end of meiosis II (*Helleborus*—Waterkeyn, 1962; *Illicium, Schisandra, Kadsura*—Hayashi, 1960; Kapil and Jalan, 1964). This type of division is exemplified by phragmoplast 'reconstruction' and appearance of secondary spindles in late meiosis II (*Laurelia*) (Sampson, 1969). Under the simultaneous type of tetrad formation, in members of a number of families (Cucurbitaceae, Plumbaginaceae and Fabaceae) an important role in microsporocyte division into microspores is played by vacuoles, which fuse to divide the microsporocyte protoplast into separate microspores.

A third type of tetrad formation, the **intermediate** one, has been described under which, following meiosis I, a furrow is laid down around the microsporocyte circumference, but towards the end of interphase II furrow growth is arrested, to be resumed after meiosis II; concurrently, initiation and growth of a second furrow occurs, resulting in the formation of four cells. The intermediate type of tetrad formation is closer to the simultaneous type.

Thus, during tetrad formation cytokinesis occurs by cell-plate formation or by furrowing. According to recent electron microscopic evidence, the vesicle-microtubular complex of the cytoplasm is actively involved in both processes (Kordyum, 1978; van Went and Cresti, 1988b). Concurrent with cell-plate formation, callose is deposited around the microspores.

Many authors regard the type of microspore tetrad formation as a taxonomic character. The simultaneous mode of tetrad formation is typical of dicotyledonous flowering plants, the successive one of monocotyledonous. However, exceptions are possible. Thus among dicotyledons, only the successive type occurs in families Myristicaceae, Monimiaceae, Lauraceae, Asclepidaceae, whereas both types are found in Annonaceae, Aristolochiaceae, Ceratophyllaceae, Cabombaceae, Berberidaceae, Ranunculaceae, Staphyleaceae and Plantaginaceae. Among monocots, only the simultaneous type occurs in Phormiaceae, Juncaceae, Cyperaceae, whereas both types are found in Butomaceae, Aponogetonaceae, Liliaceae, Iridaceae and Taccaceae.

The intermediate type has been described in dicots (Magnoliaceae, Annonaceae, Winteraceae, Nelumbonaceae and Plantaginaceae).

Which of the types of tetrad formation is evolutionally more advanced remains an open question. Some authors (Goebel, 1923; Schnarf, 1931; Sladkov, 1962, 1970) believe the simultaneous type to be more primitive since it occurs not only in flowering plants, but also in archegoniates (Gymnospermae, Lycopodiophyta, wedge-leafed plants, Polypodiophyta, and Bryophyta). Other authors (Schurhoff, 1928; Takhtajan, 1948, 1964; and others) conversely look upon the successive type as more primitive since, in their view, formation of free nuclei is a more advanced feature. Still others (Tackholm and Soderberg, 1918) hold that evolution from one mode of tetrad formation towards another could have followed different pathways in different branches of the system: in some branches one mode could have been more primitive, in others another.

Tetrads formed as a result of meiosis vary in shape and mutual arrangement of component cells according to mode of tetrad formation and microsporocyte shape.

Six types of tetrads formed have been reported: tetrahedral, isobilateral, decussate, rhombic, T-shaped and linear (Boehm, 1931; Maheshwari, 1950).

One previously advanced view (Schnarf, 1929), subsequently corroborated by other workers (Sladkov, 1989; Sladkov and Grevtsova, 1991; Rangaswamy and Janaki Subramanyan, 1992), holds that only tetrahedral tetrads are formed under simultaneous cytokinesis. Under successive cytokinesis, all types of microspore tetrads, with the exception of tetrahedral ones, are formed (Sladkov, 1989). According to Sladkov (1989), misidentification by many authors of the type of tetrad probably stems from the fact that they determine it on the basis of the already formed, but not yet released, microspores whose shape does not correspond to that of the cells formed immediately following meiosis, and fail to relate the mutual arrangements of microspores to the positions of the tetrads themselves.

Immunofluorescence of Cytoskeleton during Microsporogenesis

Knowledge concerning the organization of cytoskeleton has recently been extended through the use of indirect immunofluorescence and immunogold methods (Gunning and Hardham, 1982). By these methods, the position of cytoskeletal proteins was demonstrated in microsporogenesis and megasporogenesis.

Preparations of cells are treated with an antibody against a certain cytoskeletal protein. Once the antibody is bound to the protein, the preparations are treated with an antibody immunoglobulin, which attaches itself to the first antibody. The immunoglobulin is labelled with a marker, usually fluorochrome (FITC, rhodamine) or particles of colloidal gold. The markers show the sites of cytoskeletal proteins in the cells.

The arrangement of cytoskeleton elements changes during microsporogenesis. The microtubules are organized in several arrays: a cytoplasmic network, meiosis I and II spindles, and telophase I and II phragmoplasts (Van Lammeren *et al.*, 1985). There is a conspicuous lack of a preprophase band of microtubules which, in mitotically dividing cells, predicts the position of a cell plate formed after telophase (Hogan, 1987; Traas *et al.*, 1989).

In early prophase I microsporocytes, the cytoplasmic microtubules are randomly distributed. At diplotene, new microtubules arise at the nuclear envelope and form a dense perinuclear array. Later, in prophase I, microtubules appear at bivalent chromosomes and the bundles of these microtubules are attached to kinetochores. It may be assumed that these microtubules participate together with perinuclear ones in formation of the spindle (Brown and Lemmon, 1991), which transforms into a system of anaphase polar microtubules and fluorescent caps of kinetochore bundles. At late telophase I and early interphase, the microtubules apparently originating at the nuclear envelopes and some spindle microtubules form a phragmoplast, which reaches the surface of sister nuclei. This situation seems to be specific for meiotic cells; in mitotic cells the phragmoplasts are less developed and do not extend to the nuclear surfaces.

In the equatorial plane of the telophase I phragmoplasts, either a cell plate or an aggregation of organelles is formed. Formation of a cell plate is typical for successive cytokinesis, aggregation of organelles for simultaneous cytokinesis. In both types the phragmoplasts disappear during interkinesis and by prophase II perinuclear and kinetochore microtubules have formed again.

In plants with successive cytokinesis, the meiosis II spindles and phragmoplasts develop and act separately in both cells of a dyad. In plants with simultaneous cytokinesis, two spindles are formed, and after meiosis II two barrel-shaped phragmoplasts arise between two pairs of sister nuclei. Then the microtubules interconnect non-sister nuclei and take part in the formation of secondary phragmoplasts. The cytokinetic process is carried out by a quadripolar system of phragmoplasts, and the coenocytic microsporocyte is divided simultaneously into a tetrad.

Actin microfilaments were detected as a component of the microtubular cytoplasmic network, spindles and phragmoplasts (Van Lammeren *et al.*, 1989). They also seem to enclose the spindle in a cage-like network which extends into the cytoplasm.

During megasporogenesis of the monosporic type, the microtubule arrays are similar to those in microsporogenesis (Willemse and Van Lammeren, 1988; Webb and Gunning, 1990).

MICROSPORE

Microspore (Greek *micros* = small, *sporá* = seed, spore)—one of the cells resulting from meiosis.

The term 'microspore' is often misused as a synonym for the term 'pollen grain'. According to Schnarf (1929) and Maheshwari (1950), the microspore (more precisely, its 'germination') and the pollen grain represent different stages of the gametophyte generation. Nägeli (1842) gave the first description of microspores. Upon completion of meiosis, a tetrad of microspores enclosed in callosic walls is formed. Structural features of the microspore surface and location of pores on it are already observable at the stage of tetrads during primexine synthesis. Synthesis of primexine components within a tetrad is performed by the protoplast of each individual microspore. Primexine is composed of pectocellulosic microfibrils located between the plasmalemma and the callosic wall of the microspore. As callose undergoes enzymatic degradation, sporopollenin synthesized by tapetal cells is deposited in its stead on the microspore surface, resulting in the formation of exine. Also synthesized at the microspore stage is the inner layer of the wall of the future pollen grain, the intine, whose primary function is formation of a wall lining the aperture. In a number of genera of flowering plants (*Pseudowintera, Victoria, Philydrum, Luzula, Typha,* etc.), microspores remain within the tetrads and continue their development to form complex pollen aggregates.

Microspores vary in developmental time: from a few hours or days in tropical and subtropical plants to several months in many arctic plants. A number of classifications of microspore developmental stages have been proposed (Romanov, 1970a-c; Batygina, 1974b; Vithanage and Knox, 1980; Willemse and Reznickova, 1980).

The microspores released from a tetrad exhibit dense, non-vacuolated cytoplasm and a nucleus, with one or several nucleoli, located in the centre of the cell. An appreciable growth of microspores is normally observed, which is accompanied by cytoplasm hydration and vacuolation. Then the process of cytoplasm autophagy commences, followed by formation of new cytoplasm and ribosomes. As a result of amylogenesis, protoplasts differentiate into amyloplasts containing starch (Pacini, 1990). The starch undergoes hydrolysis during intine and aperture formation.

Subsequent amylogenesis and new cytoplasm vacuolation occur following mitosis I of the pollen grain.

During vacuolation the microspore rapidly grows in size. Thus in grasses, the microspore diameter increases more than twofold, reaching 3/4 the size of a mature pollen grain. The nucleus is pushed away from the central position towards the periphery, to a parietal position, which is often attributed to formation of a large central vacuole (Poaceae, Campanulaceae, Juglandaceae and Saxifragaceae). In some cases (*Haworthia, Juncus, Luzula, Ruppia*), no vacuole is present but the nucleus nevertheless ends up in a parietal position. With respect to the aperture, the nucleus position within the microspore prior to mitosis is strictly fixed for each particular plant species. This is probably due to the presence within the microspore of chemical and physical gradients which are determined by the architectonics and polarity of its protoplast and already observable during formation of microspore tetrads (Geitler, 1935; Maheshwari, 1950; Gerassimova-Navashina, 1951a, b; Vazart, 1958; Batygina, 1962a, b; Romanov, 1970a-c).

Microspore polarity is especially pronounced in grasses and manifests in the position of the microspore nucleus in relation to the aperture. The microspore aperture faces the tapetum. Prior to mitosis, the microspore nucleus is removed as much as possible from the aperture, whereas its position relative to the aperture may vary during microspore vacuolation and growth (Batygina, 1962a, b, 1974b, 1987a; Kihara and Hori, 1966; Romanov, 1966, 1970a, b).

In the course of development of a microspore, its cytoplasm, nuclei and nucleoli expand appreciably and the pattern of cytoplasm vacuolation undergoes changes. The condition of cytoplasmic organelles is intimately associated with the processes occurring during microspore development (Van Went and Cresti, 1989). Organelles, which are subject to particularly drastic changes, are populations of ribosomes, matrix and internal membrane systems. The ribosome number and distribution correlate strongly with the accumulation of nucleoloids in the cytoplasm. During formation of primexine, exine and intine, the shape of ER and of dictyosomes as well as the number and size of vesicles undergo changes. The abundance of ER and dictyosomes is closely associated with the intense growth of microspores and vacuolation of their cytoplasm. The microspore completes its development and enters a differentiation mitosis, which results in formation of two cells of the male gametophyte: vegetative and generative.

Microspore Cytochemistry (Plate 21)

In the course of their development, microspores greatly increase in volume. Using cytochemical and biochemical techniques, it has been possible to demonstrate that the developing microspores exhibit an increase in DNA, RNA and protein content (Bryan, 1951; Ogur *et al.*, 1951; Taylor, 1958; Woodard, 1958).

Using lily (*Lilium candidum*), sage (*Salvia sclarea*), lavender (*Lavandula angustifolia*) and coriander (*Coriandrum sativum*) as an example, microspore nuclei have been demonstrated to have an increased DNA content. In late interphase prior to mitosis (G_2 phase), the DNA content corresponds to 2C. The pattern of variation in histone content is basically similar to that of DNA content. Notwithstanding the fact that in microspores of lily, sage and lavender the increase in histone content lags somewhat behind that in DNA content, the histone content is 2C in late interphase prior to mitosis.

The process of microspore development is characterized not only by DNA and histone quantitative changes, but also by structural changes in the nucleus chromatin. Analysis of DNA susceptibility to acid hydrolysis makes it possible to determine the direction of changes in physicochemical condition of nucleus chromatin of the developing microspores of lily. The results of cytochemical studies indicate that microspore development is accompanied by changes in physicochemical condition of chromatin and by an increase in DNA content detectable upon a relatively short hydrolysis. This suggests that in the developing microspores of lily, there is a tendency towards chromatin decondensation and increased metabolic activity which reaches maximum prior to mitosis.

The results obtained enable the metabolic status of chromatin to be evaluated at different stages of interphase. The metabolic activity of chromatin appears to be relatively low during most of the G_1 phase. An increased level of metabolic activity is apparently observable prior to and during the S phase. Further increase in metabolic activity occurs during G_2, prior to microspore mitosis.

Currently available cytochemical and biochemical evidence suggests that RNA synthesis in microspores proceeds throughout the interphase, such that maximum intensity of this synthesis is observed following completion of DNA replication (Steffensen, 1966). The RNA synthesized in the course of development of microspores belongs to the ribosomal RNA type (Peddada and Mascarenhas, 1975).

RNA content of the nucleolus, nuclear chromatin and cytoplasm of the developing microspores exhibits a specific pattern of variation. This is evidenced by the results of cytochemical analysis of RNA during development of lily microspores. During the first half of interphase, RNA content increases both in the cytoplasm and in the nuclear chromatin and nucleolus. Prior to the S period in microspores, RNA content of the nucleolus declines while that of the nuclear chromatin and cytoplasm increases. With the onset of the S phase, RNA content tends to decrease in the nuclear chromatin and to increase in the cytoplasm and within the nucleolus. The results suggest a rather complex relationship between RNA synthesis and presumably, transport of the synthesized RNA from the nucleus to the cytoplasm. They simultaneously indicate that as microspores develop the overall RNA content (per cell) increases, reaching maximum prior to mitosis.

The regularities established are also observable in other plant species. Thus, in the course of development of sage, lavender and coriander microspores, an increase in RNA content occurs in both the nucleus and cytoplasm.

In developing microspores, the pattern of changes in total protein content is normally similar in many ways to that of RNA content (Moss and Heslop-Harrison, 1967). For instance, during lily microspore development total protein content is seen to increase in both the nucleus and cytoplasm.

Thus, microspore development exhibits some special features of DNA, RNA, and nuclear and cytoplasmic protein synthesis. The intensity of synthesis within microspores is probably low due to the process of microspore development being greatly extended in time. The exception is provided by late stages exhibiting higher intracellular metabolic activity, as evidenced by greater chromatin decondensation and higher RNA content.

Ultrastructural Aspects of Microspore Development (Aggregates of Endoplasmic Reticulum) (Plate 22)

Aggregates of endoplasmic reticulum (ER) are **three-dimensional structures** of a very peculiar composition. They **occupy so much of the cell cytoplasm** that they are visible even under the light microscope. They can arbitrarily be divided into two types.

The first type are so-called **chain-mail aggregates** (French *aggregates en cotte de mailles*), or 'chain-mails', reported by Eymé (1967) in cells of *Diplotaxis* and *Ficaria* nectaries. Eymé (1967) speculates that *chain-mails* accumulate lipoproteins. Precisely this term, proposed by Eymé, appears to be the most appropriate one for describing these unusual ER aggregates whose sections exhibit parallel sinusoidal (in one plane of sectioning) or linked circular (in another plane of sectioning) contours. Synonyms: pseudocrystal lattices (Wooding, 1967), labyrinths (Behnke, 1968), viewed as a reserve of membranes within the cell; convoluted ER aggregates (Esau and Hoefert, 1980) occurring during disorganizing processes within a cell; further, their presence is taken as indicative of specific cytoplasmic differentiation associated with the primary function of the cell (Esau and Charvat, 1975), with the authors reporting the presence in cisternae of these aggregates of acid phosphatases and ATPases. Evert (1984) also believes them to be sources of autolytic enzymes. Chain-mails discovered in developing microspores—**contorted ER** (Hoefert, 1969)—of species belonging to the genus *Beta* are directly bound to the nuclear envelope and, according to the author, are involved in the transport of substances from the nucleus to the plasmalemma. The same structures have been observed during rapid growth of intine in *Tradescantia bracteata* (Mepham and Lane, 1970); the authors assume ER aggregates to be involved in rapid transport to the plasmalemma of substances required for intine assembly. These and other ER aggregates like them have been found in developing phloem constituents (Telepova, 1985).

Chain-mail ER aggregates are constantly present in the cytoplasm of *Michelia fuscata* and *Manglietia tenuipes* (Magnoliaceae) microspores throughout most, if not their entire, period of development (Gabarayeva, 1986, 1987, 1991a). During the time between the early tetrad period and mid-post-tetrad, the aggregates undergo a number of changes. However, during primexine formation, there is an invariable contact between chain-mails and plasmalemma mediated by ER cisternae carrying ribosomes and being direct branches of smooth chain-mail cisternae. The fact that the intracisternal space of chain-mails is filled with an amorphous, moderately electron-dense substance and that they are permanently associated with the developing primexine suggests a synthetic function for these structures. In any case, they are undoubtedly involved in the primary function of the cell, viz. formation of a complex pollen grain wall. Considering that smooth endoplasmic reticulum (SER) is implicated in the synthesis and transport of terpenes, steroids and lipids (Vassilyev, 1977) and that moderately electron-dense material, most likely of lipid nature, is first deposited in the primexine matrix of *Michelia fuscata* and *Manglietia tenuipes* and then sporopollenin, a terpenoid substance, polymerized on this substrate, it can be suggested that precisely these substances (lipids, terpenoids) are synthesized and transported to the microspore plasmalemma by chain-mails (Gabarayeva, 1990, 1991b).

The **second type of ER aggregates** is even more unique; these aggregates penetrate the entire cytoplasm of the developing microspores of *Liriodendron chinensis* (Gabarayeva, 1990, 1991a). They consist of stacks of long tubular cisternae arranged parallel to one another. The number of cisternae in aggregates is normally odd, ranging from 3 to 13. The central cistern invariably contains an electron-dense, osmiophilic substance; in polycisternal aggregates, electron-dense cisternae usually alternate with electron-transparent ones, which is why they have been termed **zebra aggregates** (Gabarayeva, 1991a). These large aggregates form undulations in the cytoplasm, branch or form rings. Association of the ends of osmiophilic cisternae with lipid globules invariably present nearby in the cytoplasm is clearly discernible. The outer electron-transparent cisternae of the aggregates carry ribosomes, with the ends of the peripherally located aggregates lying immediately adjacent to the plasmalemma. Given the stages of *Liriodendron chinensis* microspore development at which zebra aggregates occur, the contact of the aggregate cisternae with the plasmalemma and the particularly intense accumulation of sporopollenin in the exine occurring at the time, it would appear reasonable that these structures should be involved in synthesis of sporopollenin precursors. Assumption of the lipoprotein nature of the substances synthesized, as a step in the biosynthetic pathway of this biopolymer, has been substantiated by an electron cytochemical enzymatic reaction with pronase (Gabarayeva, 1990). Structures somewhat similar to those described above (twin ER cisternae with osmiophilic material in-between and ribosomes on the outer surfaces) were detected by Fleurat-Lessard (1986) in parenchymal cells of *Mimosa pudica*. He believes them to be involved in synthesis of proteins, presumably contractile ones. ER aggregate cisternae have been found in cancer cells (Schaff *et al.*, 1972; cited in Fleurat-Lessard, 1986). Structures like these have also been reported in encysting zoospores of the fungus *Pythium aphanidermatum* (Grove and Bracker, 1978); however, their function is not discussed by the authors. The occurrence of this type of ER aggregate in organisms as dissimilar as higher plants, fungi and animal cancer cells is noteworthy. Such ER aggregates are likely to be of general cytological significance; they arise where and when there is a need for intense synthesis of particular compound proteids, the synthesis requiring spatial separation of the initial components (proteins and prosthetic groups).

POLLEN GRAIN

Pollen Grain—a multicellular structure arising by one or two mitotic divisions in the microspore, resulting in the formation of a male gametophyte. Synonym: male gametophyte[4] (Plate 23).

The microspore 'germinates', giving rise to a pollen grain which consists of two or three cells: vegetative and generative (bicellular) or vegetative and two sperms (tricellular). Three types (modes) of pollen grain development are distinguished: 'normal' type, Triglochin type, and Juncus type (Eames, 1961). Under the normal type, which is typical of most flowering plants, microspores greatly increase in volume prior to division. In the case of the Triglochin type, the microspores first undergo division and then increase in volume. This type of pollen grain development occurs

[4]Some workers extend the notion of 'male gametophyte' to the pollen tube.

in a number of monocots (*Najas, Triglochin, Ruppia, Lilaea,* and *Aponogeton*). The Juncus type is restricted to members of Juncaceae and Cyperaceae and, in Eames' (1961) view, is not clearly distinguishable from the other types. However, currently available evidence suggests that in Juncaceae (Anisimova, 1987), there is no difference in size between the microspore and the pollen grain.

The first division in the microspore is defined as an unequal or differentiation mitosis which results in the formation of two morphologically and physiologically distinct cells: **vegetative** and **generative**. A distinguishing feature of the first division is the position and orientation of the division spindle which is invariable for each particular plant species and is determined by a strictly fixed position of the microspore in the parietal layer at a particular microspore wall (outer, inner or radial, in terms of location within a tetrad) (Geitler, 1935; Gerassimova-Navashina, 1951a, b). The division spindle runs parallel to the longitudinal axis of the pollen grain (the Hofmeister-Hertwig rule) and perpendicular to the pollen wall adjoining the nucleus. The division spindle often tends to be asymmetric which is due to the nucleus position within the pollen grain. The degree of spindle asymmetry is responsible, in turn, for differential behaviour of chromosomes in late anaphase and in telophase (Geitler, 1935; Gerassimova-Navashina, 1951a, b). The presence of an asymmetric spindle is not necessarily a prerequisite for an unequal division, however, since even under a symmetric division configuration (*Asclepias, Anthericum, Adoxa, Sambucus, Cuscuta, Uvularia* and others) unequal cells are formed as well (Geitler, 1935; Maheshwari, 1950).

The resultant generative and vegetative cells are separated from each other by plasmalemmae; plasmodesmata are completely lacking (Van Aelst and Van Went, 1991). Callose has been reported to be temporarily present around the generative cell in a number of plants (*Hyacinthus orientalis, Tradescantia bracteata, Impatiens balsamina*), as well as in members of genera *Triticum* and *Chlorophytum* (Gorska-Brylass, 1967; Romanov, 1970a-c; Reznickova, 1984).

The second mitotic division occurring in the generative cell results in the formation of two male gametes (sperms). The generative cell or the sperms come to lie in the cytoplasm of the vegetative cell, the latter, as pollen matures, becoming rich in oxidative enzymes, vitamins, physiologically active and reserve substances.

According to the reserve substance composition, pollen grains can be classified into fat-containing (Juglandaceae, Eupomatiaceae, Frankeniaceae, Stachyuraceae, Lauraceae, Salicaceae, Pittosporaceae and Polygonaceae) and starch-containing (Alliaceae, Aizoaceae, Betulaceae, Butomaceae, Moraceae, Trapaceae and others). In a number of families, pollen grains accumulate starch, lipids and proteins (Brassicaceae, Grossulariaceae, Sonneratiaceae and Ruppiaceae).

The pollen wall consists of the exine (outer layer) and intine (inner layer). The exine surface exhibits furrows, pores, rimae and colpi, extending down to the intine, which are collectively called the 'aperture' and represent sites of pollen tube germination (germinal pores). Pollen grains vary widely in exine structure, aperture number and shape.

Pollen grains tend to be orbicular, ellipsoid or oblong but may also be lobate or polygonal. Filiform (Cymodoceaceae, Zosteraceae) and crescent (Ruppiaceae) pollen grains are variations from the normal pollen shape.

Normally, pollen grains occur singly, and less commonly as compound aggregates in the form of dyads (Scheuchzeriaceae), tetrads (Empetraceae, Epacridaceae, Juncaceae, Droseraceae, Philydraceae, Typhaceae, Velloriaceae and Sapindaceae), polyads (Mimosaceae) and pollinia (Asclepiadaceae, Orchidaceae).

As already noted, pollen grains may be bicellular or tricellular. Among the dicots, bicellular pollen grains are typical of 195 families while tricellular ones occurring in 98 families. Among the monocots, bicellular pollen grains occur in 41 families and tricellular ones in 28 families. Both types of pollen grains occur in orders at different levels of the phylogenetic system of flowering plants (Kamelina, 1991).

Pollen Wall (Plate 24)

The pollen wall consists of the exine and intine (inner polysaccharide layer). The exine is known to be composed mainly of sporopollenin, a highly resistant polymer of carotenoids and carotenoid esters (Brooks and Shaw, 1978). However recently, using the high resolution solid state ^{13}C Nuclear Magnetic Resonance technique, Guilford *et al.* (1988) demonstrated that sporopollenin is not a unique substance, but rather a series of related biopolymers derived from largely saturated precursors such as fatty acids. Schulze Osthoff and Wiermann (1987) showed that phenols represent integral components of exine. Wehling *et al.* (1989) assumed that p-coumaric acid is a genuine structural unit in the exine skeleton (sporopollenin).

The exine layer comprises several substrata (Fig. 5):

1. **Tectum**: partial or complete covering layer. Micropores or microchannels may perforate the tectum (tectate pollen). Tectate pollen grains are: (a) pertectate—the tectum is continuous, i.e. unbroken (e.g. *Triticum aestivum*), (b) subtectate—the tectum is not continuous (e.g. *Erucaria crassifolia*). Pollen grains without an apparent tectum are intectate (e.g. *Franklandia triaristata*, Proteaceae; Erdtman, 1969). A tectum may be smooth or carry processes of various kinds.
2. **Baculum**: bacula are generally arranged between the tectum and nexine. They are rod or rod-like elements. Bacula can vary in length and thickness. They may be straight or branched. The tectum and bacula form a stratum called the sexine (Erdtman, 1969).

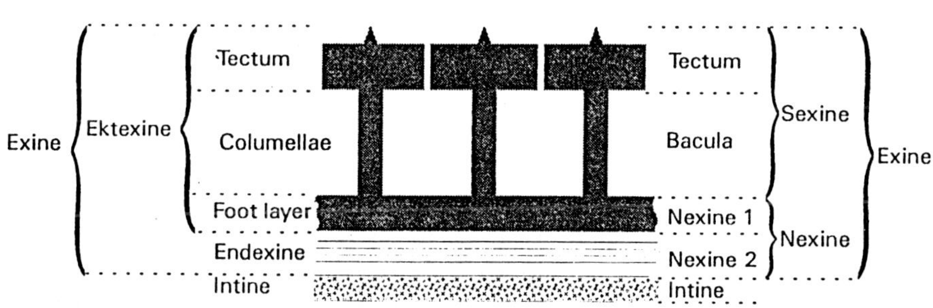

Fig. 5: Stratification of pollen wall in pollen grains of most angiosperms: according to Erdtman (right) and Faegri (left).

3. **Nexine**: appears to be more homogeneous in structure than the sexine. The upper (proximal) part of the nexine has stainability properties similar to those of the sexine. This upper part of the nexine is referred to as nexine-1. Sexine and nexine-1 together form exine-1.
4. **Nexine-2** or **endexine:** differs generally in stainability from nexine-1, nor does it have the same fine structure as nexine-1. Nexine-2 may sometimes be the only constituent of nexine, in other cases it is very reduced or even lacking.

Several terms have been used in describing the exine layers. The most common terms are those of Faegri and Inverson (1975).

The intine consists of less intergrading layers. It comprises a thick polysaccharide layer, often with radially oriented tubules, and is usually thickened at the germinal apertures. Cases are known in which the intine consists of more than one layer. These layers can be differentiated morphologically and chemically (Heslop-Harrison, 1979a, b; Kress and Stone, 1982; see also Knox, 1984).

Development of pollen wall: Following meiosis, tetrads of haploid microspores are formed. The tetrads are then enveloped by a callosic wall and disposed peripherally in the locule close to the tapetum. Callose, a β-1,3-polyglucan, can be detected around the pollen mother cells during initiation of meiosis. It forms a layer between the cytoplasm and pollen mother cell wall. Additional callose is formed after the second meiotic division and isolates the young microspores (Heslop-Harrison, 1968a,b). The plasma membrane of the tetrad microspores is coated with fibrillar material that increases in thickness gradually. The plasma membrane coat (primexine, Heslop-Harrison, 1963; glycocalyx, Rowley and Dahl, 1977) is thought to form receptors for sporopollenin precursor accumulation and polymerization. The callose wall of the tetrad starts to degrade and young microspores are released in the anther locule. Vacuolation of the cytoplasm starts soon after microspore release. In *Triticum aestivum* (El-Ghazaly and Jensen, 1986a,b), the vacuole expands as the volume of young pollen grain increases. With further development, the diameter of the vacuolate microspores increases to more than double the thickness of their wall.

Microspore walls increase gradually in thickness from young free microspores to vacuolate microspores. Development of a large vacuole in the cytoplasm of young pollen grains presumably initiates pressure on the wall of the developing pollen grains. As a result of this pressure, the layers of the wall are appressed against each other. In some cases, as in grasses, a rudimentary endexine becomes appressed against the proximal surface of the foot layer and can no longer be seen (El-Ghazaly and Jensen, 1986a). Development of the inner nexine layer (nexine-2 or endexine) and the intine may occur during the vacuolate period, as in *Vicia* (Audran and Willemse, 1982) and *Lilium* (Heslop-Harrison and Dickinson, 1969). Formation of nexine-1 (foot layer) and nexine-2 (endexine) generally takes place by formation of lamellae at the plasma membrane. These lamellae form the sporopollenin-deposition surface. 'The intine is usually laid down after the exine' (Erdtman, 1969).

Intine development could be associated with the plasma membrane, as in *Triticum aestivum* (El-Ghazaly and Jensen, 1986a, b). In some pollen types, Golgi bodies are frequent during intine synthesis, e.g. Ranunculaceae (Roland, 1971), while in others, endoplasmic reticulum and polyribosomes are abundant, e.g. *Cosmos* (Knox and Heslop-Harrison, 1970).

Ultrastructural aspects of the pollen wall have been studied in many plant species. For example, Rowley *et al.* (1981) presented the substructure in exines of *Artemisia vulgaris* following partial oxidation. They proposed a model of glycocalyx unit-complex (tuft) consisting of a supercoiled helical subunit (binder) wound around several axially straight helical subunits. Tuft units are radial in bacula and throughout the nexine.

Generative Cell (Latin *genero* = produce, generate, give rise to)—the smaller cell of a bicellular pollen grain arising by differentiation mitosis and giving rise through further division to two sperm cells. Synonyms: antheridial cell, spermatogenous cell, spermiogenous cell. Nägeli (1842) provided the first description of two unequal 'nuclei' (cells; G.A.) within the pollen grain. Strasburger (1884) elucidated their functions and designated the smaller as the 'generative' cell.

The generative cell originally abuts against the pollen wall. Its position is species-specific and is a character of taxonomic significance. It depends on the pollen grain architectonics (structural pattern), microspore protoplast polarity and is predetermined already within a tetrad (Geitler, 1935; Gerassimova-Navashina, 1951a; Vazart B., 1958; Romanov, 1970a, b). Initially, the generative cell is lenticular in shape. Electron microscopic evidence indicates that the generative and vegetative cells are only separated by their plasmalemmae (Vazart J., 1969, 1971). Gorska-Brylass (1967) reported for the generative cell the formation of an auxiliary (transient) callosic wall in two species of *Chlorophytum*, in *Hyacinthus orientalis, Tradescantia bracteata* and *Impatiens balsamina,* which presumably plays a role in generative cell differentiation. As a result of unequal mitosis, the generative cell becomes cytoplasm poor but contains the normal set of cell organelles except for the plastids (starch-free), which may be lacking in a number of plants. The generative cell mitochondria are smaller than those of the vegetative cell and possess a well-developed internal membrane system. The generative cell possesses a few dictyosomes with Golgi vesicles, rough endoplasmic reticulum (RER) cisternae, free ribosomes, spherosomes, lipid globules and small vacuoles. Microtubules arranged in stacks and oriented parallel to the longitudinal cell axis are located in the peripheral regions of the cytoplasm (Van Aelst and Van Went, 1991). The nuclear envelope of the generative cell is porous.

Upon phragmoplast resorption, RNA content in the generative cell cytoplasm decreases; then, in the course of differentiation, intense synthesis of RNA occurs and its content in the cytoplasm and within the nucleus is increased, while it concurrently declines within the nucleolus (Bolkhovskikh, 1965). DNA content in the generative nucleus is doubled relatively rapidly compared to that in the vegetative nucleus, while the increase in histone content is not so rapid (Reznickova, 1984). The generative cell cytoplasm, unlike that of the vegetative cell, lacks reserve substances, in particular starch.

An important stage in development of the generative cell is its detachment from the pollen wall and movement towards the vegetative nucleus (Batygina, 1962a, b, 1974b, 1987a; Kihara and Hori, 1966; Romanov, 1970a, b; Schlag and Hesse, 1922). Detachment of the generative cell from the pollen wall starts with loss of its lenticular shape. The central portion of the generative cell, together with the nucleus, protrudes inside the vegetative cell, the process accompanied by departure of the cytoplasm from the regions adjacent to the pollen wall; the contact surface between the generative cell and the pollen wall intine rapidly diminishes to the point of complete separation. The generative cell takes on a spherical or a radially elongated shape. Shortly after detachment of the generative cell from the pollen wall, it moves towards the vegetative nucleus, finally coming to lie in the immediate vicinity of the nucleus. In the course of further development, the generative cell is transformed and moves away from the vegetative nucleus. More often, the generative cell assumes a spindle, elliptic (ovate), occasionally crescent shape. In some plants (*Allium, Cuscuta, Ottelia, Tradescantia, Paeonia*), the generative cell is very long, such that its length may be equal to the pollen grain length or width. In *Calathea*, it is three times the diameter of the pollen wall. In *Erythronium*, the generative cell is amoeboid in shape. In the case of tricellular pollen, it is common for the generative cell to divide. The entire process of division—arrangement of chromosomes and spindle formation—is no different from a regular mitosis. Cell division occurs either by constriction or via initiation of a cell plate. Cases are conceivable wherein, in some pollen grains, a well-defined cell plate is laid down which disappears before long, and final separation of the sperm cells takes place due to formation of a constriction arising soon afterwards. In other pollen grains, the cell plate, once formed, persists, whereas formation of a constriction is delayed, albeit its initiation is conspicuous at one of the ends. If that is the case, the generative cell is divided by the cell plate prior to formation of a constriction. In the course of division of the generative cell its organelles migrate to the peripheral regions of the cytoplasm. Occurring in the vicinity of the spindle are microtubules, individual RER cisternae and numerous free ribosomes. During late telophase, dictyosome vesicles and microtubules arrange themselves in the neighbourhood of the future cell plate, contributing to its formation (Charzynska *et al.*, 1988).

Ultrastructure of Generative Cell (Plate 25)

The generative cell has been the subject of ultrastructural investigations since the early 1960s. It is an elongated cell of ovate, crescent or spindle shape. In many species, a distinct cell wall occurs between the plasma membrane of the generative cell and the adjacent inner plasma membrane of the surrounding vegetative cell. This wall probably contains cellulose and pectin (Maruyama *et al.*, 1965; Jensen *et al.*, 1974; Nakamura and Miki-Hirosige, 1985; Cresti *et al.*, 1987) and, at an early stage, may be composed of callose (Gorska-Brylass, 1967). In some species, no cell wall is present around the generative cell, as evidenced by lack of any space between the generative cell plasma membrane and the inner plasma membrane of the vegetative cell (Jensen *et al.*, 1974; De Mason and Sekhar, 1988). Cytoplasmic contents of the generative cell include endoplasmic reticulum, ribosomes, dictyosomes, numerous vacuoles, microtubules, mitochondria and, in some species, plastids (Sassen, 1964; Larson, 1965; Lombardo and Gerola, 1968; Sanger and Jackson, 1971; Jensen *et al.*, 1974; Knoth *et al.*, 1980; Russell and Mislan, 1986; De Mason and Sekhar, 1988; Schröder and Oldenburg, 1990; Zhu *et al.*, 1990a, b, 1991). However, the presence of plastids within generative cells does not guarantee that they will be inherited because they may subsequently degenerate within the generative cell, be shed from the sperm cells, be excluded from the zygote, or be sequestered into the suspensor during embryo development (Clauhs and Grun, 1977; Mogensen and Rusche, 1985; Mogensen, 1988; Sears, 1980; Hagemann and Schröder, 1989; Schröder and Oldenburg, 1990). The absence of plastids within the generative cell typically forms the basis for strict maternal plastid inheritance (Smith, 1989).

The generative cell nucleus conforms to the general shape of the generative cell and, as will be indicated in the section on the vegetative cell, the number of pores in the generative nucleus is typically much less than in the vegetative nucleus (Mepham and Lane, 1969; La Fountain J. and La Fountain K., 1973; Shi *et al.*, 1991).

Vegetative Cell (Latin *vegeto* = excite, grow, vegetate)—the larger cell of the pollen grain giving rise, upon the latter's germination, to a pollen tube. Synonyms: tube cell, siphonogenous cell. Strasburger (1884) was the first to elucidate its functions and to name it.

The vegetative cell is formed as a result of differentiation mitosis and occupies most of the pollen grain volume. Immediately following mitosis, its nucleus lies in close proximity to the generative cell, showing no difference in size from the latter's nucleus. Gradually, the vegetative cell nucleus changes appreciably in volume and starts to move to the opposite end of the pollen grain. Movement of nuclei within a developing pollen grain has received much attention and been thoroughly investigated in pollen of grasses (Batygina, 1962a, b, 1974b, 1987a; Kihara and Hori, 1966; Romanov, 1970a, b). Movement of the vegetative nucleus occurs along the pollen wall, probably together with the surrounding cytoplasm, such that the entire mass of cytoplasm that was present in the vicinity of the generative cell migrates to the opposite end of the pollen grain, the central part being occupied by a vacuole. According to Batygina (1962a, b), in grasses the vegetative nucleus migrates twice in the course of pollen development; the first transposition is into the neighbourhood of the aperture and the second to the centre of the cell. The vegetative nucleus and its nucleoli become greatly enlarged. The shape of the vegetative cell nucleus remains invariably spherical in most flowering plants, whereas in some species it undergoes a perceptible change in outlines, assuming an amoeboid shape or extending considerably, becoming filiform or breaking up into separate fragments (Maheshwari, 1950; Gerassimova-Navashina, 1951a, b; Polyakova, 1961, 1963; Batygina, 1962a, b, 1987a; Bolkhovskikh, 1965). The vegetative nucleus is assumed to exhibit a high metabolic activity. The nuclear envelope shows numerous, non-uniformly distributed pores.

DNA synthesis within the vegetative nucleus does not coincide in time with that in the generative nucleus and takes place much later, often shortly before pollen grain maturity (Reznickova, 1984). Histone content within the vegetative nucleus remains unchanged until DNA replication, with the arginine fraction prevailing in histones. The total protein content of the vegetative nucleus is higher than that of the generative nucleus, the proteins of the former exhibiting higher acidity. The vegetative nucleus is involved in RNA and protein synthesis in the cytoplasm. A correlation has been established between nucleolus size and pyronin-stainability of the cytoplasm in the vegetative cell (Polyakova and Podstavek, 1960; Polyakova, 1962). The vegetative nucleus is likely to play a part in generative cell division and in formation of sperm cells (Gerassimova-Navashina, 1951a, b; Polyakova, 1960, 1961, 1962). Prior to generative cell division, it comes to lie close to the dividing cell.

By virtue of being metabolically active, the vegetative nucleus is involved in formation and growth of a pollen tube, although in some cases it breaks down into fragments prior to pollen tube formation or fails to leave the pollen grain following formation of a pollen tube (Maheshwari, 1950; Poddubnaya-Arnoldi, 1976).

Throughout its existence, the vegetative cell exhibits the lowest nucleocytoplasmic ratio (*Lilium candidum*) (Reznickova, 1984). Intense synthesis of cytoplasm occurs within the vegetative cell so that by the time of sperm cell formation within the pollen grain or, more often, by the time of pollen grain maturity, the vacuole that has been present within the vegetative cell disappears. Prior to pollen maturity, the vegetative cell cytoplasm becomes greatly dehydrated.

The vegetative cell is a storage cell with high metabolic activity; it contains numerous reserve, energy-rich substances in the form of lipids, starch and proteins required for rapid growth of the cell and cell wall synthesis during pollen tube germination and growth.

The ultrastructure of the cytoplasm of the vegetative cell lends support to its role as a storage cell and a cell giving rise to the pollen tube (Charzynska *et al.*, 1989; Van Aelst and Van Went, 1991). In the cytoplasm of a mature vegetative cell are found numerous mitochondria with well-developed cristae, ribosomes and small vacuoles. Spherosomes, dictyosomes and large vacuoles are relatively few. Rough ER cisternae are arranged in stacks or occur singly or in pairs, associated with spherosomes and plastids. Bound ER cisternae are characteristic of mature pollen grains. Upon pollen germination, the cisternae are dispersed throughout the cytoplasm. The association of ER cisternae with spherosomes and plastids suggests lipid and starch deposition. Polyribosome regions in the cytoplasm constitute sites of active synthesis of proteids required for pollen germination. The proportion of small vacuoles and the presence of rough ER cisternae arranged in stacks might be indicative of the time required for the pollen to germinate. Abundant small vacuoles and scarcity of ER cisternae are typical of a pollen that takes a few minutes to germinate.

Ultrastructure of Vegetative Cell (Plate 26)

The ultrastructure of the vegetative cell has been well characterized. At pollen maturity (at anthesis) it is typically highly cytoplasmic without a large central vacuole (Knox, 1984). All of the common cytoplasmic components are usually present, to a greater or lesser degree, including endoplasmic reticulum, ribosomes, dictyosomes, mitochondria, plastids (often containing large starch grains), vesicles, small vacuoles, lipid bodies, protein bodies, microtubules, microfilaments and a dense background matrix (Larson, 1965; Sanger and Jackson, 1971; Jensen *et al.*, 1974; Mogensen, 1986; De Mason and Sekhar, 1988; Tiwari, 1988). In some species, particularly in grasses, in addition to the large amount of starch grains present, a significant portion of the vegetative cell cytoplasm is composed of pollen tube wall precursor vesicles (Mogensen and Ladyman, 1989). In a single rye pollen grain, these wall precursor vesicles may number over one million and occupy as much as 30% of the vegetative cell volume (Heslop-Harrison J. and Heslop-Harrison Y., 1982). These vesicles, whether formed before or after pollen germination, are of dictyosome origin and contain primarily pectin and some protein. They play a critical role in pollen tube growth as their contents are incorporated into the tip of the pollen tube wall during its elongation, and their membranes contribute to the plasma membrane of the pollen tube.

The vegetative cell nucleus is typically more rounded (though often highly lobed) and stains more faintly than the nucleus of the generative or sperm cells; it is usually closely associated with and partially surrounds these cells, resulting in the so-called male germ unit (Dumas *et al.*, 1984; Mogensen, 1992). The vegetative nucleus has a much higher density of nuclear pores than the vegetative or sperm nucleus (Mepham and Lane, 1969; La Fountain J. and La Fountain K., 1973; Shi *et al.*, 1991) and is believed to be much more active in RNA synthesis (Mascarenhas, 1975).

Sperm Cell (Greek *sperma* = seed)—a male sexual cell, or a male gamete (Plate 27). The generative cell (= spermiogenous cell, in Gerassimova-Navashina (1951a) terminology) produces two sperm cells. This event occurs either within the pollen grain or, following its germination, within the pollen tube. Sperm cells contain haploid nuclei and are characterized by a dramatically altered nucleocytoplasmic ratio. The cytoplasm of sperm cells is thinly spread along the nucleus and concentrated at the cell poles. The nuclear chromatin is condensed. The sperm nuclei usually lack nucleoli, although the latter have been reported to be present in some cases (*Triticum* species, Batygina, 1962a, b, 1987a; *Plumbago zeylanica*, Russell, 1983; *Matthiola bicornis*, Tarasenko *et al.*, 1988). In some species, the nucleolus is only observable in the nucleus of immature sperms and disappears later on (Tatintseva, 1973, 1975, 1988). The sperm cytoplasm contains the usual set of cell organelles, such as mitochondria with numerous cristae, dictyosomes associated with Golgi vesicles, ER (mostly RER) cisternae, ribosomes, lipid globules, microbodies and microtubules; in some plants plastids are absent in the sperm cells. The primary role of microtubules within the sperm protoplast is presumably that of controlling the shape of male gametes (Van Went and Willemse, 1984). Sperm cell vacuoles vary in size and shape. The structure of the male gamete plastidome is one of the most intriguing and poorly understood questions in modern plant embryology. Essential structural differences between male and female gametes of flowering plants (including the structure of their plastidome) are due to their divergent evolution. Unlike the egg cell, sperm cells have undergone a dramatic reduction in number of plastids, diminution of size and simplification of their ultrastructure. According to the concept of there being more than one plastid distribution pattern during pollen grain development and spermiogenesis in flowering plants, sperm cells in some species are aplastidic (or apoplastidic) (Hagemann, 1981; Hagemann and Schröder, 1989). However, in parallel with the ideas of a historically gradual 'physical exclusion' of male gamete plastids from the plant plastidome, a hypothesis of 'organelle alteration' has been proposed. This has been advanced on the basis of genetic evidence (Vaughn, 1981) to account for the mechanisms of plastidal inheritance in higher plants (for details, see Zhukova, 1993).

A distinctive feature of male gametes is that during spermiogenesis their cell walls gradually disappear (prior to syngamy and triple fusion, their protoplast is surrounded by the plasma membrane only). Sperm cells remain associated with one another throughout their movement along the pollen tube until contact with the egg cell and central cell. In some species, sperm cells are united by a common cell wall, in others by projections of their protoplasts. The pair of sperms is, in turn, associated with the vegetative cell nucleus. Sperms of a pair are linked to the vegetative nucleus either by a single, lengthy outgrowth enveloping the lobed vegetative nucleus or via several projections, with one sperm of a pair being so associated in some species and both in others. These complexes consisting of a pair of sperms and a vegetative cell nucleus came to be called the 'male germ unit'. However, no such complex has been found in *Helleborus foetidus* (Heslop-Harrison *et al.*, 1986). The association between sperms in *Calanthe vestita* has been demonstrated to persist after their discharge from the pollen tube. The association is assumed to be preserved in *Nicotiana rustica* as well, and not only during the discharge of the pollen tube contents into the synergid, but also at the point of divarication of the two male gametes before their movement towards the egg cell and central cell. It is conceivable that this association ensures a simultaneous contact of both sperm cells with female gametes, thus preventing either of the latter from being fertilized by two sperms (Plyusch, 1992). According to S.G. Nawaschin (1911), each of the two sperms within a single pollen tube is a mirror-image of the other, i.e., they are enantiomorphous. In some members of Liliaceae, Amaryllidaceae and Tamaricaceae, sperms of a pair differ in size, cytoplasm content and nucleus volume (Belyaeva, 1973). Difference in size between sperms of a single pollen grain has also been reported for some members of family Asteraceae, the larger sperm invariably located closer to the vegetative nucleus than the smaller one (Barnes and Blackmore, 1987). There is some evidence for sperm dimorphism in grasses as well (Batygina, 1974a). Ultrastructural analysis of sperm cells has revealed sperm dimorphism both within the pollen tube and within the pollen grain (Russell, 1984; Dumas *et al.*, 1985; Hu, 1990; Mogensen, 1990b; Southworth, 1990b; Tarasenko, 1990; Zhu *et al.*, 1990b). Sperms of a pair may differ not only in shape and size, but also in organelle composition. Thus sperms of the *Plumbago zeylanica* pollen grain have been found to differ in number of mitochondria, plastids and microbodies (Russell, 1984). No heteromorphism has been revealed for sperms of the *Hordeum vulgare* pollen grain, or for the pair of sperms within the *Populus deltoides* pollen grain (Rougier *et al.*, 1991). In flowering plants, in contrast to gymnosperms, both sperms of a pair are functional, resulting in double fertilization.

Sperm Dimorphism (Plate 28). In reviewing the early literature on male gamete morphology, Maheshwari (1950) cited reports of differences in size and shape of the two sperm nuclei within a given embryo sac as they fuse with the egg and the polar nuclei. In some cases, the smaller sperm nucleus is seen fusing with the egg nucleus while in others, the smaller nucleus fuses with the polar nuclei. Since these observations were made after syngamy (fusion of gametes), it is likely that they apply only to nuclei and not all sperm cells. In only one species (*Vinca minor*) does Maheshwari (1950) cite a case wherein sperm cells are reportedly unequal, one having a larger cytoplasmic extension than the other (Finn, 1928). Maheshwari (1950) has admitted that 'slight differences in size and shape of the two male gametes are possible'; however, he has cautioned that the differences observed may have been the result of different planes of sectioning, or a disparity in rates at which the male nuclei undergo transformations during karyogamy (nuclear fusion).

Modern investigations on sperm cell structure have employed techniques that leave no doubt about the existence of unequal sperms of a pair. Russell (1984) was the first to use serial ultrathin sections and computer-assisted reconstructions to study sperm cell structure. He discovered that in *Plumbago zeylanica*, the two sperms within the tricellular pollen grain differ significantly from each other in size, shape, and organellar content. Russell (1984) termed these cells 'dimorphic', indicating the two forms present. Sperm dimorphism has since been demonstrated in several other species (see reviews by Hu, 1990; Roeckel *et al.*, 1990; Russell, 1991; Mogensen, 1990a, 1992). Additional terminology has been employed by Russell (1991) to indicate cytoplasmic differences, 'cytoplasmic heterospermy', and nuclear differences, 'nuclear heterospermy', between two sperms of a pair. Hu (1990) considered the term 'sperm heteromorphism' synonymous with 'sperm dimorphism'.

Although sperm dimorphism is more prevalent than previously thought, it is not universal. For example, serial ultrathin sectioning and computer-generated reconstructions have shown that the sperms of a pair are 'isomorphic' (equal in form) in some species, such as barley (Mogensen and Rusche, 1985), petunia (Wagner and Mogensen, 1988) and alfalfa (Zhu *et al.*, 1992). Reviews by Hu (1990) and Russell (1991) present up-to-date lists of species that have dimorphic or isomorphic sperms.

Pollen Grain Cytochemistry (Plate 29)

First successful attempts at employing cytochemical methods to study the biochemistry and metabolism of the cells of a developing pollen grain date back to the 1950s. It was demonstrated that the generative and vegetative cells differ in nucleic acid and protein content and that the differences are in large measure responsible for the specific nature of the pollen grain metabolic status (Swift, 1950; Bryan, 1951; Woodard, 1958).

Cytochemical studies revealed the patterns of DNA content variation in the cells of a developing pollen grain. It was possible to demonstrate that in all the species examined the generative and vegetative cells acquired, immediately following an unequal mitosis, equal amounts of DNA corresponding to 1C. In the course of further development, DNA of the generative cell nucleus underwent replication, with most of the species studied showing an early entry of the generative cell into the S phase. The resultant sperm cells had the haploid DNA content (1C). However, the sperm cell nuclei of some species exhibited further increase in DNA content which was responsible for their being in the S phase or resulted in their DNA content doubling.

The pattern of DNA content variation within the vegetative nucleus of a developing pollen grain has yet to be resolved conclusively. In some species, no DNA synthesis has been detected in the vegetative nucleus, while in others the occurrence of such a synthesis has been substantiated experimentally (Bannikova *et al.*, 1985; Gorska-Brylass *et al.*, 1986).

For cytochemical detection of pollen DNA, preparations of pollen grains of lily (*Lilium candidum*) and coriander (*Coriandrum sativum*) were subjected to Feulgen staining, and those of sage (*Salvia sclarea*) and lavender (*Lavandula angustifolia*) to olivomicine staining. All the species studied exhibited replication of the generative nucleus DNA. In species with three-celled pollen grains (*Salvia, Lavandula, Coriandrum*), the generative cell divides to form sperm cells. In sage and lavender pollen grains, the sperm cells, immediately upon formation, contain a DNA amount exceeding 1C. This is probably due to considerable shortening or even omission of the G1 period and prompt entry of the generative cell into the S-phase. In mature pollen grains, at dehiscence the sperm cells exhibit a DNA content corresponding to 2C. Sperm cells of coriander pollen show no DNA synthesis, however, their DNA content staying at the level of 1C.

The pattern of DNA content variation in the vegetative nucleus exhibits some species-specific features. In lily, DNA synthesis within the vegetative nucleus is shifted towards the later developmental stages, resulting in the amount of DNA being nearly doubled. In pollen grains of sage and lavender, no DNA synthesis is detectable within the vegetative nucleus. In coriander, the vegetative cell enters the S-phase already in the early stages of pollen grain development. In mature three-celled pollen grains, the vegetative nucleus DNA content is 1.7C. The fact that DNA synthesis in the vegetative cell of pollen grains does not proceed to completion is of particular interest. Further studies have revealed that the increase in DNA content to 1.7C results from non-uniform DNA synthesis within the population of pollen grains.

A large proportion of these exhibit DNA values of 1C in their vegetative nuclei, the minority showing a 2C DNA content. This fact was demonstrated experimentally earlier for *Hippeastrum belladonna* (Pipkin and Larson, 1973).

Even comparatively early cytochemical studies sought to elucidate the patterns of quantitative variation in developing and mature pollen grains, not only in nuclear DNA, but also in histones, i.e., DNA-bound basic proteins contained in the nuclear chromatin (Rasch and Woodard, 1959; Jalouzot, 1969). Subsequently, generative nuclei were shown to contain lysine- and arginine-rich histones, whereas vegetative ones contain only arginine-rich histones (Sheridan, 1973). Of particular interest is the detection in pollen grains of an additional fraction, a so-called meiotic histone (HM). This fraction is synthesized during meiotic prophase I and persists in the generative nucleus after an unequal mitosis. Recent biochemical studies indicate that generative and vegetative nuclei of pollen grains contain five major histone fractions: H1, H2A, H2B, H3 and H4. However, vegetative nuclei showing diffuse chromatin exhibit lower levels of proteins H1 and H3 and a higher level of H4 (Taimla *et al.*, 1984). The pattern of histone content changes within generative and vegetative nuclei of pollen grains is similar to that of DNA content variation.

During pollen development RNA and total protein synthesis occurs. High levels of RNA in the vegetative cell nucleus and cytoplasm, as well as in the generative nucleus are already detectable early in the development of two-celled pollen grains (Woodard, 1958; Jalouzot, 1969; Reznickova, 1984; Khvedynich and Petrik, 1986). In the course of further development, the amount of RNA in the cytoplasm of both types of cells increases, such that its synthesis in the vegetative cell cytoplasm is intensified prior to anther dehiscence. RNA measurements in generative and vegetative cells of lily pollen (staining with gallocyanin in combination with ammonium chromic sulphate) revealed that already at the initial stages of pollen grain development the vegetative nucleolus exhibits an increased level of RNA. During further development the nucleolar RNA content of both types of cells gradually increases and then declines. The RNA content of the nuclear chromatin also tends to increase until the final stages of pollen development. It is only on the day of anther dehiscence that the RNA content of the generative nucleus chromatin decreases sharply.

The level of RNA in the generative and vegetative cell cytoplasm of a two-celled pollen grain tends to increase during early stages of its development. In the course of further development, the amount of RNA in the generative cell cytoplasm is sustained at a given value, whereas in the vegetative cell cytoplasm it continues to rise such that throughout the pollen development cycle in lily, the cytoplasmic RNA content of the vegetative cell is considerably higher than that of the generative cell.

The functional activity of pollen grain cells has been a topic in the literature for quite some years (Bolkhovskikh, 1965; Jalouzot, 1969). It does not appear pertinent to raise the question of which of the pollen grain cells (vegetative or generative) is functionally more active. Each has its unique metabolic status. Thus, in lily, comparing the functional activity of the vegetative cell, retaining the haploid DNA content for a long time, with that of the generative cell, which enters the premeiotic S-phase fairly soon, resulting in its DNA content doubling, proves a task not so easy as might appear at first sight. High functional activity of the vegetative cell manifests in the larger volume of its nucleus and in an increased nuclear and cytoplasmic RNA content, such that the vegetative nucleus exhibits more non-histone proteins per unit content of DNA. The metabolic activity of the generative cell is maintained at a level required for DNA replication during early stages of pollen development, as well as for subsequent mitosis resulting in formation of sperm cells during pollen tube growth.

Nucleus and Cell Movement in Developing Pollen Grains (Plate 30)

In studying angiospermous pollen development, an interesting phenomenon is observable, viz., the movement of nucleus within the microspore, and of the cells and nuclei within the pollen grain. Of particular interest is the phenomenon of the generative cell movement within the vegetative cell of the pollen grain. In this respect, grasses merit special attention owing to a number of developmental and structural peculiarities exhibited by their anthers, microspores, and pollen grains. The characteristic (near-ellipsoid) shape of microsporocytes determines the tangential orientation of spindles during both meiotic divisions, which, in turn, results in a monolayered arrangement of all the microspores and pollen grains, such that their pores face the tapetum. The microsporocytes, the dyad and tetrad cells all develop thick callosic walls in the form of ridges oriented towards the microsporangium cavity (Drahowzal, 1936; Carniel, 1961; Romanov, 1970a, c; Batygina, 1974b).

It is known that at a certain stage of microspore development (the stage of weakly vacuolate microspore) its nucleus migrates from the central position in the cell to the cell wall coming to lie opposite the germinal pore (Batygina, 1962a,b, 1987a; Kihara and Hori, 1966; Romanov, 1966). Then the microspore undergoes vacuolation. Following the differentiating division of the microspore (the stage of highly vacuolate microspore) resulting in the formation of a two-celled pollen grain, the vegetative nucleus migrates towards the aperture, into the outer region of the pollen grain and eventually comes to lie beneath the germinal pore.

During the course of development, the pollen grain passes through the devacuolation stage at which further migration of the generative cell and vegetative nucleus occurs. Movement of the generative cell into the neighbourhood of the germinal pore where, at the moment, the vegetative nucleus lies starts when the cell assumes an orbicular shape and detaches from the pollen wall. Subsequently, as the intensity of synthesis of the vegetative cell cytoplasm increases, the vacuole gradually decreases in size. Then the pollen grain enters the next stage of development at which the generative cell again undergoes elongation. The latter moves away from the vegetative nucleus, while still remaining within the confines of the broad parietal layer of the vegetative cell cytoplasm. Gradually the vegetative nucleus migrates, as does the generative cell itself, deeper into the vegetative cell protoplasm. According to Romanov (1970a, c), this migration of cells and nuclei occurs as passive movement.

As the generative cell divides to form two sperm cells, the vegetative nucleus stainability becomes progressively lower, its highly vacuolate nucleoli gradually disappear and the nucleus itself assumes a lobate shape. Concurrently the sperm cells undergo changes in position, shape and structure. Dramatic changes occur in sperm cells from the time of their formation inside the pollen grain until their growth in the pollen tube and entry into the embryo sac. There is some evidence for sperm dimorphism in grasses (Batygina, 1974; Mogensen and Rusche, 1985). At pollen grain maturity, the sperm cells and vegetative nucleus are most commonly located close to one another but their position in relation to the germinal pore may vary. The mature pollen grain of cereals, in particular that of wheat, is three-celled and shows two elongate sperm cells with clearly discernible ends: one blunt, the other pointed (Batygina, 1961, 1962a, b; Chu and Hu, 1981). Thus in the course of pollen grain development, the vegetative nucleus, generative cell and sperm cells change their position time and again in relation to the germinal pore.

We shall now highlight some specific features of development of the pollen grain in grasses which presumably have a bearing on migration of nuclei and cells within it. These are as follows: variable position of meiotic configurations within the microspore (normally meiotic configurations are arranged asymmetrically within the microspore), non-uniform formation of a wall between the generative and vegetative cells, which is mostly composed of callose, and its transient nature, which probably results in the generative cell being surrounded by two plasmalemmae—its own and that of the vegetative cell. A similar pattern of microspore and pollen grain development is observed in all the grasses studied embryologically to date.

Migration of nuclei is not unique to developing microspores and pollen grains. Movement of the vegetative nucleus and of sperm cells during pollen tube growth is a matter of common knowledge. There is strong evidence for migration of nuclei to opposite poles of the angiospermous embryo sac (e.g. Polygonum-type of development) at various stages of its development and in a mature embryo sac, for movement of synergid nuclei to the basal region of these cells, and of the nucleus of the egg cell to the apical region of the cell. Regular movement of nuclei is also observable in megasporocytes during female gametophyte formation in members of various taxa exhibiting other modes of embryo sac development in addition to the Polygonum-type. Of some interest is the migration of nuclei within the endosperm of *Neottia nidus-avis*, in the haustorial cells of the *Latraea squamaria* endosperm, and in the embryo of *Pyrola rotundifolia* (Teryokhin *et al.*, 1971). Nucleus movement has been reported in the developing epidermal cells of flax hypocotyl: at a particular stage of development, the nucleus 'leaves' the parietal position characteristic of the resting epidermal cells and migrates to the centre of the cell (Navashin and Makushenko, 1956). Translocation of nuclei during stomatal cell ontogeny has been studied in *Tradescantia virginiana* (Tarkowska *et al.*, 1987). Animal cells also exhibit nuclear movement. Thus, migration of nuclei is readily observable in cells involved in enamelogenesis: initially, the nucleus is situated close to the cell surface but as the cells begin to show the presence of enamel, the nucleus moves deep into the cytoplasm (Alov *et al.*, 1969). Therefore the migration of nuclei, as well as of cells, during the cell cycle should be viewed as a phenomenon of general biological significance.

Movement of gametes and organelles is presumably one of the functions of the cell cytoskeleton (microtubules and microfilaments) (Heslop-Harrison, 1987; Heslop-Harrison J. and Heslop-Harrison Y., 1989; Pierson *et al.*, 1989). A fibrillar apparatus involved in the intracellular transposition of these cellular structures has been found in the cytoplasm of the generative cell of *Haemanthus katherinae* (Sanger and Jackson, 1971) *Petunia hybrida* (Cresti *et al.*, 1976) and some other plants (Tiezzi *et al.*, 1988). Myosin and kinesin have been detected in the cytoplasm of the tobacco vegetative cell (Moscatelli *et al.*, 1988). However, in general, this most intriguing issue remains open.

Morphofunctional Status of Gametogenesis

The morphological and functional status of gametogenesis is a set of morphofunctional and phylogenetic laws of organization of male and female gametophytes of flowering plants, beginning with initiation and ending with differentiation of gametes (Chebotaru, 1983, 1985). The morphofunctional status is a guide to genetically determined aspects of structure and function underlying selective elimination of gametes, embryonal determination, and population adaptability and sustainability. Light and electron microscopic studies of gametogenesis suggest a special, crucial role for the relationship among components of the system 'maternal plant–sporogenous tissue–gametes–zygote–embryo–seed' (Chebotaru, 1960, 1968, 1972, 1978, 1983, 1985).

The basic principles of morphofunctional status are as follows:

(a) **Principle of elimination (abbreviation)** of archesporial, sporogenous and gamete formations designated by us as genetically defective structures. The occurrence and degradation of these is observable throughout meiotic and subsequent divisions of haploid structures during macrosporogenesis, microsporogenesis, gametogenesis, zygote formation and embryogenesis in which, along with production of gametes, sister or lateral formations de novo are regularly eliminated. In hybrid progeny, continuous release (diffusion) of chromatin (nuclear) material into the cytoplasm has also been observed to occur during gametogenesis. Abbreviation (a term coined by Severtsov, 1939) of haplostructures is a regular process transmitted from one generation to the next with a high degree of genetic fidelity and occurs in synchrony with the process of formation of healthy gametes. The abbreviation principle is found in most eukaryotes and acts not as a trophic inevitability of differentiation and sexualization, but as a kind of evolutionary 'sieve' of natural selection at the 'cellular', pregamete and gamete levels. The process involves a genetically determined way of identifying and eliminating the 'wasted material' (in the sense of Goldshmidt, 1923). Thus, the principle of abbreviation of genetically defective formations shows that a proportion of structures resulting from meiotic and subsequent divisions of haploid cells are, regularly eliminated in various ways, and disappear in the extensive food chain. We view gametogenesis, as a particular stage in plant ontogeny, as a kind of evolutionary 'sieve'.

(b) **Principle of gamete equipotentiality** observable in direct and back intrapopulation, intra- and interspecific crosses. Genetic equipotentiality of male and female gametes noted by many authors (Müntzing, 1967; Inge-Vechtomov, 1989; and others) is supported by embryological studies and biotechnology research (Butenko, 1970; Batygina, 1992). The observed structural differences mostly concern the nucleocytoplasmic ratio and morphophysiological state of the plastidome and chondriome (Chebotaru, 1972, 1985; Bhandari, 1984). Equipotentiality of male and female gametes as principal carriers of genotypic specificity is laid down during gametogenesis and realized during fertilization and embryogenesis. In doing so, the continuity of generations is reliably ensured through continuous meiotic recombination canalized by panmixia.

(c) **Principle of enantiomorphism during gametogenesis**, as described for male gametes, is also of general biological significance. It ensures selective fusion of gametes during syngamy. Assuming the enantiomorphic nature of sister sperm cells, Nawaschin (1927) substantiated the mechanics of biological activity and selective self-movement of male gametes during fertilization. Enantiomorphic differences may also occur between female gametes on the one hand, and their sister structures (egg cell—polar nucleus of the central cell of the embryo sac) on the other, although differences are even more pronounced during postmeiotic differentiation. Here a special role is played by the cytoplasmic component. From the standpoint of the principle proposed, it can be assumed that the strategy of syngamy as a general biological phenomenon is based on enantiomorphic synergism, i.e., fusion of mirror-image structures. Synergism promotes morphogenetic homeostasis which canalizes adaptive variation of sporogametophytes during phylo- and ontogeny.

The most important factor in the morphophysiological status of gametogenesis is an actually operating pre-gamete and gamete selection resulting not in allelic substitutions, but in differentiation and generation of equipotential reaction (Shishkin, 1984), i.e., selection of adaptive pairs of sexual cells fitting into the genotypic variability range outside of which the gametes either become eliminated or produce abortive embryos.

The morphofunctional status of gametogenesis also shows that sexual cells arise as an element of a unified system. Establishment of the morphofunctional status of gametogenesis enables one to trace the source of bioecological expediency, adaptive norm of reaction in the sense of Schmalhauzen (1940) and other important laws of sexual reproduction. The Vavilov law of homologous series in genetic variation is operative during gametogenesis.

In general, higher plant gametogenesis should be viewed as a unique, not fully understood 'botanical alphabet' in which the biological past of our planet and the evolutionary course it pursued has been written. While rock formations remind us of our planet's past by their stony silence, flowering plants do so by their perpetual ontogenetic renewal. Having evolved once in the geological past on a diploid sporophyte, the gametes reproduce themselves over the course of generations by gametogenesis.

Pollinium (Latin *pollinium* = a small clump of pollen grains)—a cohesive mass of pollen grains held together by a common envelope or by a special sticky substance. Synonyms: massula[5], compound pollen grains (Plate 31).

In angiosperms, various pollen grain configurations occur, from dyads (Scheuchzeriaceae), tetrads (Empetraceae, Epacridaceae, Juncaceae, Droseraceae, Philydraceae, Typhaceae, Velloziaceae, Sapindaceae) and polyads (Mimosaceae), to more complex pollen aggregates, such as pollinia. Pollinia have only been found in two families of flowering plants: in the overwhelming majority of monandrous orchids (family Orchidaceae, subfamily Orchidoideae) and among Asclepiadaceae.

The number of pollinia per anther may vary between 2 and 8. Pollinia exhibit variation in shape (spherical, ovoid, obovate, ellipsoidal, oblong, terete, oblate, horse-shoe-shaped, cylindrical, wedge-shaped, broadly triangular, entire, and lobed) and consistence (granulated, powdery, waxy or hard). These features are of taxonomic importance.

Pollinia develop special structures made up of a sticky substance (elastoviscin) which ensure cohesion of pollinia and serve as adaptations for insect pollination. In orchids, these structures are represented by a stalk (caudicle) and a retinaculum (viscidium). Such a pollinium, adapted for transfer by insect pollinators, is often

---

[5]By the term massula we also mean the microspores of heterosporous ferns aggregated in small clumps (Kupriyanova and Aleshina, 1967). In some orchids, pollinia are divided into massulae (Dressler, 1993).

referred to as a pollinarium. In Asclepiadaceae, the structures concerned in transfer of pollinia are represented by a translator comprising a centrally located gripping (clamping) body, or corpusculum, from which a pair of stems branch out. These attach to the pollinia in the right and left locules of two adjacent anthers. A translator can carry 2 to 4 pollinia, provided all microsporangia of an anther are well developed.

The sticky substance involved in the formation of pollinia and termed elastoviscin is similar to pollenkitt in terms of chemical composition, timing and site of synthesis, and function (Schill and Wolter, 1986). Like pollenkitt, elastoviscin is secreted by tapetal cells. In some orchids, the tapetal layer shows clusters of cells with massive tracheidal thickenings. These are specialized in synthesizing elastoviscin at the sites of formation of the aforesaid special structures carried by pollinia (Wolter *et al.*, 1988).

Formation of pollinia in orchids may occur very early in anther development. Thus in *Peristylus spiranthes*, there are clearly discernible formations comprising a few sporogenous cells and enclosed within a thicker, common wall (Zee and Siu, 1990). Prior to meiosis, these structures, composed now of microsporocytes, are invested in a thick callose wall. Callose is also deposited outside the plasmalemma of each microsporocyte. Cell division within each pollinium is strictly synchronous, which is ensured by extensive cytoplasmic connections among its component microspore mother cells. No connections like these occur among pollinia within a single anther. During tetrad formation, no callose is deposited on the surface of the microspore plasmalemma and only a few cytoplasmic connections occur among the microspores.

In most species examined, organized pollinia show an unstructured wall which, as suggested by scanning electron microscopic studies, is much thicker than the walls of component pollen grains. Pollen grains within a pollinium are no longer capable of producing their own exine due to suppression of callose deposition either around the microsporocytes and between the microspores within a tetrad (Asclepiadaceae), or around inner tetrads and between microspores (Orchidaceae). Where callose is deposited around the microsporocytes, exines are formed in pollinia. Thus, the formation of three types of exine differing in structural complexity has been reported to occur within *Peristylus spiranthes* pollinia (Zee and Siu, 1990). The first type of exine representing the outer wall of outer tetrads comprises the tectum, bacula and nexine. The second type of exine occurring in the inner walls of tetrads shows a poorly developed tectum and bacula, and a thinner nexine. The third type of exine occurs in the walls of pollen grains attached in tetrads and is even more poorly developed, composed only of small spherical bodies. In orchids, the mode of intine formation in pollen grains differs somewhat from that in other angiosperms. The orchid intine is two-layered: the outer layer designated as ectintine is darker than the inner one termed endointine.

The pollinium wall of Asclepiadaceae is differentiated anatomically and morphologically (Galil and Zeroni, 1969; Sreedevi and Namboodiri, 1982). It is composed of sporopollenin-like substances and is acetolysis resistant except at the site of pollen tube emergence, termed the germinal furrow. The germinal furrow position is a strictly genus-specific feature, and clearly defined in all 14 genera of the Asclepiadaceae family. In some genera (*Huernia*, *Gerapegia*), the germinal furrow is discernible in untreated, non-germinated pollinia. It is represented by a membrane with perforations showing variation in size, shape and position and is detectable, upon acetolysis, as a narrow, discontinuous slit varying in size between 40 mm (*Stapelia*) and 500 mm (*Calotropis*). Pollen-tube growth within a pollinium proceeds towards permeable zones. The germinal furrow itself or any artificial opening in the pollinium wall constitutes such a permeable zone.

Aggregation of pollen grains into pollinia in flowering plants is regarded as an advanced feature, a manifestation of neoteny in pollen development which can be viewed as a mechanism offering a considerable saving of material and time (Favre-Duchartre, 1984). In addition, some stages of pollen development are omitted, such as suppression of callose deposition, pollen grain exine and pore formation. Pollinia exhibit few non-functional pollen grains.

Male Gametophyte Evolution

The origin of the angiospermous male gametophyte continues to be a widely debated topic. A germinated angiospermous microspore, a pollen grain that has undergone two divisions, is in appearance more like pollen grains of the prespermaphytes (pteridophytes, gymnosperms) with their three to five divisions. This, at first sight, facilitates the task of establishing homologies, and one would believe that a consensus already exists among modern plant morphologists as to the origin of the male gametophyte. However, such is not the case.

Gaining a more or less complete understanding of the nature of the angiospermous pollen grain will primarily require a cytology-based development theory approach. As is known (Geitler, 1935; Gerassimova-Navashina, 1951a, b), the first division in angiosperms occurs within the parietal layer of the protoplast, due to a highly vacuolate state of the microspore by the time of its germination, the process of vacuolation coinciding with anther tapetum degradation. This results in a differential division, i.e., a two-celled gametophyte is formed which comprises a small parietal cell and a large highly vacuolated cell. Soon afterwards, de novo synthesis of cytoplasm begins within the larger cell, and the smaller cell, surrounded by the cytoplasm of the larger, becomes incorporated in it and together they form the entire pollen grain system. Subsequently, the two gametophyte cells appear to share the functions required for fertilization: the smaller cell develops into a spermatogenous cell and undergoes another round of division to produce two sperm cells, the larger one embarking upon a synthetic pathway producing substances required for pollen tube growth, i.e., becoming a siphonogenous cell. It is worth noting that in angiosperms, by the time of the first division of the pollen grain, pollen walls (exine and intine) have already been organized. In gymnosperms, e.g. *Larix dahurica* (Gerassimova-Navashina, 1954a, b), by the time of the first division, conditions are created within the pollen grain similar to those occurring in angiosperms; here, the tapetal layer disintegrates, resulting in a highly vacuolate microspore; the protoplast, together with the division configuration, shifts to one side, as a result of which a small lenticular cell, as in angiosperms, is detached and displaced to the wall. However, the subsequent fate of this cell is different. An inner cellulosic layer forms within the pollen grain which separates this cell from the entire pollen grain system. Thus cut off from the pollen grain system and not receiving sufficient cytoplasm of its own, this gametophyte cell, the first prothallial cell, dies.

Another cell arising under similar conditions by the second gametophyte division is destined to meet the same fate. It too is cut off from the general pollen grain system and lacking sufficient cytoplasm of its own to support its vital activities, dies soon afterwards.

Not until the time of the third gametophyte division are favourable conditions established for further development of the newly formed (antheridial) cell. The pollen grain now exhibits a lesser degree of vacuolation since conditions are created within the anther for cytoplasmic synthesis, i.e., pollen growth. By this time formation of cellulosic walls has normally been completed, thus allowing the newly formed cell to remain incorporated in the general pollen grain system and continue its development. This antheridial cell, located now in the central region of the pollen grain, undergoes another round of division to form the stalk (or basal) and spermatogenous cells. Following detachment of the antheridial cell, the apical cell becomes a siphonogenous cell giving rise to a pollen tube.

Therefore, the metabolic status and conditions of development of the cell determine its subsequent fate (differentiation). That this is the case is indicated by some abnormalities.

For example, if sprays of *Larix dahurica*, a species known to prepare for flowering in late winter, i.e., at very low temperatures, are brought indoors and fixed after a time, the microspores tend to exhibit numerous abnormalities of development. Examination of these enables elucidation of the causes of abnormal differentiation.

It suffices for as few as two developmental factors, namely vacuolation and initiation of cellulosic walls, to be displaced in time relative to each other for results to differ.

Thus if by the time of prothallial cell formation cell wall synthesis has already been completed, this cell will not be cut off from the pollen grain system and will not be eliminated.

If during its first and second divisions the pollen grain is less vacuolate, the prothallial cells will receive a larger amount of cytoplasm, sometimes quite sufficient not only for survival but for division as well.

It follows that elimination of the prothallial cells is not due to some properties inherent in them but, rather, merely associated with the conditions to which they become exposed within a particular anther during its development. The life activity of a cell is only determined by the course of its development unless, of course, it shows some intrinsic properties causing lethality.

If uniform conditions prevail within the pollen grain, i.e., no one-sided vacuolation occurs, then the first division, in both gymnosperms and angiosperms, produces two identical cells, or two identical nuclei, resulting in formation of a non-functional system.

So analysis of the course of development suggests that in angiosperms, the very first gametophyte cell is endowed with the potential to develop into a spermatogenous cell. That is, the antheridial cell (antheridium) is formed much earlier than in gymnosperms and, conversely, the first gametophyte (prothallial) cells die, strictly speaking, due to circumstances attendant on anther development. The spermatogenous cell of angiosperms cannot therefore be regarded as homologous with anything but one of the first two gametophyte cells of the gymnosperms.

Some other considerations related to the problem of origin of the angiosperm pollen grain and based on comparisons of pollen grain ontogeny in different groups of seed plants (Cycadales, Coniferales, Gnetales and Angiospermae) are worth mentioning (see Fig. 6). Microsporogenesis is known to proceed in a very similar fashion in all these groups of plants. By way of example, we shall restrict ourselves to considering the pollen grain development in *Pinus*, the process here being most complete and similar to that just described for *Larix*.

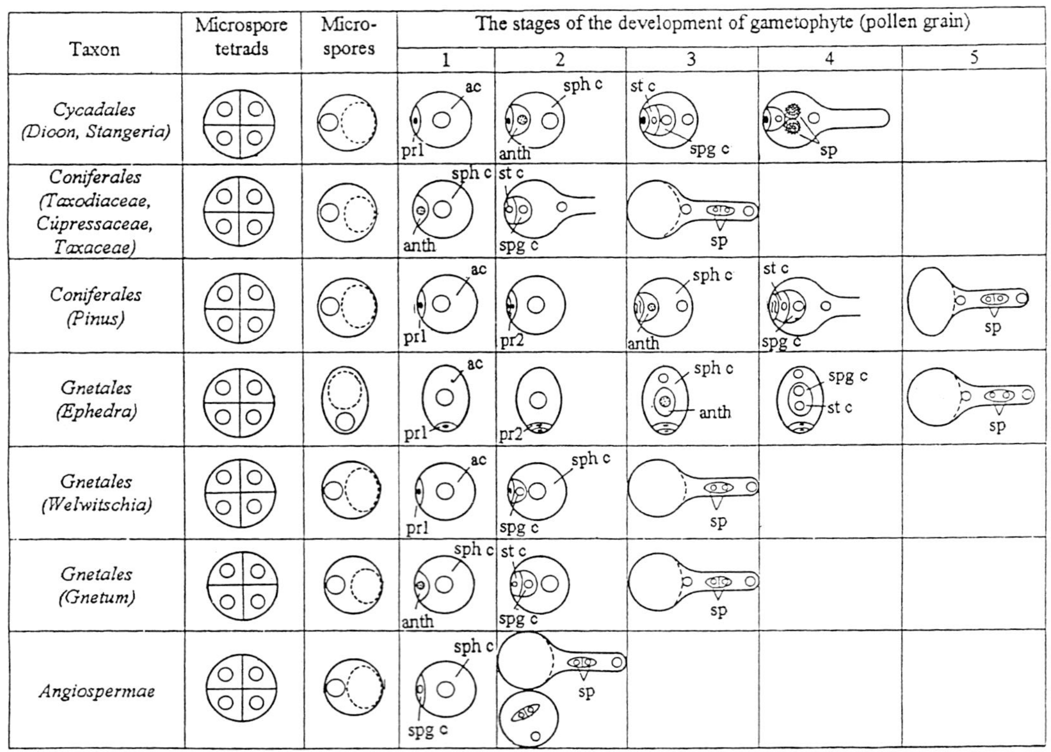

Fig. 6: Principal phases of male gametophyte development in different representatives of Angiospermae (Scheme). After Gerassimova-Navashina, 1954.

The first division of *Pinus* microspore results in formation of two cells: the first prothallial (*pr1*) cell (the vegetative cell of the prothallus) and the apical cell (*a c*) ('the youngest cell of the prothallus' according to Wettstein, 1911). During the second division, the apical cell of the organized two-celled gametophyte cuts off the second prothallial cell, *pr2*.

Neither prothallial cells, organized as small lenticular segments, show further development and degenerate, whereas the apical cell undergoes one more (a third) division. This latter cell undergoes no further divisions but assumes a number of new functions, developing subsequently into a pollen tube. From this point on, special terms are used to designate this cell: 'vegetative cell of the pollen tube', or 'tube cell'. I call it a siphonogenous cell (*sph c* in the Figure). The cell detached from the apical cell as a result of this third division is the antheridial cell ('the generative cell' in the sense of Chamberlain), designated as *anth* in the diagram. The term adequately describes the structure in question: this cell is the initial cell of the antheridium, since it divides (fourth division during gametophyte development) to form the antheridial primordium comprising a stalk cell (*st c*) ('basal cell', 'Stielzelle', 'Fusszelle') and a spermatogenous cell (*spg c*) ('body cell'). The stalk cell is virtually all that has been left of the antheridium as an organ in *Pinus*. As for the spermatogenous cell, this is already a pollen (sperm) mother cell which divides (fifth and last division in the course of development of *Pinus*) to produce sperm cells.

Some authors hold to the idea that the apical cell just prior to division resulting in the formation of an antheridial cell should be viewed as an 'antheridial initial'. According to this view, it would be logical to treat a mature pollen grain as the antheridium. However, there is no inclination to do so evident, presumably due to reported in the literature difficulties (Coulter, 1909) in establishing homologies between the cells comprising the pollen grain.

This situation led some authors (Chamberlain, 1935) to view the microspore as an antheridial initial in some Coniferales (Taxodiaceae, Cupressaceae and Taxaceae), in *Gnetum* and in Angiospermae (see figure), and to extend this interpretation to Angiospermae in general, which is hardly justifiable, as it merely lumps together two distinct morphological concepts. However, this view is still very popular (see Battaglia, 1951) and, for sure, results from failure to imagine sperm production with no antheridium involved. But as impartial analysis suggests, the latter is actually non-existent not only in angiosperms, but already in some gymnosperms as well (e.g. *Welwitschia*).

Further examination of Fig. 6 clearly indicates that the male gametophyte evolution follows different pathways in different groups of seed plants, showing marked heterochrony. Some taxa exhibit reduction of prothallial cells and persistence of the antheridial cell (and consequently of the antheridial primordium consisting of the stalk and spermatogenous cells). Thus in Cycadales, for example, there is only one prothallial cell, but an antheridial primordium is formed. In Taxodiaceae, Cupressaceae and Taxaceae, no prothallial cells are found at all, but a rudimentary antheridium persists. A situation similar to that occurring in the latter group obtains in *Gnetum*. By contrast, in *Welwitschia* one prothallial cell persists, but no antheridium is developed; instead, a spermatogenous cell is formed directly, i.e., the antheridium as an organ has ceased to exist. Thus Taxodiaceae, Cupressaceae, Taxaceae and *Gnetum* resemble the angiosperms in one way (disappearance of prothallial cells) and *Welwitschia* in another (lack of antheridium).

The male gametophyte of angiosperms (the last member of the phylogenetic series) has undergone a most dramatic reduction: it shows neither prothallial cells nor antheridial primordium. Instead, as seen above, the very first division of the microspore results in formation of a spermatogenous cell. However, there is no sufficient reason to consider the latter homologous with the spermatogenous cell of the lower groups since it originates *directly* by microspore division, instead of being cut off by a special antheridial cell.

In angiosperms therefore acceleration reaches its climax due to earlier sperm formation during gametophyte development. This has resulted in angiosperms omitting the later stages of development at which the lower groups of plants still formed the antheridial primordium. Hence the situation is essentially similar to that occurring in the female sphere where one of the seven cells of the female gametophyte develops directly into an egg cell, whereas in *Pinus*, for instance, following formation of 2048 cells of the female gametophyte only an archesporial initial is formed, which still has to undergo three rounds of division to form an egg cell. In my view, Battaglia (1951) made a serious mistake in concluding that the male gametophyte of angiosperms is supposedly 'more conservative' than the male gametophyte since it carries an antheridium, albeit reduced, while the archegonium is no longer present. Actually, the male gametophyte has undergone the same, if not higher, degree of reduction as the female gametophyte. Undoubtedly, the antheridium, like the archegonium, has long since ceased to be formed in angiosperms.

The detailed cytological analysis of pollen grain development presented above suggests that it is not unreasonable to draw the following conclusion: the process of spermatogenous cell formation in angiosperms is basically similar to prothallial cell formation in gymnosperms; it shows much more distant similarity to antheridial cell formation and is totally different from the mode of formation of the gymnosperm spermatogenous cell derived not directly from the apical cell, but as a result of division of the antheridial cell. The spermatogenous cell of angiosperms appears to be homologous with nothing but the gametophyte cell.

As for the other cell of the angiosperm pollen grain, i.e., the siphonogenous cell, the role played by the tube cell originating from it and acquiring a unique significance in angiosperms marks the highest degree of male gametophyte development at which it itself is involved in the sexual process, comprising together with the sperm cells a common fertilization system (Gerassimova-Navashina, 1951a, b, 1952, 1954a, b) and ensuring, through the process of pollination, development, along with the seed, of a fruit.

Compendium of Genes Expressed Specifically in Pollen that have been Cloned and Characterized

Pollen represents the highly reduced mature male gametophyte of flowering plants. The function of pollen is to enable the transfer of sperm cells to the female gametophyte in order for fertilization to occur. While pollen is only three-celled at maturity, it does represent the evolutionary descendant of more complex many-celled gametophytes. As such, it is not surprising to find that pollen expresses a larger number of genes during its development and germination than might be expected by its apparent simplicity. Plant biologists have begun to examine some of the genes expressed in pollen in order to learn more about the nature and regulation of the genetic program operating in pollen. Studies of this sort should provide insight into ways of manipulating genes in pollen for agricultural, biotechnological, or even medical ends, as well as to further our basic understanding of haploid gene expression.

Types and timing of genes expressed in pollen. As mentioned, the number of different messenger RNAs expressed in pollen appears to be more than might be expected for such an apparently simple organism. The total number of genes expressed in mature pollen of various species has been estimated to be around 20,000-24,000 (reviewed in Mascarenhas, 1990), which represents roughly 60% of the number expressed in the morphologically more complex shoot. Naturally, most of the genes expressed in pollen are expressed in other tissues of the plant, while only a portion are specific to pollen. Estimates of the percentage of pollen-specific genes are in the order of 10-20% (Mascarenhas, 1990). Recently, however, the suggestion has been made that many pollen-specific messages might be low in abundance and therefore difficult to detect (Hodge *et al.*, 1992). Experiments by these authors suggest a per cent closer to 30%.

Studies of the mRNA levels of genes expressed in pollen have given some indication of the timing of expression of pollen genes. In general, many pollen genes have been shown to be expressed in roughly two classes: genes expressed early in development but which decrease in expression towards maturity, and genes expressed after microspore mitosis occurs whose mRNAs accumulate towards maturity. This is clearly a simplification because several genes of intermediate expression have been characterized recently (Hodge *et al.*, 1992; Roberts *et al.*, 1993). However, most of the pollen-specific genes that have been isolated are from the 'late' class since they can be isolated from mature pollen, which is generally easier to obtain in quantity than immature pollen. Genes of this class are likely to be involved in the process of pollen maturation, germination and tube growth.

Pollen-specific genes. Table 1 contains a list of all of the genes published to date which have been shown to be pollen-specific, and which have been cloned and at least partially characterized.

There are several more genes (such as *ADH*1, *Bet*VII/profilin, *Cat*-1, *EPSP* synthase, etc.), which have been shown to be expressed in pollen, but which are not pollen-specific. These have not been included in this list, nor have anther- or tapetum-expressed genes. Most of these genes have been isolated via standard differential screening techniques, which select for RNAs which are present at much higher levels in pollen than in other tissues. As a result, most of these are presumably among the most highly expressed and therefore the easiest to recover. It must be mentioned that the putative proteins cited as having a function or enzymatic activity have not actually been tested for that activity; rather, the function has been assigned based on protein sequence homologies to known proteins.

Many of the genes listed in Table 1 represent wall-degrading enzymes and proteins involved in the cytoskeleton which could possibly have functions in pollen tube growth through the pistil. It is interesting that many of the pollen proteins that elicit an allergenic response in humans have real functions in pollen, e.g. pectate lyase, profilin, calmodulin, etc.

Thus, the angiosperm microgametophyte displays a complex genetic program during development, with apparently two main classes of genes expressed. While the earlier class of genes is presumed to be associated with immature microspore formation, the later genes probably represent those associated with pollen maturation,

Table 1: Pollen-specific genes

Gene	*Species*	*Putative function*	*References*
*Zm*58	Maize	Pectate lyase, (homology to *lat*56, *lat*59, *G*10, *Amb* a I,II)*	Hodge *et al.*, 1992
*lat*56, *lat*59	Tomato	Pectate lyase, (homology to *Zm*58,G10, *Amb* a I,II)*	Wing *et al.*, 1989
*G*10	Tobacco, tomato	Pectate lyase, (homology to *Zm*58, *lat*56, *lat*59, *Amb* a I,II)*	Rogers *et al.*, 1992
Amb a I.1-4, *Amb* a II	Ragweed	Allergens, pectate lyases, (homology to *Zm*58, *lat*56, *lat*59, *G*10)*	Rafnar *et al.*, 1991 Rogers *et al.*, 1991
LolpI, *LolpI*b	Rye-grass	Allergens	Singh *et al.*, 1991
KGB family (*Poa* p IX)	K. bluegrass	Allergens	Silvanovich *et al.*, 1991
*P*2 family	*Oenothera*	Polygalacturonase	Brown and Crouch, 1990
*PG*1	Maize	Polygalacturonase (homology to *W*2247)	Niogret *et al.*, 1991
*W*2247	Maize	Polygalacturonase (homology to *PG*1)	Allen and Lonsdale, 1993
*Bp*19	*Brassica napus*	Pectin esterase	Albani *et al.*, 1991
*Bp*10	*Brassica napus*	Ascorbate oxidase	Albani *et al.*, 1992
*NTP*303	Tobacco	Ascorbate oxidase	Weterings *et al.*, 1992
*chi*A	*Petunia*	*CHI* genes	Van Tunen *et al.*, 1990
*TUA*1	*Arabidopsis*	Alpha-tubulin	Carpenter *et al.*, 1992
*SF*3	Sunflower	DNA-binding protein (zinc finger)	Baltz *et al.*, 1992
*I*3	*Brassica napus*	Oleosin	Roberts *et al.*, 1991
*Zm*13	Maize	Unknown (homology to *lat*52)	Hamilton *et al.*, 1989
*lat*52	Tomato	Unknown (homology to *Zm*13, possible homology to Kunitz trypsin inhibitor)	Twell *et al.*, 1989
*Tpo*44, *Tpo*70	*Tradescantia*	Unknown	Stinson *et al.*, 1987
*Bp*4 (A,B,C)	*Brassica napus*	Unknown	Albani *et al.*, 1990
various	*Brassica napus*	Unknown	Mascarenhas, 1990

* These genes are also homologous to the 9612 gene which is expressed in the styles of tobacco and tomato (Budelier *et al.*, 1990).

germination and tube growth. This is supported by the observation that several pollen-specific clones of the late class have been identified whose sequences are similar to enzymes associated with cell wall synthesis and degradation—a necessary and major requirement for pollen tube growth (Wing *et al.*, 1989). These 'late' genes represent the bulk of pollen genes which have been cloned and characterized.

Work on understanding the promoter elements of pollen genes is currently very vigorous and eventually characterization of genes involved in key regulatory processes in pollen should provide insight into the nature of haploid gene expression and its relation to that of the diploid plant. Findings of this type should point the way towards practical applications of pollen molecular biology in the areas of plant breeding, biotechnology and basic science.

Anther: Histochemical and Ultrastructural Profile (Plate 32)

The anther is a specialized heterogenous cell system and is involved in protection, nutrition and dispersal of pollen. Because of the importance of anther in pollen development coupled with the advent of cyto- and histochemical techniques in conjunction with high resolution TEM and SEM techniques, it has now been possible to unravel a surge of information on different aspects of anther development. Unfortunately, most of the work is focused on monad development while there is very little information on the more complex but very promising polyad development.

Histochemistry enables identification and localization of specific substances within cells and tissues which are of great advantage in understanding the physiological function and also the sites of synthesis, storage, transport and metabolism of the tissue concerned. At each and every step of cell, tissue and organ differentiation of the reproductive cycle, there is direct involvement of either a metabolite(s) or enzyme(s) paving the way for the next tenor. Histochemical studies also aid in our understanding of the role of anther wall layer tissues in pollen development.

The young anther primordium is dominated by cell division until differentiation of various tissues within it. The archesporial cell and its derivative, the sporogenous tissue, generally exhibit a rich concentration of RNA and proteins. It is now well documented that high contents of nucleic acids and proteins play a significant role in effecting differentiation in a variety of biological systems. The sporogenous tissue in the anther is metabolically a very active tissue and forms an important growth phase since differentiation of pollen mother cells is ensured in it. The sporogenous cells enter the phase of specialized cell growth and differentiation, which culminates in formation of pollen grains. In *Sesbania speciosa* the nucleolus in the meiocytes assumes a crescent shape, moves towards the periphery and forms a prominent nucleolar-cap (Vijayaraghavan and Seth, 1991). Such a polarized nucleolar cap may represent additional adjuncts that perhaps initiate meiosis. In *Prosopis juliflora*, the cellulosic mother cell wall persists around the tetrads and stains intensely with Periodic acid-Schiff's reagent (Chaudhry, 1992). Individual microspores of the tetrad reveal PAS-positive primexine and apertures. During vacuolation in the microspore cytoplasm, the nucleus migrates towards the periphery where it undergoes an asymmetric mitotic division resulting in a large vegetative and a small lenticular generative cell. At anthesis, the generative cell becomes spindle-shaped, possesses a thin PAS positive cell wall and is embedded in the vegetative cell cytoplasm. The generative cell further divides to form a pair of elongate sperm cells, both of which stain intensely for proteins. The pollen grains are shed at the three-celled stage and are tricolporate with rugulose exine sculpturings.

Composite pollen occur in about 52 families of flowering plants (Guinet and Le Thomas, 1973). In *Albizzia lebbek* the pollen grains are shed as polyads and each polyad is a composite of 16 grains (Shah, 1992). A single sporogenous cell divides twice to form four pollen mother cells. A continuous callosic wall surrounds each anther locule. However, between the meiocytes, the callosic wall is discontinuous, thus indicating the presence of cytomictic channels. The presence of such cytoplasmic connections provides a visible means for the establishment of cohesion between adjacent microspores after dissolution of the callose special wall. Meiosis in each pollen mother cell results in the formation of four tetrahedrally arranged sets of tetrads that adhere during successive stages to form a 16-celled polyad. The exine is bilayered with the sexine confined to only the outer surface of the polyad while the nexine and intine surround each grain. Within the polyads a few PAS-positive pores are located on the outer face while others occur at the corner of the grains. The polyads are acalymmate where the grains are not united by a continuous layer of sexine. Acalymmate polyads are also observed in *Acacia* spp. (Kenrick and Knox, 1979) and many members of the tribe Ingeae (Mimosoideae: Leguminosae) (Niezgoda *et al.*, 1983). While considering the differentiation of monads/polyads, the role of anther wall layers cannot be overlooked. The wall layers consist of the epidermis, endothecium, middle layers and tapetum which function as a local reservoir for high energy source in the vicinity of the sporogenous tissue and pollen mother cells and ultimately in differentiation of the male gametophyte. It is believed that the various metabolites stored in the anther wall layers are utilized by the developing pollen. The nutritional role of the anther wall layers is indicated by the universal occurrence of plasmodesmata between these layers, the tapetum and sporogenous tissue. As the microspore mother cells undergo further differentiation, the endothecium develops radial thickenings, the middle layers become flattened and are eventually crushed and the tapetum becomes a prominent and specialized layer.

In *Prosopis juliflora* the endothecial thickenings appear as well-developed, PAS-positive fibrous bands that arise along the inner tangential walls, extend outwards and upwards and terminate near the outer tangential walls. These thickenings become more pronounced during the vacuolate stage of the microspores and help in the mechanics of anther dehiscence. The cellulosic nature of the thickenings is confirmed by a strong PAS reaction. Electron micrographs reveal a distinct fibrillar texture of these thickenings. The cells constituting the middle layers are vacuolate, rectangular and possess densely stained cytoplasm and nuclei. During meiosis in the meiocytes, degeneration sets in and the inner middle layer (next to the tapetum) is obliterated. The outer middle layer persists, however, until about the mid-maturation bicelled stage of the pollen grains. Because of the increase and decrease of starch in the middle layers and endothecial cells, the breakdown products of starch of these cells are one of the main sources of microspore and pollen development. EM studies on these middle layers reveal many chloroplasts with well-developed thylakoids and starch grains. In *Lilium* anther there are 4-5 middle layers and as a source of nutrients in the loculus, the starch of these middle layers has an important function (Reznickova and Willemse, 1980).

The host side of the sporophyte-gametophyte interface in angiosperm anthers consists of the tapetum, a tissue layer whose morphology can vary widely in different taxa from a secretory epithelium to an invasive plasmodium. Recent reports indicate that after the tetrad period a secretory tapetum may be reorganized into either (1) an amoeboid tapetum, (2) a pseudo-periplasmodium or (3) a regrouping of tapetal cells that retain clear cellular boundaries so as to essentially fill the space in loculi between microspores. Pacini (1990) and Pacini and Franchi (1991) have prepared reviews covering common tapetal types occurring in many major groups of vascular plants. A few taxa in which an invasive but non-syncytial tapetum is reported include *Canna indica* L. × *C. spec* (Tiwari and Gunning, 1986a,b) and *Cichorium intylus* (Pacini and Keijzer, 1989).

The nature, function and location of the tapetum in the hierarchy of anther wall layers makes it a tissue of considerable physiological significance as all the nutritional material entering the sporogenous cells/microspores/pollen grains passes or originates from it. In *Prosopis juliflora*, the tapetum is of the secretory type (Chaudhry, 1992). During meiosis in the pollen mother cells, prominent intercellular connections develop between the juxtaposed tapetal cells. The tapetal cells are rich in RNA and protein which implicates the presence of abundant stratified RER in the cytoplasm. During the tetrad stage, both the inner tangential and radial walls of the tapetum show dissolution. The tapetal cytoplasm becomes extremely dense and vesicular and contains copious RER profiles, mitochondria and ribosomes. Concurrent with these changes in the tapetum, the pattern elements of the pollen grain wall are laid down at the microspore surface while they are still invested within the thick callosic wall. Concomitant with the development of probacules on the microspore surface, the tapetum shows extrusion of pro-orbicules which are rapidly coated with sporopollenin and differentiate into the orbicules. Prominent embayments are formed at the surface of the tapetal cells at the loci of extrusion of pro-orbicules and profiles of ER with dilated ends are closely associated with the developing orbicules. Initially, discrete plaques of sporopollenin are laid on the pro-orbicular cores which later enlarge by accretion of more material and coalesce to form a complete coating. As the layer of sporopollenin increases in thickness, fusions occur between the adjacent bodies to form compound or complex orbicules. The pro-orbicules during the tetrad stage and even after accretion of sporopollenin are associated with rough endoplasmic reticulum, implicating its involvement in the synthesis of pro-orbicules. The origin of pro-orbicules from ER has also been reported in *Anemarrhena asphodeloides* (Chen *et al.*, 1988) and *Lilium henryi* (Herich and Lux, 1985). Sporopollenin synthesized by the tapetum condenses upon the pro-orbicular cores and the exine initials. The microspores released from the tetrads come into direct contact with the tapetal-derived sporopollenin polymer. Sporopollenin deposition progresses rapidly within and upon the patterned exine and the wall infrastructure formed in the tetrad becomes well organized. During vacuolation in the microspores, the tapetum becomes progressively more disorganized and its cytoplasm fragmented while in the microspores, pollen wall formation proceeds through endexine sporopollenin deposition and cellulosic intine formation. At anthesis the tapetum is completely lysed, leaving only the tapetal membrane studded with both single and complex orbicules. In *Albizzia lebbek* many orbicules line the inner tangential wall of the tapetum and form a 'sac'-like structure, the significance of which lies in efficient dispersal of polyads. Due to the contact pressure from the orbicules, the exine possesses a circular surface.

Anther gland. Secretory glands are widespread in plants and show great diversity in position, organization and secretion. However, secretory glands which constitute appendages on anther connectives are rare and have been reported in Myrtaceae (Beardsell *et al.*, 1989) and Violaceae (Fahn, 1979). In *Prosopis juliflora* each anther is surmounted by a gland which lies above the anther locules as an appendage on the anther connective (Chaudhry and Vijayaraghavan, 1992). The gland consists of a narrow multicellular stalk and a wide multicellular, globose head of secretory cells. The secretory cells of the globose head become densely cytoplasmic and are rich in both intracellular polysaccharidic secretions and proteinaceous inclusions. Histochemical analysis reveals that the material initially enclosed in storage vesicles of the anther gland, later accumulates on the surface. In an active and mature gland, the secretory cells lyse and through degeneration of the protoplasts (holocrine secretion) release a protein-carbohydrate exudate which passes through cuticular openings to the exterior of the gland. The heterogenous exudate that envelops the gland surface serves as a food source and attracts a wide array of foraging insects, in particular those belonging to Thysanoptera and Hemiptera. A frequent visitor is *Oxyrhachis* *tarendus* of Membracidae, commonly observed within flowers gnawing on the anther gland and other floral parts. Interestingly, an insect was even found trapped, probably while foraging, within the anther gland cells. It is suggested that besides copious amounts of pollen and nectar produced by *P. juliflora*, the protein-carbohydrate exudate of the anther gland serves as an additional food source. Insects, once attracted and during gnawing on the gland surface, are dusted with pollen and may serve as inadvertent pollen vectors when they visit one inflorescence after another, of an individual tree or even another tree to permit outcrossing within a population. The exuded secretion coheres the pollen grains, acts as a food source for the foraging insects and thus facilitates pollen dispersal. This phenomenon may significantly affect the flower-visiting behaviour of potential pollinators.

Cytomixis (Greek *cytos* = cell, *mixis* = mixing)—migration of a portion of the cytoplasm, together with the nucleus or part of it, from one cell into another, or into several adjacent cells. This occurs in both somatic and sporogenous tissues (Plate 33).

Gates (1911) was the first to notice the phenomenon of cytomixis and defined it as the process of migration of the nuclear chromatin from the donor cell into the recipient cell cytoplasm.

Various authors have since defined cytomixis as: migration of unstructured chromatin from one cell into another; exchange of cytoplasm between neighbouring cells; transposition from one cell into another of entire interphase nuclei or parts of these together with the cytoplasm; migration of the cytoplasm with the nuclear chromatin at any stage of spiralization; fusion of the protoplasts of adjacent sporogenous cells; migration between spore mother cells of the cytoplasm with individual chromosomes, chromosome segments, or chromosome associations.

Cytomixis can be induced experimentally through archesporial tissue damage (Levitskyi, 1928; Tarkowska, 1973), viral diseases (Smirnova, 1954), and by extremely adverse weather conditions (Medvedeva and Bazavluk, 1951).

Cytomixis is most common among plants with disturbed homeostasis: in hybrids, mutants, inbred lines and regenerants.

In terms of cytogenetic implications, of particular interest is cytomixis in sporogenous cells. The incidence of cytomixis is a widely varying character. In the sporogenous tissue of hybrids, it reaches maximum value during leptonema-pachynema. In late prophase-metaphase, its frequency is low. One and the same microspore mother cell may be a donor and recipient simultaneously. Mass migration of chromosomes from microspore mother cells into intercellular spaces results in completely sterile sporogenous cells.

A prerequisite to protoplast migration is the occurrence of discontinuities in cell walls resulting in the formation of cytomictic channels (Heslop-Harrison, 1966a). Cytoplasmic channels are most likely to occur where massive callose depositions are lacking and the protoplasts of adjacent meiocytes are intimately in contact with one another. No continuity of cytomictic channels with plasmodesmata has been detected, nor are their cross-section areas comparable.

The aftereffects of cytomixis complicate the pattern of meiosis. The chromatin that has migrated is identifiable by the following features: it is invariably located in the cytoplasm of the recipient cell and is never incorporated in its nucleus; it is always found in the same stage of spiralization at which it migrated. The recipient cell chromosomes therefore exhibit varying degrees of spiralization. The chromatin that has migrated may appear as homogeneous droplets of variable size and density. It can be represented by bivalents, univalents or by fragments of metaphase chromosomes.

The relationship between cytomixis and aneuploidy continues to be a widely debated topic. For the recipient meiocyte chromosome complement to be enlarged, the following requirements should be met: the chromosomes should be intact and incorporated in the mitotic apparatus of the recipient cell, otherwise the excess chromatin will be eliminated. According to numerous reports, including those substantiated by electron microscopic studies, the migrating chromosomes tend to be extensively damaged (Weiling, 1965; Romanov and Orlova, 1971; Tarkowska, 1973). Foreign chromatin is normally rejected by the recipient cell. The rejection is effected by cell wall initiation resulting in the formation of small, abnormal "companion" cells, with both donor and recipient cells degenerating. In almond and peach, cytomixis caused extensive pollen sterility (Soodan and Wafai, 1987).

The probability of occurrence of meiocytes with increased chromosome number following cytomixis can hardly be a high one since it depends on a combination of events. Undoubtedly, the chromatin-donating cells that have lost most of their chromosome complement are incapable of further development. There is no evidence by which to judge the viability of meiocytes that have been depleted of a small portion of their nucleus.

The phenomenon of cytomixis has been erroneously attributed by some authors to the existence of an S cytoplasmic gradient within microsporocytes of hybrids and to differential rates of nuclear division in parental stocks.

Some modern authors, following Bleier (1930) who took the microsporocytes showing varying degrees of chromosome coiling as an indication of the parental genomes being isolated in hybrids of grasses, misinterpreted the consequences of cytomixis. Thus, Shkutina and Khvostova (1966) attributed the presence within a single meiocyte of chromosomes showing varying degrees of coiling to the separation of parental chromosome sets. Citing as evidence in support of their hypothesis, the authors refer to cases wherein, in addition to normally coiled chromosomes, prophase chromosomes occur. The authors believe this to be the extreme manifestation of non-coordinated functioning of the wheat and rye genomes, i.e., the prophase-like chromosomes are the rye chromosomes lagging behind the wheat chromosomes in the degree of coiling.

Cytogenetically, it is essential to know whether the chromosomes showing a lower degree of coiling belong to the microsporocyte nucleus and what the numbers of chromosomes in the nuclei of these meiocytes are. According to the aforesaid authors, a meiocyte of the hexaploid *Triticale* should exhibit 28 wheat chromosomes (= 14 bivalents) as more tightly coiled and 14 rye chromosomes (= 7 bivalents) as prophase-like chromosomes. In fact, the meiocyte nucleus comprises 21 bivalents composed of wheat and rye chromosomes showing equal degrees of coiling, with less tightly coiled chromosomes, like the more concentrated masses of chromatin, being the supernumerary ones. They have been incorporated in the recipient meiocyte as a result of cytomixis.

It should be emphasized that cytomixis never occurs simultaneously and regularly in all the cells of sporogenous complexes. Its manifestation is highly sporadic. These features of cytomixis suggest that it constitutes a disturbance of the normal process of microsporogenesis and that genetic imbalance is a factor considerably enhancing the rate of cytomixis, if not primarily responsible for it. Homeostatic disturbances create conditions for highly negative responses to any adverse factors such as mechanical injury of the anther, extreme environmental conditions and hybridity.

The view of cytomixis as a regular phenomenon promoting genetic variability in nature and under experimental conditions held by some authors (Shneider, 1988) appears to be an unconvincing one. Cytomixis is a disturbance of the cell reproduction process more or less frequently accompanying microsporogenesis in plants. The most significant effect of cytomixis is partial sterility of sporogenous complexes of the anther, especially in hybrids.

Cytoplasmic Male Sterility (CMS)—male sterility resulting from interactions between the specific S cytoplasm and the *rf rf* homozygous recessive genes (Kaul, 1988) (Plate 34). This phenomenon was first described as contabescence by Corrence in 1904 in summer savory (*Satureja hortensis*). Laser and Lersten (1972) reported the presence of CMS in 140 species, 47 genera and 20 families of angiosperms; Kaul (1988) presents evidence for 288 species in which this type of male sterility has been found.

CMS is characterized by normal, regular meiosis with no disturbances. Developmental abnormalities occur most frequently during the postmeiotic period, at the tertrad stage or shortly after release of microspores from tetrads, during development of a microspore into a pollen grain, the male gametophyte (Oryol, 1968; Simonenko, 1982). In some cases, development proceeds up to a two- or three-celled stage of pollen grain.

Abnormalities appear as progressively higher vacuolation of the microspore protoplast, abnormal development of mitochondria, plastids and sporoderm which fails to form an intine.

A characteristic feature of anthers is not only abnormalities in the development of sporogenous tissue, but also malfunctioning and abnormal development of the tapetum which is presumably the site of expression of the CMS genetic factors (Flawell, 1974). This is manifested either in premature degeneration and lysis of the tapetum or, most commonly, in its long persistence or extensive growth following formation of microspores.

Anthers of the cytoplasmic male sterile plants exhibit hypertrophied cells of the tapetal layer, transformation of the secretory tapetum into a false parietal intratapetal plasmodium following the breakdown of radial walls of the tapetal cells, and abnormal development of orbicules or their complete absence (Shivanna and Johri, 1985). The micrographs show two variants of abnormal development of anther tapetum in maize: with the Texas or CMS-T cytoplasm exhibiting local formation of a false plasmodium and with the S-type of CMS where each tapetal cell is hypertrophied, shows a thickened wall, is highly vacuolate and persists until the complete degeneration of young microspores.

The fundamental role of tapetum in pollen sterility has also been demonstrated using molecular biological techniques. Transgenic plants of tobacco and rape carrying a dominant *TA29* RNase gene specifically expressed in the tapetum, exhibit complete pollen abortion resulting in complete male sterility. It has been suggested that the transfer of this gene into other plants may be used for the practical purposes of mass production of hybrid seeds (Mariami *et al.*, 1990).

The genetic mechanisms of CMS have been most extensively studied in maize. Jones (Jones, 1950; Jones *et al.*, 1957) originated the concepts of plasmagenes and chromogenes. He came to the conclusion that pollen sterility in maize results from the interaction between a special cytoplasm carrying a heritable mutation and recessive nuclear factors showing Mendelian inheritance. The dominant state of these factors results, with the same CMS cytoplasm, in restoration of pollen fertility. In a sterile analogue of the maize line, pollen fertility restoration is due to the presence of two dominant alleles, *Rf1* and *Rf2*, all the known stabilizers of sterility show the *rf1rf1Rf2Rf2* genotype, the sterile line exhibiting two pairs of recessive alleles, *rf1rf1* and *rf2rf2*, and the S cytoplasm (Duvick, 1959). Similar genetic patterns are found in other plant species and genera (Edwardson, 1970).

Mitochondria are presumed to be cytoplasmic determinants of CMS in some plants (Peterson *et al.*, 1975; Barrat and Flawell, 1975). However, there is evidence suggesting that plastids are responsible for this character (Ahokas, 1978).

CMS is used extensively in breeding work with maize, sorghum, sunflower, sugar-beet and other crops, for mass production of F_1 hybrid seeds.

Development of Embryo Sacs in Anthers (Plate 35): The phenomenon of reversal of sexual state—formation of 'embryo sacs' within pollen grains—was first detected by Nemeč (1898) in *Hyacinthus orientalis* and subsequently called the 'Nemeč phenomenon'. More recently (De Mol, 1923; Stow, 1930, 1934; Naithani, 1937), similar patterns were observed following special treatment of bulbs of various *Hyacinthus* varieties whose anthers were mainly at the stage of meiosis. The Nemeč phenomenon has also been reported in *Ornithogalum nutans* and *Leptomeria billardierii* (Geitler, 1941; Ram, 1959b).

Experimental evidence on the formation of 'embryo sacs' within *Heuchera micrantha* anthers (Vijayaraghavan and Ratnaparkhi, 1977) is worthy of special attention. Considered in this work are also possible patterns of behaviour of microspores within tetrads and at post-tetrad stages, as well as issues pertaining to the genesis of such 'embryo sacs'. The authors found that the 'embryo sac' formation here can conform to the Polygonum-, Plumbago- or Penaea-type, although quite often no walls are initiated in this case.

Mature pollen grains with a normal seven-celled embryo sac (three cells of the egg apparatus, three antipodal cells and the central cell with two polar nuclei) have also been reported in grasses (*Hordeum vulgare* and *Triticum aestivum*—Batygina, 1987a).

Davis (1968) presented evidence for the possible formation of 'embryo sacs' (embryo sac-like structures) from individual cells of the anther tapetum in *Calotis lappulacea* (Asteraceae). In describing this phenomenon, Davis points out the morphological and structural variability of these structures as well as their increased numbers in anthers exposed to various treatments.

Despite the fact that opinions differ as to the origin of embryo sac-like structures, analysis of the available data suggests that these can be formed by four morphogenetic pathways: as a result of division of a microsporocyte, microspore, vegetative cell or generative cell. It appears to be all but an established fact that these embryo sac-like structures should be preceded by coenocytic structures from which, depending on the number of cell divisions and the pattern of distribution of nuclei, 'embryo sacs' of either Polygonum-, or Plumbago- or Penaea-type can be formed. Not all the coenocytic structure formed during anther ontogeny develop into 'embryo sacs' since for this to occur, specific factors determining establishment of polarity along the micropylar-chalazal axis of the so-called pollen grain are required.

Most of the authors reporting the Nemeč phenomenon in various members of Liliaceae, Santalaceae, Saxifragaceae and Poaceae, point out that anthers developing embryo sac-like structures differ in pattern of development of each sporangium (its wall and sporogenous complex). Thus, occasionally the microspores degenerate completely within three microsporangia of a single *Heuchera micrantha* anther, whereas an embryo sac-like structure develops within the fourth microsporangium (Vijayaraghavan and Ratnaparkhi, 1977). Different patterns of tapetum differentiation (formation of 'palisade-like cells') have also been observed in different microsporangia of a single anther. That is, there appears to be a certain correlation between formation of embryo sac-like structures and the wall structure of the microsporangium within which these develop. It has been substantiated experimentally that there is a direct relationship between tapetum development and pattern of microspore and pollen grain development (Zenkteler, 1962; Hayward and Manthriratna, 1972). There is general agreement among authors studying this phenomenon that formation of embryo sac-like structures is usually due to extensive (up to 90%) degeneration of pollen grains. This has led some authors to assume the presence in the anther of a specific necrohormone inducing the formation of embryo sac-like structures (Haberlandt, 1921; Stow, 1934).

A comparative morphophysiological study of pollen grains developing embryo sac-like structures and normal embryo sacs of *Hyacinthus* has revealed a close similarity between these two types structure. That the former possess normal 'female potency' is, in Stow's opinion, evidenced by the fact that in in-vitro culture, pollen tubes of normal grains fertilize such embryo sac-like structures. An increase in number of nuclei within these structures following fertilization is taken by the author to be indicative of the process of endosperm nuclei formation. Unfortunately, it is unclear whether or not a zygote, and subsequently an embryo, is formed in the process.

Whether such 'pollen grains' with 'female potency' are completely enclosed in exine, as reported in *Heuchera,* or only develop it at one end of the embryo sac-like structure, is still an open question. Nor is there consensus of opinion among authors as to the pole at which the exine is formed. Some authors believe the exine to be localized at the site of the egg apparatus formation (Geitler, 1941), others position it in the region of the antipodals (Stow, 1930; Naithani, 1937). The exine is disrupted later in development of embryo sac-like structures.

The factors inducing formation of such embryo sac-like structures within the anther are still unclear. Geitler argued that the phenomenon is genetically controlled. Naithani, while also explaining this phenomenon in terms of Mendel's laws, believed temperature to be responsible for the changes occurring the pollen. The necrohormone hypothesis has not been completely validated either.

Gerassimova-Navashina (1958) reasoned that metabolic disturbances occurring in anthers on exposure to different temperatures promote expansion of the microspore and make possible further divisions within it. Where embryo sacs develop from microspores, the process of embryo sac formation does not just bear a superficial resemblance to that of the embryo sac development from megaspores, as assumed by Gerassimova-Navashina, but rather the two processes are essentially identical as a result of identical operation of fundamental laws of cell development in similar conditions. Therefore, it is not quite correct to attribute to the microspore any 'male' or 'female' potency, as is done by Stow (1934), since any embryonic cell has but a single 'potency', i.e., to develop and divide under appropriate conditions. If a change in conditions occurs, it invariably brings about changes in the pattern of development, and this is observed during the formation of various types of embryo sacs (for details, see further: Embryo Sac: Origin and Development).

The above findings have acquired a new meaning with the discovery of another phenomenon allowing the production of haploid plants from microspores and pollen grains (see Batygina, 1987a).

Anther as Model for Studying Morphogenetic Potencies and Pathways—Stamens, particularly anthers, are the structures currently most often used in experimental embryology, primarily for production of haploid plants. Major morphogenetic processes occurring in the plant as a whole (for example, alternation of generations) are represented in anther morphogenesis.

To gain a more complete understanding of the anther as a model system it is necessary, first and foremost, to develop methods of studying this structure involving unification of the nomenclature of developmental stages; formulation of the main problems touched upon in the studies; a unified approach to studying morphogenetic pathways of the anther in in-vitro culture in various species of angiosperms; an integrated study of all the structures of the stamen and female reproductive sphere. Despite the fact that the anther is a complex integrated system, in terms of methods employed in studying it, it is undoubtedly a simpler experimental subject, a compared with the ovule, for in-situ, in-vivo and in-vitro studies of morphogenesis, since in anther studies it is easier to identify and simulate individual morphological processes and correlations (Batygina, 1983a, c; Reznikova, 1984).

In view of this, a particular investigation strategy probably needs to be developed which would involve the following steps:

1. Studying stamen development taking into account the biology of plant development.
2. Studying stamen development in species contrasting in mode of pollination (e.g. chasmogamy vs. cleistogamy).
3. Comparing the kinetics of morphogenetic and physiological and biochemical processes occurring in the stamen.
4. Comparing evidence from in-situ and in-vitro studies of stamen development.
5. Simulating conditions for each stage of anther development taking into account morphogenetic and morphophysiological correlations.

The above approach would make it possible to identify critical periods in stamen development determining specific development of its component structures. This would enable control of the in-situ, in-vivo and in-vitro development of the anther. Furthermore, supplementing the culture medium with particular 'triggers' and inducers at different stages of anther culture in various species of angiosperms would enable assessing the morphogenetic potency of different anther cells as well as identifying morphogenetic pathways in in-vitro culture: embryoidogenesis, organogenesis and histogenesis.

PART THREE—OVULE

OVERVIEW

Ovule—a seed plant structure formed on the placenta. It is the organ in which the processes of megasporogenesis, female gametophyte formation and fertilization take place and which, through complex transformations, gives rise to a seed. The ovule is a modified megasporangium of seed plants enclosed in sterile envelopes having a protective function.

Göthe (1790) had already used the term 'Samenknospe' and regarded the ovule as a metamorphosed bud, viewing the nucellus as a shoot apex and the integuments as a modified ultimate pair of leaves. The term ovule (Latin *ovulum*) came to be most widely used in English literature as well as French literature. The corresponding term in German literature is Samenanlage.

The ovule consists of the central body—**nucellus** enclosed in one or two protective envelopes, **integuments**, and attached to the placenta by a stalk, the **funiculus**. The portion of the ovule where the nucellus fuses with the integuments is termed the **chalaza** (Esau, 1977). At the opposite (micropylar) pole of the ovule, the free ends of the integument(s) extend to form a more or less lengthy tube, the **micropyle**, or micropylar canal. Pollen in gymnosperms (except in the Araucariaceae) and pollen tube in flowering plants find their way through the micropyle to the egg apparatus; it is through the micropyle that eruption of the radicle occurs during seed germination. Most workers are agreed that the nucellus represents the homologue of the megasporangium of heterosporous pteridophytes. The tendency of the archesporial cells within the megasporangium to become sterile during changeover to heterospory, already apparent in higher spore-bearing plants, reaches its peak in spermatophytes. However, in contrast to spore-bearing plants, spores are not liberated from the megasporangium in seed plants, with not only male gametophyte development, but also early stages of the newly formed sporophyte occurring within the megasporangium. Hofmeister (1851) was the first to advance this hypothesis.

The ovule of seed plants, including many angiosperms, is massive, crassinucellate. Its nucellus develops from numerous placental cells, dermal ones in the gymnosperms and dermal and subdermal in flowering plants (Schnarf, 1929; Allen, 1946, 1947; Takhtajan, 1964). On this basis, seed plants are assumed to have originated from ancient eusporangiate pteridophytes. Among the flowering plants, the most massive nucellus is found in unitegmic dicotyledonous species belonging to Hamamelidae as well as in primitive taxa of dicotyledons (Rosaceae, Ranunculaceae, etc.) and in monocotyledons. In these, the dermal cells of the nucellar apex normally divide periclinally to form a massive nucellar cap which sometimes projects into the micropylar base. This distinguishes flowering plants from gymnosperms, especially from their more primitive representatives in which, by contrast, the microsporangium unfurls as it were, resulting in the formation of a pollen chamber—a depression in the nucellar apex—in the region where its cells have been lysed. Unfurling of the nucellus can be regarded as a feature inherited from spore-bearing progenitors. The micropylar canal cells secrete a fluid (the so-called pollination drop) which entraps airborne pollen grains and, as it evaporates, draws them into the pollen chamber where they germinate.

In the more advanced angiosperms, the ovule is tenuinucellate with a very small thin nucellus consisting of one or two cell layers and developing from one or a few subdermal and dermal cells of the placenta.

In some members of the subclasses Dilleniidae, Rosidae and Caryophyllidae, the nucellar apex extends a great deal to form a nucellar beak projecting through and sometimes even beyond the micropyle. Where the nucellar beak adjoins the obturator, it facilitates the entry and growth (passage) of the pollen tube (Cooper and Brink, 1945; Favre-Ducharte, 1963; Savchenko, 1973). The beak-like appendage resembles the outgrowth projecting beyond the ovule and serving to entrap pollen in Araucariaceae, Saxegothaea, *Cedrus* and some other gymnosperms. The presence of a beak-like outgrowth in flowering plants may be indicative of their potential ability to entrap pollen. This has been confirmed in experiments by Maheshwari and Kapil (1963) and Savchenko (1973). Martens (1957) regarded this structure as a malformation. Interesting insights into the function of the beak-like outgrowth are provided by its formation in Betulaceae where during anthesis the ovary is still not completely formed, the ovules are 'naked', thus enabling one to speak of a kind of 'gymnospermous' condition. It is likely that the beak-like appendage is a manifestation of anachronism (Savchenko, 1960, 1973).

According to the type of nucellus and the number of integuments, five types of ovule are distinguished: crassinucellate, uni- or bitegmic; tenuinucellate, uni- or bitegmic; and lastly 'naked' (ategmic) with a reduced nucellus (Schnarf, 1929).

In the ovular primordium of some taxa, a single archesporial cell and a megasporocyte are formed, the latter giving rise to megaspores. In a number of taxa, a large proportion of the archesporial cells prove to be fertile (Casuarinaceae, Betulaceae, Fagaceae, Ranunculaceae, and some members of Rosaceae, Asteraceae, etc.). Normally, a multicellular archesporium is considered to be a primitive feature; its presence in evolutionally advanced taxa may be either a manifestation of heterobathmy, or a derived condition. Meiotic division results in the formation of megaspores. The female gametophyte most commonly develops from a single megaspore, the one normally located closer to the chalazal end, i.e., to the nutrient supply, although potentially, any megaspore within a tetrad can develop. The other megaspores of the tetrad undergo reduction at certain developmental stages.

Germination of a megaspore results in formation of a female gametophyte. The female gametophyte exhibits different patterns of development in different divisions of seed plants. In most spermatophytes it is monosporic, being formed from a single megaspore which, in the process, expands a great deal due to the consumption of circumjacent nucellar cells. In gymnosperms, repeated nuclear divisions initially occur within the spore, resulting in the formation of 260 or more nuclei arranged around the large central vacuole in the parietal cytoplasm. Subsequently, the nuclear phase gives way to the cellular one. Cells are formed centripetally and then collapse centrifugally under the influence of the proembryo. At the micropylar end of a multicellular prothallium, archegonia are formed: one (*Torreya*), two (the majority of gymnospermous taxa) or more (Cupressaceae), up to 200 (*Widdingtonia*). In Cupressaceae, the archegonia form a cluster separated from the prothallium by a special wall. If archegonia are numerous in Cycadaceae, they are scattered over the entire surface of the prothallium. The archegonia are normally embedded completely in the gametophytic tissue, the archegonial wall is, as a consequence, greatly reduced, the collar becoming shorter in many taxa (as in Cycadopsida where it is composed of as few as two cells). Finally, no archegonia in the classical sense are formed in genera *Gnetum* and *Welwitschia* belonging to the division Gnetophyta.

In flowering plants, archegonia are lacking, the number of nuclear divisions is greatly reduced and no multicellular prothallium is formed. The megaspore gives rise to an embryo sac. The 'normal' eight-nucleate embryo sac—the Polygonum type of development—has been found in 80% of the flowering plant species studied. Other types of embryo sac are also distinguished.

Many hypotheses have been advanced in the literature on the origin of the embryo sac. Most authors regard this as a greatly modified and reduced female gametophyte comprising as few as 7 cells and 8 nuclei and showing an early tendency for syngamy (sexual fusion). One of the most interesting hypotheses is that proposed by Porsch (1907); it bridges the gap between flowering plants and the other spermatophytes. Porsch believed the angiospermous embryo sac to be a result of further reduction of the spermatophyte prothallium. Here the principle of material economy and acceleration of developmental rates, observable throughout higher plant evolution, reaches its climax. One of the factors accelerating development appears to be light, which in passing through the translucent carpels that are more delicate than the scales of strobili of other spermatophytes, could enhance the rate of gamete maturation. Experimental studies have shown anthesis and light intensity to be positively correlated (Golenkin, 1927; Senyaninova-Korchagina, 1959).

The **integument** encloses the nucellus of an ovule. In most **spermatophytes**, including the majority of sympetalous and some dialypetalous dicotyledonary angiosperms and certain monocotyledonary ones, there is one integument (unitegmic condition). Most dialypetalous dicotyledons and monocotyledons have two integuments, an inner and an outer (bitegmic condition), with the outer often being more conspicuous than the inner.

Some authors regard arils and arillodes as a third integument arising as a result of splitting of the outer integument. Sometimes the third integument is considered a structure *sui generis*. The origin and homology of the integument remains an open question. Corner (1949a) has described the expanded tissue between the outer and inner integuments involved in the formation of the seed coat (testa) in three genera of the family Annonaceae as a fourth integument, taking the aril expanding after fertilization and developing into fruit pulp for the third integument. However, many morphologists question this view (e.g. Savchenko, 1973; and others).

In most spermatophytes, the integument fuses with the nucellus, remaining free, at most, at the micropylar end (Eames, 1961), and it is only in seed ferns and in most relatively primitive angiosperms (*Cleome, Julianiania, Casuarina, Juglans, Myrica*) that both integuments are distinctly free, not fused to one another or to the nucellus. In members of some taxa, the entire process of fusion can be followed. Thus in some members of Rosaceae, Fabaceae, Myristicaceae, Ochnaceae, Capparaceae, Ranunculaceae, Fagaceae and in many monocots, free integuments are tightly adpressed to one another without fusing while in some genera of Ranunculaceae, Fabales, Palmales and some others, all degrees of fusion can be followed up to complete union when the borders between the integuments are no longer distinguishable. Where the integuments are not fused to the nucellus, the latter may have a stalk lifting it over the integument. This can be considered a primitive feature.

It can be inferred from cases of united integuments that the simple (single) integument in some angiosperms has arisen from the double one secondarily, congenitally, as a result of fusion, with the tendency towards unitegmy occurring independently and heterochronously in different families (Coulter and Chamberlain, 1903; Maheshwari, 1950; Eames, 1961; Poddubnaya-Arnoldi, 1964; Takhtajan, 1964; Savchenko, 1973; and others). The fusion of two integuments gives rise to a single, massive integument. Such a fusion may be due to the evolutionary trend towards concrescence, common to all floral organs of angiosperms, or, according to Joshi (A. Joshi, 1939; P. Joshi, 1960), they may have fused as a result of pressure of the ovary wall on the ovules.

The primary function of the integument is to protect the nucellus against adverse environmental factors. It develops in the seed into the seed coat. In the ovule, it also functions as a trophic (storage) and partially secretory tissue.

The origin of the integument in spermatophytes remains an enigma. Many hypotheses have been advanced but none is wholly satisfactory. For instance, the integument has been interpreted as a mere outgrowth of the sporophyll, a modified indusium in pteridophytes, a recently evolved structure, or a modified leaf. Currently most workers entertain the hypothesis of its synangial origin, proposed by Benson (1904). According to this hypothesis, the nucellus evolved from a transformed single fertile megasporangium once located in the central portion of the sorus of ancestral Lyginopteridophyta, the integuments arising by fusion, sterilization and transformation of 1-2 circles of sporangia surrounding it. Some authors believe the two integuments of flowering plants to have been derived from the two circles of sporangia; others consider the outer integument to have originated from the cupule of ancestral Lyginopteridophyta. It is worth noting that the integuments of different taxa are probably not always homologous.

In flowering plants, morphological features of the ovule and funiculus, their shape and relative positions as well as the position of the ovule in the ovarian cavity, and primarily the orientation of micropyle, are all important taxonomic characters. Numerous studies have shown these characters to reflect particular biological features of the taxa exhibiting them, their evolutionary trends.

One of the most important features of the ovule is orientation of the micropyle. Depending on this, fertilization is either hampered or facilitated, the pollen tube pathway either longer or shorter, and the transport of water to the seed (embryo) required for initiation of growth processes either facilitated or hampered. Occasionally, micropyle orientation also affects the process of embryo germination, the position of the radicle tip in the fruit. The cause of this or that orientation of the micropyle should in each particular case be ascertained on the basis of biological properties of the plant. Therefore, the type of ovule and its orientation in the ovary are an important indicator of the plant's biological properties and in combination with other features, can shed some light on the possible direction of evolution of a given plant group and the evolutionary level attained (Korchagina, 1969).

The first attempt at classifying angiospermous ovules was made by Mirbel (1829), the founder of the French school of plant anatomists. Studying the ovule development in a number of families, he came to the conclusion that the straight ovule, termed by him orthotropous or atropous, is the original, primitive type of ovule which occurs in Juglandaceae, Polygonaceae and Urticaceae. He considered the inverted ovule, which he termed anatropous, to be the most common type of ovule. In addition to these two types, two more types were described: curved or campylotropous, and double-curved (horseshoe-shaped) or amphitropous. The former is found in Cruciferae and Caryophyllaceae and the latter in Chenopodiaceae. Since then, the main features used in classifying angiospermous ovules have been the shape of ovule and funiculus, and the position of the nucellar and micropylar axes in relation to the placenta and funiculus (Warming, 1878, 1913; Kozo-Polyansky, 1922; Sahni, 1925; Netolitzky, 1926; Golenkin, 1927; Goebel, 1933; Maheshwari, 1950; Eames, 1961). Based on these criteria, five basic (classical) types of ovule have been recognized (Goebel, 1933) (Plate 36).

1. **Orthotropous** or **atropous ovule—straight ovule**. The micropyle lies in the same axis with the hilum and funiculus. Thus the ovule is at a right angle to the placenta, with the chalaza and micropyle lying opposite each other. This type of ovule has also been termed radial or symmetrical. The funiculus is normally short and sometimes absent altogether. Such an ovule occurs in members of dicotyledonous families belonging to orders Centrospermae, Polygonales, Juglandales, Piperales and Enantioblastae. Among the monocotyledons, it is found in some members of Najadaceae, Commelinaceae, Restionaceae, Araceae and Smilacaceae.
2. **Anatropous** or **inverted ovule**—an ovule turns over 180° such that the micropyle, facing the placenta, and the hilum come to lie in parallel, with the chalaza and micropyle lying opposite each other. The anatropous curvature causing the nucellus to turn 180° is due to unequal, intercalary growth of the funiculus which curves beneath the chalaza. The apical portion of the integument forming the micropyle tends to develop on the funiculus side. This is the most common type of ovule among angiosperms, especially in primitive orders and families belonging to subclasses Magnoliidae and Alismatidae.
3. **Hemitropous ovule**. The nucellus with integuments forms an angle of 90° with the placenta and funiculus, the micropyle and chalaza opposite each other. This is an intermediate form between the orthotropous and anatropous condition and is found in members of Primulaceae, Scrophulariaceae and other families.
4. **Campylotropous ovule**—an ovule with one-sided curvature. The nucellus and integuments show unilateral expansion. The nucellus curvature brings the micropyle closer to the placenta, almost level with the chalaza. This type of ovule occurs in some members of the order Caryophyllales and in the Capparaceae family.
5. **Amphitropous ovule**—an ovule with two-sided curvature. The ovule curves in such a way that the longitudinal section of the nucellus is horseshoe-shaped. The micropyle is level with the chalaza. Bending of the nucellus is accompanied by the formation, near its ventral side, of a parenchymatous outgrowth (basal body) which extends into the arch of the nucellus. This ingrowth may develop from nucellar cells, from the inner or outer integument, or from both. The basal body cells accumulate nutritional material. This type of ovule is found in memberes of the superorder Lilianc, in the order Caryophyllales and families Alismataceae, Brassicaceae and Fabaceae.

The last four types of ovule (anatropous, hemitropous, campylotropous and amphitropous) are also termed dorsoventral, or asymmetric, by Goebel (1933).

The above classification, although incorporated in many hand- and textbooks of botany, does not appear to cover adequately the existing diversity of morphogenetic types of ovules and their transitional forms.

A need has arisen for incorporating other characteristics into the existing classification(s) or developing a new one based on entirely different principles. Thus, the **classifications of Takhtajan** (1964) and **Eames** (1961) **also distinguish ovules according to their orientation in the ovary**. More detailed classifications have been and are increasingly being proposed which incorporate additional features, such as vascular supply of the funiculus, ontogenetic aspects of the ovule, etc. Most likely this is primarily due to the increasing number of well-studied plant species among which plants occur with ovules exhibiting more complex shapes and a wider variety of ovule orientations within the ovary, making it impossible to fit them into any of the existing classifications. This creates a pressing need to expand the 'range' of ovule types, to revise and refine existing concepts and definitions, since one and the same concept can often be taken by different authors to mean different things (e.g. 'campylotropous' in the sense of Mirbel, Warming, and Bocquet) and, conversely, different terms can be used to describe the same type of ovule.

Bocquet (1959) made a major contribution to further elaboration of the existing classifications of ovule types. In doing so, he was taking into account not only relative positions of the ovule and funiculus with its vascular supply but, primarily, the features of the funiculus itself as well. Based on these criteria and combining such a feature as a 'straight' or 'curved' funiculus (the 'ortho-' or 'ana-'group respectively) with the conventional characters of the ovule and 'factoring out' as it were the funiculus characteristic, **Bocquet recognized two series of ovule types:** (1) ovules with a straight ('ortho') funiculus (funicular bundle): **orthotropous, ortho-campylotropous, ortho-amphitropous** and (2) ovules with an inverted, curved ('ana') funiculus (funicular bundle): **anatropous, ana-campylotropous, ana-amphitropous**.

In critically reviewing the classification of Bocquet, it should be noted that the underlying principle is indeed a very important one. However, as Bocquet himself admits, this classification is far from comprehensive and not perfectly consistent, with the definitions of concepts provided by Bocquet being rather restrictive at times. For example, he need not have specified, while quite correctly recognizing epitropous and apotropous ovules, that these assume the respective position due to the upward or downward bending of the distal portion of funiculus. Epitropy and apotropy can be due to other factors also, such as bending of the ovule, of the funicular and other regions of the ovule, etc.

A similar approach to classifying ovules by shape of the ovule and the funiculus was adopted by Savchenko and Petrova (1963), Savchenko and Komar (1965) and Savchenko (1973).

A number of authors consider the position of ovule in the ovary, along with the shape of funiculus, to be a reliable criterion for classifying ovules. On the basis of this, they recognize the following types of ovule: erect (located at the base of the ovary, with the micropyle pointing upward), and horizontal or lateral (located on the side walls or on the central placenta). Judging by the definitions in the literature, we are dealing here with ovules borne on straight funiculi. But in what category should ovules with curved funiculi, e.g. anatropous ones, be placed remains unclear.

The drawback to these classifications is that they lump together characteristics of different categories (position within the ovary, type of funiculus, etc.).

In connection with classification of morphological types of ovules, the work of Sterling (1963) is of interest. She distinguished ovules not only on the basis of ovular and funicular shape, but also considered the position and orientation of ovule in the ovary very important features. Her approach to ovule classification is of interest in that she does not lump different properties under a single concept (category) but classifies ovules in several groups according to different features. In order to identify the type of a given ovule, Sterling first determined, separately for each character, the group to which the ovule belongs and then, by combining the names of these several groups, established the type of ovule. Classifying ovules of genus *Prinsepia*, for example, she took into account the positions of obturator, funiculus and ovule, as well as the course of vascular bundles. Based on these criteria, she distinguished, using the recognized terminology of Lindley (1836, 1853), the following types of ovules: erect, pendent and pleurotropous.

Based on the funiculus curvature along the dorsal or ventral side of the carpel (determining the position of ovule within the ovary), **Sterling recognized the following types of ovules: ascendent and descendent.** In the former case, the funiculus ascends along the side of the ovule to the dorsal side of the carpel, while in the latter case it descends along its ventral side. Considering the fact that the ovule can, in this case, show a variety of orientations, Sterling distinguished, depending on the micropyle orientation (upward or downward relative to the upper region of the ovary) respectively epitropous and apotropous ovules, both among erect (erect-epitropic; erect-apotropic) and pendent (pendent-epitropic; pendent-apotropic) ovules, and so on. Thus, Sterling classified ovules on the basis of at least four characters (criteria).

Therefore, the gradation of ovules according to their position and orientation in the ovary proposed by Sterling shows a higher degree of flexibility and covers a wider diversity of forms. Moreover, it takes into account the fact that the position of ovule is affected not only by the unilateral growth of integuments in the chalazal region causing the ovule to turn, but also by the bending (curvature) of the apical portion of the funiculus and, in other cases, of its basal portion as well. Thus, due to a one-sided curvature, the funiculus can become S-shaped.

A similar approach to classifying ovules was suggested by us (Korchagina, 1969). The classification is in the form of a table listing **major ovular features which are of great biological and morphological significance and essential for distinguishing at least groups of types: the orientation of micropyle relative to the base of the stylar canal, position of the ovule in relation to its point of attachment, position of the funiculus with respect to other ovular structures, angle of attachment of the ovule to placenta, degree of curvature of the longitudinal axis of the ovule, and presence or absence of funiculus.** We suggest that the type of ovule should be established by grouping ovules according to each particular feature, with the name of the type being a combination of names of the resultant groups. An approach like this allows, if necessary, additional characteristics to be incorporated into the description of type.

The type of ovule is of phylogenetic importance. Some plant morphologists consider the orthotropous or atropous ovule to be primitive, while others believe the anatropous ovule is the original type. Various modifications of the basic types of ovules were probably arising as a result of variations in ovule structure, number, placentation type, degree of development of the transmitting tissue, etc. All these factors were, to be sure, instrumental in bringing about modifications of the original type of ovule found in prespermaphytes. There is reason to believe that orthotropy is a primitive feature, but it could be a derived character as well. In Betulaceae (genus *Duschekia*), for example, a changeover from the anatropous condition to the orthotropous one is observable which could, as a result of reduction of the placental base, give rise to a basal ovule of the Juglandaceae type (Korchagina, 1965).

After fertilization, the process of embryo and endosperm development begins which is accompanied by ovule structural changes typical of the developing seed. Since the shape of ovule and funiculus undergoes drastic changes during plant ontogeny, it is essential that characterization of ovules be made at embryo sac maturity, since prior to this stage the ovule is still not organized completely, whereas following fertilization it is no longer an ovule, but a young seed.

Structure and Functions of Campylotropous Ovule—Mirbel (1830) was the first to classify the different shapes of the angiospermous ovule in more detail. He recognized orthotropous, campylotropous and anatropous ovules. In most hand- and textbooks of the late nineteenth and early twentieth century Mirbel's three main types were used. They were illustrated by Engler and Prantl (1889) in 'die natürlichen Pflanzenfamilien'. Consequently this figure was incorporated in a number of botanical books, as those of Wettstein (1911) and Coulter and Chamberlain (1903).

Confusion in the classification of ovules started again with Goebel in 1923. Goebel pointed to the differences between the development of ovules in Leguminosae and in Cruciferae or Caryophyllaceae. He introduced the term amphitropy for ovules with embryo sacs, or embryos, curved in their middle to two sides and becoming more or less horseshoe-shaped. The term campylotropy was restricted to ovules with a weaker and one-sided curvature. In the second edition Engler (1926) followed Goebel and extended the original figure accordingly. Notwithstanding the wide distribution and general acceptance of the campylotropous and amphitropous types, practical application of the ovular classification met considerable difficulties.

Bocquet (1959) tried to solve the problems by introducing a more detailed classification. He clearly demonstrated that campylotropy and amphitropy are independent processes in respect to anatropy. Combining the basic series of ortho-, hemi- and anatropy, he distinguished 6 different types of ovules with curved embryo sacs. Furthermore, he characterized the amphitropous ovule by the formation of a mass of cells which extends into the arch of the nucellus. This structure was called the central body by Bocquet.

However, it did not solve the practical problems encountered in classification of ovules with curved embryo sacs. A comparison of the systematic data in some, more recent handbooks of Davis (1966), Corner (1976), and Cronquist (1981) shows still many contradictory records for families having ovules with curved embryo sacs. This holds for Acanthaceae, Cneoraceae, Crossomataceae, Dilleniaceae, Menispermaceae, Oleaceae, Portulacaceae and others. The existing literature and our own observations clearly show that campylotropous ovules are highly diverse in structure. The campylotropous curvature may start at different stages of ovular development. The central body over which the embryo sac becomes curved may be of quite different origin and position. From inside outwards it may be formed by endosperm, the nucellus or chalaza, inner integument, raphe, funiculus or even placenta. The growth processes involved are cell division, cell stretching and resorption of the nucellar tissue by the embryo sac, or a combination of these. Campylotropy is of polyphyletic origin, widely distributed within the angiosperms and found as a rule or as a an exception in more than 65 families belonging to different superorders of both dicotyledons and monocotyledons. It is best understood in the orders Caryophyllales, Capparales and Fabales. In many other cases campylotropy is restricted to the family level, genus or even species level. Several families have campylotropous ovules varying in shape, e.g. Caryophyllaceae, Papillionaceae and Myrtaceae.

Moreover, a fundamental difference exists between the functions of anatropy and orthotropy on the one hand, and campylotropy on the other. The anatropous ovule is widespread among angiosperms. According to Davis (1966), 204 of the 315 families are exclusively anatropous. Most probably anatropy is the original type of ovule in angiosperms. By the anatropous curvature the micropyle is brought near the placenta, thus shortening the pathway for the pollen tube. Orthotropy is found in 20 angiospermous families and is often associated with mono-ovulate pistils with a single, erect ovule, the micropyle directed towards, or even contacting the stylar transmitting tissue (e.g. Piperaceae, Polygonaceae, Myricaceae, Juglandaceae, Urticaceae and Zosteraceae).

The main function of campylotropy lies not in the ovule, but in the seed. Seed ecologists have demonstrated a correlation between life span, habit, habitat and seed size. Seed size involves a compromise with seed number and changes in seed size have repercussions on aspects of seed dispersal, seed germination and seedling establishment. Campylotropy enables seeds to contain embryos up to twice as long as the seed itself. Long embryos have longer roots, hypocotyls and/or cotyledons and are supposed to have better opportunities in seedling establishment.

In conclusion we can state that campylotropy is polyphyletic in origin and diverse in morphology and that there is no sound basis for distinguishing separate categories of campylotropy and amphitropy.

Campylotropy is a character of the seed, although often already expressed at the ovular stage. Functionally, campylotropy should be likened to embryo 'packing', that is, to a means of enlarging the size or surface of the embryo or of its parts (in a manner reminiscent of folded cotyledons), rather than to anatropy or orthotropy.

Ovular Primordium (Latin *primordium* = rudiment, eminence)—an undifferentiated, young ovule (Plate 37). The morphology of ovule primordium, a hemispherical structure, was described by Mirbel (1829).

Based on the histogen theory proposed by Von Hanstein (1868), the ovular primordium histogenesis was followed for the first time in members of the Piperaceae family (Schmitz, 1872). A number of authors (Guttenberg, 1960; Fagerlind, 1961; and others) treated the ovular primordium from the standpoint of the tunica-corpus theory proposed by Schmidt (1924). Based on the pattern of initiation of placenta, two groups of ovular primordia are recognized (Kordyum, 1967, 1968, 1978; see below: Megasporangium). In group B, the ovular primordium is formed predominantly by periclinal divisions of cells of the subdermal placental layer and is composed of a one-layered tunica and a corpus. In group A, periclinal divisions occur in subjacent layers of the placenta and the ovular primordium comprises a two-layered tunica and a corpus. Bouman (1984) suggested that ovular primordia should be divided into three-zonate (group A according to Kordyum) and two-zonate (group B). In more recent works (Venturelli and Bouman, 1986; Grootjen and Bouman, 1988; and others), the terms 'tunica' and 'corpus' are used to describe the ovular primordium.

Early investigators had already noted that the ovular primordium arises by periclinal divisions of cells of the subdermal layers of the placenta (Schmitz, 1872; Strasburger, 1872; Warming, 1874). Warming (1878) recognized three modes of ovular primordium formation: (1) by division of subdermal cells, (2) from cells of the subjacent layer and (3) by simultaneous divisions of cells of the subdermal and third layer. He did not consider these differences in ovular primordium histogenesis of crucial importance since a number of plants exhibit various combinations of these pathways.

Most workers fail to trace the ovular primordium genesis and merely outline the patterns of cell division in different layers of the placenta.

During the formation of ovules lacking the chalazal region of the nucellus (generally, these are typical tenuinucellate ovules), periclinal divisions occur in the third placental layer. It is only in cases like these that it is possible to speak of dermal and subdermal layers being common to the placenta and the ovular primordium. For a great number of taxa of flowering plants, it is periclinal divisions of cells of the subdermal placental layer resulting in the formation of the greater part of the ovule (nucellus, the outer region of the chalaza and funiculus, occasionally the integument) which are probably most important in ovule histogenesis. Cells of the subjacent layers of the placenta divide periclinally to form tissues providing a contact between the ovule and the placenta. Thus, periclinal divisions of cells of the subdermal and subjacent placental layers result in the ovular primordium layers being 'shifted' relative to the initial placental layers. The outer derivatives arising by divisions of subdermal placental cells form an archesporium and initials of the lateral nucellar region and outer chalazal and funicular regions, while the inner ones, along with derivatives of the third and subjacent placental layers, form longitudinal tiers of cells. This results in the formation of ovular primordia composed in most flowering plants, of a dermal, subdermal and one or a few longitudinal layers of cells.

As shown earlier for the shoot apex, the tunica-corpus theory provides some insight into the arrangement and enlargement of cells in the apical meristem proper. The role of meristem in differentiation of tissues and structures becomes clear if one proceeds from the cytohistological zonation concept of shoot apex (Foster, 1938; Popham and Chan, 1952; Vasilevskaya and Kondratyeva-Melvyl, 1958; Gifford and Corson, 1971; Mignotte *et al.*, 1989; and others). According to this concept, the following regions can be distinguished in an organized ovular primordium: **peripheral** (epidermal layer), **apical** (region of differentiating archesporial cells in the subdermal layer), **lateral** (cells of the subdermal layer below the level of the archesporial cells), **basal** (longitudinal tiers of cells below the epidermis), and **transitional** (this is differentiated centrifugally and extends through the basal and lateral regions up to the peripheral one). The apical, lateral and basal regions differ in structure. Normally, at the level of the transitional region integuments are formed concurrent with or shortly after its differentiation; above this level the nucellus is differentiated, and below it the chalaza and funiculus. Each region of the ovular primordium gives rise to a particular ovular region whose degree of development varies among taxa, resulting in a wide diversity of ovules. The peripheral region differentiates into the nucellar, chalazal and funicular epidermis and, often, integuments. The apical region gives rise to a megaspore mother cell, or to a megaspore mother cell and a parietal tissue. The lateral region contributes to the formation of the lateral nucellar region and the outer chalazal and funicular regions, the basal region giving rise to the basal (axial) nucellar region and to procambial cell layers in the chalaza and funiculus. The transitional region forms the ovular hypostase proper.

MEGASPORANGIUM

Nucellus (Latin *nucellus*)—a major ovular structure wherein initiation and development of sporogenous cells and female gametophyte occurs. Synonyms: megasporangium, macrosporangium (Plate 38). Brown (1826b) established the presence in the ovule of a central body, or 'nucleus' (core), and integuments. Mirbel (1829) was the first to apply the term 'nucellus' (French *nucelle*) to the central part of the ovule. Hofmeister (1851) convincingly demonstrated the nucellus of angiospermous ovule to be homologous to the megasporangium of other higher plants. This assumption was supported by some outstanding botanists (Warming, 1878; Schnarf, 1929; Takhtajan, 1964, 1980b; and others). Aspects of nucellus genesis were described for the first time by Warming (1878) in his seminal paper 'De l'ovule', based on results of studies of a great number of plant species.

The nucellus develops from the apical portion of an **ovular primordium**. The initial cell layers of the ovule primordium—tunica (epidermal and subepidermal layers) and corpus—undergo further development to form specific nucellar tissues. The morphological appearance of the nucellus starts with integumentary initiation at its base which is continuous with the chalaza. At the nucellar apex, one or several subepidermal cells organize into an **archesporium**. From this point on the nucellus develops as a non-uniform structure composed of sporogenous cells and somatic tissue viewed as a megasporangium wall. In flowering plants, in contrast to gymnosperms, the sporogenous tissue and megasporangium wall have undergone drastic reduction.

The archesporial cell directly (without division), or following detachment of the **cover (parietal) cell** organizes into a megaspore mother cell, a **megasporocyte**. The latter undergoes **meiosis** to form **megaspores**. These, upon germination, develop into a **female gametophyte** termed the **embryo sac**. The number of megasporocytes within a nucellus may range from just one to a few tens. When the archesporial cell does not cut off a parietal cell, the megasporocyte is situated directly below the epidermis. If the parietal cell is cut off, it can divide (anticlinally or periclinally) to form a sterile **parietal tissue**. The nucellus enlarges in volume most rapidly during megasporogenesis and early stages of megagametogenesis, primarily due to divisions of cells of the subepidermal and subjacent layers. Differences in mitotic activity of initial layers of the nucellus result in considerable variation in size, shape and ratio of its parts in members of various groups of flowering plants. The sterile portion of the nucellus enclosing the megasporocyte may be multi- or monolayered, with an entire gamut of intermediate forms, and its persistence in the ovule may be short or long term.

The nucellus, when present in a mature ovule, varies greatly in shape. In anatropous and orthotropous ovules, it is straight, cylindrical (*Sedum, Houttuynia*), spherical (*Magnolia, Peperomia*), or conical (*Polygonum, Nelumbo*). In campylotropous ovules or derived forms of these, the nucellus may exhibit varying degrees of curvature (*Portulaca, Allium*). The nucellus is often divided into three regions: micropylar (apical) region situated over the megaspore mother cell or embryo sac; lateral (middle) region flanking the megaspore mother cell or embryo sac; and chalazal (basal) region lying below the level of the megaspore mother cell and embryo sac. The nucellus shows varying degrees of development of its individual regions. In some taxa (Casuarinaceae, Haloragaceae), all three nucellar regions are well developed, in others it is one of the regions that is predominantly developed. For instance, a massive micropylar region may occur in combination with a poorly developed or undeveloped chalazal region (Ulmaceae, Polygalaceae) and, conversely, a bulky chalazal region can be found together with a weakly developed micropylar region (Brassicaceae, Crassulaceae, Saururaceae). In a number of taxa (Magnoliaceae, Poaceae, Liliaceae, Nelumbonaceae), a multilayer, massive lateral portion is formed. In members of some families (Euphorbiaceae, Rosaceae, Trapaceae, Cucurbitaceae and others), the apical nucellar region expands to form a beak-like outgrowth projecting into the micropyle. No detailed classification of types of nucellus which would adequately reflect the great diversity of its structure and functions has been developed. Based on the extent of development of nucellus, Van Tieghem (1898) classified ovules under two types: **crassinucellate**, with a massive, multilayered nucellus and **tenuinucellate**, with a thin, one- to two-layered nucellus. These types have become established in the literature even though there is no clear dividing line between them and a variety of intermediate forms exist (Schnarf, 1929; Eames, 1961; Kordyum, 1978; Bouman, 1984). It has also been proposed that nucellus longevity, number of archesporial cells, presence of parietal cells, ability of epidermal cells to divide periclinally, level of integument initiation and a number of other features should also be used as additional classification criteria (Van Tieghem, 1901b; Asplund, 1920; Dahlgren, 1927; Schnarf, 1929; Maheshwari, 1950; Kordyum, 1978; Bouman, 1984). Davis (1966) proposed that the presence or absence of parietal cells be used as a sole criterion in distinguishing crassi- and tenuinucellate ovules. In addition to two basic types, she recognized the 'pseudocrassinucellate' type representing nucelli lacking parietal tissue but with a multilayer epidermis in their apical region. In our opinion, her approach to classifying nucelli is one-sided since it takes into account only the structure of the apical region of the nucellus and is at variance with the recognized basic types of nucellus (in their original sense). The classification of **megasporangia** proposed by Kordyum (1978; see further: Classification of Megasporangium Types), taking into account mainly the contribution of placental layers to the formation of megasporangium and the characteristics of archesporium, is restricted to an early developmental stage (megasporogenesis) and therefore is not sufficiently comprehensive.

For a more adequate description of the nucellus, we suggest that while preserving its classical types, additional properties reflecting its structural and developmental features be taken into consideration.

Type I nucellus. In crassinucellate ovules, the nucellus can be described as a **thick-walled, massive** structure (Latin *nucellus crassus*). It is comprised of an epidermis and derivatives of the subepidermal layers arising by periclinal divisions. This type of nucellus carries one to a few sporogenous cells and persists in a fertilized ovule for some time during embryogenesis. Type II nucellus. Tenuinucellate ovules have a **thin-walled, poorly developed** nucellus (Latin *nucellus tenuis*), represented by one or two cell layers—epidermis and subepidermis—that has lost its ability to divide periclinally. It normally contains a single sporogenous cell. The megaspore mother cell and embryo sac, when poorly developed, are flanked by cells of the subdermal layer from which they have been derived; however, when these undergo intensive growth, they extend beyond the subdermal layer and appear to be enclosed in the epidermis only. The thin-walled nucellus degenerates during megagametogenesis and does not persist in the fertilized ovule.

In addition to typical features, some supplementary characteristics related to various aspects of nucellar development have been proposed. **A. Degree of development of nucellar regions** (micropylar, lateral, and chalazal): well-developed, moderate, poorly developed, or completely reduced. **B. Degree of tissue differentiation**: presence or absence of parietal tissue, multilayer epidermis, lateral parenchyma, hypostase, postament and other tissues. **C. Longevity of nucellus** in relation to megasporogenesis, gametogenesis and embryogenesis. For instance, the nucellus of *Myriophyllum* is massive, with three well-developed regions; a nucellar cap and a parietal tissue differentiate in the micropylar region, and a vascular strand composed of elongate cells in the chalazal region; the nucellus begins to degenerate following the globular proembryo stage. In *Lathraea*, the nucellus is thin-walled, with poorly developed chalazal region; it degenerates at the two- to four-nucleate stage of the embryo sac.

The nucellar structure has been studied with the light microscope in a great number of species and with the electron microscope in few species. Ultrastructural studies have revealed the meristematic nature of cells of the developing nucellus, plasmodesmal connections between adjacent cells, as well as developmental patterns of nucellus. A massive nucellus normally exhibits histological differentiation: an **epidermis** and a parenchyma can be distinguished in it. In most angiosperms, the epidermis is one-layered, since its cells undergo only anticlinal divisions. In some species, they can divide periclinally to form a multilayered nucellar cap (Polygalaceae, Haloragaceae, Asparagaceae and others). The epidermal surface is covered with a cuticle. In the course of development, the nucellus can become differentiated into tissues varying in cell wall thickness, organelle composition and pattern of cytoplasm vacuolation (Jensen, 1965a; De Boer-de Jeu, 1978; Noher de Halac, 1980; Plisko, 1980, 1982; Wilms, 1980; Folsom and Cass, 1988). However, highly variable composition of tissues and lack of unified nomenclature for these make comparative analysis difficult.

In the micropylar region of nucellus, a **parietal tissue**, a derivative of the primary archesporium, organizes. This varies in number of layers (from a single to 20 and more), becoming heteromorphic in some species (*Alcea rosea*—Plisko, 1980). Members of many families (Hamamelidaceae, Malvaceae, Chenopodiaceae, Sonneratiaceae) develop a 'strand' composed of elongated cells extending from the micropyle to the egg apparatus which functions as a pollen-guiding tissue ensuring pollen tube transfer.

In the well-developed chalazal region of the nucellus of some taxa (Saururaceae, Droseraceae), a central strand differentiates from elongated cells or, in some instances, from tracheid-like cells (Casuarinaceae, Euphorbiaceae). In *Capsella* (Schulz and Jensen, 1971) and *Spinacia* (Wilms, 1980), two types of tissue have been distinguished in the chalazal region of the nucellus: the tissue next to the antipodal cells is composed of thick-walled cells in the former and thin-walled cells in the latter, both types containing in their cytoplasm well-developed ER, abundant mitochondria and dictyosomes, and a few plastids with starch grains and also lipid droplets. Cells of this nucellar region are progressively compressed and destroyed by the growing embryo sac, while cells of the nucellar base undergo intensive divisions to form a 'chalazal proliferating tissue' during early embryogenesis. In the chalazal region of the nucellus of many species, a morphologically well-defined tissue termed **hypostase** is organized. Some taxa may during embryo sac development form a **postament**, as an axial part of the nucellar base, and a **podium**, a cup-like remnant of the nucellar base.

In the multilayered lateral region of the nucellus of some plants (*Oenothera, Spinacia, Alcea*), the cell layers become differentiated into inner and peripheral layers (Noher de Halac, 1980; Plisko, 1980, 1982; Wilms, 1980). It is believed that transformations occurring in the inner layers are due to their earlier degeneration. However, in *Glycine max* (Folsom and Cass, 1988) the inner nucellar tissue is differentiated, starting at the megasporocyte stage, due to wall thickening, abundant ribosomes in the cytoplasm, numerous plasmodesmal connections of cells with one another and with the coenocytic gametophyte. This tissue bears no resemblance to degenerating inner nucellar tissues of other species.

Nucelli of different types vary considerably in their degeneration pattern and timing. The multilayered nucellus usually degenerates after fertilization, at various stages of embryo development (Papaveraceae, Polygalaceae, Malvaceae, Linaceae). Normally, degeneration starts at the inner cell layers of the lateral region adjoining the embryo sac (*Alcea*). Cells of the micropylar and chalazal regions of the nucellus can persist much longer than those of the lateral region (*Sedum, Colchicum*). It has been demonstrated that degeneration of the nucellar cells adjoining the megasporocyte and the young embryo sac may commence concurrent with activation of cell divisions in peripheral nucellar layers (Norstog, 1974; Tilton and Lersten, 1981a, b; Willemse and De Boer-De Jeu, 1981). Formation of a lytic zone encompassing several nucellar layers around the periphery of the developing embryo sac has been observed (Norstog, 1974; Plisko, 1982; and others). Degeneration of nucellar cells in *Oenothera* (Noher de Halac, 1980) and *Alcea* (Plisko, 1982) precedes accumulation in them of storage material (lipids, polysaccharides, proteins). Degeneration of the nucellus presumably occurs in two ways (Plisko, 1982): (1) by cytoplasm autolysis and aggregation of cell walls near the embryo sac wall (*Aquilegia, Alcea*); (2) without cytoplasm autolysis but with disappearance of the typical cell wall (*Helianthus, Calendula, Hordeum, Oenothera*). It has been suggested that autophagic vacuoles, lysosomes, multivesicular bodies and ER are involved in cytoplasm autolysis.

A thin-walled nucellus generally has a short development cycle. Its cells show few organelles and lack reserve material; they usually degenerate in an unfertilized ovule during development of the embryo sac (most members of Sympetalae), although there are some exceptions. For example, a one-layered nucellus persists in mature ovules of *Sambucus* and *Gymnadenia*. Degeneration of the nucellus most commonly starts in the micropylar region, then extends to and includes the middle region and, subsequently, the chalazal region whose remnants may persist around the antipodal cells for some time after fertilization. In some taxa (*Impatiens, Reseda*), degenerating epidermal cells in the micropylar region may become thick-walled and serve as a kind of storage tissue.

The nucellus of an unfertilized ovule may accumulate storage material in the form of starch (Caryophyllaceae, Eupomatiaceae), as well as a few lipid droplets and protein bodies. In some taxa (Piperaceae, Caryophyllaceae), the nucellus functions as a storage tissue for the developing embryo and persists in mature seeds as a perisperm. As an intensively growing organ, the nucellus contains physiologically active substances (ascorbic acid, SH groups, heteroauxin). The composition of the physiologically active substances and their distribution in various nucellar tissues have been found to vary among species. Histochemical studies of the crassinucellate ovule of *Nelumbo nucifera* have shown the presence in the nucellus of specialized tissues varying in cell structure and nucleic acids content (Vasilyeva *et al.*, 1988). The nucleic acids content has proven to be highest in the chalazal region of the nucellus, lower in the peripheral layers of the micropylar region of the nucellus, and the lowest in the cell layers adjoining the embryo sac. Evidence on the nucellar cell structure in various plant species has been summarized by D'Amato (1984): in some species, the nucellar cells are diploid (*Campanula*); in most species they undergo chromosome endoreduplication, resulting in varying levels of ploidy (8 *n* in *Pedicularis*, 16 *n* in *Cucumis*, 32 *n* in *Daphne*, and 256 *n* in *Trapa*).

In some taxa, nucellar cells appear to be capable of being organized into **nucellar embryos** which may arise in any region of the nucellus (Euphorbiaceae, Poaceae) or, more frequently, in the micropylar region (Buxaceae, Cactaceae, Anacardiaceae), and then grow into the embryo sac.

An unusual behaviour of the chalazal region of nucellus has been observed in members of Podostemaceae in which local degeneration of nucellar cells below the developing embryo sac results in formation of a coenocytic nucellar plasmodium which functions as an endosperm (Mukkada, 1962a, b).

The primary function of nucellus is that of producing megaspores and supplying nutrients to the female gametophyte. Nutritive and transport functions of the nucellus are evidenced by the presence of plasmodesmata connecting the nucellar cells (which together comprise the symplast) and of special structures between the chalazal vascular supply and the embryo sac (axial rows of cells, podium, postament, hypostase, procambial cell layers). The dynamics of sugar content support the transport function of the nucellus (Willemse and Franssen-Verheijen, 1983). It has been experimentally substantiated that nutrients from the vascular supply are transported to the nucellus: in *Zephyrantes*, the ^{14}C isotope was observed to accumulate rapidly in the nucellar layers surrounding the embryo sac (Coe, 1954). It has been demonstrated experimentally that the developing embryo sac obtains its nutrition from both the intact and the degenerating nucellus.

In addition, the massive micropylar region of the nucellus has a function in the guidance and presumably growth of the pollen tube (Chaban and Yakovlev, 1974).

Various taxa exhibit different patterns of nucellar development due to close interactions between the nucellus and other ovular structures, such as embryo sac, integument, chalaza, etc. Nucellar behaviour is probably determined by developmental and nutritional patterns of the embryo sac. For example, degeneration of the nucellus during early developmental stages correlates with formation in the integument of an endothelial layer which comes into contact with the embryo sac and endosperm. The nucellus may take over the functions of the endosperm (Podostemaceae) or act as a food-storage tissue in mature seeds in which the endosperm is poorly developed (Piperaceae).

It is generally recognized (see Schnarf, 1929) that evolution of the nucellus has been towards its reduction from a massive central body, within which multiple sporogenous cells develop, to a small eminence when a single megaspore mother cell develops under cover of one epidermal layer of the nucellar wall. Reduction of the sporogenous and somatic tissues may not have proceeded concurrently, with the processes resulting in sterility of sporangia probably having played a significant part in the evolution of the megasporangium (Kordyum, 1978). Instances of progressive reduction of the nucellus are found not only among members of related families, but even within the same family (Ranunculaceae, Caryophyllaceae, Liliaceae, Asparagaceae), in which both crassinucellate and tenuinucellate ovules have been reported. The best-known example of progressive reduction of the sterile nucellar tissue to a few epidermal cells is found in members of Apocynaceae (Guignard, 1917) and Rubiaceae (Fagerlind, 1937). These have undergone a series of evolutionary transformations from a morphologically well-defined nucellus in the form of the central part of the ovule to a few epidermal cells over the megasporocyte which are embedded in the chalazal tissue of the ovule. An even greater reduction of the nucellus, associated with general reduction of the ovule, is observed in members of highly specialized families—Loranthaceae, Balanophoraceae, Viscaceae (Eames, 1961; Teryokhin, 1987a, c). In these, formation of megaspores and female gametophyte occurs within the undifferentiated placental tissue or in the ovarian base tissues.

The nucellar structure is a stable diagnostic feature of the species and often of higher taxonomic categories and is viewed as an indicator of the evolutional level of the ovule morphogenetic organization. Distribution of nucellar types within the system of flowering plants is reviewed in a number of publications (Dahlgren, 1975, 1980; Kamelina, 1991). A well-developed thick-walled nucellus is typical of most monocotyledons and of some dicotyledons occurring at the base of the system of flowering plants, whereas a thin-walled and early degenerating nucellus is characteristic of more advanced taxa (Schnarf, 1929; Takhtajan, 1964).

Nucellar Epidermis (Greek *epi* = on, and *derma* = skin)—a surface layer of nucellar cells. Sometimes this is the only megasporangium wall layer surrounding the megaspore mother cell (Orobanchaceae, Orchidaceae, Acanthaceae). The outside of the nucellar epidermis is covered with a layer of cuticle. The epidermal cells are connected with one another and with the cells of inner layers by plasmodesmata. Where the nucellar wall is one-layered, plasmodesmata are present only on the radial cell walls, as in *Calendula* (Plisko, 1982). Generally, cells of the nucellar epidermis undergo only anticlinal divisions. In the micropylar region of some taxa they can divide periclinally. Periclinal divisions at early stages of ovular development were observed by Warming (1878) in *Aristolochia* and *Ficus*, and by S.G. Nawaschin (1894) in *Betula*. More recently, periclinal divisions of epidermal cells have been reported in members of many families (Aceraceae, Lauraceae, Rosaceae and others). In the lateral region of the nucellus, periclinal divisions of epidermal cells are less common (Portulacaceae—Batygina, 1983b; Maranthaceae—Kamelina, 1990a, b; Haemodoraceae—Alimova, 1990).

The multilayered cap-like structure in the nucellar apical region derived by periclinal divisions of the epidermal cells and overhanging the parietal tissue or the micropylar end of the embryo sac, has been termed the **nucellar cap**. Understood in this manner, the term has been recognized by the majority of modern workers (see Davis, 1966; Bouman, 1984; *Comparative Embryology of Flowering Plants*, 1981, 1983, 1985, 1987, 1990). However, the term 'nucellar cap' is also used to refer to the entire apical part of the nucellus situated above the embryo sac (Eames, 1961; Savchenko, 1973), especially where the genesis of the epidermal and parietal tissues is difficult to follow. The nucellar cap may comprise two or more layers. The degree of its development is a species-specific feature. The multilayer nucellar cap is found in members of Caryophyllaceae, Calycanthaceae, Phytolaccaceae, Rosaceae and other families. A nucellar cap can give rise to an **epistase** (Van Tieghem, 1901), a tissue with thickened cell walls which persists in mature seeds (Potamogetonaceae, Ranunculaceae).

At the beginning of nucellar development, the epidermis is meristematic in nature and is not dissimilar from the inner layer. During pre- and post-fertilization stages the cells of the nucellar epidermis can undergo dramatic changes in shape, size, cytoplasm density, cell wall structure, etc.

The most common changes in epidermal cells are their elongation parallel to the nucellar surface. For example, in *Oryza sativa* the epidermal cells are hexahedral. They stretch out considerably along the periphery of the developing kernel, their anticlinal walls showing ribbed thickenings (Ellis and Chaffey, 1987). Less commonly, epidermal cells are stretched out radially, perpendicular to the nucellar surface, as in *Begonia* (Anisimova, 1983b) and *Drosera* (Teryokhin, 1985). The epidermis may assume the form of a secretory layer composed of densely cytoplasmic cells and simulating the integumentary tapetum (*Papaver* and *Roemeria,* Papaveraceae—Ilyina, 1981; *Scilla,* Liliaceae—Savchenko and Komar, 1965; *Melo* and *Cucumis,* Cucurbitaceae—Dzevaltovskyi, 1983). The nucellar epidermal cells that undergo the most drastic changes are those in the micropylar region of the nucellus. In some plants, a few cells on the nucellar apex stretch out considerably towards the micropyle and become more densely cytoplasmic, as reported in members of the Agavaceae (Tilton and Mogensen, 1979), Portulacaceae (Batygina, 1983b), Cactaceae (Naumova, 1983a), Haloragaceae (Nikiticheva, 1985b), Marantaceae (Kamelina, 1990a) and other families, and is probably due to the pollen tube transmission.

Specific changes in epidermal cells of the micropylar region of nucellus occur during fertilization in members of the Piperaceae family: the outer cell wall acquires a tortuous outline, thus increasing the surface of contact of the nucellar epidermal cells with the adjoining integumental cells (Nikiticheva *et al.*, 1981). In the micropylar portion of the *Reseda* nucellar epidermis (Chaban and Yakovlev, 1974), cell walls become thickened owing to deposition of mucilage which begins at the stage of megasporogenesis. The cell cytoplasm shows abundant dictyosomes, extensive rough ER and numerous vacuoles. By the time of fertilization, the epidermal cell cytoplasm will have been completely resorbed and replaced by mucilaginous substances. In the epidermal (and the only) layer of the *Impatiens* nucellus, dramatic cell wall thickening occurs during embryo sac development accompanied by cytoplasm compression (De Boer-de Jeu, 1978). As suggested by the aforesaid authors, the thick-walled cells of the nucellar epidermis are presumably involved in the protection of the embryo sac against the pressure exerted by the pollen tube. They can also serve as an additional food-storing reservoir.

The longevity of the nucellar epidermis varies in taxa with different types of ovule. In tenuinucellate ovules, the nucellar epidermis as the only layer degenerates even before fertilization (Orobanchaceae, Scrophulariaceae and others). There are exceptions, however. Thus in *Gymnadenia* (Shamrov and Nikiticheva, 1992) and *Sambucus* (Kamelina *et al.*, 1987), the epidermal layer persists in the mature ovule whereas it normally degenerates in early stages in most species of the Orchidaceae and Caprifoliaceae families. In crassinucellate ovules, the epidermal layer normally disintegrates following degeneration of inner tissues at various stages of seed development. Cases are known in which upon disintegration of most of the nucellus, the micropylar portion of the nucellar epidermis above the developing embryo persists for a long time: this has been reported in some members of the Crassulaceae (Nikiticheva, 1985a), Resedaceae (Chaban, 1983), Berberidaceae (Kapinos and Kagramanova, 1966) and other families. In *Ceratophyllum*, the nucellar epidermis persisting in the seed forms a dense layer surrounding the embryo and possibly serving as a seed coat (Shamrov and Batygina, 1988).

Parietal Tissue of Nucellus (Latin *parietalis* = parietal)—a tissue in the micropylar region of the nucellus which lies under the epidermis and originates from the parietal (cover) cell. The parietal cell is cut off upon periclinal division of the archesporial cell and can, in turn, undergo periclinal and anticlinal divisions to form a more or less massive tissue. The presence of parietal tissue is taken by most workers to be the diagnostic feature of the crassinucellate ovule (Asplund, 1920; Davis, 1966). However, this view is open to discussion.

The peak of parietal tissue formation tends to coincide with megasporogenesis and embryo sac development (Juncaginaceae, Polygalaceae, Smilacaceae). The number of parietal tissue layers formed varies widely. The presence and degree of development of the parietal tissue is a fairly consistent taxonomic feature.

A multilayered parietal tissue (up to 10 and more layers) is typical of many families belonging to the orders Magnoliales, Laurales, Casuarinales, Caryophyllales, Paeoniales, Violales, Fagales, etc. A massive parietal tissue may be differentiated into a number of structurally distinct cell layers. Thus between the micropyle and the egg apparatus a specialized 'transmitting tissue' may be formed which is composed of elongated cells and serves as a pollen-guiding tissue. It adjoins the parenchyma-type cells, as demonstrated in *Alcea rosea* (Plisko, 1980), *Spinacia oleracea* (Wilms, 1980) and other plant species.

In some families (Liliaceae, Alliaceae, Phormiaceae, Asparagaceae), the presence of parietal cells in the nucellus is an inconsistent feature: they are formed in some species but not in others. Here the nucellus appears to be undergoing a progressive evolutional reduction. In a number of species a parietal cell is cut off but degenerates soon afterwards (*Houttuynia cordata*—Nikiticheva, 1981b; *Calathea louisae*—Kamelina, 1990b). In taxa with tenuinucellate ovules, the archesporium does not cut off a parietal cell (e.g. families belonging to Sympetalae).

A popular view holds that evolution of the nucellus, including its parietal tissue, has been toward progressive reduction. A nucellus with a multilayered parietal tissue is regarded as a more primitive condition compared with a nucellus whose parietal tissue is reduced to a single cell or lacking (Netolitzky, 1926; Schnarf, 1929; K. Dahlgren, 1928; Kordyum, 1978). Thus in evolutionally advanced families, the parietal tissue is normally absent in the nucellus (R. Dahlgren, 1980).

Megasporogenesis—a process in which megasporocytes undergo meiosis giving rise to megaspores (Plate 39). There are three types of megasporogenesis: monosporic, bisporic and tetrasporic.

Monosporic development occurs in the great majority of angiosperms. A megasporocyte after meiosis I is divided into two cells called a dyad; following meiosis II four megaspores are formed. They are usually arranged in a row, a linear tetrad. One of these megaspores, called functional, is capable of developing into an embryo sac. The three remaining, degenerating megaspores undergo hydrolysis and are presumably utilized by the developing functional megaspore. The functional megaspore is either the chalazal or the micropylar one in a tetrad. The chalazal position is characteristic for the Polygonum-type of embryo sac development, the micropylar for the Oenothera-type.

Bisporic development. Meiosis I leads to formation of a dyad, but after meiosis II the mechanism of cytokinesis is blocked and dyad cells remain binucleate. The chalazal binucleate megaspore is functional in the Allium-type of embryo sac development, the micropylar one in the Endymion-type.

Tetrasporic development. The mechanism of cytokinesis in the megasporocyte is not active and meiosis results in the formation of one tetranucleate megaspore (coenomegaspore) as in *Lilium, Fritillaria, Plumbago,* etc. During meiotic prophase I, the megasporocyte cytoplasm and cell wall undergo structural and chemical changes. Mitochondria and plastids dedifferentiate becoming almost indistinguishable, except in species in which megasporocyte plastids contain starch grains (*Oenothera, Epipactis, Iris*). Some processes in a prophase I megasporocyte are similar to those described in microsporogenesis: the nuclear envelope extends and produces vesicles and the nucleolus gives rise to nucleoloids which after metaphase I appear in the cytoplasm.

Cytoplasmic changes have mostly been studied in *Lilium* megasporocytes. A dramatic increase in endoplasmic reticulum takes place, coinciding with growth of the megasporocyte which attains about 200 µm in length and 50 µm in width. These megasporocytes are considered to be among the largest in angiosperms. ER cisternae lie in stacks at the nucleus, but later in prophase I become arranged into rings isolating portions of cytoplasm. Presumably the isolated cytoplasm compartments are protected from extensive hydrolytic destruction occurring outside the ring. However, hydrolytic enzymes are also present in the ringed cisternae (Rodkiewicz and Mikulska, 1965a, b; Dickinson and Heslop-Harrison, 1977; Fougère-Rifot, 1987b; Kapil and Bhatnagar, 1981).

Distribution of organelles. Plastids and mitochondria of the early prophase I megasporocyte in an orchid are aggregated in the micropylar region, in the later prophase I they are evenly distributed or, in some plants, are concentrated in the chalazal region. This results in an unequal megasporocyte division and promotion of the chalazal, functional megaspore.

Plastids in megasporocytes of Onagraceae prior to meiosis aggregate in micropylar and chalazal regions, while other organelles are present throughout the cytoplasm. During two divisions of meiosis plastids remain in the same regions and are eventually enclosed in the micropylar and chalazal megaspores of a linear tetrad. These two megaspores are capable in some Onagraceae of developing into embryo sacs (Rodkiewicz and Bednara, 1986).

Megasporocyte wall. From early prophase I a new component—polysaccharide callose—is laid down in the entire wall of a megasporocyte that follows monosporic or bisporic type of development. By middle prophase I callose has disappeared from the chalazal region of Oenothera megasporocytes. A callose layer in the chalazal wall of an orchid megasporocyte (Polygonum-type embryo sac development) becomes perforated. After meiosis I and II, megasporocytes are divided by callose walls which isolate degenerating megaspores (Rodkiewicz and Bednara, 1986).

It has been reported that callose fluorescence is not present in semithin preparations of megasporocytes of some species. However, weak fluorescence becomes visible when the squash preparations of these megasporocytes are treated with aniline blue, a reagent for callose.

According to some experiments, callose walls are not permeable for medium size and large molecules; therefore, megasporocytes for a part of prophase I, and degenerating megaspores are isolated from some factors acting in the nucellar tissue.

This impression of isolation increases when we consider the lack, or very small number of plasmodesmata in side walls of a megasporocyte. Plasmodesmata are quite numerous only in the apical region where a functional megaspore is to be developed (Bell, 1989).

There is an obvious correspondence between the site of callose disappearance of the megasporocyte in prophase I and the position of the functional megaspore in a tetrad. In addition, this region of the wall is perforated by plasmodesmata and in some plants the cell wall labyrinth is formed. Megasporocytes and megaspore tetrads are strongly polarized systems in which the position of functional megaspores is determined early in development (Willemse and Bednara, 1979; Rodkiewicz and Bednara, 1986).

Archesporium (Greek *arche* = beginning, and *sporá*)—a cell or a few cells of subdermal layer in the apical region of the nucellus primordium producing a sporogenous and a parietal (cover) cell or differentiating directly into a sporogenous cell (megaspore mother cell) giving rise to megaspores.

The archesporium is formed in both the ovule (female archesporium) and the anther (male archesporium), the two types having many features in common in terms of development. It was first described by Warming as 'initial (mother) pollen cells' (Warming, 1873) and as 'initial (mother) embryo sac cells' (Warming, 1878). Either cells of the subdermal layer or its inner derivatives resulting from periclinal divisions turn into 'initial cells'.

Goebel (1880) proposed a more concise term—'archesporium'—for Warming's 'initial cells of sporogenous tissue' of higher plant sporangia. Initially he viewed it as a 'hypodermal cell cutting off a tapetal cell', as is evident from his description of sporangium development, say, in leptosporangiate ferns. However, he failed to provide a precise definition of the term, resulting in a rather loose interpretation of the concept of archesporium by the author himself later (Goebel, 1898-1901) and by other workers.

Schnarf (1929) believed that subdermal cells can be viewed as archesporial when they differentiate among the surrounding cells and become the initials giving rise to an embryo sac. What is probably meant here by archesporium is the cells of the subdermal layer which develop directly into megaspore mother cells. At the same time, his archesporium typification treats the archesporium as subdermal initial cells giving rise to a sporogenous complex and a sterile megasporangium wall tissue (cover cells) (archesporium types I, II, III and IV).

Dahlgren (1928) distinguished primary archesporial cells located within the subdermal layer of a young ovule and developing directly into embryo sac mother cells and secondary archesporial cells which are not formed until the parietal cells have been cut off by subdermal cells. Some workers (Sladkov and Grevtsova, 1988, 1989) think that subdermal cells in the nucellar apex should be referred to as 'initial cells of the sporangium wall and archesporium' and the term archesporium reserved for cells giving rise to spores alone.

Ambiguity in interpreting the term archesporium could have stemmed from lack of conclusive evidence on genesis of the archersporium and from comparing different stages of its development. However, an archesporium (uni- or pluricellular) can only be identified at early stages of its development since sporogenous cells in some taxa (Causarinaceae, Ceratophyllaceae, Rosaceae) may later undergo mitotic divisions.

Archesporial cells can be differentiated from surrounding nucellar cells by larger size and more prominent nuclei, denser cytoplasm, and histochemical and ultrastructural characteristics. The archesporial cell cytoplasm of many plants has been found to contain high levels of RNA and proteins (Petrovskaya, 1955; Vlasova and Serebryakova, 1963; Panchaksharappa and Syamasundar, 1975; Bhandari *et al.*, 1980; Shamrov, 1991; and others). In some plants (*Cypripedium insigne*—Poddubnaya-Arnoldi, 1959; *Gentiana cruciata*—Shamrov, 1990d) it shows low RNA and protein levels but high starch content. Ultrastructural studies (various *Dendrobium* species—Israel and Sagawa, 1964) have revealed that compared with the surrounding cells, the archesporial cell cytoplasm exhibits more abundant mitochondria, plastids and ER, numerous lipid droplets and gradual disappearance of plasmodesmata. In any case, archesporial cells begin to elongate either during their direct development into sporogenous cells or when proceeding to differential periclinal division to produce a parietal and a sporogenous cell. The latter pathway is characteristic of most primitive families with a massive nucellus in which sporogenous cells are found to lie beneath the parietal cells, embedded in the nucellus. The former pathway is associated with the absence of periclinal divisions and direct development of the subdermal archesporial cell into a sporogenous cell and subsequently into a megaspore mother cell. This is a shorter pathway of megaspore mother cell development and is typical of highly specialized families with tenuinucellate ovules. Where several archesporial cells are differentiated, the resultant sporogenous cells may undergo further divisions which are responsible for multiplication of cells of a single—sporogenous—type of tissue whose cells develop into megaspore mother cells.

The first classification of archesporia to be proposed was that by Schnarf (1929); the criteria underlying this classification are the number of archesporial cells and their capacity to divide periclinally to form a cover and a sporogenous cell. He recognized 6 types of archesporium. In types I, II and VI, a large number of archesporial cells differentiate: the types differ in presence (I, II) or absence (VI) of differentiation divisions producing sporogenous and parietal cells. A single archesporial cell (types III, IV and V) can cut off a parietal cell (III, IV) or directly develop into a megaspore mother cell (V). Intermediate conditions have been reported which are characterized by initiation of several archesporial cells, with only the central one undergoing further development.

All later classifications of archesporia (Poddubnaya-Arnoldi, 1964, 1976; Kordyum, 1978; Kamelina, 1991) use the principles proposed by Schnarf and offer more numerous examples. The Sladkov and Grevtsova (1989) classification is based on the stage of sporogenous cells prior to their transformation into megaspore mother cells rather than on the stage of archesporium development (sensu Schnarf).

Thus analysis of the literature available enables two basic types of female archesporium to be distinguished: pluricellular (Casuarinales, Myricales Paeoniales and others) and unicellular (Ericales, Orchidales, Scrophulariales and others). The main characteristic feature of both the pluri- and unicellular archesporium is their capacity to undergo a differentiation division to produce a parietal and a sporogenous cell. Each type is subdivided into two subtypes: subtype A—the archesporial cell divides to form a parietal and a sporogenous cell; subtype B—the archesporial cell develops into a megaspore mother cell without undergoing a differentiation division. We see no need for including further divisions of these cells resulting in their multiplication in the classification of archesporia.

Major assumptions as to the evolution of the archesporium were made by Schnarf (1929, 1931) and corroborated by subsequent studies (Maheshwari, 1950; Poddubnaya-Arnoldi, 1964, 1976; Kordyum, 1968, 1978; and others). Evolution of the archesporium was towards its reduction and proceeded along two different pathways, and by a combination of these: by reduction in number of archesporial cells and by reduction in number of divisions in derivatives of the archesporium (parietal and sporogenous cells) occurring during the developmental period that elapses between an archesporial cell and a megaspore mother cell.

Sporogenous Cell—an initial cell producing spores. It is formed early in ovular development from the subdermal archesporial cell in one of two ways: either as a result of its transformation without division, or following a differentiation division producing a sterile parietal cell and a sporogenous cell.

If an archesporial cell is defined as a meristematic cell and megasporocyte (megaspore mother cell) as a highly specialized cell with a particular state of its wall and cytoplasm and its nucleus entering meiosis, then in our opinion at any stage of transformation of an archesporial cell into a megasporocyte a cell can be termed a sporogenous one. It is probably for this reason that well-defined morphological criteria for identifying a sporogenous cell are lacking in the literature and its structure is poorly understood. Sporogenous cells can undergo divisions to form a multicellular sporogenous tissue; however, this process is difficult to distinguish from that of pluricellular archesporium initiation. Where several archesporial cells are laid down, each or only a proportion of these can embark upon the developmental pathway culminating in formation of a sporogenous cell. A multicellular sporogenous tissue may consist of a single (Papaveraceae Rubiaceae and others) or a few (Casuarinaceae, Loranthaceae, Paeoniaceae and others) layers of cells. Conclusive evidence on the origin of these layers is lacking. Development of a sporogenous cell into a megaspore mother cell is accompanied by certain morphological changes. It becomes conspicuous among meristematic cells of the nucellus by its larger size, dense cytoplasm, prominent nucleus, thicker cell wall, loss of plasmodesmata and preparation of the nucleus for meiosis. The structure of a megasporocyte—the final stage of sporogenous cell development—has been extensively studied using various investigative techniques. Some authors make no terminological distinction between the archesporial, sporogenous and megaspore mother cell (megasporocyte). This does not seem to be at variance with the facts where the archesporium develops into a sporogenous cell, and the latter into a megasporocyte, without division.

A single sporogenous cell normally develops in a one-layered nucellus (the Sympetalae families) but not infrequently in multilayered ones as well (Cynomoriaceae, Haloragaceae, Scheuchzeriaceae). Development of two or more sporogenous cells is characteristic of ovules with a massive, multilayered nucellus (Caryophyllaceae, Paeoniaceae and others).

Megasporocyte (megaspore mother cell)—a sporophytic cell which, following meiosis, is transformed into a megaspore giving rise to a female gametophyte (embryo sac) (Plate 39). In diplosporous apomicts, however, an embryo sac develops directly from the megasporocyte in which the mechanism of meiosis is suppressed.

A megasporocyte is formed from an archesporial cell differentiated in the subepidermal layer of nucellus, within the presumptive micropylar region of the ovule. Differentiating archespores increase in size, growing along the micropylar-chalazal axis. The somewhat elongated archespore of some species enters meiosis and becomes a megasporocyte. In other species, the archespore at first undergoes asymmetrical mitosis, dividing into a small subepidermal cell (a cover cell) and a larger cell constituting a megasporocyte.

The megasporocyte develops in a polarized environment of the ovule and itself displays certain features of polarization: elongated shape, the nucleus usually shifts towards the micropylar apex and an uneven distribution of organelles in the cytoplasm and callose in the megasporocyte wall often accurs.

Appropriate conditions for inducing and carrying out meiosis occur only in the micropylar area of the nucellus. The nature of meiotic inducers is still unknown; the mechanism of induction must be stable and precisely balanced since in many species only one nucellar cell acquires competence to respond to meiotic induction. In some species, a multicellular archesporial tissue differentiates constantly or occasionally. Therefore, a certain number of megasporocytes enter meiosis, resulting in the formation of several megaspore tetrads (Bouman, 1984; Fougère-Rifot, 1987b).

Megaspore Tetrad (Greek *tetra* = four, *megas* = large, and *sporá*)—structure composed of four haploid cells resulting from two meiotic divisions of a megasporocyte.

Archesporial cells give rise to megaspore mother cells. The megaspore mother cell of the angiosperms, like that of heterosporous Lycopodiaceae, pteridophytes and gymnosperms, undergoes two further divisions to produce, similar to the formation of a microspore tetrad, a tetrad of megaspores with haploid chromosome numbers.

During megaspore formation meiosis proceeds in much the same way as during formation of microspores but is somewhat delayed.

The transverse walls between the megaspores are laid down sequentially (in a successive manner). The types of tetrads formed are: (1) linear, (2) T-shaped, (3) ⊥-shaped, (4) tetragonal, (5) tetrahedral. During megasporogenesis, as with microsporogenesis, callosic walls are formed whose functions are similar to those of microspores. The distinguishing feature of the megaspore (like any other cell, including the archesporial) is its polarity resulting from cytoplasmic and nuclear dissimilarity between the upper and lower portion of the cell. Embryo sac polarity is a consequence of the polarity of archesporial cells and megaspores.

Most flowering plants (if embryo sac development is of the Polygonum-type) produce four megaspores, of which the lower one enlarges in size and develops into an embryo sac, the upper three megaspores degenerating.

In addition to this type of megaspore formation, there are others: formation of three, two or a single megaspore. In the latter case, the megaspore mother cell develops directly into the embryo sac mother cell. Although the embryo sac is normally formed from the lower megaspore, occasionally it develops from the upper megaspore or from the middle ones. In all members of the Onagraceae family and in some Asteraceae, the embryo sac develops from the upper megaspore. There are cases, albeit very rare, wherein the embryo sac is formed from the middle megaspores, as in *Limnocharis emarginata, Aralia racemosa* and *Colchicum autumnale*. The functional megaspore is distinguished from the very beginning of its initiation from the degenerating megaspores by larger size. Subsequently, the difference becomes even more pronounced due to rapid growth of the former at the expense of the latter. Occasionally there is more than one functional megaspore. In *Phellodendron amurense*, all the megaspores within the tetrad were observed to germinate. Even so, typically as few as one megaspore becomes fully developed; one of the developing embryo sacs pushes out the others which degenerate at various, primarily early stages of development (Starshova and Solntseva, 1973).

Classification of Megasporangium Types (Plate 40)—Studies of ovule differentiation in members of various families of angiosperms have shown the megasporangium structural pattern to vary according to number of central cells of the subepidermis (second layer of the tunica), number of differentiating archesporial cells, rate of division of vegetative and, in some cases, sporogenous cells of the subepidermis, timing and site of integument initiation, as well as degree of curvature of the ovule as a whole. Based on these features, a classification of megasporangium types was proposed (Kordyum, 1968, 1978). **Group A**: An ovule primordium originates as a bulge arising from the placenta by (predominantly) anticlinal divisions of cells of the two peripheral layers of the placenta—epidermal and subepidermal—and by periclinal and anticlinal divisions of cells of the subjacent layers. Thus, the undifferentiated ovule is composed of cells of two layers of the tunica and cells of the corpus. **Group B**: An ovule primordium arises as a bulge from the placenta by anticlinal divisions of cells of the epidermal layer of the placenta and by periclinal divisions of cells of the subdermal layer. The undifferentiated ovule is composed of cells derived from a single outer layer of the placenta (tunica) and the subdermal layer.

Within the A group, five types are distinguished. **Type I**—*Casuarina*. A few central cells of the second layer of the tunica differentiate into primary archesporial cells, which divide periclinally to form parietal (cover) cells located below the epidermis and secondary archesporial cells. Repeated periclinal (and anticlinal) divisions also occur in somatic cells derived from the second layer of the tunica and in parietal cells. Following 1-2 rounds of periclinal divisions, the secondary archesporial cells enter meiosis, i.e., become megaspore mother cells. Periclinal divisions of epidermal cells, giving rise to an integument, begin below apical cells of the corpus (species belonging to families Casuarinaceae, Corylaceae, Hippocastanaceae, Lauraceae, Paeoniaceae, Rhamnaceae, Rosaceae, etc.). The number of corpus cells within a megasporangium varies according to the level (position) of integument initiation. **Type II**—*Malabaila*. A few central cells of the second layer of the tunica differentiate into primary archesporial cells, which develop directly into megaspore mother cells. Periclinal divisions of epidermal cells giving rise to an integument begin below the apical cells of the corpus (species belonging to Apiaceae, Brassicaceae and Rubiaceae). **Type III**—*Ribes*. Only one of the central cells of the second layer of tunica differentiates into a primary archesporial cell; this divides periclinally to form a parietal cell and a secondary archesporial cell, which develops into a megaspore mother cell. Periclinal (and anticlinal) divisions take place in somatic cells derived from the second layer of the tunica and in the parietal cell. Periclinal divisions of epidermal cells giving rise to an integument normally start at the level of apical cells of the corpus or somewhat below it (species belonging to families Magnoliaceae, Saxifragaceae, Betulaceae, Polygonaceae, Plumbaginaceae, Liliaceae, Amaryllidaceae, Chenopodiaceae, Fabaceae, Myrtaceae, Portulacaceae, Caryophyllaceae, Urticaceae, Capparidaceae, Zingiberaceae, etc.). **Type IV**—*Anisum*. Only one of the central cells of the second layer of tunica differentiates into a primary archesporial cell, which directly (without division) develops into a megasporocyte. Only somatic cells of the second layer of tunica undergo periclinal divisions. Divisions of the epidermal cells giving rise to an integument occur at the level of apical cells of the corpus or primary archesporial cell (species belonging to Actinidiaceae, Ranunculaceae, Papaveraceae, Apiaceae, Solanaceae, Boraginaceae, Gentianaceae, Poaceae, Alismataceae and Hydrocharitaceae). **Type V**—*Helianthus*. Asymmetry of the ovule is pronounced at early stages of its differentiation. All the central cells of the second layer of the tunica undergo periclinal divisions. Repeated divisions only occur in one pair of cells such that the basal cell of a pair divides in different directions and the apical cell only divides periclinally to form a row of cells, the upper one of which develops into a megaspore mother cell. Periclinal divisions of epidermal cells giving rise to an integument occur at the level of apical cells of the corpus. This results in the formation of a nucellus composed of a few epidermal cells and cells derived from the second layer of tunica. The megaspore mother cell is only surrounded at the base by cells derived from the second layer of the tunica, the rest of it surrounded by epidermal cells (a number of species belonging to Asteraceae and Poaceae).

Types VI—VII fall within the B group. **Type VI**—*Achillea*. The undifferentiated ovule comprises a single-layer tunica and 3 to 5 layers of cells derived from the subdermal layer of the placenta. One to two layers of the ovular apical cells located below the epidermis differentiate into archesporial cells, which develop directly into megaspore mother cells. Periclinal divisions of epidermal cells giving rise to an integument occur at the level of the archesporial cell base or somewhat below it. The nucellus is predominantly formed from epidermal and archesporial cells, which develop directly into megaspore mother cells (some species belonging to Asteraceae). **Type VII**—*Antennaria*. The undifferentiated ovule is composed of a single-layered tunica and three to four layers of cells derived from the placental subdermal layer. The apical cell of the central row differentiates into an archesporial cell, which develops directly into a megaspore mother cell. Periclinal divisions of epidermal cells giving rise to an integument take place at the level of the archesporial cell base (species belonging to Asteraceae). **Type VIII**—*Himantoglossum*. The undifferentiated ovule comprises a single-layered tunica and a row of cells derived from the placental subdermal layer. The apical cell of the subdermal layer differentiates into an archesporial cell. Periclinal divisions of epidermal cells giving rise to an integument occur at the level of the archesporial cell base or slightly below it. The archesporial cell, surrounded only by epidermis, develops directly into a megaspore mother cell (species belonging to Orchidaceae).

To establish the type of megasporangium, it is necessary to follow its formation from the earliest stages up to the stages of megaspore tetrad and uninucleate embryo sac formation. Later on, the ovule morphological features associated with specialization and formation of a female gametophyte become more conspicuous. A large megasporangium with a multilayered wall, showing a parietal tissue, and a multicellular archesporium (type I) is viewed as an original, primitive type of megasporangium. Megasporangia of this type normally show corpus cells as well. Differentiation of archesporial cells is similar to that in microsporangia (a few central cells of the second layer of tunica acquire the appearance of primary archesporial cells which undergo periclinal divisions to form parietal and secondary archesporial cells capable of one or a few mitotic divisions). Evolution of the megasporangium was towards its reduction, in both portions—somatic and sporogenous, but the two processes may not have proceeded concurrently. Sterilization processes appear to have played an important role in the evolution of the megasporangium. One and the same evolutionally more advanced type of megasporangium could have originated from different, more primitive types (Fig. 7).

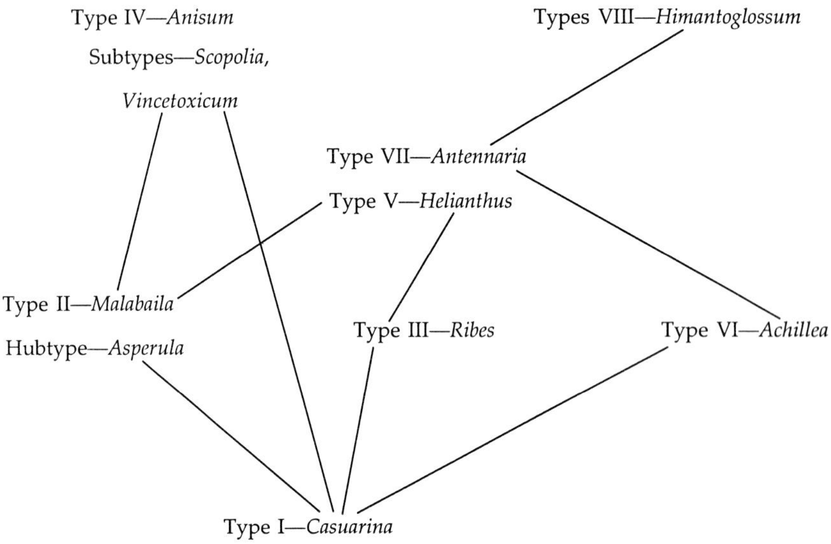

Fig. 7: Classification of megasporangium types.

Types IV, VII and VIII are regarded as the more advanced types of megasporangium, showing a single-layered wall, unicellular archesporium and lack of parietal cells. These types of megasporangium have been reported primarily in families exhibiting a number of features of advanced organization.

INTEGUMENTS

INTEGUMENT (French *in* for un-/in- and *tégument* for coat, envelope)—an ovular structure surrounding the megasporangium (= nucellus) (Plate 41).

The integument was first described in the seed as an 'outer membrane' (Grew, 1672; Malpighi, 1675). Closer examination of the ovular structure (Brown, 1818, 1826b, 1834) revealed the presence of two 'membranes' or 'envelopes' in it (French *tégument*). It was suggested that the inner membrane is the true one. The outer membrane is auxiliary and 'covers', as it were, the developing ovule; this latter membrane was termed 'integument'. Subsequently (Schleiden, 1837) the term integument was used to denote both ovular membranes (inner and outer integument respectively).

Angiospermous ovules show a variable number of integuments (two, one or none) and are accordingly called **bitegmic** (having two integument)—in nearly all monocotyledons and many dicotyledons; **unitegmic** (having one integument)—predominantly in sympetalous taxa of dicotyledons; and **ategmic** (having no integument)—in some parasitic plants (such as Misodendraceae and some members of Olacaceae). Among flowering plants, bitegmic ovules have been reported in 237 families and unitegmic ones in 114 (Kamelina, 1991). The integument(s) can show varying degrees of development. The highest degree of its reduction to initial cells is characteristic of some parasitic plants (e.g. Santalaceae—Ram, 1959a, b; Raj, 1970; some members of the Gentianaceae—Figdor, 1897; Bouman and Louis, 1989). A total absence of integument has only been observed in undifferentiated ovules. In some flowering plants, three and even four integuments have been reported. Some authors (Gaertner, 1788; Warming, 1878; Corner, 1954a, b; and others) consider the **aril** to be the third integument, and the middle integument, as described in Annonaceae (Corner, 1949a), the fourth one. Most authors (Schweiger, 1905; Alexandrov, 1954; Eames, 1961; Komar, 1973; and others) view the aril not as a de novo formation but as specialization of individual structures of an already formed ovule.

Internal (inner) integument (Latin *internus* for internal plus integument)—an envelope proximal to the nucellus in bitegmic ovules. In monocotyledons, it is normally two- or three-layered. However, the inner integument of some dicotyledons can be composed of a higher number of cell layers: 4 to 5 (Burseraceae, Stachyuraceae) or 7 to 10 (Neuradaceae). Its cells can become the initials of somatic embryos (embryoids) (Orchidaceae, Ochnaceae, Celastraceae—Afzelius, 1928; Chiarugi and Francini, 1930; Copeland, 1966; Naumova, 1970, 1983b). In the course of development, the inner integument degenerates completely in some taxa (Polygonaceae, Zingiberaceae) and forms the seed tegmen in others (Amaranthaceae, Berberidaceae).

External (outer) integument (Latin *externus* for external plus integument)—an envelope distal to the nucellus in bitegmic ovules. It varies in thickness between 2 (Papaveraceae, Saururaceae, some members of Liliaceae) and 20 to 25 layers (Paeoniaceae, some members of Nymphaeaceae). In members of nearly 80 families of dicotyledons and monocotyledons (Bouman, 1984), a vascular strand is differentiated within the outer integument; this may be simple (unbranched) (Hemerocallidaceae, Magnoliaceae) or branched (Connaraceae, Meliaceae) which extends up to the level of micropyle (Savchenko and Komar, 1965; Boesewinkel and Bouman, 1984). In the course of seed development, the outer integument in most species transforms into a testa.

As growth proceeds, the integument gradually surrounds the nucellus. At the ovular apex, free margins of the integument form a channel for pollen tube passage and entry into the embryo sac, the **micropyle**. Normally, the inner integument forms the micropyle. Occasionally, both integuments contribute to its formation: the inner integument forms the **endostome** and the outer one the **exostome**. In anatropous ovules, a portion of the integument is normally fused to the funiculus to form a **raphe.**

The integuments may give rise to various structures: **obturator** (both integuments), **operculum** (inner integument), **aril** (outer integument), **integumentary tapetum** (inner or single integument). Within the integument (single or outer), one or more vascular bundles can differentiate (see further: Vascular Supply of the Ovule).

In bitegmic ovules, the first integument to develop is usually the inner integument. In some taxa, the outer integument is formed first (Euphorbiaceae), or both integuments differentiate simultaneously (Juncaceae). As already apparent in early illustrations of ovule genesis by Mirbel (1829), the integuments are initiated as ridges and usually arise by periclinal or oblique divisions of dermal or subdermal initials of the ovular primordium (Warming, 1878). In the region of the forming integument, a zone (normally one- or two-layered) of actively dividing cells is observable, differing from surrounding cells by large size and often prominent nuclei. As an exception (e.g. Ceratophyllaceae—Shamrov, unpublished), these divisions occur in both dermal and subdermal layers. In some taxa (Brassicaceae—Roth, 1957; Liliaceae—Bouman, 1971; Orchidaceae—Shamrov and Nikiticheva, 1992), the integument may comprise only the outer derivatives of epidermal initials, whereas normally all derivatives of initials contribute to formation of an integument. The inner integument is almost invariably of dermal derivation (Warming, 1878; Bouman, 1974; and others), while the outer integument, like the single one, can be either dermal (Netolitzky, 1926; Shamrov, 1990d; Shamrov and Anisimova, 1993a, b) or subdermal (Warming, 1878; Pèchoutre, 1902; Bouman, 1974; and others). The pattern of integument initiation may be used as a taxonomic feature (Bouman, 1984).

In plurilayered integuments, an epidermis and a parenchyma are distinguished. Formation of a plurilayered integument normally occurs by periclinal divisions of cells of middle layer(s). In some cases, periclinal divisions of epidermal cells on the outside (Cucurbitaceae) or inside (Magnoliaceae) or on both sides simultaneously (Paeoniaceae) also contribute to formation of a plurilayered integument. Ovules with plurilayered integuments have been reported in both primitive and more advanced taxa. Thus, plurilayered integuments are characteristic of many primitive taxa (Calycanthaceae, Magnoliaceae, Myristicaceae), while two- to three-layered integuments are typical of more advanced taxa (Didieraceae, Caryophyllaceae). On the other hand, a number of taxa at the base of the system of flowering plants (Aristolochiaceae, Papaveraceae, Saururaceae) show two- to three-layered integuments, with the more advanced taxa (Elaeagnaceae, Rhamnaceae, Sympetalae) exhibiting plurilayered ones. The single integument is usually more massive than integuments in bitegmic ovules.

The phylogenetic significance of the multilayer condition of integument remains an open question. Some authors consider this character a primary feature (Bouman, 1974; Boesewinkel, 1981; Goldberg, 1986), others a secondary one (Netolitzky, 1926; Savchenko, 1973). The evolutionary history of the integument should probably be considered in conjunction with other codeveloping structures of the ovule as well as with the seed-coat structural features, and the mode of dispersal and germination.

The number of integuments in the ovule is a species-, genus- and, often, family-specific taxonomic character. In some families (Fagaceae, Monimiaceae, Piperaceae, Rosaceae), both unitegmic and bitegmic species occur. The bitegmic condition is considered a primitive feature. The evolutional changeover from bitegmy to unitegmy must have occurred many times, independently and heterochronously in different phylogenetic lines of angiosperms (Eames, 1961; Takhtajan, 1948, 1964; Bouman, 1984).

It was long supposed that the unitegmic condition arose in two ways: (1) by suppression of development of one of the integuments (some Hydnoraceae, Piperaceae, Rafflesiaceae), or (2) by fusion of primordia of the two integuments to form a common structure (Coulter and Chamberlain, 1903; Takhtajan, 1948, 1964; Maheshwari, 1950; Eames, 1961). Detailed studies (Bouman, 1974, 1984; Bouman and Calis, 1977; Bouman and Schrier, 1979) have explained the nature of congenital fusion, which is only possible when both integuments are of the same derivation. In addition, another developmental pattern for the unitegmic condition has been observed, viz., by shifting of the initials during development. The single integument arose as a result of shifting and gradual elimination of the primordium of the inner dermal integument in ovules where the outer integument is of subdermal derivation (Bouman and Calis, 1977) (some Juglandaceae, Ranunculaceae, Rutaceae; and Ulmaceae—Dottori, 1991), or of the primordium of outer integument in the ovules when both integuments are of dermal origin (*Impatiens* from Balsaminaceae—Boesewinkel and Bouman, 1991). In *Ceratophyllum* no complete elimination of one of the integuments occurs. Probably, the integument has arisen by a long and gradual drawing together of the initials of two integuments of different derivation. The upper portion of the single integument represents the inner integument formed from dermal initials and its middle and lower portions represent the outer integument developed from subdermal initials (Shamrov, 1995).

The protective function is the primary function of both the integument and its derivative, the seed coat. In addition, the integument is the site of accumulation of storage reserves (starch, proteins and lipids) and therefore provides nourishment to the embryo sac and, at post-fertilization stages, to the developing embryo.

The problems of integument evolution have always been the subject of much discussion. The integument is homologized with the leaf organs of the vegetative bud, with modified parts of the leaf and with fern indusium (see Worsdell, 1904). Several theories on ovule origin exist: the synangial theory (Benson, 1904; Meeuse, 1964), telomic theory (Haan, 1920; Walton, 1952; Long, 1966), nucellus modification concept (Andrews, 1961), and double integument theory (Camp and Hubbard, 1963a, b). Differences in these concepts centre on the origin of the integument. None of the theories suggest homology of the integument with fertile structures. A point of view exists that of the two integuments, only the inner is the true integument corresponding to the single integument of gymnosperms; the outer integument is of cupular origin (Stopes, 1905; Walton, 1952; Stebbins, 1974). An alternative view holds that the inner integument has originated de novo, or during transformations of an already appearing megasporangium (Meeuse and Bouman, 1974).

Integumental Epidermis (Greek *epi* = on, over and *derma* = skin, + integument)—a surface layer of cells of the ovular integument. At early stages of ovule development, the epidermal cells are normally small, isodiametric, thin-walled, with a centrally located nucleus. In a number of taxa (Plantaginaceae, Polemoniaceae), epidermal cells enlarge dramatically after fertilization (by a factor of 60 to 80 in Plantaginaceae). This is accompanied by thinning down of the cell walls (Hyde, 1970) or by straightening of their outer and radial walls (Zhukova, 1987), as well as by enlargement of the nucleus and nucleolus and changes in organelle distribution. Ultrastructural studies suggest that these cells exhibit the characteristics of meristematic cells: presence of vacuoles, dense hyaloplasm showing numerous ribosomes, and poorly developed ER (Vassilyev and Plisko, 1971a; Plisko, 1971). By the time of embryo sac formation, the epidermal cells have increased in size and become vacuolated, such that the nucleus migrates to the outer tangential wall. Their radial walls are crossed by plasmodesmata, which are more numerous in the outer integument. Epidermal cell specialization is accompanied by disappearance of plasmodesmata from the inner tangential walls (Mogensen, 1973; Wilms, 1980; Belyaeva, 1983; and others). In the course of development, epidermal cells often develop on the outside of the integument, wall thickenings, and can become tangentially (Begoniaceae, Datiscaceae, Gentianaceae) or radially (Grossulariaceae, Solanaceae) elongated, taking on the appearance of a palisade tissue. In most plants, they persist in a mature seed as a component of the seed coat. The outer integument (Amaryllidaceae, Liliaceae, etc.) shows the presence of stomata (Petrova, 1977).

The epidermal cells of the integument(s) are covered by a cuticular pellicle. When the integuments are in close contact with one another or with the nucellus, the cuticular pellicles may come together to form a structurally heterogeneous cuticle, which generally persists in a mature seed. Epidermal cells can accumulate various reserves: starch, tannins, calcium oxalate crystals, etc. In some taxa, epidermal cells on the outside of the integument (the single in Plantaginaceae and the outer in Juncaceae) become filled with mucilaginous substances of polysaccharide nature at final stages of seed development. Secretion of mucilaginous substances is accompanied by disappearance of starch and progressive degeneration of the protoplast (Hyde, 1970; Shamrov and Anisimova, 1993a, b).

Integumentary Tapetum (Latin *tapetum integumentale*)—a portion of the integumental epidermis on the embryo sac side (Plate 42). The term was coined by Palm (1915). Synonyms: endothelium, epithelium, endodermis. Of these, the term endothelium introduced by Schwere (1896) is the most commonly encountered in modern literature.

The integumentary tapetum is normally formed in tenuinucellate ovules, typically in unitegmic ovules (most members of the Sympetalae) and less commonly in bitegmic ones (from the inner integument, as in Primulaceae and some members of the Linaceae). It has also been found in crassinucellate ovules with an ephemeral nucellus: unitegmic (Alangiaceae, Davidiaceae, Griselliniaceae) or bitegmic (as in some members of Linaceae). Formation of integumentary tapetum is characteristic of many dicotyledonous families, while it has been reported in as few as 7 monocotyledonous families (Araceae, Arecaceae, etc.).

The endothelium shows a number of features common to all taxa and some taxon-specific features. It is formed as a result of transformation of integumental epidermis into secretory cells on the nucellus side (subsequently coming into contact with the embryo sac) and represents a layer of secretory cells. In rare instances, where the integumental epidermis degenerates early, an endothelium can develop from the subdermal layer (as in *Melampyrum* belonging to Scrophulariaceae) (Arekal, 1963; Nikiticheva, 1979a).

Endothelium differentiation occurs during intensive growth of the integument: its cells tend to exhibit orderly patterns and homogeneity. Their shape, initially isodiametric, varies with respect to the long axis of the embryo sac: in most taxa, cells elongate at a right angle to the axis, whereas in some taxa (Asteraceae, Gentianaceae and others) they do so parallel to the axis; occasionally, the integumentary tapetum cells retain their isodiametric shape (*Lathraea,* Scrophulariaceae). In some species, cells of the single-layered endothelium become in the course of development, non-uniform in size: groups of small cells alternate with larger cells, resulting subsequently in a ruminated endosperm (species of tribe Verbasceae belonging to Scrophulariaceae).

On the basis of ultrastructural features, three stages of integumentary tapetum development have been distinguished (using *Calendula officinalis* as an example—Vassilyev and Plisko, 1971a). **During embryo sac development**, the endothelial cells are palisade in shape, contain small vacuoles and show the characteristics of meristematic cells (poorly developed endoplasmic reticulum, high density of hyaloplasm, abundance of free ribosomes) (stage 1). **At the stage of mature**

embryo sac, endothelial cells flatten along its surface and show the characteristics of differentiating cells (disappearance of typical vacuoles, denser hyaloplasm, extensive rough ER, active Golgi apparatus and numerous ribosomes) (stage 2). **During embryogenesis**, endothelial cells exhibit tangential elongation, become highly vacuolate and acquire structural features typical of differentiated parenchymatous cells (stage 3). Similar structural changes occurring in endothelial cells during embryo sac development have also been reported for other species (Newcomb, 1973a, b; Zhukova, 1987; Van Went, 1989; and others). Normally, the endothelium is one-layered but in some members of Asteraceae, Orobanchaceae and Gentianaceae it is multilayered (Schnarf, 1929; Tiagi, 1965; Shamrov, 1990d).

The extent of the integumentary tapetum along the embryo sac varies from family to family, and within a single family (Ericaceae, Gesneriaceae, Asteraceae, Scrophulariaceae, etc.): it can be well developed along the entire length of the embryo sac or just along a certain portion of it, surrounding its central or chalazal, or less commonly micropylar region. These features of the endothelium presumably correlate with the structure of the endosperm and its haustoria (Nikiticheva, 1979b).

The endothelial cells are usually uninucleate and only in some taxa (Asteraceae, Balanitaceae, Solanaceae) they are bi- or multinucleate; the nuclei may be polyploid. High ploidy of nuclei of endothelial cells has been reported in *Pedicularis palustris* (Steffen, 1955). Ultrastructural studies have shown the integumentary tapetum cells to be connected to one another by numerous plasmodesmata crossing the radial walls, resulting in the whole endothelial layer forming a single symplast (Vazart B. and Vazart J., 1965; Vassilyev and Plisko, 1971a; Zhukova, 1987; Van Went, 1989; and others). In tangential walls, there are fewer plasmodesmata on the integumental parenchyma side and no plasmodesmata at all on the embryo sac side. On the endothelial cell surface, that is, between the integument and the embryo sac, a cuticular layer is formed which is continuous at first but then, by the time of embryo sac maturation, becomes discontinuous. During seed development, the tangential and radial walls of endothelial cells can develop secondary thickenings of variable composition. The most thickened tangential walls are generally those adjoining the embryo sac; these are impregnated with electron-dense material, probably sporopollenin, in *Impatiens valeriana* (Van Went, 1989). In *Nymphoides humboldianum* (Batygina and Kolesova, 1987), the radial cell walls are the more thickened ones. In some plants (*Vicia* and *Lathyrus* of Fabaceae), the endothelial cells show the properties of transfer cells as evidenced by the presence of outgrowths on tangential walls (Gunning and Pate, 1974). It has been found that in *Petunia hybrida* (Esser, 1963), pollination promotes deposition of callose in the integumentary tapetum cell walls, which disappears after fertilization. In *Impatiens*, callose deposition accompanies degeneration of endothelial cells and appears as a layer coating the endosperm (Van Went, 1989).

In many species of the Solanaceae family, endothelial cells proliferate following incompatible crosses, resulting subsequently in the death of hybrid embryos (somatoplasmic sterility) (Satina *et al.*, 1950; Rappaport *et al.*, 1950).

Histochemical studies have shown endothelial cells to contain carbohydrates (often in the form of starch) and lipids as well as physiologically active substances: ascorbic acid and some enzymes (esterase, peroxidase) (Zinger, 1958; Van Went, 1989; Shamrov, 1990d; and others).

Combined morphological, ultrastructural and histochemical studies of the integumentary tapetum have revealed that this tissue exhibits a high physiological activity and presumably is multifunctional.

On the hypothesis of secretory function of the integumentary tapetum, the cells of this tissue produce hydrolytic enzymes which break down and 'digest' the integumental parenchyma cells (Guignard, 1893, Goebel; 1933; Zinger; 1958; and others). The hydrolysates, according to most authors, are transported to the embryo sac. Opinions differ as to the pathways of their transport (Netolitzky, 1926; Schnarf, 1929; Zinger, 1958; Vazart W. and Vazart J., 1965; Newcomb, 1973a, b; Bhat and Vijayaraghavan, 1980). Some authors believe that nutrients are supplied to the embryo sac directly from the integumentary tapetum, others regard the endothelium as a barrier tissue preventing the direct transport of nutrients. Metabolites are supplied via the endothelium, to the chalazal portion of the ovule and thence through specialized tissues, to the embryo sac. It has also been suggested that the endothelium determines allometric growth of the embryo sac and integument (Vassilyev and Plisko, 1971a; Kapil and Tiwari, 1978a). But this does not appear to be its specific function since similar growth occurs in ovules lacking an endothelium. Furthermore, it has been demonstrated in *Impatiens*, that the endothelium develops independent of the embryo sac (Van Went, 1989).

Where the endothelium or its remnants in the form of tangential cell walls persist in a maturing seed (some species belonging to Linaceae, Scrophulariaceae, Solanaceae and others), it may have a protective function with respect to the endosperm and the embryo, and possibly other functions in germination. The evolutional changes in endothelium are closely associated with transformation of the entire ovule and, in particular, with nucellar reduction. Its specialization within the integument represents a higher level of organization of ovules of flowering plants.

Integumental Parenchyma (Greek *par* = equal, *enchyma* = filled, + integument)—a parenchymal tissue situated below the epidermis of plurilayered integument. It may comprise 1 to 20 and more layers: 1 to 2 layers in members of Gesneriaceae (Alimova and Yakovlev, 1982) and Cneoraceae (Kamelina, 1985a) families, 4 to 7 in Staphylleaceae (Alimova, 1985) and Linaceae (Nikiticheva, 1985c) and 10 and more layers in Bignoniaceae (Freiberg, 1987a), Gentianaceae (Shamrov, 1988) and Nelumbonaceae (Batygina, 1981). Ultrastructural studies suggest that the integumental parenchyma cells, like the cells of the entire integument, exhibit the characteristics of meristematic cells at early developmental stages. During embryo sac formation, the cells undergo specialization and accumulate polysaccharides, proteins and lipids (Zinger, 1958; Mogensen, 1973; Newcomb, 1973a, b; Wilms, 1980; Van Went, 1989; Shamrov, 1990d; and others). In some taxa (Asteraceae, Valerianaceae), cells of the inner layers adjoining the endothelium synthesize and secrete mucilaginous substances, resulting in activation of the Golgi apparatus, cytoplasm vacuolation, disappearance of starch and gradual protoplast degeneration (Vassilyev and Plisko, 1971a; Kamelina and Zhukova, 1987). Integumental parenchyma cells may be ultrastructurally distinct from epidermal cells: for instance, in representatives of Polemoniaceae the chloroplasts are 2 to 3 times as large, contain starch and show a more extensive system of thylakoids (Zhukova, 1987). In some plants (e.g. *Heliophila amplexicaulis*, family Brassicaceae—Rodionova, 1972, 1992), large multinucleate cells have been observed to form within the integumental parenchyma which perform a nutritive function. In many taxa, integumental parenchyma cells partially or completely degenerate during seed development.

MICROPYLE (Greek *micros* = small, fine, and *pyle* = gate, entrance)—a tiny canal above the nucellar apex formed by an extension of the integument(s) overarching the nucellus and ensuring entry of a pollen tube into the embryo sac under porogamy. Synonyms: germ pore, germinal pore, germinal aperture (Plate 43). Turpin (1806) introduced the term micropyle, although Grew (1672) was the first to describe it.

In unitegmic ovules, the single integument forms the micropyle. In bitegmic ovules, it can be formed by both integuments (Malvaceae, Grossulariaceae, Moraceae) or by the outer integument (Salicaceae and others). The portion of a bitegmic ovule formed by the inner integument is termed the **endostome** and that by the outer integument the **exostome** (Greek *stoma* = mouth, *endon* = inside, *exo* = outside); the terms were coined by Mirbel (1829). The micropylar canal may be short or long. A micropyle formed by both integuments may be straight (Ruppiaceae, Centrolepidaceae, Musaceae and others) or zigzag (Hamamelidaceae, Papaveraceae, Berberidaceae and others). Occasionally the upper portion of the integument expands to form folds or twists, giving the micropyle the appearance of a long winding canal (*Malpigia, Cissus*). In some plants, no micropyle is formed due to underdevelopment of the integuments (*Torenia, Philadelphus*) or as a result of complete fusion of their margins (*Alchemilla, Ficus*). Some angiosperms exhibit penetration of the embryo sac into the micropylar canal (Flacourtiaceae, Loranthaceae) or formation of a beak-shaped outgrowth of the nucellus, the nucellar beak (Euphorbiaceae), thus preventing the margins of the integuments from drawing together to form a typical micropylar canal. The resultant wide opening (aperture) was termed 'acropyle' by Baillon (1885).

Cells lining the micropylar canal normally exhibit dense cytoplasm and relatively large nuclei (*Linum*). They may be covered with hairs (a number of species of Asteraceae), papillae or, as in *Begonia, Nelumbo*, etc., have a smooth surface often showing wax depositions. Cells lining the micropyle are involved in active secretion of chemotropic substances aiding in passage of the pollen tube through the style and its entry into the embryo sac.

While assisting the pollen tubes in finding their way to the nucellus and the embryo sac, the micropyle is concomitantly the last barrier (in the stigma-ovary-micropyle series), limiting the number of penetrating pollen tubes. In many plants, cells lining the micropylar canal become filled with tannins immediately after passage of the pollen tube.

OPERCULUM (Latin *operculum* = lid)—a lid-like structure formed by cells of the apical portion of the inner integument. The term was proposed by Hegelmaier (1868) but the structure was already described earlier (Gaertner, 1788-1807).

Genesis of the operculum has been discussed in the literature for years. It was suggested that both the inner and the outer integument contribute to its formation (Caldwell, 1899). Detailed studies of the operculum in members of Lemnaceae by S. Maheshwari (1954, 1956) confirmed Hegelmaier's view of the operculum as a specialized tissue of the inner integument. It is in this meaning that the term operculum has been used in the book *Comparative Embryology of Flowering Plants* (Yakovlev, 1981). Loose interpretation of this term should probably be avoided whenever, under the name 'embryo lid' or 'seed lid' (embryotega or embryostega), one deals with structures of different origin which occur in the form of a lid (stopper or plug) over the root end of the embryo (Boesewinkel and Bouman, 1984; Danilova and Kirpichnikov, 1985; and others). Opercula have been reported in both monocots (Lemnaceae, Liliaceae, Zosteraceae, etc.) and dicots (Haloragaceae, Nymphaeaceae, Penaeaceae, etc.), the operculum in monocots (as in the order Zingiberales) often being formed concurrent with the **micropylar collar**.

Formation of the operculum commences during embryo sac development and is normally preceded by an increase in number of cell layers at the inner integumental apex. Subsequently epidermal cells on the outside (Costaceae—Panchaksharappa, 1962; Lemnaceae—Kapil and Vasil, 1963) or on the inside (Liliaceae—Shamrov, unpublished data) dramatically increase in size, especially in the radial direction. Either the entire apical portion of the inner integument develops into an operculum (Lemnaceae—Kapil and Vasil, 1963; and others) or only its individual layers (Zosteraceae—Teryokhin and Shibakina, 1985). In doing so, the cell walls increase in thickness and the subjacent integumentary cells are gradually crushed and lignified. The operculum reaches maximum development in a mature seed, providing protection for the micropylar region of the seed and aiding in germination.

Micropylar Collar (Latin *collare* = collar, + micropyle)—an annular structure at the apex of the outer integument. Humphrey (1896) coined the term. The micropylar collar has only been found in monocots belonging to orders Commelinales and Zingiberales. It is differentiated within the ovule prior to fertilization. Periclinal divisions of the cells and local increase in number of layers of the outer integumental parenchyma result in formation of an annular structure determining a peculiar shape of the micropylar end of the ovule and seed. In the micropylar collar region, the inner integument and nucellus develop folds (Mauritzon, 1936, Chikkannaiah, 1962; Grootjen and Bouman, 1981; Grootjen, 1983; Kamelina, 1990b).

The micropylar collar persists in a mature seed. In this case, its inner layer of cells may become sclerotic. Often developing concurrently with the operculum, the micropylar collar forms the massive mechanical tissue of the seed coat, which provides an additional protection for the micropylar region of seed and assists in seed germination (Boesewinkel and Bouman, 1984).

OTHER SPECIAL STRUCTURES OF THE OVULE

Aril (Latin *arillus* = aril)—a seed appendage in the form of specialized structures of different origin which are initiated within the ovule and continue to develop in the seed. Synonym: caruncle (Plate 44). In mature seeds, arils are fleshy, often coloured structures serving as adaptations for seed dispersal.

The term 'aril' was proposed by Linnaeus (1751) to denote a saccate envelope enclosing the seed. Subsequently, as new evidence accumulated, the term acquired a slightly different meaning. Thus, Gaertner (1788-1807) defined aril as an extra seedcoat of funicular origin. Richard (1808) established that the aril originates in the hilar region of the ovule and does not reach full development until after fertilization. Lastly, Planchon (1845), in his major work, subdivided the arils into **true** (funicular appendages at the base of the ovule) and **false** (various overgrowths of the outer integument) called arillodes (Latin *arilloid*). He also introduced the concept of strophiole (various outgrowths along the raphe). These terms continue to be the subject of discussion to the present day since the structures they refer to are of taxonomic significance. The true aril develops from funicular tissues at the site of the hilum over the obturator. It is initiated in the ovule at the two- to four-nucleate stage of embryo sac and subsequently remains unfused to the seedcoat. True aril plugs, in a manner somewhat similar to that of the obturator, the micropyle in anatropous ovules. **Arillodes** (false arils) and their variations (strophioles and caruncles) never plug the micropyle since they are normally laid down from the margin of the exostome or are outgrowths (extensions) of other regions of the outer integument (Gaertner, 1788-1807; Richard, 1808; Planchon, 1845; Pfeiffer, 1891; Van Tieghem, 1901a, Van der Pijl, 1955, 1957; Takhtajan, 1964; Komar, 1965 a,b; and others).

Baillon (1876) took a somewhat different approach to the problem. He proposed that a single term 'aril', collectively referring to both true and false arils, be reserved for this structure, while specifying, in addition, its localization. He also suggested that the concepts of 'common' and 'specific' aril, i.e., localized in a particular region (raphe, chalaza, funiculus, micropyle) should be introduced: for example, funicular aril (*Nymphaea*), micropylar aril (*Euphorbia, Polygala* and others), etc. Baillon's (1876) definition of aril received support from many of his contemporaries (Pfeiffer, 1891; Schweiger, 1905). Later, however, other authors, in particular Netolitzky (1926), could not help noticing that Baillon unduly expanded the concept of aril, unjustifiably extending it to piliform structures and pappi. Goebel (1933) also pointed out this fallacy of Baillon.

Among the works appearing subsequent to Baillon's (1876) publications, the most exhaustive was that of Pfeiffer (1891) who, while embracing the ideas of Baillon and criticizing those of Planchon, considered it necessary that the definition of aril should include its location in and connection to the seed and further held that a true aril is fused to the seed coat at the site of its initiation only. Pfeiffer (1891) was against the then rather popular interpretation of aril as a third integument, considering the fact that an aril arises on the already fully developed and fertilized ovule. Among those sharing Pfeiffer's opinion on the issue was Schweiger (1905), who had written a monograph on the development of caruncles and obturators in Euphorbiaceae and emphasized that the caruncle (a micropylar arillode) is as such completely distinct from the third integument.

Maheshwari (1950) cited an example of formation de novo of an aril as a third integument in *Ulmus americana* by splitting of the outer integument. A very interesting case of formation of four integuments in some members of Annonaceae has also been reported: two normal integuments, an aril (a well-developed portion of the outer integument), and a middle integument, a formation de novo between inner and outer integuments (Corner, 1949a, b). Eames (1961) was against viewing the aril as a third integument; he considered the aril to be a part of the ordinary integument, the result of its specialization associated with zoochory. The nature of aril as a third integument is still an open question since what the opponents of this view were usually dealing with was the arillode, i.e., an outgrowth of the outer integument.

Attempts have been made to examine the nature of the aril from the evolutional standpoint. Thus Corner (1949b, 1953, 1954a, b) in his durian theory (which takes its name from the fruit of the Indo-Malay tree *Durio zibethinus,* family Bombaceae) relates the presence of an aril to fruit evolution. He noticed that the true aril is uncommon among herbaceous plants and is more commonly found in tropical trees (an observation which needs to be corrected nowadays). Corner concluded that fruit showing arils are a primitive type. In advanced forms, especially those with an inferior ovary, arils are rare. Corner suggested that in early zoochores the aril could have been the third, fleshy integument which had undergone reduction in the course of evolution, its rudiment in extant plants being represented by various structures. It is worth noting here that small arillodes, according to Corner, are indicative of a higher degree of specialization and a more advanced level. The durian theory is very intriguing and merits special consideration. It attracted considerable attention at one time. One of the strongest opponents of this theory appears to be Van der Pijl (1955, 1957) who pointed out that Corner ignored the ecological factors and the cases where an aril was formed de novo, failed to analyze critically the accumulated evidence and, in some cases, confused arillodes with arils. Van der Pijl himself draw a clear-cut distinction between arils and arillodes. Unlike Corner, he considered the undifferentiated, tough sarcotesta a primitive feature and initial structure for formation of more specialized structures, arillodes. According to him, the arillode is an incomplete, reduced sarcotesta ecologically better adapted for zoochory. From this point the evolutional transition from sarcotesta to arillodes of variable type and position and finally, as the most advanced evolutional stage, to aril can be followed. At the same time, Van der Pijl did not rule out the possibility of independent origin of aril in some instances. Nor did he claim his view to be generally applicable; he believed that not all plant groups evolved adaptations for zoochory by developing such structures as the sarcotesta and arillodes. Other developmental pathways are also possible.

Moreover, it was later been noticed (in particular, Corner (1976) in contrast to an earlier theory) that there exist mixed structures: a combination of a micropylar arillode and a true aril. These are called arillodia (Latin *arillodium*), a term proposed earlier by Planchon (1845). In our view, compound structures of the arillodium type, which previously received little attention, are fairly common. Despite the lack of consensus regarding the nature of false and true arils, the taxonomic and evolutional significance of this feature should not be ignored. The formation of only strophioles or only caruncles has been reported to occur in all members of some families, as in Euphorbiaceae (Takhtajan, 1964; Van der Pijl, 1969; Corner, 1976; Kapil *et al.*, 1980; Boesewinkel and Bouman, 1984).

Obturator (Latin *obturator* for plug, stopper)—a tissue of secretory type comprised of elongated epidermal cells of the ovary or ovule which grow toward the micropyle and often plug entry into it (Plate 45). The term was proposed by Baillon (1858), the structure had earlier been described under the name **'transmitting tissue'** (Brongniart, 1827).

Savchenko (1973) proposed a classification of obturator types. Depending on the structure on which the obturator differentiates, the author recognized the following types of obturator: **funicular, integumental, placental** and **carpellary**. Veselova (1989, 1991) suggested that, in addition, **nucellar** and **septal obturators** should be distinguished. In our opinion, the term 'carpellary obturator' does not describe the origin of the obturator adequately. Since in this case the obturator originates from epidermal cells of the ovarian wall it should be called **parietal** (Latin *parietis* for wall). Analysis of the evidence available suggests that, depending on origin, two major types of obturator should probably be distinguished: (1) **ovular** (Latin *ovulum* for ovule) obturators formed from ovular cells and (2) **ovarian** (Latin *ovarium* for ovary) obturators arising from ovarian cells. Depending on the obturator position in the ovule or ovary, the following variants are distinguished: in the ovular type: **integumental, funicular** and **nucellar**; in the ovarian type: **placental, parietal** and **septal** obturators. There also exist obturators of intermediate ovular-ovarian type which are formed from both ovular and ovarian cells.

Among ovular obturators, the most common is the funicular obturator. This forms at the base of the funiculus of predominantly anatropous ovules (Molluginaceae—Narayana and Jain, 1962; Liliaceae—Savchenko and Komar, 1965). The integumental obturator forms as a result of expansion of the apical margin of the normally internal integument which may project into the stylar canal (Araceae—Fagerlind, 1944b; Poaceae—Savchenko and Petrova, 1963) and less commonly the outer integument (Nymphaeaceae—Winter, 1987). The nucellar obturator originates from apical cells of the nucellar epidermis and as a multilayered beak-like structure may protrude beyond the micropyle coming into contact with obturators of the ovarian type (Nyctaginaceae—Bhargava, 1932; Rosaceae—Savchenko, 1959; Polygonaceae—Solntseva, 1983; Trapaceae—Batygina and Kolesova, 1985; Caryophyllaceae—Veselova, 1989).

Ovarian obturators have been reported in rather few taxa. The parietal obturator most commonly forms at the base of the style (Rosaceae—Arbeola and Herrero, 1987), but occasionally at the ovarian base (Thymelaeaceae—Fuchs, 1938). The placental obturator is derived from the placental cells adjoining the ovular cells (Liliaceae—Berg, 1962; Alliaceae—Shamrov, 1990a). The septal obturator is formed from cells of the ovarian septae in the syncarpous gynoecium (Caryophyllaceae—Veselova, 1991).

The type of obturator depends on structural features of the funiculus. In funicular ovules, the obturator normally develops from cells of the funiculus, and less commonly from those of the funiculus and integument (*Victoria amazonica,* family Nymphaeaceae—Shamrov and Winter, 1991). In sessile ovules and in funicular ovules with a poorly developed funiculus, obturators are derived from placental cells (Juncaceae, Paeoniaceae—Shamrov and Anisimova, 1993a, b; Shamrov, pers. comm.), or from placental and funicular cells (*Gagea stipitata,* Liliaceae—Shamrov, 1992b).

Obturators are of epidermal derivation and comprise fairly large cells with thickened outer walls, dense cytoplasm and a prominent nucleus with a few nucleoli. They show the features of transfer cells (Johnson *et al.*, 1975, Peterson *et al.*, 1979; Singh and Walles, 1992; Walles, 1992; and others) and like stigmatic and stylar canal cells, are a variety of transmitting tissue (Tilton and Horner, 1980; Tilton *et al.*, 1984; Clifford and Orvens, 1990). They synthesize fat and sugar droplets and mucilaginous substances which can be excreted onto their outer surface (Schweiger, 1905; Savchenko, 1973; Arbeola and Herrero, 1987; Shamrov and Anisimova, 1993).

Integumental, nucellar, funicular and parietal obturators normally arise during the embryo sac formation and become fully developed by the time of entry of the pollen tube into the ovarian cavity (Bhargava, 1932; Savchenko and Komar, 1965; Savchenko, 1973; Winter, 1987; Shamrov and Winter, 1991; and others). The timing of placental obturator differentiation varies: this begins at the ovular primordium stage in some taxa (Liliaceae–Shamrov, unpublished data) and at the binucleate stage of embryo sac in others (Euphorbiaceae—Maheshwari and Johri, 1941; Alliaceae—Shamrov, 1990a; Juncaceae–Shamrov and Anisimova, 1993). Pollen tube growth is believed to occur along the surface of the obturator whose cells excrete a substance;

however, the mechanisms triggering the process are poorly understood. It has been found that in *Prunus persica* (family Rosaceae) pollination has no effect on secretion production and that pollen tube growth does not start until disappearance of storage starch from the cells of the parietal obturator (Arbeola and Herrero, 1987). Obturators normally disintegrate after fertilization but occasionally a placental obturator persists in a mature seed (*Luzula pedemontana,* family Juncaceae—Shamrov and Anisimova, 1993).

Accumulation of various plastic substances by the obturator cells showing the characteristics of transfer cells as well as their peculiar growth towards the micropyle suggest a function for obturator as a secretory structure providing nourishment for the pollen tube and ensuring its growth and entry into the embryo sac (Maheshwari, 1950; Savchenko, 1973; Johnson *et al.*, 1975; Tilton and Horner, 1980; Tilton *et al.*, 1984; and others).

Chalaza (Greek *chalaza* for nodule)—a basal region of the ovule where the base of the nucellus and that of the funiculus unite to form a single structure (Plate 46). The term was proposed by Treviranus (1805).

The chalaza is not clearly delimited from other ovule parts. An arbitrary boundary of the chalaza could be the site at which the cuticle terminates (Netolitzky, 1926) or the level at which the epidermal layers become united (Zinger, 1958). We believe that the position of hypostase within the ovule can be used as a criterion in identifying the chalazal boundary.

Developmental and structural patterns of chalaza vary according to the morphological type of ovule. The first classification of chalaza types was proposed by Periasamy (1962b) and based on two criteria: (1) the proportion between the chalaza and other ovular structures and (2) the contribution of chalaza to formation of the seed coat. According to this, the following types of chalaza are distinguished: (1) normal chalaza comprising a relatively small portion of the ovule and seed and making a minor contribution to formation of the seed coat; (2) massive chalaza, or pachychalaza by Corner's (1976) definition, accounting for the bulk of seed and arising by intercalary growth in every direction during the post-fertilization period; it forms most of the seed coat; and (3) perichalaza (the term was proposed by Corner, 1949a) accounting for the major portion of the seed and arising by intercalary growth in the raphal and antiraphal regions; makes up the bulk of the seed coat.

We propose a different classification of the chalaza types based on the following criteria: (1) position in the ovular primordium and size in relation to other ovular structures, (2) vascularization pattern, and (3) the extent of contribution to seed coat formation. The following types of chalaza are distinguished: **pachychalaza** (Greek *pachys* for massive), **mesochalaza** (Greek *mesos* for middle; normal chalaza in Periasamy's terminology), **perichalaza** (Greek *pery* for around), and **leptochalaza** (Greek *leptos* for thin; term proposed by us).

Pachychalaza has been reported in a number of taxa from different levels of the phylogenetic system of flowering plants. We suggest that three pachychalaza variations be recognized on the basis of spatiotemporal development of ovular structures. **Typical** variation—from the very beginning the chalaza (exo-endopachychalaza) occupies the major portion of the ovule and is completely formed in the mature seed (Cannaceae, Rhizophoraceae, Tropaeolaceae). The chalaza may arise below the site of attachment of only the inner integument (endopachychalaza in Rhizophoraceae—the term proposed by Boesewinkel and Bouman, 1984), or of the outer integument (exopachychalaza in Nelumbonaceae— the term proposed by Shamrov, 1995). **Juvenile** variation (Latin *juvenilius* for young)—the chalaza is initiated as in the typical variation; however, prior to fertilization (Ceratophyllaceae) or immediately thereafter (Scheuchzeriaceae, some members of Santalaceae), the micropylar portion of the seed becomes the more massive part. **Mature** variation (Latin *maturus* for ripe, mature)—the chalaza becomes massive following fertilization, due to intercalary growth below the site of attachment of the integument(s) (some members of Lauraceae, Myristicaceae). In pachychalazal ovules, the embryo sac starts development at the site surrounded by nucellar tissues and the resultant embryo undergoes further development while being surrounded by chalazal tissues. The funicular vascular bundle generally ramifies to form a network of postchalazal bundles some of which extend into the integuments. In a mature seed, the pachychalaza derivatives constitute the bulk of the seedcoat.

Mesochalaza has been found in ovules of the majority of flowering plants. The chalaza in this case occupies a small part of the developing ovule and seed and makes a minor contribution to formation of the seedcoat. The vascular bundle usually terminates in the chalaza.

Perichalaza is characterized by unidirectional growth of the chalaza in the raphal and antiraphal regions. The postchalazal vascular bundle ramifies into the integument. The seed coat is formed predominantly from integumentary cells (Annonaceae, Monimiaceae, Vitaceae, some Lauraceae).

Leptochalaza is characteristic of the tenuinucellate ovules of a number of taxa occupying the upper level of the system of flowering plants (Begoniaceae, Gesneriaceae, Orchidaceae, and others). It is represented by a small cluster of cells degenerating in the course of seed development. The leptochalaza derivatives make practically no contribution to the seed coat formation. Procambial cells normally formed within the funiculus are absent from the chalaza.

Early stages of chalazal development are poorly understood. Our data suggest that the layers involved in the formation of particular chalazal regions are identifiable very early in ovular development. In the subdermal layer of an ovule with the mesochalaza, below the forming integument (the outer one in bitegmic ovules), periclinal divisions of cells occur, resulting in formation of the outer chalazal region. In anatropous ovules (Campanulaceae, Gentianaceae, Nymphaeaceae), periclinal divisions only occur on the dorsal side, while in orthotropous ovules (Barclayaceae—Winter, 1987) they take place on both sides of the ovule. The axial row of cells of the ovular primordium located below the differentiating hypostase transform into the inner chalazal region in which a vascular bundle differentiates.

The position of the chalaza in the ovule as well as one of its structural features (the ovular vascular bundle normally terminates in the chalaza) suggest that the chalaza serves as a kind of 'depot' for metabolites coming from both the vascular bundle and the ovular regions adjacent to the chalaza. Another function of the chalaza—the protective one—is most typical of pachychalazal seeds whose coat is primarily made up of chalazal derivatives.

Evolutionary transformations of chalaza within the angiospermous ovule are still poorly understood. It is conceivable that the ovule with a mesochalaza is the original one. The emergence of ovules with leptochalaza can be regarded as reflecting a major direction in the evolution of structures—from large, massive to small, few-layered ovules—which is manifested in reduced structure size. Emergence of pachychalazal ovules and seeds, found in both primitive and advanced taxa, must have occurred many times in the course of evolution and appears to be an adaptation for dissemination and germination.

FUNICULUS (Latin *funiculus* for cord)—a structure attaching the ovule to the placenta (Plate 47). Within the funiculus, a vascular supply differentiates which translocates nutrition from the ovary to the ovule and the developing seed. Synonym: seed stalk. The term was proposed by Mirbel (1829).

Depending on the presence or absence of funiculus, two types of ovules have been recognized (Korchagina, 1969): with a seed stalk and without it or sessile (the ovule is practically without a funiculus, showing a broad base whose tissue grades directly into the placental tissue to form a placentochalaza). We propose a different interpretation of sessile ovules and a classification of ovules according to their structure and degree of funiculus development which includes 3 types. On the basis of degree of funiculus development, ovules may be classified into: **funicular** (ovules with a funiculus), **afunicular** (funiculus as a structure lacking), and **sessile** (Latin *sessilis* for assidenous, or attached to something by its base; the ovule is attached to the placenta by the basal portion of the raphe).

In **funicular ovules**, the funiculus may be either long or short, straight or curved, twisted or rugose (plicate), which largely determines the morphological type of ovule and its position within the ovary (Maheshwari, 1950; Korchagina, 1969; Savchenko, 1973; Danilova and Kirpichnikov, 1985; and others). The funiculus is conspicuous in ovules occupying a lateral position within the ovary (e.g. Nymphaeaceae, especially in *Nuphar lutea* whose ovule is hypertropous—Winter and Shamrov, 1991). A long funiculus is common in pendulous ovules (Amaranthaceae—Kajale, 1937; Brassicaceae, Capparaceae—Rodionova, 1972, 1983). In circinotropous ovules, the long funiculus is badly twisted (Cactaceae—Archibald, 1939; Plumbaginaceae—Korobova, 1983). In some plants, it can aid in expulsion of seeds from the fruit (Fabaceae—La Rue, 1954; Magnoliaceae—Earle, 1938).

In **sessile ovules,** which are usually reflexed (anatropous, campylotropous and amphitropous), at some point in their development the funiculus fuses to the integument to form a raphe. In cases like this, the funiculus proper is not morphologically well defined, the ovules being attached to the greatly expanded placental outgrowth by the basal portion of their raphe (Campanulaceae, Juncaceae, Paeoniaceae). In some plants (Cactaceae), the funiculus has been reported to form branches (Buxbaum, 1953; Kapil and Vasil, 1963) which appear to be a kind of placental outgrowth rather than the basally branching portion of the funiculus.

In **afunicular ovules**, the funiculus is completely reduced, the placenta and chalaza forming a single structure, the placentochalaza (Poaceae—Alexandrov *et al.*, 1944; Batygina, 1974a; *Melanthesa rhamnoides,* family Euphorbiaceae—Singh, 1968; Ceratophyllaceae—Shamrov, 1981; *Santalum album,* Santalaceae—Shamrov *et al.*, 1992).

The funiculus may develop various types of hairs. The obturator and aril may be formed from its tissues.

NEW APPROACH TO INTERPRETING THE OVULE BASAL STRUCTURES (Plate 48)—The ovule of flowering plants develops a number of special structures (**hypostase, podium, postament**) whose genesis, position and functions have been the subject of wide discussion.

The concepts of **hypostase, podium** and **postament** differ markedly among various authors and sometimes are even contradictory.

These concepts date back to the ideas of Van Tieghem (1901b, 1903). He proposed the term 'hypostase' to designate the tissue in the form of a disc, cup or column below the embryo sac and composed of cells with lignified walls and acting, according to him, as a barrier tissue to prevent encroachment of the developing embryo sac on the ovular base. In introducing the term, he did not take into account the type (mode) of nucellus development (crassi- and tenuinucellate ovules) or the stage of embryo sac development (pre-fertilization: organized, mature; post-fertilization: during formation of the embryo, endosperm, perisperm, *etc.*)

Dahlgren (1940) coined the term 'podium' and acknowledged the need for retaining the term 'postament' pioneered by Westermaier (1890). According to these authors, the postament is a column of cells beneath the antipodal cells of the embryo sac, the podium being a persistent, cup-like remnant of the nucellar base. Dahlgren and some other authors (Mauritzon, 1933; Maheshwari, 1950; Savchenko, 1973; Poddubnaya-Arnoldi, 1976; and others) believed that the hypostase, podium and postament can undergo interconversion during their development. This resulted in ambiguous interpretation and confusion of these concepts.

Since the introduction of the terms hypostase, podium and postament, a considerable body of evidence has been amassed on the ovule structure and development. Studies of these structures from the earliest stages of ovule development onwards have led us to propose a somewhat different interpretation of them. The hypostase, podium and postament develop as a single structure from the earliest stages of ovule development. In addition to the dermal and subdermal layers, one or more tiers of cells are organized in the forming ovular primordium which run parallel to its longitudinal axis. The cells in these axial rows are of dual origin: in the apical region they arise by periclinal divisions of subdermal cells of the placenta, whereas in the middle and basal regions they do so by periclinal divisions of cells of the third and subjacent layers. They actively divide, their initial divisions being predominantly transverse, to form in the aggregate, the basal meristem.

Longitudinal (axial) rows of cells below sporogenous and gametophytic structures were reported earlier as well (Warming, 1878; Heusser, 1915; Schnarf, 1929; Nawaschin, 1894). However, the formation of axial rows of cells was not examined in relation to the functions of these structures. Our data (Batygina, 1974b; 1987a; Shamrov, 1990d; Shamrov and Nikiticheva, 1992; Shamrov and Anisimova, 1993; Tobe and Batygina, in press; and others) suggest that the axial rows of flat or elongated cells play a most important role in the ovule: transport of nutrients to sporogenous and gametophytic structures from the placental and ovarian vascular supply, starting at the earliest stages of ovular development.

Axial rows of cells of the ovular primordium may give rise to various special structures. In crassinucellate ovules, cells of the proximal (to the archesporial cell) region develop into a postament, those of the distal region into a vascular bundle and those of the middle region into a podium and a hypostase (the central region). In tenuinucellate ovules, only a hypostase and vascular bundle are formed: the hypostase (central region) from cells of the proximal region and the vascular bundle from those of the distal region. Cells of the podium and hypostase have a dual origin and the lateral regions of these structures, unlike the central ones, develop from cells of the lateral meristem of the ovular primordium. The timing of initiation, longevity and structural features of the hypostase, podium and postament vary between taxa, show close association with other developmentally coupled structures of the ovule and seed, and are determined by the pattern of megasporogenesis and of the embryo sac, embryo and endosperm development. **In our view, the term hypostase should be considered a broad term referring collectively to the hypostase proper, the podium and the postament.** However, all these structures, while performing similar functions, differ in pattern of genesis.

Hypostase (Greek *hypo* = under, below, and *stasis* = standstill, stagnation)—a tissue organized at the base of the nucellus and integuments. In tenuinucellate ovules (Gentianaceae, Lamiaceae), the hypostase is located immediately below the embryo sac. In crassinucellate ovules (Ceratophyllaceae, Elaeagnaceae, Nymphaeaceae, Resedaceae), between the hypostase and the embryo sac, there is a nucellar tissue which may disintegrate during seed development, as a result of which the hypostase ends up below the endosperm. The hypostase is initiated at the level of the forming integument (inner one, in bitegmic ovules), almost concurrent with the latter's initiation. Occasionally it is organized prior to differentiation of other ovular structures (Juncaceae). Differentiation of the hypostase proceeds centrifugally. This structure may be represented by a disc-like cluster of cells or, less commonly, by 1 to 3 cells (some members of Orchidaceae and probably, Orobanchaceae). Initially, these cells are isodiametric, thick- walled, densely cytoplasmic, with prominent nuclei showing a few nucleoli each. In some plants, the hypostase cells are binucleate. Moreover, the nuclei may fuse (Campanulaceae, Liliaceae, Scrophulariaceae). In the fully formed hypostase, the cells are discoid and vacuolate.

The nature of the hypostase cells varies between species. Thus, for example, the hypostase of *Anethum graveolens* shows two types of secretory cells: peripheral cells with a prominent nucleus and middle cells with thin walls and small nuclei (Szujkó-Lacza, 1978). In *Retzia capensis,* however, the peripheral radially elongated cells of the hypostase exhibit thick, lignified walls, the central portion of the hypostase being composed of thin-walled cells (Engell, 1987). The cell walls show variation in structure. In *Delphinium camptocarpum,* the hypostase cells exhibit thick, laminated walls lacking plasmodesmata (Belyaeva, 1983) whereas cell wall protuberances have been reported in some species belonging to the Solanaceae family (Ermakov and Goryaeva, 1982) . The hypostase cell walls may become lignified (Polygonaceae, Rosaceae, Malvaceae, Amaryllidaceae, Urticaceae and others) or suberized (Liliaceae, Ulmaceae, Geraniaceae and others). Thickening of the hypostase cell walls may occur via callose, cutine, suberin and lignin deposition (Netolitzky, 1926; Savchenko, 1973). The cells are often filled with tannin-like substances (Bouman, 1984). In *Allium odorum,* the hypostase cells are densely cytoplasmic and bear similarity to epidermal cells of some hydathodes (Haberlandt, 1923, 1925).

A characteristic feature of the hypostase at later developmental stages in some flowering plants is formation of air pockets resulting from anastomosis of some of its cells (Maheshwari, 1950) as well as of schizogenous reservoirs filled with various substances including tannins. In some species (e.g. *Scilla sibirica*), the hypostase cells secrete substances of unknown composition which diffuse in a fan-like manner (Gerassimova-Navashina and Batygina, 1958; Savchenko and Komar, 1965).

Depending on the plant species and developmental stage, the hypostase cells can accumulate various storage reserves: starch, protein, fats, tannins, etc. For instance, carbohydrates, proteins and fats are normally accumulated by *Agave*, only carbohydrates by *Ornithogalum* (Tilton, 1980), and dextrins, starch and proteins by *Gentiana cruciata* (Shamrov, 1990d).

Longevity of the hypostase is believed to be correlated with patterns of nucellus, integument and embryo sac development, and tends to be reduced in evolutionally more advanced taxa. Thus in the Sympetalae with tenuinucellate unitegmic ovules, the normally thin-walled hypostase cells degenerate very early during embryogenesis. In taxa with crassinucellate ovules, the hypostase is more massive, often cup-like, its thick-walled cells either degenerating at late stages of embryo development or persisting in the mature seed (Cephalotaxaceae, Ceratophyllaceae, Tuncaceae).

Of particular interest is morphogenetic correlation between the presence of hypostase and the type of endosperm established in both crassinucellate and tenuinucellate ovules of various families (Tohri and Seghal, 1963; Szujkó-Lacza, 1978; and others). It has been found that in some species hypostase survival is longer with a cellular endosperm than with a nuclear.

The hypostase is a multifunctional structure. Its location near the vascular bundle, the structure of its cells resembling that of the secretory cells and accumulation by its cells of various reserve substances have led most authors to regard the hypostase as a tissue ('glandular tissue', 'nutritive apparatus') concerned with secretion of enzymes and growth regulators (Håkansson, 1923; Paliwall, 1950; Adatia and Shah, 1952; Szujkó-Lacza, 1978; Tilton, 1980; Boesewinkel and Bouman, 1984; and others).

Furthermore, the hypostase location next to the vascular bundle and its persistence in the mature seed in some plant species suggest that it may play a role in maintaining the water balance of dormant seeds during long periods of heat and dryness (Goebel, 1933; Tilton, 1980; Engell, 1987; and others). Other functions attributed to the hypostase are mechanical and protective, in particular that of a barrier tissue precluding excessive expansion of the embryo sac, embryo and endosperm (Eames, 1961; Venkateswarlu 1972; Engell, 1987; and others). In *Retzia capensis* for example, peripheral, concentrically arranged layers of the thick-walled cells of the hypostase prevent the aggressive chalazal haustorium of the endosperm from invading the ovular base (Engell, 1987).

Still another function of the hypostase is that of haustorial activity: together with the chalazal parenchymatous tissue and antipodal cells, it forms a haustorial apparatus (Zinger, 1958). However, the primary function of the hypostase in most species studied appears to be that of supplying nutrients: its cells specialized in the nutritive function (regardless of their shape and wall structure) are involved in transport of metabolites to sporogenous and gametophytic structures.

Postament (Latin *postament* for column)—a tissue of nucellar origin occurring as a column below sporogenous or gametophytic structures. Synonyms: column, pedestal, base for the antipodals (Plate 49).

The term postament was pioneered by Westermaier (1890) in reference to a column of elongated cells located below the antipodal end of embryo sac. The column cells are distinguished from the surrounding cells by denser cytoplasm and more intensely staining walls. According to Westermaier, the primary function of the postament is to provide nutrition. Dahlgren (1939, 1940) extended the meaning of the term and believed the postament to be present both at the embryo sac stage and at various stages of seed development. He suggested that the postament could be formed in one of two ways: (1) by the tissue situated below the antipodals growing into the embryo sac cavity or (2) through transformation of the hypostase occurring as a column beneath the embryo sac and composed of thick-walled cells. Some authors support the former assumption (Komar, 1965a; Savchenko, 1973; and others), other authors the latter (Palm, 1915; Schnarf, 1929; Mauritzon, 1933; Poddubnaya-Arnoldi, 1976; Freiberg and Kamelina, 1981).

In our view, it is the upper cells of the longitudinal (axial) rows of the ovular primordium cells located below the archesporial cell that develop into a postament. In a mature ovule, the postament represents the axial part of the nucellar chalazal region. In grasses, this structure has been described as 'a kind of hypostase' (Batygina, 1974b, 1987a).

Thus, a postament is formed in ovules with a differentiated chalazal region of the nucellus which can be few-layered (Araceae, Orchidaceae, Ranunculaceae) or comprise a fairly large number of layers (Costaceae, Grossulariaceae, Zingiberaceae). The postament is composed of rows of elongated or flat cells extending to the hypostase and structurally resembling the procambial cells (Mücke, 1908; Joshi and Venkateswarlu, 1935; Kamelina and Proskurina, 1987; Alimova, 1990a; Shamrov and Nikiticheva, 1992; and others). The postament cells, as in *Quercus gambelii* (Fagaceae), show an elongated nucleus, numerous mitochondria, free ribosomes and extensive ER, and are interconnected by a few plasmodesmata. Around the periphery of these cells, the plasmalemma forms invaginations of varying size (Mogensen, 1973).

Postament longevity varies in different plants and is correlated with the pattern of development and functioning of other nucellar structures. Most commonly, the column of cells in the chalazal region of the nucellus disintegrates shortly after fertilization. In some taxa (Taccaceae—Håkansson, 1921; Poaceae—Batygina, 1974b, 1987a; Asparagaceae—Satarova, 1990), the column cells collapse even before fertilization. These are gradually replaced by the growing embryo sac, resulting in a more intimate contact with the vascular supply of the ovule and ovary. Members of some families (Calycanthaceae—Kamelina, 1981b) develop in place of the degenerating cells of the axial part of the nucellar base, a lysigenous cavity above the hypostase.

Whenever the postament persists for a long period of time, the lateral layers and subsequently the dermal layer in the nucellar chalazal region as well, break down and are replaced in the course of development by the chalazal haustorium of the central cell of the embryo sac (subsequently endosperm) (Araceae—Mücke, 1908; Bull, 1938; Geissolomataceae—Stephens, 1909; Pontederiaceae—Ono, 1928; Fagaceae—Brown and Mogensen, 1972; Mogensen, 1973; Grossulariaceae—Shamrov, unpublished data). In some taxa (Araceae, Costaceae, Geissolomataceae, Zingiberaceae), it is composed of lignified cells and persists in the mature seed. Cell wall lignification may start as early as during megasporogenesis and embryo sac development (Costaceae—Panchaksharappa, 1962; Magnoliaceae—Freiberg and Kamelina, 1981; Sterculiaceae—Venkata Rao, 1953; Zingiberaceae—Panchaksharappa, 1966; Gurudeva and Somashekar, 1985; and others).

The structure of postament cells and their arrangement in rows, as well as the dynamics of accumulation in them of proteins, starch, dextrins and other reserves, all suggest that the primary function of the postament is that of transport of metabolites arriving from the hypostase to sporogenous and subsequently gametophytic structures. Another function whenever the postament persists in the mature seed would appear to be mechanical protection against unfavourable environmental factors. Besides that the postament, as a hypostase, supports water balance in maturing and dormant seed.

Podium (Greek *podion* = for base)—a cup-like structure formed in the nucellar base (Plate 50).

The term podium was introduced by K. Dahlgren (1940) to designate the remnant of the chalazal (basal) region persisting in the nucellus. On the basis of the transformation pattern of this nucellar region, **columnar** and **cup-shaped podia** can be distinguished.

The cup-shaped podium (term proposed by us) is formed in crassinucellate ovules by the persistent lateral region of the nucellus following fertilization. It differentiates as a cup-shaped structure in the chalazal region of the nucellus early in ovule development. In the mature ovule, its upper boundary lies at the level of antipodal cells. The cup-shaped podium cells are dual in origin and arise by periclinal divisions: lateral regions—predominantly from subdermal cells, and central region—from a portion of cells of the middle region of the ovular primordium basal meristem. Intervening between the podium and the embryo sac is the nucellar parenchyma which disintegrates during seed development, the podium gradually coming into contact with the endosperm. The structure and timing of differentiation of the cup-shaped podium correlate with the degree of massiveness of the crassinucellate ovule. For example, in Ceratophyllaceae the podium begins to form at the sporogenous cell stage and comprises 8 to 12 layers of cells in a mature ovule. In less bulky ovules of the Juncaceae, the podium starts to differentiate either in early stages of megasporogenesis (*Luzula*) or late in embryo sac development (*Juncus*) and is 3 to 5 layers thick. Normally, the podium comprises several layers of flat, densely cytoplasmic cells. Initially, its cells are thin-walled but later on generally become thick-walled. The podium persists, completely or partially, as a cup-like structure in the mature seed (Ceratophyllaceae, Grossulariaceae, Juncaceae, presumably Cyperaceae, Degeneraceae).

The columnar podium or the chalazal proliferating tissue is predominantly formed in ovules with poorly differentiated and early degenerating micropylar and middle regions of the nucellus (Alangiaceae, Liliaceae, Zosteraceae). The entire chalazal region of the nucellus transforms into the podium (Dahlgren, 1940; Komar, 1982; Alimova, 1987; Shamrov, 1992b; and others) which initially appears as a column and then, following disintegration of a portion of its cells, gradually assumes the shape of a cup.

We suggest that a cup-shaped podium starting to develop from the earliest stages of ovule formation and readily distinguishable from other nucellar structures should be considered a true podium. Contrarily, the columnar podium combining features of the postament and podium (Shamrov, 1992a) and comprising the entire chalazal region of the nucellus could probably be termed the postamento-podium. In the course of seed formation, it gradually disintegrates, starting at the apical part.

Podium cells have been found to contain proteins, dextrins and tannins. Structural and histochemical studies suggest that the podium is involved in translocation of metabolites arriving from the vascular bundle via the hypostase. Its central part provides nutrition for the embryo sac, while its more massive lateral portion nourishes the lateral region of the nucellus persisting for a long time after fertilization. The podium may have a protective function in the mature seed.

EMBRYO SAC

EMBRYO SAC (FEMALE GAMETOPHYTE): Origin and Development (Plates 51 and 52): It is generally believed that one of the main features distinguishing angiosperms from phylogenetically lower groups of plants is the presence in them of a unique embryo sac and double fertilization. Based on the striking similarities in embryo sac structure (the so-called 'normal' or Polygonum-type embryo sac development) exhibited by the overwhelming majority of angiosperms and universal occurrence among them of double fertilization, it appears very likely that precisely these features could have evolved but once in the evolutional history of angiosperms (Schnarf, 1929, 1936; Poddubnaya-Arnoldi, 1930, 1951; Maheshwari, 1937, 1948, 1950; Romanov, 1944). Taxonomists consider this to be an irrefutable argument for a strictly monophyletic origin of angiosperms (Boush, 1944; Grossheim, 1945; Zhukovskyi, 1949). As far as other special types of embryo sac found in some angiosperms are concerned, these are viewed as derivatives of the 'normal' type which is thought to be the original, primitive one (Schnarf, 1929; Poddubnaya-Arnoldi, 1930, 1951; Maheshwari, 1937, 1948, 1950; Romanov, 1944). A few plant embryologists accept the possibility that some other types of embryo sac might have evolved independently but their line of reasoning is not sufficiently convincing (Joshi, 1938; Modilewski, 1953).

Ruling out formal homology of the embryo sac structures to structures of mature gametophytes of lower forms, we suggest that the 'normal' type of embryo sac be examined not only based on some regularities peculiar to it, but also on the basis of most general laws of the life cycle of the cell and cell components. Such an approach provides not only a better insight into the origin and development of the 'normal' type of embryo sac, but also 'abnormal' types. Moreover, if analysis is based on general laws of vital cell activity, primarily rules of cell division, an insight is gained not only into the evolution of embryo sac structure, but also into double fertilization as the result of the same process of megagametophyte development abbreviation.

Let us recall the basic laws operating in any embryonic cell (Hofmeister, 1867; Sachs, 1877, 1882; Hertwig, 1894; Gerasimoff, 1890; Wilson, 1925; Navashin, 1947; Gerassimova-Navashina, 1951a, b).

1. The nucleus and consequently mitotic configuration tend to be positioned in a specific 'dynamic centre' of the cell (presumably due to repulsion from the protoplast walls). The location of this centre often coincides with the geometric centre of the cell; if the cell is asymmetrically vacuolated (as with a young uninucleate microspore), the nucleus lies in the centre of the highest cytoplasm concentration.
2. The mitotic configuration axis runs parallel to the long axis of the polarized protoplast to divide the elongated cell transversely (this is typical of organ meristems showing longitudinal growth, e.g. root apical meristem).
3. Each new plane of division tends to cross the preceding one at a right angle (typical of microsporogenesis).
4. Sister nuclei moving apart during division repel each other; this mitotic repulsion is terminated by formation of cell walls or by sufficiently strong development of a vacuole between the nuclei. That is, according to Rule 1, the nuclei tend to occupy the dynamic centres of their protoplasts (e.g. spermiogenesis in angiosperms).

5. In bi- and multinucleate cells (coenocytes), the nuclei divide synchronously. Mutual repulsion occurs among all the nuclei and not only between the coderived ones (Rule 4), resulting in all the nuclei lying at maximum separations, which determines their distribution pattern. For this reason, the nuclei come to lie at opposite poles in a binucleate coenocyte (e.g. the first meiotic division in the microsporocyte); at the vertices of the tetrahedron or at the corners of the square (due to rotation of the plane of division, Rule 3) in a tetranucleate coenocyte (e.g. the second meiotic division in the microsporocyte); in multinucleate coenocytes, all the nuclei become quite uniformly distributed throughout the cytoplasm or within its parietal layer if the coenocyte is highly vacuolated (e.g. the secondary nuclear endosperm of angiosperms).
6. In postmitotic coenocytes, cellularization does not start until a particular level of cytoplasm 'saturation' with nuclei is attained. Where the distance between nuclei is very long or there is a very large vacuole between them, such an interaction is not established, the cell wall is not formed, and the nuclei continue to lie in the common cytoplasm.
7. After a complete termination of division processes within the coenocyte, the nuclei cease to influence one another (e.g. mutually repel each other) and where unobstructed by cell formation, tend to move, according to Rule 1, towards the common dynamic centre.

Operation of these rules varies according to a particular situation created as a result of differentiation of cells and tissues (vacuolation, changes in shape, direction of pressure, unidirectional transport of nutrients, etc.). Thus, in a tissue composed of closely packed cells and in highly elongated cells, Rule 3 may not be operative; strong development of the vacuole prevents formation of a cell wall following division and so on. These factors frequently interact with one another in various combinations. However, a key factor producing a particular effect can always be identified among them.

Turning to the well-known pattern of development of the 'normal' type of embryo sac, it is immediately apparent that the leading role in this developmental pathway is played by the above-listed basic cellular mechanisms. In this sense, **there is nothing specific in the embryo sac which could be interpreted as an inherited structure that had once evolved in a single progenitor.** The megaspore nucleus divides to form two daughter nuclei which move, as with any division, to opposite poles of the polarized cell (Rules 1 and 2). The second and third divisions are perfectly synchronous and perpendicular to each other (Rules 3 and 5), resulting in 'puzzling' quartets of nuclei invariably forming at each pole. The nuclei in each quartet repel one another (Rule 4) to become invariably positioned in such a way that three come to rest against the 'roof' of the coenocyte, the fourth nucleus coming to lie opposite its middle portion occupied meanwhile by an expanded vacuole. This moment coincides with nuclear 'saturation' (Rule 6) of the polar aggregations of cytoplasm where cell walls are formed between the nuclei, i.e., formation of the egg apparatus and antipodal cells takes place. The highly vacuolated state and embryo sac expansion prevent formation of a wall in-between the two fourth nuclei of each nuclear tetrad, resulting in formation of a large central cell with two polar nuclei. Soon afterwards, this develops a single dynamic centre at which the polar nuclei converge (rules 1 and 7).

Thus, the 'normal' type of embryo sac is the net result of cell growth and development under the given conditions. Any embryonic cell placed in similar conditions will invariably follow this developmental pathway, which is determined by general laws of the cell life cycle rather than total dependence on some specific laws unique to the megaspore. Such specific laws could modify this development only to a certain extent.

The aposporic development of embryo sacs reported in many apomicts (as in a number of species of Compositae) is, in our opinion, unassailable proof of the correctness of precisely this view. In apomictic plants, the 'normal type' of embryo sac is known to be formed not from a megaspore but directly from a megasporocyte which, notwithstanding its failure to follow a specific developmental pathway leading to formation of a megaspore as a gametophyte initial, does produce an embryo sac. This is difficult to explain by any means other than the operation of general laws of cell development under specific conditions.

A dramatic proof of the foregoing is also provided by some special cases, primarily the well-known amazing phenomenon of young **pollen grains** (microspores) developing into embryo sac-like structures under extremely high temperatures (Nemeč, 1898; Stow, 1930).

In view of the above, the 'normal' type of angiospermous embryo sac can, in principle, be envisioned as having originated from the embryo sac of some phylogenetically less advanced plant group which might even be similar to extant gymnosperms. It is obvious that the difference in gametophyte development between gymnosperms and angiosperms is due to a most general cause, namely, alteration of the entire metabolic pattern which could have been brought about in the progenitors of angiosperms by nothing other than a radical change in their environmental conditions.

As already noted (Rule 6), cellularization, i.e., complete formation of an embryo sac, is dependent on a particular ratio between the rate of nuclear divisions and accumulation of cytoplasm within the coenocyte. Angiosperms exhibit overwhelmingly higher nuclear division rates compared to gymnosperms and consequently attain the above ratio far in advance of the latter. Another characteristic feature of angiosperms is that the young embryo sac begins to show longitudinal growth very early in development, resulting in its protoplast being essentially divided into two by the vacuole; it is for this reason that polar groups of nuclei, and subsequently cells, are formed. Therefore, changes in the original shape of the embryo sac combined with its rapid development determine the ultimate arrangement of the nuclei formed in it and hence the final cellular structure of the embryo sac. A completely different situation occurs in gymnosperms, as in conifers, wherein the embryo sac retains a more or less rounded shape, which (in combination with the low developmental rate) invariably results in the nuclei being uniformly distributed throughout the parietal layer of the cytoplasm (according to Rule 5). Therefore, dissimilar though they seem to be, the eight-nucleate stages of gametophyte development in one group or the other are actually nothing more than variations—resulting from exposure to different sets of environmental conditions—**of one and the same fundamental process.** In one case, the net result is a polarized structure of the angiospermous embryo sac, in the other a uniform distribution of nuclei within the parietal layer of the cytoplasm. This difference does not reflect the presence of any specific organizational pattern, however, but is due instead to non-specific factors.

Even the early gymnosperms appear to show adequate prerequisites for subsequent emergence of the angiospermous embryo sac. Therefore, the embryo sac of angiosperms could have evolved more than once during evolutional history from a few of their diverging ancestors which, however, showed essential prerequisites for such an evolution in the form of comparatively similar structures (the phenomenon of parallelism here, with respect to the emergence in flowering plants of an embryo sac of the 'normal' type).

Impressive evidence in support of these inferences also comes from studies of other types of angiospermous embryo sac which can be used to demonstrate that their development is also governed by the same general (fundamental) laws of cell cycle; operation of these laws is only modified in response to peculiar features of ontogeny of the generative organs of a particular plant species. However, among these factors there is always a critical one which determines the pattern of embryo sac ontogeny; identification of this is essential, in most cases, for proper understanding of the origin of the type of embryo sac in question.

We revert to bi- and tetrasporic embryo sacs which, *ab initio*, show the 'normal' structure (the Allium-type and Adoxa-type; Plate 51, D-E). Their special feature is that they develop respectively from two and four megaspores forming coenocytes and represent the highest level of megagametophyte development abbreviation. These types give further support to our view that the 'normal' type of embryo sac structure is not a genetically pre-programmed organization realized through the development of a megaspore of a particular origin, such as the one resulting **directly** from one or both meiotic divisions without delimitation of megaspores (i.e., without formation of cell walls between megaspores). It is obvious that the original binucleate coenocytes of mono-, bi- and tetrasporic embryo sacs (Plate 51 C′, D′ and E′) **are not homologous** to one another, yet the net result in all three cases is much the same, namely, an embryo sac of 'normal' structure ('convergent development' according to Modilewski, 1953, or 'inherited structure of final stages' in Romanov's terminology, 1944). It appears that our view offers the most plausible explanation, namely, that regardless of its origin the coenocyte follows similar developmental pathways under similar conditions, its development being governed by the same general rules of cell cycle. Therefore, everything that has been said about development of the 'normal' type of embryo sac holds true for these types too and, in fact, is actually supported by the results from studies on these.

Another, very important conclusion follows from the foregoing, namely, there is no reason to believe that bi- and tetrasporic embryo sacs were invariably bound to evolve from the 'normal' type assumed to be the original one. On the contrary, it is not unreasonable to leave room for independent origins of bi- and tetrasporic condition in the progenitors of angiosperms. The occurrence of tetraspory in *Gnetum* appears to be a most convincing proof of this.

Within each of the three natural subdivisions of embryo sacs (mono-, bi- and tetrasporic), a given number of types is distinguished which varies according to which structural and developmental features are considered essential. Therefore, different authors recognize different numbers of embryo sac types. As we have seen, most authors agree that the embryo sac owes its origin to the occurrence in a single progenitor of angiosperms of some specific laws of megagametophyte development whose operation resulted in the emergence of the 'normal' type of embryo sac which subsequently gave rise to all the other types. However, further examination of the development of various types of embryo sacs, apart from those discussed above, suggests that in this case too it would be unreasonable to assume the operation of some specific laws (a fixed number of divisions, a particular polarity, etc.) since the existing spectrum of embryo sac organization types is actually nothing more than variations of the same general laws resulting from specific conditions under which they operate, the conditions in each case being established during the ontogeny unique to each particular plant species.

Consider first the so-called Oenothera-type of embryo sac which is unique within the monosporic group. This is known to exhibit the distinctive feature of being composed of as few as four cells (instead of seven), i.e., of the egg apparatus and the central cell containing only one polar nucleus. This kind of half-composition of embryo sac results from the fact that during its development the embryo sac undergoes only two divisions instead of the three observed normally. Another characteristic feature of this type is that it develops from the micropylar megaspore, in contrast to the 'normal' type in which the embryo sac is known to develop normally from the chalazal megaspore.

Authors taking a formal approach to interpreting this type of embryo sac development assume that the progenitors of Onagraceae (where no type other than this one has been reported) were bound to acquire another three peculiar features of embryo sac development in addition to and independent of the genetically determined survival of the micropylar (instead of chalazal) megaspore.

1. Migration of the nucleus of the megaspore to its micropylar end where it divides to form the egg apparatus ('establishment of unipolarity').
2. Localization of the vacuole in the chalazal region of the megaspore.
3. Reduction of the number of divisions from three to two.

But, in our opinion, rather than assume the occurrence of four genetically determined features, it would suffice to make a single, fairly reasonable assumption to explain, taking into account the general laws of cell development, the features of the Oenothera-type of embryo sac development and their interconnection. For all the above-listed features of embryo sac development in Onagraceae to be perceived, one needs only to assume the establishment in this family of a special order of nutrient supply to the developing megaspore.

Indeed, if the development of a single megaspore within a tetrad depends primarily on the fact that from its very inception it obtains better nutrition, then the source of nutrient supply might be expected to be located on the side of the developing megaspore. In plants with the 'normal' type of monosporic embryo sac development, the flow of nutrients is therefore bound to be from the chalaza, as is usually assumed. In plants with the Oenothera-type embryo sac development, the flow is bound to be reversed, resulting in the developing (functional) megaspore being the one lying immediately next to the micropyle. This is conceivable, for example, where sporogenesis proceeds at a higher rate compared to formation of the ovular conductive system which fails to reach the level of the chalaza by the time of completion of the microspore tetrad formation. However, in addition to assimilates delivered to the ovule by the ovular vascular supply, of great importance are undoubtedly the plastic substances the developing megaspore obtains from the adjacent cells and primarily from the absorbed products of disintegration of the three abortive megaspores. With the 'normal' type of embryo sac development, the source of supply of these substances is located opposite that of assimilates flowing through the chalaza. With the Oenothera-type embryo sac development, both sources of nutrient supply are located on the same side since the supply of assimilates from the chalaza is presumably already established during disintegration of the three abortive megaspores. This one-sided (i.e., with both sources of supply located on the same side) supply of nutrients to the developing megaspore, which has considerably expanded by this time, cannot but produce asymmetry in its metabolism, readily apparent in this case from the one-sided vacuole formation at the chalazal end and in the migration of the major portion of the cytoplasm along with the nucleus to the opposite, micropylar end of the embryo sac. Therefore, the megaspore nucleus apparently finds itself there in conditions identical to those in which under the 'normal' type of development, one of the two coderived nuclei resulting from the first division in the megaspore is found. Hence, it is only natural that the same processes should occur as in the micropylar end of the 'normal' embryo sac, i.e., two successive divisions resulting in formation of an egg apparatus and a polar nucleus of the central cell (here 'central' only in name, of course). Thus, the number of divisions is reduced by one, due to the fact that in the Oenothera-type the first division to occur **corresponds physiologically** to the second of the 'normal' type, the second division corresponding to the third in the 'normal' type; the division physiologically equivalent to the first in the 'normal' type is not at all possible here.

Thus, using the Oenothera-type as an example, it can be seen that, though this type of embryo sac might seem unique, its dissimilarity from the 'normal' type can be reduced to the operation of such non-specific factors as the ratio between ovule developmental rate and rate of development of megaspores which affects the order of nutrient supply.

The largest number of embryo sac types is known to occur in the tetrasporic group where, given formation of the initial tetranucleate coenocyte, there is from the outset the greatest diversity of possible developmental pathways. However, analysis of these types of embryo sac development reveals that underlying each developmental pattern is, as with any embryo sac, a single key factor **of non-specific nature**. First and foremost, mention should be made of instances wherein the governing factor is the **shape** of the original cell.

Thus, comparing the spherical embryo sac of the Peperomia-type with the very elongated embryo sac of the Pyrethrum-type, it can readily be seen, if one takes the general laws of cell development into account, that the shape of the original coenocyte was the governing factor in determining emergence of these tetrasporic types of embryo sac.

In the case of the Peperomia-type embryo sac, the orbicular shape of the megasporocyte determines the pattern of arrangement of nuclei, namely, the nuclei are uniformly distributed around the periphery, resulting in more or less evenly spaced cells rather than two polar groups of cells. The remaining free ('polar') nuclei come together in the dynamic centre of the globular embryo sac.

If we turn our attention to the Pyrethrum-type embryo sac, here too the behaviour of nuclei can be predicted based on the coenocyte shape alone: rapid longitudinal growth of the coenocyte prevents the nuclei undergoing the second division from changing their division plane, resulting in a chain-like arrangement of the nuclei along the coenocyte; division plane rotation occurs during the third division as soon as this is made possible by spatial relations within the expanding coenocyte.

The above examples give a good indication of the extent to which the type of embryo sac development is dependent on the ovule developmental pattern, the latter being directly responsible for the shape of the initial cell (coenocyte) of the embryo sac: the *Peperomia* ovule exhibits radially symmetrical growth while that of *Pyrethrum* shows longitudinal growth. There is no doubt that of prime importance also are such non-specific factors as type of placentation, structure of the ovary, and possibly of the entire flower, let alone the ratio between the rate of development of the ovular conductive system and that of the megaspore (see above). Unfortunately, these important factors were completely ignored in establishing embryo sac type, the latter considered not only without regard for ovule type, but also represented quite arbitrarily as an oval or even a circle. If we briefly consider some other tetrasporic types of embryo sac that appear at first glance to be highly individualistic, we shall see that they are actually nothing more than variations resulting from the same general laws operating in various combinations under specific conditions created by the overall course of ontogeny of an organ or even the entire organism.

First and foremost we shall dwell on the Fritillaria-type embryo sac which commands universal attention by its developmental features. Some plant embryologists (Joshi, 1938) used it as evidence for the polyphyletic origin of angiosperms. This type of embryo sac is primarily characterized by asymmetrical arrangement of nuclei within the tetranucleate coenocyte. Instead of the usual arrangement of two nuclei at each end (2+2), as is the case in the familiar Adoxa-type (i.e., tetrasporic type embryo sac with a normal ultimate structure), here the nuclei are arranged in such a way that only one of them stays at the micropylar end, the other three ending up at the chalazal end (the 1+3 arrangement).

However, the principal features of the Fritillaria-type development, namely, 'migration' of three nuclei to the chalazal end, can in reality be reduced to a single process of suppression (of functional activity) of the chalazal end of the embryo sac, an early manifestation of which is a unique asymmetrical (1+3) arrangement of nuclei.

The arrangement of nuclei is governed by **dynamic** forces; hence any change in conditions affecting the mutual repulsion of nuclei is bound to have an effect on distribution of nuclei.

Based on the above general physiological laws, the unusual 1+3 arrangement of megaspore nuclei within the *Fritillaria* coenocyte can probably be most reasonably explained by the fact that the two lower nuclei begin to show early signs of suppressed activity. Therefore, when the upper pair of nuclei become most widely separated by virtue of their mutual repulsion, the lower pair of nuclei do not prevent (by their repulsion activity) the lower nucleus of the upper pair from moving farther towards them, i.e., towards the chalazal pole. As a result, the lower nucleus of the micropylar pair moves far away, towards the chalazal end, joining the chalazal pair of nuclei to form a single group, thus producing a characteristic 1+3 arrangement of nuclei.

The next step in this depression process manifests in fusion of the three nuclei in division at the chalazal end, resulting in formation of a 'secondary tetranucleate' coenocyte. The subsequent course of events at the chalazal end of the embryo sac and the type of ultimate structure of the embryo sac are, as already mentioned, dependent on the extent and pattern of that same depression, commencement of which should, in our opinion, be referred to the tetranucleate coenocyte stage, i.e., to very beginning of embryo sac development. However, commencement of suppression of the chalazal end activity has until now been referred to a much later period, i.e., the time of final division, the degree of suppression of which determines the ultimate number of nuclei in the embryo sac ('depression' according to Romanov or 'failure of the nuclei to divide' according to Fagerlind).

Another feature of the Fritillaria-type embryo sac is that under normal development it undergoes one additional division compared to the 'normal' tetrasporic type (Adoxa-type). Due to the fact that only one nucleus remains at the micropylar end of the primary tetranucleate coenocyte (by virtue of 1+3 distribution of the nuclei), conditions are apparently created there which are comparable to those of the preceding developmental stage, i.e., to the binucleate coenocyte. Therefore, two divisions occur unobstructed which normally take place at the micropylar end and produce a micropylar tetrad of nuclei, and subsequently the egg apparatus with an upper polar nucleus. In other words, using the statement made while discussing the Oenothera-type embryo sac development, the second division of nuclei in the Fritillaria-type megasporocyte is **physiologically equivalent** to the first division in the Adoxa-type, the third division being equal to the second one in the Adoxa-type.

The foregoing is borne out by the mode of embryo sac development in *Tulipa tetraphylla* which has been described as a peculiar type of development and, in a certain respect, is the opposite of the Fritillaria-type, since here a reverse arrangement of nuclei (3+1 instead of 1+3) occurs at the tetranucleate stage. Of crucial importance here appears to be the shape of the tetranucleate coenocyte whose chalazal portion is extended along the longitudinal axis, the micropylar part being greatly expanded. This feature is furthermore accompanied by weak depression during development. The chalazal pair of nuclei is thereby oriented along the longitudinal axis of the coenocyte while the micropylar one is located perpendicular to it, the inevitable result of the principal laws of division operating under these conditions. Being repelled from its sister (coderived) nucleus, the upper nucleus of the chalazal pair migrates upwards approaching the micropylar pair of nuclei which, by virtue of their lateral position (transverse position), have no room to move farther apart. The subsequent longitudinal extension of the coenocyte resulting from vacuole development makes the 3+1 arrangement even more pronounced. Since the micropylar end turns out to be relatively saturated with nuclei already at the tetranucleate stage, the number of divisions is reduced by one compared with the Fritillaria-type, in which the Tulipa tetraphylla-type is similar to the Adoxa-type, differing from the latter only by the ultimate structure, which is due simply to the original shape of the coenocyte.

Thus, the three characteristic tetrasporic types of embryo sac discussed above are in effect the product of variation of one and the same basic process of development of the tetranucleate coenocyte. Variations in this process are due to non-specific factors, such as shape of coenocyte and differential physiological state of the micropylar and chalazal ends resulting presumably from different patterns of nutrient supply of these embryo sac regions. The point is just when and in what combination do these factors operate. It is natural that all these types, as well as a number of others, should occur among closely related species, and not to wonder that sometimes the pattern of embryo sac development of these types varies within a species or even in the same individual (Romanov, 1936, 1938, 1939b, 1944).

While considering the developmental pathway of the Fritillaria-type embryo sac, we found it necessary to assume a very early occurrence in it of physiological inequality of the chalazal and micropylar ends of the young embryo sac, which initially appears to be responsible for the specific 1+3 distribution of nuclei and for the entire subsequent course of development. In our opinion, this is attributable exclusively to differential nutrient supply of the chalazal and micropylar regions of the developing embryo sac, i.e., to the same non-specific factor which is undoubtedly also responsible for the suppression of three megaspores under the monosporic type of development. If such is the case, then the depression occurring in the tetrasporic embryo sacs can be seen as an indication of their secondary derivation from monosporic types of embryo sac. This assumption is also supported by the widespread occurrence of the original 1+3 distribution of nuclei found, apart from the Fritillaria-type, in a number of tetrasporic embryo sacs (the Drusa-, Anthemis- and Plumbagella-types, according to Romanov).

This raises the question of why in the emergence of tetrasporic condition the megaspores that came to be suppressed were those in the chalazal region and not in the micropylar one as seen in the majority of the original monosporic types. An obvious answer appears to be as follows: the tetrasporic types of embryo sac with suppression of the micropylar end could not persist since such a developmental pattern would not lead to formation of a functional egg apparatus, selection being only for embryo sacs with the chalazal end suppressed. Unassailable proof of the very possibility of such a 'reversal' of the process in monosporic progenitors of the tetrasporic types of embryo sac comes from the Oenothera-type. Moreover, normal embryo sacs have been reported, which usually develop from a micropylar megaspore (Fagerlind, 1944a, b).

We have barely touched upon the crucial question of why the number of cell divisions occurring during embryo sac development is relatively constant (the question going back almost to the time of discovery of embryo sacs). Were it not for this constancy of number of divisions (especially at the micropylar end of embryo sac), the entire sequence of events leading to fertilization in angiosperms would be impossible. This constancy seems to be so mysterious that even those plant embryologists who came near to generalizing the factors of embryo sac development (Romanov, 1944; Fagerlind, 1944a) were baffled by the peculiarity of embryo sac organization, taking it to be a kind of standard unique to the female sphere of angiosperms. Fagerlind even accepted idealistic definitions and spoke of a certain 'purposefulness' in embryo sac development. Attainment of this genetically determined 'purposefulness' may, according to Fagerlind, be only postponed or precipitated by developmental factors, but divisions in the embryo sac 'are supposed' to continue until the purpose (i.e., formation of the egg apparatus and primary endosperm nucleus) is achieved.

The explanation for this should apparently be sought not in the admission of a genetically predetermined developmental 'sequence' but, once again, in examining the most fundamental laws of cell development, i.e., in the **cytological approach**.

It is well known that in a cell undergoing division, protoplasm synthesis all but ceases. It is for this reason that with divisions occurring in rapid succession, as during sporogenesis or in early stages of embryo development, the cells diminish in size, failing to reach normal size between two successive divisions. In embryonal tissues (meristem), a particular rate of cell divisions is established as well as a closely coupled to it rate of new protoplasm synthesis taking place in-between divisions, resulting in the mean cell size stabilizing at a constant value. Moreover, a more or less definite relation is established between the nucleus mass and cytoplasm known as the nucleocytoplasmic ratio; in effect, we implied the same thing earlier when we used the expression 'saturation of cytoplasm with nuclei'. The same relations are, in principle, bound to occur in coenocytes resulting from repeated nuclear divisions without cells being organized. However, with the nuclear division rate exceeding a given 'standard' value, cytoplasm synthesis will invariably lag behind, the nucleocytoplasmic ratio will shift in favour of the nuclear fraction, the division process eventually coming to a halt; obviously, other things being equal, the divisions will terminate sooner the higher the rate of division. Therefore, the number of successive divisions leading to formation of a coenocyte and consequently the ultimate number of nuclei in it, is bound to depend on the one hand, on the reserve of plastic substances in the initial cell and on the other, on the rate of nuclear division which, in turn, governs the amount of new plastic substances synthesized in-between divisions. As for the cytoplasm or, more cautiously expressed, plastic substances required for a cell to be able to divide, their supply in each cell is undoubtedly not large. In the vast majority of embryonal cells, i.e., in meristems, this reserve appears to be just sufficient for a single round of division, otherwise repeated divisions would occur in sister cells without their growth in-between divisions. This, however, is not the case. Only in special cases where repeated divisions do occur, in rapid succession, is this reserve of plastic substances presumably large enough, having been accumulated by the cell as a result of specific conditions of its development, which seem to be primarily characterized by a delay in division.

A mother cell undergoing two meiotic divisions is known to be such a cell undergoing repeated divisions without a 'rest pause' to allow replenishment of the aforesaid reserve. This is particularly true for microsporocytes undergoing two divisions in rapid succession, with no cell enlargement in-between divisions. Why do microsporocytes undergo two and only two divisions? It is extremely unlikely that the division process comes to a halt on completion of the second division due to some external inhibitory factor. Strictly speaking, an assumption like this would be entirely improbable, as is evidenced by large anthers in which divisions proceed 'in a wave': the first meiotic division is still in progress at one end while the second one is already underway at the other end. One would have to admit then that the proposed external factor arrests the second division without interfering with the first one, which is absolutely incredible. Therefore, the most reasonable assumption would be that the mother cell has its own 'reserve' which is sufficient for as few as two divisions and that this reserve has been accumulated by it during the premeiotic period in which the cell exhibits prolonged growth without division. The same, in principle, is true of megasporogenesis.

Each of the four megaspores resulting from the two meiotic divisions is bound to undergo a certain enlargement without which further division, i.e., initiation of gametophyte formation, is impossible. It is very likely that the megaspore destined to undergo regular mitoses accumulates during its enlargement a reserve identical to that acquired by an ordinary cell meristem, i.e., sufficient for one division only. This raises the question of why the megaspore (under normal type development) undergoes two successive divisions rather than a single one.

A tentative answer to this question might be as follows. A total of four megaspores are formed, which upon a single division of each, would produce an aggregate of eight cells. To put it differently, together they apparently possess the *reserve* required and sufficient for completion of all the divisions leading to formation of a normal embryo sac. And when it is remembered that during development of a normal monosporic embryo sac one megaspore consumes the remaining three, it is reasonable to assume that in this way reserves of all the four megaspores of a tetrad come to be concentrated in a single (functional) megaspore. From this it follows inevitably that the developing megaspore is bound to undergo precisely three divisions—no less, no more.

Indeed, we proceed from the assumption that for a cell division to occur a particular amount of reserve material is required. Therefore, cell components being doubled during cell division, it is not unreasonable to suggest that for a division to take place a double amount of reserve material is required compared to that of each sister cell resulting from division. Thus, taking the relative reserve of a cell ready to divide to be numerically equal to two, the contents of each sister cell should evidently be assumed to be equal to unity. If the functional (surviving) megaspore really absorbs the essential reserve substances of the remaining three, identical to it, megaspores of the tetrad, then it is bound to contain a **quadruple** reserve which, by our hypothesis, will be equal to eight. Since the amount of reserves is assumed to be halved upon division, the indicated megaspore contents will have been depleted after **three** successive divisions, resulting in formation of a young eight-nucleate embryo sac, which can be viewed as eight, still undifferentiated great-granddaughter cells originating from the initial megaspore.

This assumption is borne out by development of a tetrasporic embryo sac of normal structure (the Adoxa-type) (Plate 51, E) where only **one** division occurs in the tetranucleate coenocyte, the latter being nothing more nor less than four united megaspores. Consequently, in our hypothesis, the monosporic type embryo sac development is actually differentiated from the tetrasporic one solely by the fact that a single division available to each of the four megaspores fits here, as it were, in a single megaspore and is therefore 'fractionated' in time; contrarily, in tetrasporic type development it is completed in one step. On the assumption of a genetically determined developmental sequence, according to which the division process 'comes to a halt' upon the third division under the normal monosporic type embryo sac development, the tetrasporic type assumably exhibits its own peculiar developmental sequence under which halting occurs upon the first division.

By our hypothesis, under the bisporic type embryo sac development (Allium-type; Plate 51, D), a binucleate (bisporic) coenocyte is bound by virtue of absorbing the other pair of megaspores, to be capable of undergoing precisely two divisions instead of one (as with tetrasporic type development). Otherwise one would probably have to assume here the existence of still another, new developmental pattern, namely, cessation of divisions upon the second division, as opposed to monosporic and tetrasporic types of development wherein halting occurs respectively upon the third and first divisions.

It follows from the above assumption that the whole peculiarity of the angiospermous female gametophyte here too reduces to no more than a highly increased rate of cell division compared with that in gymnosperms—an acceleration resulting in the divisions occurring in the megaspore leaving no time for synthesis of substances required for cell division. In gymnosperms with their extremely low rates of division in megaspores, the latter are in effect closely similar in this respect to ordinary meristematic cells, i.e., in-between divisions they have time, following each division, to synthesize new reserve material, resulting in a greater number of divisions in them.

In summary, just as the 'normal' type of embryo sac could have originated more than once during a particular evolutional period, so must have the other types and, what is more, they did so independent of the 'normal' type. However, we do not by any means rule out the possibility, which in many cases has came to be a proven fact, that different types of embryo sac could have evolved from one another and primarily from the 'normal' type.

Embryo Sac: Organized and Mature: Given the lack of distinction in earlier embryological literature between the concepts of 'organized (formed) embryo sac' and 'mature embryo sac', the following definitions of these notions were proposed (Batygina *et al.*, 1982). The **organized embryo sac** is an embryo sac in which cellularization has been completed. It is usually characterized by taxon-specific number and disposition of cells. It was found experimentally that at this stage of development the embryo sac is not yet ready for fertilization (Batygina, 1974b). The **mature embryo sac** is defined as an embryo sac in which differentiation and specialization of elements exhibiting taxon-specific features have been completed; the embryo sac is ready for fertilization. The duration of different embryo sac developmental stages, ranging from several hours to a few days, is taxon-specific.

In conjunction with the above concepts, a brief mention should probably be made of the terms 'poorly differentiated' and 'undifferentiated' egg apparatus which are used in to both light and electron microscopic studies. Unfortunately, the properties allowing a line of demarcation to be established between these concepts are lacking in the literature. Admittedly, it has been pointed out that in defining these concepts morphological features such as polarity, topography, shape, size, presence or absence in the synergids of a filiform apparatus should be taken into consideration (Gerassimova-Navashina, 1962; Petrova, 1977; Plyusch, 1992). However, all these features are exhibited by a mature embryo sac in which the processes of differentiation and specialization have been completed. Lack of the filiform apparatus and other features in many taxa are by no means indicative of the egg apparatus being 'poorly differentiated' or 'undifferentiated'. The structural similarity of the three cells of the egg apparatus to one another, as in *Lilium*, does not provide sufficient grounds for describing it as 'poorly differentiated' or, as in *Ulmus*, viewing it simultaneously as 'differentiated' (with respect to some features) and 'poorly differentiated' (with respect to others) (Plyusch, 1992). In a mature embryo sac, all sexual elements, including all the cells of the egg apparatus, are differentiated, but the degree of differentiation may vary. It is not quite correct to speak of undifferentiated or poorly differentiated egg apparatus in the strict sense of the term 'differentiation'. Even provisional use of these terms appears illegitimate since neither describes the phenomenon adequately.

Egg Cell—a female gamete which, upon fertilization, develops into an embryo (a new plant organism) (Plate 53). The presence of an embryonal initial in the embryo sac was first reported by Brongniart (1827) who termed it the embryonal vesicle (French *vesicle embryonnaire*). In 1858, the German botanist Hartig coined the term egg cell for it.

The egg cell is one of the three cells of the egg apparatus located at the micropylar end of the embryo sac. In some species with Plumbago- and Plumbagella-types of embryo sac development, only one cell—the egg cell—occurs in the micropylar region (*Plumbago capensis, P. zeylanica, Plumbagella micrantha*). Cases of an increased number of cells in the egg apparatus have been reported. For instance, the egg apparatus of

Crepis capillaris and *Ochna serrulata* contains up to five cells, two to three of these being the egg cells (Gerassimova, 1933). An egg apparatus composed of 5 to 7 cells is a common feature of *Tulipa tetraphylla*- and *Eriostemones*-type embryo sac development (Romanov, 1939, 1945, 1971b).

In flowering plants, the egg cell is comparatively large, being normally larger than the synergids (Elaeagnaceae, Gentianaceae, Loranthaceae, Malvaceae, Rhamnaceae, Rosaceae and others). Most commonly, it is a pear-shaped cell. The egg apparatus may occupy a minor portion of the embryo sac (Paeoniaceae, Portulacaceae and others) or more than one-third or even half of it (Aizoaceae, Cactaceae, Clusiaceae, Nyctaginaceae). The egg cell of a mature embryo sac exhibits a distinct morphological polarity. This is most clearly indicated by the relative positions of the egg nucleus and the large 'central' vacuole. The egg nucleus is normally located at the chalazal pole of the cell and the vacuole at the micropylar pole. The polarity of the egg apparatus, like differentiation of its cells, becomes apparent during embryo sac maturation. Initially, all three cells of the egg apparatus exhibit some structural similarity to one another (Guignard, 1882; Romanov, 1936; and others). In *Gossypium*, for example, this similarity is manifested in similar dispositions of nuclei and vacuoles in all the three cells of a young egg apparatus (the vacuoles occupying the chalazal portion of the cell, with the nucleus located above the chalazal region). Moreover, the egg cell is appreciably smaller than the synergids. Subsequently, the egg cell enlarges; this enlargement presumably occurs primarily due to the vacuole increasing in size (Romanov, 1936). Several cases of deviation from the typical polarity of the egg apparatus have been reported. The position of the egg nucleus in *Plantago lanceolata* is variable: it may be centrally or apically located (Vannereau, 1978). Nor is the pattern of arrangement of nuclei and vacuoles consistent in the egg apparatus of members of the Papaveraceae family (Ilyina, 1981). *Samadera indica* (Simaroubaceae) normally exhibits a typical polarity of the egg cell; however, in some embryo sacs unusual disposition of the egg nucleus and vacuole has been noted, which is viewed as resulting from abnormal differentiation of cells of the egg apparatus (Solntseva, 1985). Egg apparatus have been reported with all the three cells differentiating as synergids (*Eryngium planum, Osmorhiza aristata*—Grevtsova, 1987) or as egg cells (*Sambucus*—Kamelina *et al.*, 1987). It is not inconceivable that some of the aforesaid examples of atypical polarity of the egg apparatus cells observed in some species are indicative of embryo sac immaturity, rather than taxon-specific features. Ultrastructural characteristics of deviant types of egg apparatus structural organization are given in the works of Plyushch (1991, 1992).

The egg cell protoplast contains a complete array of cell organelles. The egg nucleus is generally larger than the synergid nuclei (exceptions are seen in members of Resedaceae in which the egg nucleus is smaller than the synergid nuclei) (Chaban, 1983).

Comparative ultrastructural analysis of egg cell plastids based on evidence in the literature and our own observations suggests that the plastids of the angiospermous egg cell are represented by leucoplasts (these being amyloplasts in most species studied). In the angiospermous egg cell, chloroplasts have been found neither in leuko- nor in chloroembryophytes. This distinguishes it from the egg cell of many algae and that of some higher plants (Zhukova, 1983, 1993). Plastids are mainly distributed around the egg nucleus and are amoeboid in shape. Leucoplasts of the *Aconitum napellus* egg cell exhibit extremely deep invaginations (Zhukova and

Sokolovskaya, 1983). The inner membrane system of egg cell plastids is poorly developed in the species studied. Only a few thylakoids occur in their stroma. In plastids of *Capsella bursa-pastoris* (Jensen, 1968), *Epipactis atrorubens* (Savina and Zhukova, 1983) and *Aconitum napellus* a peripheral plastid reticulum is observed. The plastid stroma is very dense. However, in many species, plastoglobules and ribosomes are observable in it. Because of their poor internal differentiation, angiospermous egg cell plastids came to be regarded as a form of proplastids, or eoplasts (Thomson and Whatley, 1980). At the same time, angiospermous egg cell plastids have been found to accumulate reserve carbohydrates in the form of starch. Starch has not been found in the egg cell plastids of some species, such as *Crepis capillaris* (Gerassimova-Navashina and Gulyaev, 1973), *Cytinus hypocistis* (Ponzi and Pizzolongo, 1976), *Aconitum napellus* (Zhukova and Sokolovskaya, 1983), *Agave parryi* (Tilton and Mogensen, 1979), *Ornithogalum caudatum* (Tilton, 1981) and *Nicotiana rustica* (Plyushch, 1989). Among possible taxon-specific features of the egg cell plastid ultrastructure, along with lack of amylogenesis, are formation of concentric membrane structures (*Aconitum napellus*—Zhukova, unpublished data); associations of plastids with ER membranes (*Epidendrum scutella*—Cocucci and Jensen, 1969); deposition in the plastid stroma of a substance of unknown nature in the form of a few globules (*Gossypium hirsutum*—Jensen, 1968); and accumulation of numerous plastoglobules (*Lycopersicon esculentum*—Ermakov and Goryaeva, 1982).

Egg cell protoplasts of most angiospermous species exhibit a large vacuole. In some species, this vacuole occupies most of the egg cell volume, as in *Alcea rosea* (Zhukova, unpublished data), *Gossypium hirsutum* (Jensen, 1964, 1965b, 1968), *Capsella bursa-pastoris* (Schulz and Jensen, 1968b), *Hibiscus costatus, H. furcellatus* (Ashley, 1972), *Ornithogalum caudatum* (Tilton, 1981). In addition to the large vacuole, some small vacuoles occur. These are located closer to the chalazal pole of the protoplast in some species (*Capsella bursa-pastoris, Plumbago zeylanica*) and to the micropylar pole in others (*Linum catharticum*). In some species, no large vacuole has been found (Diboll and Larson, 1966; Yoffe, 1968; Cocucci and Jensen, 1969; Chaban, 1975; Maze and Lin, 1975). The absence of a central vacuole is regarded as a structural feature of the egg cell in the species studied. These observations may be the result of immature embryo sacs having been used in the studies, however (see above). Further studies will show which of these assumptions is the more realistic one.

The egg cell wall is non-uniform in thickness. It is thickened in the micropylar region but almost completely devoid of projections characteristic of other embryo sac cells. An exception is the egg cell of *Plumbago zeylanica* and *P. capensis* which shows a filiform apparatus-like structure at its micropylar pole (Cass, 1972a, b; Cass and Karas, 1974). In the overwhelming majority of species studied, the polysaccharide cell wall at the chalazal pole of the egg cell of the embryo sac ready to be fertilized is absent. The egg cell wall becomes incomplete during differentiation of this cell. Initially, upon completion of cytokinesis in the coenocytic embryo sac, all the three cells of the egg apparatus are completely enclosed in cell walls. During embryo sac maturation and preparation for fertilization, the chalazal portion of the cell walls disappears. This has been demonstrated in a number of studies (Cass, 1981; Fougère-Rifot, 1981; Wilms, 1981; Savina and Zhukova, 1983; Gvilava, 1987; and others). Therefore, the absence of cell walls from the chalazal pole of the egg apparatus cells can serve as one of the indicators of degree of embryo sac maturity. The selective loss of a portion of the cell wall is, in the view of Cass (1981), an adaptation for a more efficient gamete transmission and probably for 'recognition' via specific membrane contacts.

In some plant species, the egg apparatus cells are completely devoid of cell walls (*Potamogeton*—Wiegand, 1900; *Peperomia blanda*—Plyushch, 1983, 1988; *Gagea lutea*—Plyushch, 1992). Upon completion of plasmogamy, the egg cell is completely surrounded by the cell wall, thus precluding further penetration of sperm cells and providing a relative metabolic isolation of it from the surrounding cells and tissues. Comparative morphological studies using light and electron microscopy showed that the egg cell structure varies considerably among various species of flowering plants. The common features of egg cell organization are: the chondriome, plastidome, free ribosomes and vacuome are well developed; organelles such as the ER and Golgi apparatus are very poorly developed; the cell wall is absent from the chalazal pole of the cell (Vassilyev and Plisko, 1971b; Jensen, 1972; Kordyum, 1978; Tilton, 1981; Zhukova, 1983, 1984; Willemse and Van Went, 1984; Plyushch, 1991, 1992).

The egg cell of flowering plants is viewed as a highly specialized cell (Gerassimova-Navashina, 1971; Batygina, 1987a)[1]. Comparative ultrastructural analysis of the egg cell lends support to this view. Among the features indicative of a high degree of the egg cell specialization are: peculiar combination of structural features such as presence of amyloplasts characteristic primarily of storage tissues (leucoplasts lacking starch only occur in certain species), their predominant distribution around the nucleus, extensive chondriome and vacuome, a small amount of ER and Golgi apparatus, abundant ribosomes (mainly in the form of monosomes) and selective partial loss of the cell wall (Zhukova, 1983, 1984). The egg cell closely interacts with the maternal organism. Accumulation of starch in it is indicative of activation of not only biosynthetic processes controlled by amylosynthase, but also of the processes ensuring supply of monosugars to the cell. Indirect indication of activation of nutrient transport to the egg cell is provided not only by starch (as well as storage lipid and protein) synthesis, but also by the presence of plasmodesmata in the egg cell wall and the degree of development of a vacuome capable of accumulating inorganic compounds as well as soluble sugars, amino acids, organic acids and other metabolites, i.e., of building up a pool of substrates for numerous synthetic processes (Zhukova, 1983). The latter ensure egg cell growth (including growth by elongation) during its preparation for fertilization, disintegration of the cell wall at the apical pole of the egg cell (involvement of enzymes in the process is not inconceivable) and an increase in amount of plastidome and chondriome. Therefore, during differentiation of the egg cell not only is its polarity established, but also a high energy potential as well as a temporary storage of reserve substances (starch in amyloplasts, lipid inclusions) and possibly soluble metabolites in the vacuome. The specialized nature of the egg cell makes it capable of performing specific functions such as acceptance of the sperm cell, fusion with it and preparation for division, i.e., for transition of a newly formed organism from the unicellular to the multicellular stage of its existence.

Synergids (Greek *syn* for together, *ergon* for work)—satellite cells closely associated with the egg cell (Plate 54). Schacht (1857) introduced the term. These cells (normally two), form with the egg cell a single complex in the embryo sac—the egg apparatus.

---

[1]According to some authors, the egg cell is a non-specialized parenchymatous cell (Russell, 1993).

Synergids are lacking in species with Plumbago- and Plumbagella-types of embryo sac development.

Synergids show considerable variation in size, shape, structure and viability in different taxa. Normally, they are smaller than the egg cell in some taxa they are larger or equal in size to the egg cell, the egg apparatus cells distinguishable from one another only by their disposition in the embryo sac. In some plants, as in *Allium* species, the synergids differ in size, one of them capable of becoming a giant cell or undergoing a number of divisions. The morphology of synergids is highly variable. They may be funnel-shaped as in *Crepis tectorum* (Gerassimova-Navashina, 1957), dentate as in *Triticum aestivum* (Batygina, 1962a), with a broad base (Anacardiaceae, Davidiaceae and Valerianaceae) or with a greatly extended and tapering terminal portion (Asteraceae, Calyceraceae and Solanaceae). In a number of *Allium* species (Haberlandt, 1922; Modilewski, 1925) and in *Eucalyptus cinerea* (Polunina, 1959), they often resemble the egg cell. In some species, the synergids, especially their basal portion, extend such that their tips project deep into the micropyle and even into the ovular tissues. The presence of synergids with haustorial outgrowths is a characteristic feature of some taxa of flowering plants (e.g. a number of tribes of the Asteraceae family and some genera of the Poaceae family). The most lengthy and narrow synergid haustoria projecting into the ovary and extending up the style to penetrate it to 1/3 of its length have been reported in *Quinchamalium chinense* of the Santalaceae family (Johri and Agarwal, 1965).

Synergids exhibit a well-defined polarity manifested in both their cell wall structure (see below) and protoplast organization. A common feature of these cells is the basal location of the nucleus (in close proximity to the filiform apparatus) and the apical location of the vacuole. Such a disposition persists up to and after fertilization (*Epipactis atrorubens and Valeriana officinalis*). The opposite arrangement of these organelles has been observed in species of Anacardiaceae, Apiaceae, Combretaceae, Myrtaceae. The synergid nuclei are lobate in shape (spherical in synergids of an immature embryo sac), smaller than the egg nucleus or equal to (Lecythidaceae) or larger than it (Convolvulaceae and Resedaceae). Normally, the synergids are uninucleate but occasionally they are binucleate as in *Anthemis alpina*, *Carya illinoensis* and *Dictamnus albus*, or even trinucleate as in *Santalum album*. The nucleus is usually haploid. Giant polyploid nuclei and polytene chromosomes have been observed in the synergids of some species. In the Tropaeolaceae members studied, synergid nuclei have been found to undergo division such that mitoses remain incomplete (Walker, 1947; Batygina, 1985b). The synergid plastidome is represented by leucoplasts showing a poorly developed system of inner membranes. Many taxa show the presence of amyloplasts but the starch content of these is very low (*Alcea rosea*—unpublished data; *Proboscidea louisianica*—Mogensen, 1978; *Valeriana officinalis*—Kamelina and Zhukova, 1987). No starch has been found in the synergids of *Zea mays* (Diboll and Larson, 1966), *Epidendrum scutella* (Cocucci and Jensen, 1969), *Petunia hybrida* (Van Went, 1970a, b), *Hordeum vulgare* (Cass and Jensen, 1970), *Aquilegia formosa* (Vijayaraghavan *et al.*, 1972) and *Agave parryi* (Tilton and Mogensen, 1979). Association of plastids with ER has been reported (*E. scutella*). Plastids are distributed throughout the protoplast. Mitochondria, on the contrary, are predominantly concentrated near the filiform apparatus (FA). The vacuome is represented by a single large vacuole or numerous small vacuoles in the apical part of the synergids. Ribosomes are usually abundant. Dictyosomes are moderately active and are generally located in the FA region. The endoplasmic reticulum (ER) has been reported to be both rough and smooth. Due to the Golgi apparatus and ER activity, the synergid protoplast has an alveolar appearance. Synergids are generally conspicuous by the highest density of hyaloplasm and by accumulation during development of osmiophilic inclusions in vacuoles and between membranes of ER, as well as those of mitochondrial, plastid and nuclear envelopes. In most species, the synergids are partly surrounded by a cell wall which thins down towards the chalazal end of the cell and finally disappears. The apical part of the synergids is bounded only by plasmalemma. Species are known in which the egg apparatus cell walls are completely lacking in a mature embryo sac, the cells being surrounded by plasmalemma alone. (*Potamogeton* species—Wiegand, 1900; *Peperomia blanda* and *Gagea lutea*—Plyushch, 1988b). A structure known as the filiform apparatus (FA) differentiates in the basal portion of synergids of most plants. Schacht (1857) was the first to describe this cell wall structure and introduce the term while studying embryo sacs in *Campanula, Watsonia* and *Zea*. Hofmeister (1857) concurrently described the filiform apparatus in *Crocus*. Jensen (1963-1964) was the first to describe the FA ultrastructure in cotton 100 years later, and Van der Pluijm (1964) followed its development in detail in *Torenia fournieri*. The structure is a system of twisted (tortuous), highly branched cell wall projections at the micropylar pole of the synergid. These wall projections increase the plasma membrane surface, thus imparting to the synergid the functions of a transfer cell effecting a short-distance transport of substances. Species with a poorly developed FA are also known (*Valeriana officinalis*). The FA has been found to be highly variable in ultrastructure and chemical composition (Van Went and Linskens, 1967; Vijayaraghavan *et al.*, 1988). In *Jasione montana*, considerable cutin depositions in the cell wall at the micropylar pole of the synergids have been reported (Berger and Erdelská, 1973). Classifications of FA types have been proposed (Van Went and Linskens, 1967; Cocucci, 1981; Vassilyev and Lobkov, 1990). However, species lacking the filiform apparatus occur. These are *Petunia hybrida* (Van Went, 1970b), *Helianthus annuus* (Newcomb, 1973a), *Calendula officinalis* (Plisko, 1977), *Adoxa moschatellina, Crepis tectorum, Gagea lutea, Lilium regale, L. candidum, Peperomia blanda, Rudbeckia hirta* (Plyushch, 1992) and others. In some species, a specific cell wall thickening has been found at the micropylar region in the form of a layer of mucilaginous substances (polysaccharides or proteins) only discernible under an electron microscope (Vassilyev and Lobkov, 1990).

The structure of synergids is highly unstable. Changes occurring during differentiation and specialization of the egg apparatus cells (increase in size, variation in nucleus shape, protoplast vacuolation, formation of a filiform apparatus and disappearance of the apical portion of the synergid cell wall) are accompanied by the so-called degeneration of synergids. In majority of species investigated degeneration of one of the synergids occurs prior to the arrival of the pollen tube. In some species, both synergids may degenerate, as in some *Triticum* species (Batygina, 1974b), in *Spinacia oleracea* (Wilms, 1981), *Hordeum vulgare* (Mogensen, 1984), *Triticum aestivum* (You and Jensen, 1985), *Brassica campestris* (Van Went and Cresti, 1988a) and *Peperomia blanda* (Plyushch, 1992). In *Galanthus nivalis*, ultrastructural differentiation of the mature embryo sac synergids with respect to number of actin microfilaments has been established. Degeneration occurring following pollination has been observed in a synergid saturated with actin (Bednara *et al.*, 1994). Using a number of species as an example, it has been demonstrated that the synergid into which the pollen tube penetrates is the one that shows signs of degeneration (Jensen and Fisher, 1968a; Schulz and Jensen, 1968b; Cocucci and Jensen, 1969; Cass and Jensen, 1970; Mogensen, 1972, 1978; Maze and Lin, 1975; Mogensen and Suthar, 1979). The time of degeneration is variable: in some species it occurs before penetration of the pollen tube into the synergid, in others after penetration. Most workers believe the process to be triggered by pollination promoting the transformation of one of the synergids (rarely of both) into a receptive one. In *Panicum* sp., however, synergid structural changes similar to apoptosis were found to occur independent of pollination, during pistil maturation. These proceed in both sister synergids, but do so asynchronously, and are accompanied by secretion of calcium from these cells. The calcium secreted by the synergids initially accumulates in the apoplast, between FAs and in the embryo sac wall. Then calcium is transported towards the micropyle and farther along the outer integument of the ovule (Chaubal and Reger, 1994). Synergids are usually short-lived and degenerate shortly after fertilization. Cases of persistence of both or one of the synergids after fertilization and for a long time afterwards are known (Arecaceae, Limnanthaceae, Loranthaceae, Malvaceae, Orchidaceae, Papaveraceae, Parnassiaceae and Valerianaceae), with the persisting synergid increasing in size in some taxa (Berberidaceae, Bixaceae, Liliaceae, and Saxifragaceae). The giant synergid characteristic of *Allium* species is particularly long-lived. In *Allium nutans,* the giant synergid does not disappear until formation of a multicellular embryo (Stozharova, 1973). The nucleus of such a synergid is very large and is a product of endomitosis.

Synergids are multifunctional cells. They play a vital role in the attraction of pollen tubes to embryo sacs, discharge of their contents into them and in ensuring contact and fusion of the discharged male gametes (sperm cells) with female gametes. The mechanisms underlying the processes continue to be the subject of debate. One view holds that the synergids are involved in synthesis and secretion of substances chemotropically active with respect to pollen tubes (Van der Pluijm, 1964; Jensen, 1965c; Van Went and Linskens, 1967; Fougère-Rifot, 1975, 1987a; Plisko, 1977; Plyushch, 1988a, 1992). According to other authors, the attraction of the pollen tubes is a feature of the ovule rather than of the embryo sac. In the Malvaceae species studied, 'dissolution' of the cells in the micropylar region of nucellus has been observed prior to entry of pollen tubes, indicating the latter's route to the embryo sac. Such a transformation of the nucellar cells may start even when the pollen tubes are still in the style (Jensen, 1969; Chandra and Bhatnagar, 1975). There is also evidence suggesting lack of the influence of ovules on pollen tube growth (Stosser, 1986). Electron microscopic studies show that along with a secretory activity, synergids exhibit characteristics of transfer cells, allowing them to be regarded as multifunctional cells. They are involved in promoting directional growth of the embryo sac towards the micropyle (Plyushch, 1992). The synergids (including persistent after fertilization cells) are involved in absorption, storage and transfer of nutrients from the nucellar tissue and/or the integuments toward the embryo sac, playing an important role in the nutrition of the egg cell, the central cell, the zygote, the endosperm and the embryo. Electron microscopic studies have revealed that the pollen tube normally penetrates the embryo sac through the filiform apparatus of the receptive synergid where its growth is arrested (in species lacking FA, the place of entry of the pollen tube has not been established). Formation of synergid embryos has been reported in Anacardiaceae (Alexandrovskyi and Naumova, 1985), Apiaceae (Kordyum, 1967), Betulaceae (Korchagina, 1981), Gentianaceae (Maheshwari, D. and Lakshminarayana, 1977), Hypericaceae (Bugnicourt, 1969), Nyctaginaceae (Kajale, 1938), Poaceae (Batygina and Mametyeva, 1979), Zygophyllaceae (Kapil and Ahluwalia, 1963; Masand, 1963) and Amaryllidaceae (Vorsobina and Solntseva, 1990).

CENTRAL CELL of the Embryo Sac—a cell that derives its name from its location in the embryo sac (Plate 55). Synonyms: endosperm primordium, endosperm mother cell. It generally occupies most of the central portion of the embryo sac, between the egg apparatus and the antipodal cells (or the nucellus). Like the egg cell, this cell fuses with one of a pair of sperm cells delivered by the pollen tube in the embryo sac during double fertilization whereby it is transformed into an endosperm initial giving rise to the seed endosperm.

The central cell is generally larger than (Davidiaceae, Paeoniaceae, Trochodendraceae and members of many other families), rarely equal to or somewhat smaller than (Apocynaceae, Asclepiadaceae, Caprifoliaceae, Cynomoriaceae, Dipsacaceae and Valerianaceae) any other cell of the embryo sac. Its shape varies among taxa. Central cells may be markedly elongate, being many times longer than wide (Magnoliaceae, Malvaceae, Nelumbonaceae and Trochodendraceae), expanded at the micropylar end and tubular at the chalazal pole (Linaceae and Malvaceae), oval (Araliaceae, Gentianaceae, Limnanthaceae and Monimiaceae), oval-elongate (Elaeagnaceae and Rafflesiaceae) and isodiametric (Primulaceae, Ranunculaceae and Valerianaceae). Curved central cells have been observed in embryo sacs of some members of Caryophyllaceae, Chenopodiaceae and Papaveraceae. In members of Alismataceae, such as *Alisma natans*, this cell assumes a U-shaped appearance. During embryo sac maturation and preparation for fertilization, the central cell of many taxa greatly expands. This is for the most part due to a higher rate of its longitudinal growth towards the chalaza in some species (*Veltheimia viridiflora*—Hofmeister, 1859; *Beta vulgaris*—Oksiyuk, 1927; *Marcgravia purpurea*—Swamy, 1948b; *Carallia lancaefolia*—Nikiticheva and Yakovlev, 1985) and in the opposite direction, with the embryo sac protruding into and beyond the micropyle in other species (*Phaseolus vulgaris*—Weinstein, 1926; *Torenia* species—Krishna Iyengar, 1941). In some taxa, branching of the central cell occurs due to its local lateral growth (*Olax stricta*—Agarwal, 1963).

The central cell contains two, less commonly one or a few nuclei, termed polar nuclei. One of the most dramatic structural transformations of this cell during differentiation of the embryo sac cells is the formation of a so-called secondary nucleus of the central cell—a product of fusion of two (occasionally more) polar nuclei (see Polar Nuclei below).

The central cell plastidome in an overwhelming majority of species studied is represented by large leucoplasts (more commonly amyloplasts). Considerable starch reserves prior to fertilization have been observed in the central cell of *Arachis, Tilia, Penstemon, Acacia, Nicotiana* and *Calendula*. Starch is present in plastids of *Gossypium, Capsella, Spinacia* and *Valeriana*, but in lower amounts than in the egg cell. No starch has been found in plastids of the central cell of *Helianthus* (Newcomb, 1973a), *Linum* (Vazart B. and Vazart J., 1966), *Aconitum* (Zhukova and Sokolovskaya, 1977, 1983) and *Alcea* (Zhukova, unpublished). Chloroplasts have been found in the central cell of Brassicaceae (Jensen, 1972; Schulz and Jensen, 1973; Sumner, 1989; Mansfield *et al.*, 1991) and Resedaceae (Chaban and Yakovlev, 1974; Chaban, 1979). The central cell is the only cell of the embryo sac in which the presence of chloroplasts has been reliably established for the aforesaid taxa (Zhukova, 1984). Mitochondria are generally abundant and highly differentiated. The central cell is highly vacuolate; its vacuome is primarily represented by large vacuoles located in the central part of the cell. The bulk of its cytoplasm accumulates near the egg apparatus and the antipodal cells, the remaining part being pushed by the vacuoles to the cell periphery where it forms a thin parietal layer along the lateral wall of the embryo sac. Extensive ER, predominantly RER, is characteristic of the central cell. ER membranes are especially abundant near the egg apparatus. Parallel ER profiles are also numerous near the antipodal cells and along the cell wall projections. Dictyosomes are generally concentrated at the periphery of the cell, close to its wall. Their activity is most likely associated with the biosynthesis of the substances of the rapidly growing cell wall and wall projections (at places where the central cell borders the synergids and the nucellus in *Aquilegia vulgaris*—Fougère-Rifot, 1978). The central cell cytoplasm is rich in ribosomes, both as monosomes and polysomes. An exception is the *Aconitum napellus* embryo sac in whose central cell no free but only membrane-bound ribosomes have been found.

The central cell, like any other cell of the embryo sac, shows distinct polarity. Like the egg apparatus cells, it is only surrounded by a partial cell wall. The cell wall is absent at the micropylar pole of the cell, at the place of contact with the egg apparatus. The central cell wall bordering the antipodal cells contains plasmodesmata and shows wall projections in most species, such as *Stipa elmeri* (Maze and Lin, 1975), *Aquilegia vulgaris* (Rifot, 1973), and *Aconitum napellus* (Zhukova, unpublished). In *Eschscholtzia californica,* cases of complete disappearance of this partial cell wall during the embryo sac maturation have been reported (Fougère-Rifot, 1981). The lateral wall of the central cell is normally thickened where it borders nucellar or integumental cells. It shows well-developed protuberances. In some species (*Crepis capillaris*—Godineau, 1971; *Stellaria media*—Newcomb and Fowke, 1973; *Plantago lanceolata*—Vannereau, 1978; *Spinacia oleracea*—Wilms, 1981) the micropylar part of the lateral wall is studded with protuberances, in others (*Linum usitatissimum*—Vazart, 1971) at the chalazal one. In *Gossypium hirsutum* (Schulz and Jensen, 1977) and *Aquilegia vulgaris* (Fougère-Rifot, 1978), such outgrowths cover most of the lateral wall of the central cell. In *Capsella bursa-pastoris* (Schulz and Jensen, 1971), *Helianthus annuus* (Newcomb and Steeves, 1971) and *Euphorbia helioscopa* (Gori, 1977), finger-like projections of the cell wall occur at both the micropylar and chalazal pole of the central cell. In the mature embryo sac of *Stellaria,* the cell wall labyrinth is present at the micropylar end of the central cell, whereas in the formed embryo sac small outgrowths occur along the entire wall of the central cell where it borders the nucellus (Kudlicka *et al*., 1981). In a number of species (*Plantago lanceolata*—Vannereau, 1978; *Valeriana officinalis*—Kamelina and Zhukova, 1987; and *Nicotiana rustica*—Plyushch, 1992) formation of projections has also been observed at the boundary between the central cell and the synergids. No cell wall projections have been found in the central cell of the mature embryo sac of *Capsella bursa-pastoris.* They form after fertilization, in early embryogenesis (Schulz and Jensen, 1973).

The contact area of the central cell is maximal with the surrounding nucellus and/or integuments. The cell is assumed to play an important role in absorption, transport and secretion of metabolites. It serves as a storage of food reserves, such as starch, lipids (particularly large lipid reserves have been observed in *Sonneratia caseolaris*—Kamelina, 1985d) and possibly proteins. It also secretes some enzymes involved, presumably, in lysis of the surrounding tissues—the nucellus and integuments. Thus high-rate synthesis of constitutional and reserve substances and catabolic processes occur concurrently in this cell. All these functions stem from the principal biological role of the central cell as the endosperm mother cell.

POLAR NUCLEI (Greek *polos* for axis, pole)—nuclei of the central cell (endosperm mother cell) of the embryo sac of flowering plants (Plate 56). The term 'polar nuclei' was proposed by A. Fischer (1880) who distinguished the 'upper polar nucleus' (German *oberer Polkern),* the 'lower polar nucleus' (*unterer Polkern*), and the product of their fusion—the 'central nucleus' (*Centralkern*). Guignard (1881) used the terms polar nuclei and secondary nucleus of the embryo sac[2]. Subsequently these terms gained general acceptance. Most species have two polar nuclei—a micropylar (upper) one and a chalazal (lower) one. They separate from the polar groups of nuclei of the coenocytic embryo sac during its transition to cellularization in the course of formation of the egg apparatus and antipodal cells. In most cases, these nuclei are identical in size and structure, making it difficult, upon their drawing together, to tell the difference between the micropylar and the chalazal nucleus. Where the nuclei are not equal in size, the larger nucleus is normally the micropylar one, occasionally the chalazal one (Fritillaria-type embryo sac development). Based on comparative studies of species with the Polygonum, Oenothera, Allium and Fritillaria type embryo sac development, a hypothesis was advanced on the stimulating role of the chalazal polar nucleus in the second link of double fertilization (Gvaladze, 1976). According to the author, this nucleus promotes preferential development of the endosperm over the embryo early in seed development.

Depending on the type of development of the female gametophyte, the number of the polar nuclei in its central cell may vary, as opposed to the egg cell which normally contains a single nucleus. Occasionally the central cell contains just one—micropylar polar nucleus (e.g. in Oenothera- and Eriostemones-types of embryo sac development; according to Romanov's (1971b) classification). Most commonly, the central cell is diploid, less commonly haploid (see above) or polyploid, depending on type of embryo sac development. Thus, for example, in embryo sacs of Penaea-type development, four nuclei fuse to form a secondary nucleus. In *Peperomia* species (with Peperomia-type embryo sac development), the nuclei present in the central cell (14 in *P. hispidula,* 8 in *P. pellucida,* 4 to 8 in other species) come into contact with one another to form a polyploid complex corresponding to the secondary nucleus. Eventual fusion of these nuclei occurs following contact with the sperm cell. In apomictic embryo sacs, the number of polar nuclei may vary and increase considerably.

In some taxa, in particular those of genus *Allium,* such a character as identity of the polar nuclei in size and structure has also proven variable. In *A. schoenoprasum, A. cepa* and *A. sativum,* the two polar nuclei are indistinguishable from each other in both shape and size. Prior to fertilization, the nuclei come together. At this point in time they have no particular, well-defined location. They can occur both in the micropylar end of the embryo sac in close proximity to the egg apparatus and in the central, less commonly chalazal part of the embryo sac. Shortly before the fertilization, they fuse to form a secondary nucleus of the central cell, or just come into contact with each other. In *A. atroviolaceum* and *A. rotundum,* on the contrary, the polar nuclei differ not only in size but also in condition of their chromatin. The micropylar polar nucleus is 2 to 3 times larger than the chalazal one, contains a few nucleoli and

[2]A more appropriate term would be 'the secondary nucleus of the central cell of the embryo sac' (Ed.).

generally gives a negative reaction after Feulgen staining. The smaller chalazal polar nucleus, unlike the micropylar one, is Feulgen-positive (Gvaladze, 1962, 1966, 1973). A similar situation was reported earlier for members of genus *Gagea* (Romanov, 1961).

The time of fusion of polar nuclei varies among different taxa: they can fuse prior to, during or following fertilization. Thus, for instance, in *Capsella bursa-pastoris* (Schulz and Jensen, 1968b, 1973), *Helianthus annuus* (Newcomb, 1973a) and *Stipa elmeri* (Maze and Lin, 1975), fusion of polar nuclei is completed before fertilization. In cotton (Jensen, (1963) 1964; Schulz and Jensen, 1977) and barley (Mogensen, 1982) fusion starts before pollination and is not completed until the sperm nucleus approaches them. In spinach, polar nuclei fuse during fertilization (Wilms, 1981). Such plants are in the minority among the flowering plants studied. In some species, this character is relatively unstable. In wheat, for example, polar nuclei normally come close together but do not fuse before fertilization. Fusion commences after one of them comes into contact with the sperm nucleus. There is reason to believe that in some ovules fusion of polar nuclei may occur even before fertilization. Upon fusion of polar nuclei in wheat, fragments of cytoplasm entrapped by them during their drawing together are often discernible (Batygina, 1974b). A similar phenomenon has been observed during fusion of polar nuclei in *Impatiens glandulifera* (Steffen, 1951) and *Arachis hypogaea* (Gerassimova-Navashina, 1954). The presence of cytoplasmic 'entrapments' in *Triticum aestivum* may indicate that fusion of polar nuclei has already commenced but is not yet completed, since upon complete fusion the entrapped cytoplasmic fragments are no longer visible. Electron microscopic studies in grasses have provided support for these observations (Mogensen, 1982). Jensen ((1963) 1964) was the first to describe the fusion of polar nuclei ultrastructurally in cotton. Fusion of nuclei starts with formation of evaginations by their nuclear envelopes and subsequent contact between them at several points along the region of drawing together of the nuclei. Initially the outer nuclear membranes come into contact and fuse, followed by contact and fusion of the inner membranes of the nuclear envelopes. This results in formation of a number of narrow bridges between the two nuclei; these gradually enlarge and coalesce. Small portions of cytoplasm become trapped between these bridges. The two nuclei and their nucleoplasm become continuous and nuclear fusion is thus completed. Studies of fusion of polar nuclei in *Adoxa moschatellina* indicate interruptions in evaginations of the envelopes of fusing polar nuclei, whereby integrity of the envelopes is ensured since the nuclear membranes fuse at the places of these discontinuities (Bannikova and Plyushch, 1990). Nucleoli fusion follows nuclear one, indicating completion of fusion at the level of chromatin. In polar nuclei of *T. aestivum*, shortly after their fusion, a number of additional nucleoli (as many as 20 or more) often appear. These differ dramatically in size and colour from the nucleoli released by the sperm cell. Similar nucleoli occasionally occur in unfertilized polar nuclei (Batygina, 1974b). No fusion of polar nuclei takes place in mature embryo sacs of members of such families as Alangiaceae, Coriariaceae, Gentianaceae, Loasaceae, Nitrariaceae, Rhamnaceae and others.

Polar nuclei and the secondary nucleus of the central cell are the largest nuclei of the embryo sac. The haploid nucleus of the Onagraceae central cell reaching 25 μm in diameter is no exception. Normally, polar nuclei contain a small amount of heterochromatin and a large nucleolus showing a few nucleolar vacuoles. The nuclear envelope is deeply invaginated and in many places connected to rugose ER cisternae; the nucleus is lobate. In some species, the secondary nucleus (or contacting polar nuclei) is located near the egg cell and is separated from the chalazal pole of the cell by a large vacuole. In other species, it lies in the central portion of the embryo sac and is connected to the egg apparatus by a well-defined strand of cytoplasm. Siting of the secondary nucleus at the chalazal end is much less common and is typical of plants characterized by the Helobial type of the endosperm development. Cases of a similar disposition of the secondary nucleus are also known in species with a cellular endosperm.

Antipodal Cells (Greek *anti* for opposite and *pous* for foot)—a group of cells (usually three) comprising the chalazal, occasionally the chalazal-lateral portion of a mature embryo sac (Plate 57). Hofmeister (1849) first characterized these cells while describing the most typical embryo sac organization and termed them *'gestreckten Zellen'*. In subsequent articles, he defined them as *'Gegenfussler'*, *'Gegenfusslerzellen'* and *'Gegenfusslerinnen'*. The term 'antipodal cells (antipodals)' introduced into botanical literature is a translation of the German term *'Gegenfussler der Keimblaschen'* into English (Henfrey, 1853). The presence of three antipodals is characteristic of Polygonum, Allium, Adoxa and some other types of embryo sac development and observed in the overwhelming majority (over 80%) of the flowering plant species studied embryo-logically. Embryo sacs of the Drusa-type of development have 8 antipodal cells while those of the Plumbagella-type show only a single (triploid) antipodal cell. In some taxa, cytokinesis does not take place during formation of the antipodal complex, the chalazal pole of the embryo sac retaining the free-nuclear state. Under the Oenothera-type embryo sac development, no antipodals are formed at all.

Antipodal cells are highly variable in structure and longevity, with variation observed both at the family (Amaryllidaceae, Arecaceae, Gentianaceae, Ranunculaceae and others) and species level (*Parnassia* species and others). Antipodal cell polymorphism is due to the existence of more than one type of embryo sac development as well as to variation among taxa of the entire system of interactions of the ovule components.

Cell arrangement in a three-celled antipodal complex may be T-shaped, linear (parallel or transverse to the chalazal-micropylar axis), botryoidal, etc. Antipodal cells also show variation in shape, observable even within a single antipodal complex. In some species, the antipodals are morphologically similar to the egg apparatus, as in *Allium paniculatum* (Modilewski, 1928a), *Rudbeckia* species (Maheshwari and Srinivasan, 1944; Battaglia, 1947; Movsesyan, 1968; Solntseva, 1971, 1987), in some members of Asclepiadaceae (Davis, 1966) and *Sambucus* species (Tuchina, 1985). Using *Scilla sibirica* as an example, such a structure of antipodal cells has been shown to persist after the fertilization (Gerassimova-Navashina and Batygina, 1958).

Most commonly, antipodal cells are ephemeral structures. According to many authors, their disintegration occurs as early as during preparation of the embryo sac for fertilization or shortly after fertilization. At the same time, in many taxa of flowering plants, antipodal cells have been found to be capable of not only increasing in size and/or proliferating into a multicellular tissue, but also of persisting for a long time during seed development. Among such taxa are many members of Arecaceae, Berberidaceae, Nelumbonaceae, Papaveraceae, Poaceae, Ranunculaceae, Rubiaceae and others. Thus in *Sasa paniculata* (Poaceae), the number of antipodal cells increases up to 300 (Yamaura, 1933). However, exceptions are known in this very family: as few as three antipodal cells are present in the embryo sac of *Echinochloa frumentacea* and *Panicum miliare* (Anton and Cocucci, 1984). Cell arrangement in the multicellular antipodal complex of members of Nelumbonaceae and Gentianaceae is quite unique. In many representatives of the Alangiaceae, Asteraceae, Betulaceae, Morinaceae and Poaceae, division of each of the initial three antipodal cells results in formation of embryo-like structures. The genesis of these has been followed in detail in 4 species of *Stipa*. It has been demonstrated that the mitotic activity of the embryo-like antipodal structures gradually declines, the identity of the three cellular groups is lost and the cellular complex extends along the embryo sac wall and progressively degenerates (Yakovlev and Solntseva, 1965). A case of formation of an antipodal embryo reaching maturity, however, has been reported (*Rudbeckia laciniata*—Solntseva, 1987). In many taxa, expansion and/or proliferation of the antipodal cells continues into the post-fertilization period. Thus, in *Aconitum napellus,* the three expanded antipodals may persist until the mature seed stage (Osterwalder, 1898). Case of a similar antipodal cell vitality have been reported for Cynomoriaceae (Teryokhin *et al.*, 1975).

Antipodal cells are normally uninucleate, with an initially haploid (in species with Fritillaria- and Plumbagella-types of embryo sac development, triploid) chromosome set. However, in many flowering plants the number of antipodal nuclei may increase up to 60 during embryo sac differentiation. Evidence from electron microscopic studies suggests that in some *Linum* species the inner cell walls in the antipodal complex are incomplete (Vazart J., 1968). In *Oryza sativa,* such cell wall formation during amitotic nuclear divisions occurs by 'free' growth (Dong and Yang, 1989). Antipodal nuclei of some species undergo endoreduplication; giant endopolyploid nuclei and polytene chromosomes have been reported (Tschermak-Woess, 1956; Turala, 1966; Petrova *et al.*, 1985; Ishioka and Sato, 1990). An increase in nuclear ploidy also occurs in multinucleate antipodal cells due to fusion of individual nuclei, resulting in a secondary reduction of their number and formation of large nuclei of irregular shape. Examples are known of antipodal complexes varying in ploidy of their nuclei (Sato, 1990 and others). In some maize varieties, by contrast, persisting antipodals constitute a massive haploid tissue (Randolph, 1936).

Antipodals of a mature embryo sac exhibit a well-defined morphological polarity and are extremely rich in organelles, such as primarily plastids (including amyloplasts with small starch grains) and mitochondria, as well as ribosomes and rough ER. No chloroplasts have been found in antipodal cells. Polarity with respect to the micropylar-chalazal axis of the embryo sac is established due to differences in composition and structure of the cellular organelles, in cytoplasm vacuolation and cell wall structure (Zhukova and Sokolovskaya, 1977). Along with these features, common to antipodals of nearly all the objects studied and differentiating them from other embryo sac components, some differences have been found. Almost all the species studied exhibit rough ER: some as extensive ER in the form of ergastoplasm or concentric circles around small vacuoles, as in *Linum usitatissimum, Helianthus annuus, Aquilegia vulgaris, Stipa elmeri* and *Aconitum napellus,* others as long ER profiles located near the plasma membrane, as in *Epidendrum scutella, Linum catharticum* and *Zea mays.* There is also an appreciable polymorphism in the structure of antipodal plastids. Antipodal cells, as opposed to egg apparatus cells, are completely surrounded by a cell wall (there are, however, some exceptions). The antipodals are organized like transfer cells. Depending on the region of the antipodal cell, the antipodal cell walls show high variability in thickness, electron-optical density and chemical composition as well as in topography, size and shape of cell wall protuberances and presence or lack of plasmodesmata. In some species, antipodals are capable of forming haustoria (Rubiaceae—Maheshwari, 1954; Poddubnaya-Arnoldi, 1976; Aristolochiaceae-Teryokhin and Anisimova, 1981; Asteraceae—Solntseva, 1987 and *Sedum fabaria*—Nikiticheva, 1985a).

Antipodal cells are multifunctional structures. Their principal function is selective absorption and directional translocation of nutrients and metabolites from the maternal organism via the chalazal region of the ovule to the embryo sac. Apparently, due to this fact they are generally the first to differentiate in the embryo sac and in some taxa, are superior to the egg apparatus cells in size (or in number). Antipodals serve as centres of both nutrient absorption and transformation. The structure of the nucleus (and in particular its nucleolus), abundance of ribosomes, extensive rugose ER and its associations with vacuoles in the antipodals, suggest predominance of processes concerned with protein metabolism. It is not inconceivable that the antipodals are also involved in secretion of metabolites. The antipodals of many species are capable of storing reserve proteins and lipids taken up during post-fertilization development. This attests to their function of storing nutrients for the developing endosperm and embryo. Starch, however, is rarely found in antipodal plastids; it has been reported in *Epidendrum scutella, Linum usitatissimum, Ornithogalum caudatum, Valeriana officinalis* and *Zea mays*. A notable feature of antipodal cells of the flowering plants is that they combine the traits of specialized and meristematic cells (Zhukova, 1984).

Fine Structure of Female Gametophyte Development (Plate 58): Development of the megagametophyte (= female gametophyte or in angiosperms, embryo sac) is divided into two connected developmental phases: *megasporogenesis*, which entails formation and maturation of the initial products of meiosis, followed by *megagametogenesis*, which begins with mitotic division of meiotic products and continues through cellularization and maturation of the megagametophyte. Although this review provides an overview of the process of embryo sac development, specialized reviews on the fine structure of the female gametophyte are available elsewhere (Rodkiewicz, 1978; Kapil and Bhatnagar, 1981; Willemse and Van Went, 1984). Despite the number of published works in this field, not all of the myriad types of embryo sac development (Maheshwari, 1950) have been described at the ultrastructural level.

Megasporogenesis. The process of megasporogenesis is initiated in the paternal sporophytic tissue with formation of one or multiple archespores. These cells are characterized by larger size, prominent nucleoli and small vacuoles; ultrastructurally, this is an unpolarized cell with abundant and poorly differentiated organelles (Eymé, 1965; Kapil and Bhatnagar, 1981). A single archesporial cell will become conspicuous. The archesporial cell either divides mitotically before meiosis or directly serves as a megasporocyte (= megaspore mother cell) (Maheshwari, 1950). The megasporocyte is defined by its entry into the meiotic pathway. In the event that the archesporial cell divides, the inner (chalazal-facing) cell becomes the megasporocyte, whereas the outer (micropylar), parietal cell forms additional layers of vegetative tissue through continued mitotic divisions.

The megasporocyte undergoes dramatic cytoplasmic changes accompanying meiosis and transition between sporophyte and gametophyte generations. The fine structure of the megasporocyte is similar to that of meristematic cells at the initiation of meiosis (Eymé, 1965; Dickinson and Heslop-Harrison, 1977; Kapil and Bhatnagar, 1981; Schulz and Jensen, 1981). Mitochondria and plastids are not highly differentiated at this stage. Ribosomes are abundant, as well as ER. Polarity of the cell is variably expressed; although in *Myosurus* all organelles are reportedly present in future cytoplasm of the functional megaspore (Woodcock and Bell, 1968), in most plants a weaker degree of polarization occurs (Rodkiewicz, 1978; Russell, 1979; Medina *et al.*, 1981; Schulz and Jensen, 1981, 1986). The cell wall of the megasporocyte is somewhat thicker than that of surrounding cells and in many cases appears specialized (Kapil and Bhatnagar, 1981). If the plant forms a megagametophyte from only one of the nuclei produced by meiosis (= monosporic development) or two of the products of meiosis (= bisporic development), callose (a β-1,3 polyglucan) is expected to be present in addition to the normal cell wall constituents. This may be visualized using 0.005% decolorized aniline blue in 0.1 M KH_2PO_4, pH 8.2. Callose typically accumulates specifically at the abortive pole of the megasporocyte in monosporic and bisporic development (Rodkiewicz, 1970) even when polarity of megasporocyte is the opposite to normal (Polygonum) type development, as in *Oenothera* (Jalouzot, 1978). However, in cases in which all four of the nuclei generated during meiosis form the female gametophyte (= tetrasporic development), callose is absent from the cell wall (Rodkiewicz, 1970). Plasmodesmata typically occur only in areas of the cell wall that do not accumulate callose. In tetrasporic embryo sacs, they occur principally in the chalazal cell wall and otherwise in the cell wall of the future functional megaspore.

During prophase I, mitochondria and proplastids undergo dramatic structural changes and dedifferentiate, becoming remarkably similar in appearance, which led early cytologists to question whether heritable organelles persisted throughout meiosis or were regenerated from other organelles or the nuclear envelope (Dickinson and Heslop-Harrison, 1977). Ultrastructural and biochemical studies have confirmed that DNA-containing organelles remain present but are highly modified during prophase. Ribosome density is also reduced at this stage (Rodkiewicz and Mikulska, 1964, 1965a; Dickinson and Heslop-Harrison, 1977; De Boer-De Jeu, 1978; Rodkiewicz, 1978; Willemse and Franssen-Verheijen, 1978; Russell, 1979; Schulz and Jensen, 1981; Willemse and Van Went, 1984). Autophagic vacuoles generated from the endoplasmic reticulum and pre-existing vacuoles are frequently found in the cytoplasm and occasionally contain membranes or organelle-like objects (Russell, 1979; Schulz and Jensen, 1981). Two relatively non-specific reclamation enzymes have been localized in these vacuoles—acid phosphatase (Willemse and Bednara, 1979; Schulz and Jensen, 1981) and esterase (Willemse and Bednara, 1979), suggesting that selected cellular materials are degraded during this phase and broken down into their precursor components.

Dedifferentiation of the meiotic cytoplasm reaches its most extreme level at the middle of meiosis I; this reflects the same trends observed during microsporogenesis (= male formation of meiotic products; see Anther, Part II) and appear to be specific to the same phases of meiosis (Dickinson and Heslop-Harrison, 1977). In male sporogenesis, decreases in level of nucleic acids and proteins closely correlate with general increases in hydrolytic enzyme levels. Ultrastructural data confirm that similar changes are likely in the megasporocyte during meiosis (Dickinson and Heslop-Harrison, 1977). Different species of angiosperms appear to express these characteristics to variable degrees. The most extreme case appears to be the developing female gametophyte of *Lilium* (Dickinson and Heslop-Harrison, 1977; Rodkiewicz and Mikulska, 1964, 1965a, b), in which areas of concentric lamellar ER appear to be retained as reservoirs of protected cytoplasm sequestered from the general hydrolytic activity present in the cytoplasm. In plants with monosporic embryo sacs, however, autophagic activity appears to be restricted to specific compartments, fewer cytoplasmic reservoirs occur and elimination of ribosomes is only partial (De Boer-De Jeu, 1978; Jalouzot, 1978; Willemse and Franssen-Verheijen, 1978; Russell, 1979; Willemse and Bednara, 1979; Medina *et al.*, 1981; Schulz and Jensen, 1981).

Products from the first cycle of division during meiosis (= meiosis I) are termed dyad cells and those completing both cycles (= meiosis I and II) megaspores. In plants with either monosporic or bisporic female gametophyte development, abortion of either dyad cells or megaspores cells is apparently genetically preprogrammed. Non-functional megaspores and dyad cells are often characterized by slight impoverishment of plastids, whereas dictyosomes, RER and vacuoles may be more frequent than in the functional cell (Willemse, 1981). Abortive cells are frequently invested in a conspicuous callose-rich cell wall (Rodkiewicz, 1970; Russell, 1979; Schulz and Jensen, 1986) and lack plasmodesmata (De Boer-De Jeu, 1978; Jalouzot, 1978; Russell, 1979; Schulz and Jensen, 1981, 1986; Willemse, 1981). Callose is evident in both the side and crosswalls of the abortive megaspores. Dense areas of hydrolytic activity appear in the degenerating cells, including peroxidases and esterases (Willemse and Bednara, 1979). Acid and alkaline phosphatases are not restricted to these cells and are frequently observed in the functional cell as well (Willemse and Bednara, 1979; Schulz and Jensen, 1986). Degenerating cells are initially similar to the megasporocyte but eventually the breakdown of materials becomes general within the cell as vital intracellular membrane systems are broken down (Cass *et al.*, 1985; Schulz and Jensen, 1986). Eventually the plasma membrane is lost and lytic activity becomes general (Russell, 1979; Cass *et al.*, 1985; Schulz and Jensen, 1986). Presumably the degraded products of these abortive cells are ultimately resorbed, as are those of the surrounding cells of the nucellus (Russell, 1979; Schulz and Jensen, 1986). Unlike the breakdown of nucellar cells, however, the degenerate megaspores and dyad cells abort rapidly prior to encroachment of the developing megagametophyte (Rodkiewicz, 1978).

The functional megaspore (in all modes of development) is characterized by the redifferentiation of the mitochondria and plastids and the ultrastructural appearance of increased synthetic activity, including the increased accumulation of cytoplasmic ribosomes, multiplication of organelles and cellular enlargement (Eymé, 1965; De Boer-De Jeu, 1978; Russell, 1979; Schulz and Jensen, 1986). Storage products may also reappear at this stage (Willemse and Franssen-Verheijen, 1978), including lipids and starch grains (Willemse and Bednara, 1979). In *Capsella* small microvillus-like projections of the plasma membrane become conspicuous and may facilitate resorption of material from abortive megaspores (Schulz and Jensen, 1986). In the cell wall, small inclusions are conspicuous in callosic cell walls at this stage (Russell, 1979; Schulz and Jensen, 1986).

Megagametogenesis. Completion of megasporogenesis is marked by mitotic division of the megaspore nucleus or nuclei. In monosporic development, the functional megaspore has a single nucleus, whereas in bisporic and tetrasporic development, the megaspore contains two and four megaspore nuclei respectively. These different nuclear complements imply that monosporic megaspores require three cycles of mitotic division, bisporic megaspores require two cycles and tetrasporic megaspores may require as few as one cycle to reach their mature nuclear complement (Maheshwari, 1950).

In monosporic and bisporic types of megagametophyte development, a significant percentage of the megasporocyte cytoplasm will be apportioned into abortive cells: in monosporic development 67 to 75% of the megasporocyte may abort and in bisporic development, half of the meiocyte may abort. Since tetrasporic megagametophytes retain their entire cytoplasmic volume as they enter megagametogenesis, they appear precocious in terms of size, degree of enlargement and cytoplasmic maturation (De Boer-De Jeu, 1978).

Cytoplasmic organization entails several prominent changes during female gametogenesis, including formation of the central vacuole, replication of heritable organelles, one or more cycles of mitotic division and ultimately, cellularization of the megagametophyte (Kapil and Bhatnagar, 1981; Willemse, 1981). Central vacuole formation typically involves production of small vacuoles, de novo or from ER, that coalesce to produce the prominent vacuole characteristic of the central cell at maturity (Russell, 1979; Kapil and Bhatnagar, 1981; Schulz and Jensen, 1986). In *Zea* (Russell, 1979), mitochondria undergo dedifferentiation and redifferentiation during the four-nucleate megagametophyte stage; this coincides with completion of intense lytic activity in the central vacuole and transition from vegetative to reproductive differentiation. Callosic walls are evident in the walls of abortive megaspores and may remain evident through the two-nucleate megagametophyte.

Free nuclear division (i.e., formation of nuclei through mitotic division without intervening cell wall formation) continues to occur until the mature complement of nuclei is achieved. In most cases, the mature embryo sac contains eight nuclei; this is changed in many cases by the occurrence of 'strike' (= non-division of cell is expected to divide according to typical plants) or subsequent multiplication of cells (Maheshwari, 1950). The origin of the two synergids from the singlemost micropylar nucleus of the four-nucleate megagametophyte and the sisterhood of the egg and polar nucleus have been reported in numerous light microscopic studies (Maheshwari, 1950) and confirmed by electron microscopy (Cass *et al.*, 1985, 1986). Cellularization of the megagametophyte in the micropylar region is initiated through formation of separate phragmoplasts between the two pairs of sister nuclei (Cass *et al.*, 1985, 1986). The cell wall originating between the two synergids forms a common synergid partition, while the cell plate arising between the egg and the central cell forms the chalazal wall of all the cells the egg apparatus (= egg cell and two synergids) and then branches towards the micropyle to complete separation of the egg and synergids. The cells of the egg apparatus subsequently expand, causing expansion of the beaded cell walls to be locally discontinuous at the chalazal end of the cells (Cass *et al.*, 1985, 1986)—a structural modification that later facilitates sperm deposition and gametic fusion in some angiosperms (Kapil and Bhatnagar, 1981). Development of the so-called 'hooks' of the synergids, ubiquitous in angiosperm megagametophytes, is the result of chalazal expansion of the synergids relative to their initial attachment point (Cass *et al.*, 1986). Cells of the egg apparatus develop the cytoplasmic organization appropriate to their cell type during final maturation. Polar nuclei (= central cell nuclei) migrate to a common location where they contact each other, partially fuse, or completely fuse, depending on the type of megagametophyte development expressed (Maheshwari, 1950). Antipodals mature, may subsequently persist, divide further or degenerate, also depending on the species (Maheshwari, 1950).

Cellularization of Free-Nuclear Female Gametophyte in Angiosperms (Plate 59): An initial free-nuclear stage invariably occurs during development of the female gametophyte in all monosporic, bisporic and tetrasporic types of embryo sacs. The process of cellularization in the free-nuclear female gametophyte delimits the female gamete or the egg cell. One of the least studied aspects of megagametophyte development in angiosperms is the organization and differentiation of egg cell and other related components of the embryo sac. Although angiosperm embryology has been under investigation for almost a century, still details on wall formation in the free-nuclear megagametophyte are meagre. Only recently have some publications dealing with this aspect contained information on the formation and extension of the walls and cellularization (Cass *et al.*, 1986; Bhandari and Chitralekha, 1989). Simple reasons for this lacuna in our knowledge have been (1) lack of sophisticated techniques to resolve the initially formed thin walls and (2) the rapid rate at which cellularization takes place. Increased use of electron microscopy, thin and semithin sectioning of resin-embedded material, and publication of a few articles (Newcomb, 1973; Kapil and Bhatnagar, 1981) highlighting the lack of knowledge on cellularization revived interest in this particular aspect. In the last decade detailed information on the process has been provided by Cass *et al.* (1985, 1986) for *Hordeum vulgare*, Bhandari and Chitralekha (1989) for *Ranunculus sceleratus*, and Folsom and Cass (1990) for *Glycine max*. However, the information is still too meagre and further investigations in many more species are desirable before a general account on the precise mechanism of wall formation and subsequent cellularization can be presented. Hence in this article we merely elaborate the salient features of different mechanisms of cellularization of the megagametophyte proposed to date.

The process of female gametophyte cellularization has been studied primarily in gametophytes which develop into eight-nucleate seven-celled embryo sacs. In such cases the eight-nucleate stage is preceded by the four-nucleate one in which two nuclei are disposed at each of the two (micropylar and chalazal) poles of the megagametophyte. The four nuclei divide mitotically and form eight nuclei, four at each pole. Walls form between the four nuclei at each pole resulting in an embryo sac with two groups of three uninucleate cells at either pole separated by a large central cell containing two polar nuclei. The three cells at the micropylar end comprise the egg apparatus while the three at the chalazal end give rise to three antipodals. Bhandari and Chitralekha (1989) consider that (1) the precise positioning of the nuclei and (2) the plane of nuclear division in the four-nucleate megagametophyte are prerequisite to the actual act of cellularization. At the narrow chalazal region the two nuclei are positioned one above the other along an oblique axis; in the broad micropylar region the two nuclei are juxtapositioned in a horizontal plane. However, a chalazal movement of one of the two micropylar nuclei has been reported by Folsom and Cass (1989). In *Ranunculus sceleratus* the plane of division of the two nuclei in each set is perpendicular in the set but oblique to the long axis of the embryo sac. This results in positioning of the daughter nuclei in a distinct 1+2+1 arrangement in the chalazal region (Bhandari and Chitralekha, 1989). In the broad micropylar region, the arrangement is 2+1+1 but appears to be 3+1 due to the distribution of nuclei within the thin layer of cytoplasm here.

Wall formation at the two poles of the gametophyte is simultaneous. According to Bhandari and Chitralekha (1989) and Folsom and Cass (1990), the walls arise as cell plates across phragmoplasts during the last stage of division of the four nuclei. A few early reports also mention involvement of cell plates in cellularization of the megagametophyte (Cass *et al.*, 1986; Bhandari and Chitralekha, 1989; Folsom and Cass, 1990). Contrarily Cass *et al.* (1985, 1986) opine that wall formation takes place in the eight-nucleate embryo sac independent of and a little after nuclear division. These cell plates are thin and disc-like.

Formation of three distinct cell plates at each pole of the megagametophyte has been reported by Cooper (1935, 1941), Rees-Leonard (1935), Smith (1943) and Bhandari and Chitralekha (1989). Folsom and Cass (1990) after studying the position and orientation of cell walls in a cellularized embryo sac, also concluded that three cell plates must be involved in cellularization. Of the three cell plates, two are formed across primary phragmoplasts following nuclear division. One cell plate (CPI) is formed perpendicular to the long axis of the embryo sac. After extension it fuses with the embryo sac wall along its circumference and cuts off the polar nucleus from the other (three) nuclei. The second cell plate (CPII) between the second pair of sister nuclei is laid oblique to the long axis of the embryo sac; it then actually curves around one of the nuclei enclosing it to separate from the two non-sister nuclei. The third cell plate (CPIII, smaller than the other two), arising slightly later is parallel to the long axis of the gametophyte and positioned between the two non-sister nuclei. On extension, this plate comes into contact with and joins the oblique cell plate on one side, with cell plate I on the side facing the central vacuole and with the megagametophyte wall along the rest of the margin. Cooper (1935) and Smith (1943) considered that this cell plate (CPIII) arises across the spindle persisting from the nuclear division preceding formation of the four-nucleate embryo sac—an observation awaiting confirmation from ultrastructural studies. Interestingly, the two polar nuclei at the two poles are cut off diametrically opposite to each other.

According to Cass *et al.* (1986), initially only two walls are formed. The one formed between the two synergid nuclei is vertical and the other formed between the micropylar polar and egg nucleus horizontal. This horizontal wall branches into two: one branch continues to extend horizontally and separates the polar nucleus from the egg apparatus nuclei while the other branch grows vertically towards the embryo sac wall between the egg and synergid nuclei and fuses with the vertical wall between the synergid nuclei, which separates the egg cell from the synergids. One of the two vertical walls is highly curved.

In ultrastructural studies these cell plates look beaded and apparently are formed and extended by fusion of dictyosome-derived vesicles. Distinct microtubules are associated with the vesicles at the growing ends of the walls (Cass *et al.*, 1986). We (Bhandari and Chitralekha, unpublished) were unable to find any active dictyosomes or dictyosome-derived vesicles or microtubules at the extending ends. On the other hand, we found a large number of RER cisternae next to the cell plates. Therefore, we believe that these cisternae have an important role in the extension of the plates. The cell plates exhibit an electron-dense central core (middle lamella) surrounded by an electron-lucent periphery (primary cell wall).

From the three cells formed at the micropylar region of the embryo sac, one cell differentiates into the female gamete or egg cell and the other two into synergid cells. During cellularization and just after inception these cells are structurally very similar; neither structure nor position differentiate the egg cell from the synergids. Differences between the cells manifest themselves subsequently. However, from the position of the nuclei and spindles, most earlier workers and recently Cass *et al.* (1986) and

Folsom and Cass (1990) agree with the concept proposed by Porsch (1907) that the egg and micropylar polar nuclei are sister nuclei as are the nuclei of the two synergid cells. We found no evidence in *R. sceleratus* to substantiate such a concept.

Living Angiosperm Embryo Sac (Plate 60): Unlike the male gametophyte, the embryo sac is relatively inaccessible to structural and functional study. However, living embryo sacs are interesting and important for several reasons. Firstly, they are important for verification and precision of observations obtained from fixed material. The temporal sequence of events occurring in embryo sacs before, during and after fertilization can be characterized by the investigation of living embryo sacs. In addition, it is possible to recognize an exactly defined developmental phase of living embryo sacs for immediate fixation of objects for light and electron microscopy (Sidorova, 1985). Isolated embryo sacs are also used in cytoskeleton studies (Bednara *et al.*, 1990).

Biotechnological manipulations, in particular controlled fertilization, microinjections of genetically transformed sperm cells or foreign DNA, induction of different types of apomixis etc. have been using living embryo sacs for the past decade (Huang and Russell, 1989; Cass, 1990; Kranz *et al.*, 1991).

The models and methods of cultivation or preparation selected play an important role in the study of living embryo sacs. In choosing a suitable plant species, attention should focus on those ovules in which natural characteristics enable observation without isolation of the embryo sac from the ovule.

Original characteristics of ovules enabling direct observation of the living embryo sac were utilized in early reports. For example, an embryo sac growing out from the ovule is characteristic of genus *Torenia* as well as several species from families Loranthaceae, Santalaceae and some others. *Torenia fournieri* was recently used in a study of fertilization in vitro (Keijzer *et al.*, 1988).

Small and tenuinucellate ovules without integument or with only one transparent integument represent suitable objects for investigation of living embryo sacs too. *Monotropa hypopytis,* whose embryo sac had already been observed in the XIXth century by Strassburger, belongs to the classical objects from this group of plant species. The structural and functional characterization of different elements of embryo sacs of some orchids (*Calanthe veitchii, Dendrobium nobile* and *Cypripedium insigne*) before and after fertilization was accomplished by Poddubnaya-Arnoldi (1958, 1960). Her successful and admirable work encouraged research of living embryo sacs using another species with new methods of observation. Small, tenuinucellate and unitegmic ovules of *Jasione montana* were used for microcinematographical study. The transparent integument of *Jasione montana* ovules enabled observation of nucleolar vacuole pulsation in the egg cell, secondary nucleus and also in different cells of the embryo and endosperm after fertilization (Erdelská, 1973). The contraction cycle ranged from 7 minutes to several hours depending on the volume of arising vacuoles as well as on the total cell activity and duration of the mitotic cycle. The microcinematographical analysis also included direct measurements of the cell cycle duration in early endosperm development beginning with fusion of sperm with the secondary nucleus of the embryo sac. The duration of the cell cycle in the early cellular endosperm of *Jasione montana* was 5-8 hours at 25°C. Formation of the cell wall lasted 20-40 minutes in similar conditions. Time parameters of the linear embryo development indicated the shortest mitotic cycle of the basal cell of two- and four-celled linear embryos compared with the apical one. The apical cell of the linear tetrad had the longest mitotic cycle. Observation of living embryo sacs enabled temporal and spatial relations between early embryo and endosperm development of *Jasione montana* to be precisely determined.

Only ovules with larger embryo sacs and solid embryo sac wall are suitable for mechanical isolation. Embryo sacs are isolated under a microscope using entomological pins or a micromanipulator.

Some processes associated with fertilization and postfertilization development were observed in partly isolated embryo sacs of snowdrop (*Galanthus nivalis* from family Amaryllidaceae). Movement speed of sperm nucleus in the central cell of the embryo sac is about 3 μm min^{-1}. Formation and growth of the male nucleolus in the secondary nucleus during triple fusion take about 150 minutes to complete and fusion of mature nucleoli about 10-15 minutes at 25°C. This object also enabled investigation of early endosperm formation in species with a nuclear-type endosperm, as well as growth and first division of the zygote (Erdelská, 1983).

Development of techniques for enzymatic isolation of both fixed and living embryo sacs represents significant progress in embryo sac investigation. Ovules are macerated in the mixture of enzymes (pectinases, cellulases, snailase, driselase, etc.) degrading cell wall components. Species with smaller ovules and smaller embryo sacs better tolerate maceration procedures, repeated washing and centrifugation. Embryologists Enaleeva, Tyrnov and Khokhlov (1972) were among the first to succeed in developing the technique of enzymatic isolation of embryo sacs. Subsequently Chinese scientists, namely Zhou and Yang (1985), improved this technique and succeeded with embryo sacs of *Antirrhinum majus, Helianthus annuus, Nicotiana tabacum* and other species. The technique of enzymatic separation permits isolation of embryo sacs in sufficient quantity for direct observation, cultivation, gamete isolation and other manipulations. Laboratories successful in isolation of embryo sacs have recently increased in number. The number of species used for isolation has likewise increased (Theunis *et al.*, 1991).

Media improving the transparency of integuments (such as mineral oils, silicone fluids, etc.) are suitable for observation of an embryo sac inside ovules. Media and conditions used for cultivation of isolated cells or protoplasts in vitro are the most suitable for the enzymatically isolated embryo sacs.

Future studies of living embryo sacs will no doubt include long-term cultivation of all embryo sac development phases. Such isolations are also preconditions for successful isolation of individual cells or protoplasts.

TYPES OF EMBRYO SAC DEVELOPMENT

Principles of Embryo Sac Classification

The concept of type of embryo sac development is defined by three features: (1) number of megaspores forming an embryo sac, (2) number of mitotic cycles taking place after meiotic divisions, and (3) nuclear behaviour which determined peculiarities of the construction (structure of an embryo sac). Nuclear behaviour refers foremost to the definitive distribution of nuclei (4 + 0, 1 + 3, 2 + 2, etc.)[3] and also the facts of

[3]We mean nuclei distribution among the embryo sac poles: the first figure indicates the number of nuclei at the micropylar pole, the second, at the chalazal. While characterizing the types we use such terms as uni-, bi-, tetra- and polypolar embryo sac instead of one-, two-, four- and multipolar embryo sac (Romanov, 1981).

secondary redistribution and fusion of chalazal nuclei, which change the final result of the definitive distribution. It is irrational to consider the depression of the chalazal end of an embryo sac a feature of a development type since it is usually changeable in ontogenetic development. It determines the appearance of reduced forms within different types. The general number of nuclei in a mature embryo sac should also not be considered a feature of a type, as it varies according to chalazal depression intensity.

The names of different types of embryo sacs repeat the names of those taxa (mainly genera) in which these types were initially discovered or described in the most exact way. There are no standard principles of nomenclature of embryo sac development type, but 13 types can be recognized in flowering plants, characterized by sexual reproduction, and studied from the point of view of embryology. Each type mentioned represents a quite stable embryo sac ontogenesis, typical for the given taxon. As ontogenetic deviations for a given species some types of embryo sacs have been found which are normal for other congeneric species but may also occur as deviations only. Many of the latter reveal rather interesting ontogenetic and phylogenetic aspects (Fig. 8).

The several various types of embryo sac development described are probably the result of evolution within flowering plants. If one assumes that the embryo sac of flowering plants has descended from a female gametophyte of Gymnospermae of the type known in Cycadales, Ginkgoales and Coniferales, then **the primitive features of the type** of embryo sac development will be **formation from one megaspore** and **the largest number of mitoses**. Only the Polygonum-type possesses a combination of these features and hence should be considered the primitive type. Usually the megaspore tetrads of the Gymnospermae taxa mentioned are so similar to the megaspore tetrads of flowering plants that it is impossible to distinguish them. It is the final phase of ontogenetic evolution, i.e., the mature embryo sac, which is new and characteristic of only flowering plants. Among the quite considerable variety of final phase constructions, the primitive construction for flowering plants will be the one finishing in the evolution of the primitive, i.e., Polygonum-type, embryo sac.

Polygonum-type Embryo Sac Development—monosporic, bipolar, undergoing three mitotic divisions. The first information about the embryo sac structure of this type was gathered by Hofmeister (1847). However, the complete picture of its development, with *Polygonum divaricatum* taken as an example, was given by Strasburger (1877). This type of embryo sac is the most widespread among Angiospermae (more than 80%), which probably explains the fact that its initial name given by Palm (1915) was 'normal'. Maheshwari (1948) later gave it the name Polygonum-type to designate the genus in which it was first discovered.

In the classical variant of development, the Polygonum-type embryo sac is normally formed from a chalazal cell of the microspore tetrad. However, in some families (Chenopodiaceae, Cucurbitaceae, Flacourtiaceae and others) cases of its development from the epichalazal, micropylar or submicropylar cells of a tetrad are known. During the process of megaspore development into an embryo sac, its volume and size of the nucleus expand considerably, and the number of organelles in the cytoplasm rapidly increases. In the chalazal region the callosic wall which usually completely surrounds the megasporocyte and its daughter cells during meiosis dissolves (Rodkiewicz, 1970, 1978; Russell, 1979).

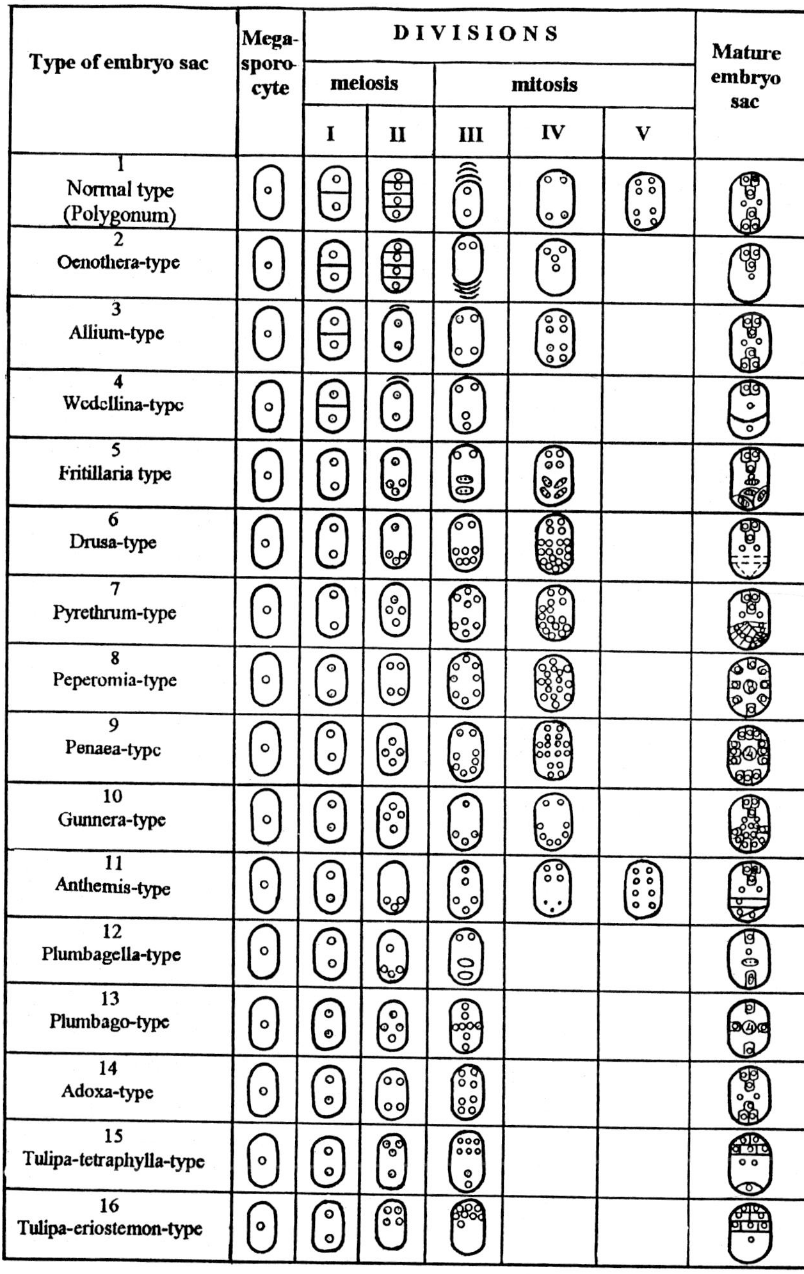

Fig. 8: Principles of embryo sac classification (Romanov, 1944). Author's later ideas presented in text.

After the first mitotic division of a megaspore nucleus, newly-formed nuclei come to lie at opposite poles in the 2-nucleate coenocyte and a large vacuole appears between them. As a result of the next two mitotic divisions an 8-nucleate embryo sac forms with four nuclei at each pole, followed by cytokinesis. In the micropylar part an egg apparatus differentiates, which consists of the ovule and two synergids, and in the chalazal part, three antipodes; the central cell consists of two polar nuclei separated from the micropylar and chalazal poles (and hence are called the micropylar and chalazal nuclei).

The embryo sac formed is bipolar, 7-cellular and 8-nucleate.

The number of cells in a mature embryo sac can vary. In a typical case it is 7-cellular, 7-8-nucleate. Change in number of cells can occur at the coenocyte stage of embryo sac development due to falling out of one or two divisions during its development as well as fusion of the chalazal group of nuclei (for example, among representatives of families Orchidaceae and Trapaceae). These deviations result in a decrease in number of cells in the formed embryo sac to four, five or six. Change in number of cells may be a secondary character, that is, taking place during embryo sac maturation. So, in consequence of destruction of the formed antipodals, a decrease in number of cells in the mature embryo sac occurs (for example, in families Alismataceae and Menyanthaceae). Contrarily, additional divisions of the cells (nuclei) of the antipodes both before and after fertilization lead to a secondary increase in number of cells (nuclei) in the embryo sac (Poaceae, Typhaceae and many others).

Based on the number of mitotic divisions as well as nuclear behaviour in the course of embryo sac development for representatives of family Orchidaceae, Abe (1972a, b) proposed four types of development.

The form and size of an embryo developing according to the Polygonum-type as well as the structure of the egg apparatus and antipodals, vary greatly and are taxon-specific.

Light and electron microscopic analysis of the structure of Polygonum-type embryo sacs during the course of their development in various species of plants revealed no features specific to this type compared with other types (Jensen, 1973; Kapil and Bhatnagar, 1981; Willemse and van Went, 1984; Rodkiewicz and Bednara, 1986; Fougère-Rifot, 1987b; Plyushch, 1992).

The question of evolutional interrelation of the Polygonum-type embryo sac with other types of embryo sacs of flowering plants is debatable. Many researchers (Palm, 1915; Chiarugi, 1927; Schnarf, 1936; Maheshwari, 1937; Gerassimova-Navashina, 1954a, b, 1971; Poddubnaya-Arnoldi, 1964, 1976; Romanov, 1971b; Kordyum, 1978; Coulter, 1984; Teryokhin, 1984, 1991; Plyushch, 1992) consider it to be initial with respect to other types of embryo sacs. These authors base their main arguments in support of this point of view on the fact that this type is widespread among Angiospermae and therefore phylogenetically ancient. Furthermore, as a result of meiosis, formation of isolated spores versus coenospores of bi- and tetrasporic development, may also be considered historically initial. A number of authors (Campbell, 1901; Schürhoff, 1919; Fagerlind, 1944a; Smirnov, 1982) do not consider a monosporic Polygonum-type embryo sac to be initial, however, basing their stand on the fact that this type of embryo sac has no homologue among Gymnospermae gametophytes, whereas a tetrasporic embryo sac is more readily homologous with a gametophyte of some Gymnospermae, for example *Gnetum*. According to another opinion (Modilewski, 1928b, 1953), all types of Angiospermae embryo sacs are unique and equivalent variants of a common ancestral type; an 8-nucleate embryo sac is not primitive but simply the most widespread and perfect variant among others.

OENOTHERA-TYPE EMBRYO SAC DEVELOPMENT—monosporic, unipolar, undergoing two mitotic divisions (Plate 61). First discovered among representatives of family Onagraceae (Hofmeister, 1847, 1849). After tracing its evolution in detail in *Oenothera lamarckiana* (Geerts, 1908), *O. biennis, Epilobium angustifolium, E. dodonaei, Circaea lutetiana* (Modilewski, 1909), it was established that **it descends from a micropylar megaspore of a tetrad.** The Oenothera-type embryo sac development was subsequently described for a great number of species of this family (Gachechiladze, 1981; Wieliunska-Woitas, 1984; Batygina, 1985a) and considered to be characteristic of the Onagraceae family only (Maheshwari, 1948) except for genus *Trapa* in which the Polygonum-type embryo sac development has been demonstrated (Maheshwari, 1947; Batygina and Kolesova, 1985). After lengthy discussion the genus was placed in a separate class, viz. the monotypic family Trapaceae (Takhtajan, 1980a).

Embryo sac development similar to the Oenothera-type was discovered in *Schisandra chinensis* (Yoschida, 1962; Swamy, 1964) as well as in *S. grandiflora* and *S. repanda* (according to illustrations in the works of Kapil, 1962; Hayashi, 1963; Kapil and Jalan, 1964). However, in the species of family Schisandraceae studied, it is the **chalazal** and not the **micropylar megaspore** (as in Onagraceae) which usually becomes the initial cell of an embryo sac. Apparently because the antipodals are ephemeral in species of genus *Schisandra*, embryo sac development was considered to be Polygonum in type (Hayashi, 1963; Kapil and Jalan, 1964; Davis, 1966) or modifications of the Polygonum-type (Yoschida, 1962). According to Swamy (1964), 50% of the ovules of *S. chinensis* exhibit Oenothera-type embryo sac development.

Since the question of type of embryo sac development for *Schisandra* species remains debatable to date, the publication by Battaglia (1986) 'Embryological questions. Do new types of embryo sac occur in *Schisandra*?' is of interest as a comparative analysis of works concerning embryo sac evolution for various *Schisandra*[4] species. The author, **based solely on the fact of embryo sac origin from a chalazal megaspore concludes that it is necessary to distinguish a new, Schisandra-type embryo sac development.**

However, most researchers while defining the type of embryo sac development do not attach decisive importance to the position of the functional megaspore in a tetrad or triad, but consider the three main features (Romanov, 1971b):

1. number of megaspores forming an embryo sac;
2. number of mitotic cycles taking place after meiotic divisions;
3. nuclar behaviour determining the structure (organization) of an embryo sac.

Further, it should be mentioned that for some species of flowering plants, for example those of Cucurbitaceae (Dzevaltovskyi, 1983) and Flacourtiaceae (Alimova, 1983), any of the cells in the triad or tetrad of megaspores can become functional regardless of the type of embryo sac development.

The character of embryo sac development for some representatives of Onagraceae and Schisandraceae is similar with reference to the main features (Table 2). So Battaglia's proposal to distinguish a special Schisandra-type embryo sac development

---

[4]Battaglia, 1951; Yoschida, 1962; Kapil, 1962; Hayashi, 1963; Kapil and Jalan, 1964; Swamy, 1964; Davis, 1966; Bhandari, 1970.

Table 2. Characteristics of Oenothera-type and Schisandra-subtype embryo sac (ES) development

	Oenothera-type	*Schisandra subtype*
Main Features		
Number of megaspores forming ES	Usually one	Usually one
Number of mitoses taking place after meiotic divisions	Two	Two
Nuclear behaviour determining ES structure	4 + 0	4 + 0
Secondary Features		
Number of megaspores forming	Tetrad (less often triad)	Triad (less often tetrad)
Position of functional megaspore	Micropylar (usually)	Chalazal (usually)
Morphology of ES elements	Synergids with filiform apparatus	Synergids without filiform apparatus (usually)

according to one feature (initiation of an embryo sac from a chalazal megaspore) does not seem very convincing at the moment.

Broad comparative embryological researches on separate Nymphaeles sensu lato taxa with respect to their systematic position, volume and interrelations have established the special nature of their embryo sac development. Batygina and coworkers (Batygina *et al.*, 1982; Batygina, 1985a) were the first to establish that in *Cabomba caroliniana* the embryo sac is **unipolar** in genesis and descends from the **chalazal megaspore of the triad**. These peculiarities of embryo sac development made it possible to define the Cabomba subtype of the Polygonum-type. More detailed researches on various Nymphaeaceae (*C. caroliniana, Nymphaea alba, N. gigantea, Nuphar lutea, Victoria amazonica* and *Barclaya (Hydrostemma) longifolium*) have confirmed the fact of embryo sac origin from the chalazal megaspore of the triad (less often, from the tetrad), its unipolarity and also two mitotic divisions (Winter, 1987; Titova, 1988, 1990; Winter and Shamrov, 1991a). According to Winter (1987), the embryo sac development of the Nymphaeaceae species studied by her is of the Oenothera- or Schisandra-type according to the Romanov (1971b) and Battaglia (1986) classifications. The embryo sac development in *Nyphaea heudelotii* is considered an intermediate type between Oenothera- and Polygonum-types (Van Miegroet and Dujardin, 1992).

The debatable character of the question concerning Oenothera- and Schisandra-type embryo sac development as well as results of the latest embryological researches on various taxa of families Onagraceae, Schisandraceae, Cabombaceae, Nymphaeaceae and Barclayaceae make it necessary to revise the Oenothera-type embryo sac development once again. Taking into account the similarity of embryo sac development of these taxa with respect to the three main features, we consider it incorrect to distinguish an independent type, namely Schisandra. Concomitantly, such distinctive features as number of megaspores forming, position of the functional megaspore, different nature of specialization of embryo sac elements and probably some other features allow examining it now as a **subtype**[5]. Let us briefly discuss therefore the main features of embryo sac evolution for representatives of the taxa mentioned above, as already shown in Table 2, which confirm the expediency of distinguishing the Schisandra-subtype of the Oenothera-type.

[5]The term 'subtype' is legitimate and has been used in a number of classifications (see: Romanov, 1971b; Poddubnaya-Arnoldi, 1976; Teryokhin, 1991).

It is universally recognized that the Oenothera-type embryo sac development—**monosporic, unipolar, undergoing two mitotic divisions**—is intrinsic to representatives of family Onagraceae only (Maheshwari, 1950). Usually the **micropylar cell** of a linear, T-shaped or oblique **megaspore tetrad** becomes the **initial cell of an embryo sac.** There are cases of absence of cellular wall formation in the micropylar cell of the dyad and development of a **megaspore triad** in some species (*Stenosiphon linifolium* and *Jussiaea repens*). Instances are also known (*E. hirsutum, O. odorata, O. graveolens, O. pycnocarpa, O. lamarckiana*) of simultaneous development of micropylar and chalazal megaspores, resulting in formation of two embryo sacs (Hoeppener and Renner, 1929) and also non-synchronous functioning of all the four megaspores giving rise to 4- and 2-nucleate embryo sacs (Subramanyam and Govindu, 1948).

Degeneration time of the remaining three megaspores differs. For most species of Onagraceae they often persist up to the stage of fully-formed embryo sac (e.g. for *Ludwigia palustris*, Wieliunska-Woitas, 1984). Sometimes degeneration features are observed before the micropylar megaspore has started division, for example *E. hirsutum* (Gachechiladze, 1981). Detailed researches on the first stages of megasporogenesis for various representatives of Onagraceae carried out using fluorescence, interference and electron microscopy, have established the specific polarity and arrival of plasmodesmata in the micropylar zone of the megasporocyte (Jalouzot, 1971, 1978; Rodkiewicz *et al.*, 1976).

The nature of laying, time of appearance and disappearance of callose in the megasporocyte and its absence at the apical end of the micropylar megaspore predetermine only functioning of the micropylar megaspore and degeneration of the other three megaspores (Kuran, 1972; Bednara and Rodkiewicz, 1975; Sniezko, 1976; Rodkiewicz and Sniezko, 1978; Sniezko and Harte, 1984; Wieliunska-Woitas, 1984). The authors think that callose plays an important role in formation of the female gametophyte. The characteristics of structure, distribution and number of organelles in the cytoplasm of the megasporocyte and megaspores have been delineated (Jalouzot, 1978; Rodkiewicz *et al.*, 1976; Gvilava, 1987; Plyushch, 1992). For example, for *E. palustre* the functional micropylar megaspore contains the largest number of plastids and the least number of dictyosomes compared to other megaspores (Bednara and Rodkiewicz, 1974). The characteristic feature of embryo sac development of the Oenothera-type is formation of a large vacuole in the chalazal part of the functioning megaspore with the nucleus moving from the central part of the cell to the apical. As a result of two mitotic divisions (usually) taking place in inter-perpendicular directions, a 4-nuclei embryo sac is formed. During this the embryo sac expands both in micropylar and chalazal directions. The form and size of the fully-formed and mature embryo sac vary (Wieliunska-Woitas, 1984; Batygina, 1985a).

The process of cell formation starts in a 4-nucleate embryo sac. There is a 3-cell egg apparatus in which the cells do not differ much morphologically and a central cell in the fully-formed embryo sac. Disposition of the polar nucleus with respect to the egg apparatus is variable.

In the mature embryo sac of representatives of family Onagraceae, the elements of the egg apparatus are quite large and specialized. Typically the egg cell is located lower than the synergids and may be highly vacuolated, especially in its apical part. The synergids are also vacuolated and usually have characteristic dentate outgrowths and a filiform apparatus in the basal part.

The embryo sac of the Schisandra-subtype of Oenothera-type development—**monosporic, unipolar, undergoing two mitotic divisions**—differs from the embryo sac of the Oenothera-type per se in that it usually forms from the **chalazal cell of the triad**, and less often from the tetrad of megaspores.

Embryo sac development of representatives of Schisandraceae, Cabombaceae, Nymphaeaceae and Barclayaceae should be ascribed to this subtype. Special attention should be paid to cases of formation of monosporic (50%) and bisporic (50%) embryo sacs in one *S. chinensis* flower (Swamy, 1964).

In spite of certain differences in megasporogenesis, embryo sac development in representatives of Onagraceae and Nymphaeales sensu lato is fundamentally similar. However, egg apparatus elements in the mature embryo sac of the studied species of Cabombaceae and Nymphaeaceae are less specialized and there is often no filiform apparatus in the synergids.

For some species of families Cabombaceae and Nymphaeaceae cases of anomaly in embryo sac development have been observed:

1. Increase in number of divisions of embryo sac nuclei. For example, the polar nucleus of *Anogra pallida* repeatedly divides, forming up to 140 nuclei, and the synergid contains up to 20 nuclei (Johansen, 1931).
2. Changing of polarity—formation of bipolar embryo sac of Polygonum-type development. For example *L. octovalvis, E. nervosum, Nuphar luteum* (Batygina, 1985a; Winter and Shamrov, 1991a).
3. Differentiation disruption of separate elements of the embryo sac. For example the egg apparatus of *L. octovalvis* contains one egg cell and one synergid.

We propose that the aforesaid anomalies make it possible to distinguish the Schisandra-subtype of Oenothera-type embryo sac development but do not constitute an independent Schisandra-type (as proposed by Battaglia, 1986).

There are two points of view on the origin of Oenothera-type embryo sac development. In accordance with the hypothesis of monophyletic origin of embryo sacs, the Oenothera-type descends from the Polygonum-type (Battaglia, 1986). Information is available on embryo sac development according to the Polygonum-type among representatives of Onagraceae, Schisandraceae and Nymphaeaceae, which supports this opinion.

According to Gerassimova-Navashina (1954, 1958), embryo sac development proceeds in compliance with the main regularities intrinsic to any vegetative cell, and the nature of formation of an embryo sac of this or that type depends on external conditions. Hence the author concludes that all types of embryo sac development, including the Oenothera, could appear in the evolutionary process repeatedly and independent of each other (polyphyletic orgin).

Discovery of the Oenothera-type embryo sac development in taxa of different phylogenetic levels augments our knowledge of the spread of this type among Angiospermae. Probably more detailed comparative researches on the complex of morphological features at different levels of organization will provide clues on the origin of various types of embryo sac development.

Allium-type Embryo Sac Development—bisporic, bipolar, undergoing two mitotic divisions (Plate 62). The mother cell of megaspores undergoes division and forms a dyad of cells. The upper cell of the dyad dies off and the lower one becomes the initial cell of the embryo sac. After two successive mitotic divisions 8 nuclei have formed, which are situated uniformly at the two poles of the sac.

Allium-type embryo sac development was first described for *Allium fistulosum* (Strasburger, 1879a, b). Subsequently it was established that this type of development is characteristic of all representatives of genus *Allium* (Weber, 1929; Messere, 1931; Ustinova, 1944; Glushenko, 1958; Gvaladze, 1962; Sokolov, 1968).

Allium-type embryo sac development has also been described for representatives of family Alismataceae—*Sagittaria guayanensis* (Johri, 1936) and *Alisma plantago* (Johri, 1936), and for some representatives of other families (Amaryllidaceae, Balanophoraceae, Butomaceae, Datiscaceae, Liliaceae, Orchidaceae, Viscaceae, Podostemaceae).

The embryo sac formed is 7-cellular and consists of the egg cell, two synergids, the central cell with two polar nuclei (micropylar and chalazal) and three antipodes that are usually not large and degenerate soon after formation. Typical embryo sacs of Allium-type are 8-nucleate (Schaffner, 1896; Murthy, 1935) as for example, *Sagittaria guayanensis* (Johri, 1936) and *S. sagittifolia* (Nikiticheva, 1990). Their egg apparatus is 3-cellular and the synergids have tapering ends and a filiform apparatus.

During maturation of this type of embryo sac a decrease in number of antipodals often occurs. As a result, the mature embryo sac of Allium-type when ready for fertilization, is 4-cellular and consists of the egg cell, two synergids, the central cell with two polar nuclei and no antipodes. Rarely, antipodals remain after the fertilization process, imitating an egg apparatus.

Allium-type development was described for *Agraphis patula (= Scilla hispanica)* (Treub and Mellink, 1880). But it was later established that in *Scilla* species, it is the micropylar cell of the dyad which gives rise to an embryo sac and not the chalazal one typical for the Allium-type.

The bisporic Allium-type embryo sac is usually derived from the monosporic one (Poddubnaya-Arnoldi, 1964; Plyushch, 1992).

Podostemum-type Embryo Sac Development—bisporic, unipolar, undergoing one mitotic division (Romanov, 1944) (Plate 63). Due to uncertainty about the actual existence of Podostemum-type embryo sac development, it was excluded from some embryo sac classifications, in particular that proposed by Willemse and Van Went (1984). The question of type of embryo sac in representatives of family Podostemaceae has long been debated and is still not resolved (Maheshwari, 1950; Battaglia, 1971, 1987b; Arekal and Nagendran, 1976). The problems are mainly three: manner of embryo sac formation in Podostemaceae, reality of the Podostemum-type and interpretation of the structure of the formed embryo sac. Embryo sacs of all the studied representatives of Podostemaceae belong to the bisporic series (bisporic type, according to Teryokhin, 1991). Two pathways of embryo sac development have been described for representatives of this family. In the first case, after division of the nucleus of the chalazal cell of the dyad and nuclear movement towards the poles, the destinies of the nuclei differ. The lower (chalazal) nucleus is pycnotized and then becomes completely lysigenous. Sometimes this nucleus manages to pass one division before destruction sets in. Regardless, these nuclei are usually destroyed by the time of embryo sac maturity (Razi, 1949). Rarely, their remnants are visible in the chalazal cell of the mature embryo sac. The micropylar nucleus divides twice, after which cytokinesis begins. The characteristic peculiarity is variability in location of the four cells of the formed embryo sac. Mostly, two cells are situated one beside the other at the micropylar end, and lower, two more cells lie one below the other. Sometimes an inverted structure is formed when two of the chalazal cells lie one beside the other. In some cases, the cells are arranged in a row, one above the other. Sometimes different variants of cell disposition can be found in the limits of the same species, e.g. in *Hydrobryopsis sessilis* (Arekal and Nagendran, 1975a). So, the peculiarities of manner of embryo sac formation described above consist firstly of just one (micropylar) nucleus of the two formed during the first division in the chalazal cell of the dyad functioning during the developmental process and in the course of embryo sac development, three nuclear divisions take place. The formed structure of the embryo sac is always 4-cellular. Battaglia (1971) considered this pathway of formation by right of priority the Apinagia-type. It was first described in Van Went's works (1908, 1910).

For some representatives of the family, in particular *Podostemum subulatus*, another pathway of embryo sac development was described (Magnus, 1913; Chiarugi, 1933). Here the nucleus of the chalazal cell of the dyad divides twice and the same 4-cellular structure is formed, but without additional (degenerating during development) nuclei. This type of development was termed the Podostemum-type and precisely this manner of embryo sac formation often raises doubts. Based on analysis of the latest researches on embryo sac development in representatives of Podostemaceae, carried out mainly by embryologists of the Indian school (Arekal and Nagendran, 1975a, b, 1976, 1977a, b), the Italian embryologist Battaglia (1987b) put forward the suggestion, which appears well grounded to us, that for all the representatives of the family studied that very type of embryo sac development is characteristic which was first described for *Apinagia* with more or less early disappearance of the degenerating chalazal nuclei. In this connection, it should be admitted that the name Apinagia-type is more priority-driven and correct than Podostemum-type.

Discussion about the faithful interpretation of embryo sac structure in family Podostemaceae (see: Battaglia, 1987b; Teryokhin, 1991) is considered quite important relative to problems pertaining to regularities and directions of female gametophyte evolution of seed plants. In our opinion, regardless of cell disposition three structures are present in the embryo sac of representatives of Podostemaceae: synergids (one or two), egg cell and haustorial cell (one or two). By its origin the haustorial cell is a homologue of the embryo sac central cell which only contains the upper polar nucleus. However, this cell is analogous to antipodals and maintains their function. Such interpretation is confirmed by two considerations. Firstly, in Podostemaceae a specific 'liquid' perisperm located under the embryo sac substitutes the absent endosperm by the time embryo development commences. Secondly, as in family Orchidaceae (in *Ophrys*), the absence of endosperm leads to 'double' fertilization becoming singular. Apparently these changes are associated with the peculiarities of the reproductive biology of these highly specialized water plants. In particular, they are associated with the need for very quick and econonomical seed ripening in specific life conditions. The Apinagia (= Podostemum) type embryo sac obviously requires further and more detailed research.

Balsamita (Anthemis)-type Embryo Sac Development—tetrasporic (pseudomonosporic with degeneration of three nuclei of megaspores), bipolar, undergoing three mitotic divisions (Romanov, 1944; Fagerlind, 1944a) (Plate 64). Discovered only in *Balsamita vulgaris* (= *Chrysanthemum balsamita*) from family Asteraceae. The sequence of development was described by Fagerlind (1939c) and Harling (1951). In the megasporocyte both meiotic divisions take place without cytokinesis and the 4 formed nuclei are located according to the formula 1 + 3. The three nuclei that move to the chalazal pole soon become pycnotic. At this stage the coenocyte cell forms an outgrowth directed towards the micropyle into which the micropylar nucleus enters. It increases in volume and participates in mitosis. The two daughter nuclei move away from each other and become separated by a large vacuole. After three mitoses 4 nuclei are formed at the micropylar and chalazal poles of the embryo sac. Thus the embryo sac is formed through 5 nuclear divisions (2 meiotic and 3 mitotic). The process of cellularization is completed by formation of the 7-cellular embryo sac with a 3-cellular egg cell apparatus. In the central cell both polar nuclei fuse before fertilization. The antipodals are represented by three cells: the two adjoining the central cell can become polynucleate while the chalazal antipodal cell possesses a lateral outgrowth and contains one nucleus and remnants of three pycnotic sporic nuclei. Fagerlind (1939c) found embryo sac development in *Ch. balsamita* similar to that described earlier for *Anthemis tinctoria* (Holmgren, 1915). However, it was later discovered that *Anthemis* has 2-mitotic (2-phasic) embryo sac development (Harling, 1950). According to Harling (1951), tetrasporic, pseudomonosporic, 3-mitotic (3-phasic) embryo sac development had been found to date (1950) only in *Chrysanthemum balsamita (= Balsamita vulgaris)*. The specific manner of embryo sac development in this species is considered by most researchers to be a special type of tetrasporic gametophyte.

Drusa-type Embryo Sac Development—tetrasporic, unipolar, undergoing two mitotic divisions (Maheshwari, 1948) (Plate 65). First described in *Drusa oppositifolia* (Håkansson, 1923). As a result of meiosis a 4-sporic coenocyte is formed. In the interphase of divisions II-III, the megaspore nuclei dispose according to the scheme 1 + 3 (one nucleus at the micropylar pole, three at the chalazal). In contrast to the Fritillaria-type, the chalazal nuclei do not draw together. In the Pyrethrum-form, the megaspore nuclei dispose according to the scheme 1 + 1 + 1 + 1 and the final polarization according to the scheme 2 + 6 is observed only in the interphase of divisions III-IV. Division III is synchronous with division figures of the three chalazal nuclei disposed at considerable distance from each other. As a result, 8 nuclei form with the distribution 2 + 6. In the interphase of divisions III-IV, a depression of the chalazal end usually takes place. At the same time, one-three of the lowest nuclei, being smaller, stain much more intensively than the others. Division IV of such nuclei passes heteronomously with pronounced chromosome adhesion or chromatin agglutination and, not infrequently, with no division whatsoever. Mitotic irregularity grows in the direction of the chalazal end and constitutes a manifestation of the so-called cytoplasmatic gradient (Romanov, 1965b).

After division IV the embryo sac forms with 4 nuclei at the micropylar end and 8-12 at the chalazal end. The formed embryo sac consists of a 3-cellular egg cell apparatus, the central cell with 2 polar nuclei and the antipodals with 7-11 cells. The number of the latter can decrease as a result of formation of 2-nucleate cells (Pyrethrum-form) or failure of cell walls to form between 4-6 upper nuclei (*Tulipa*). Distortions in development are observed during polarization of the megaspore nuclei and at the time of division III and are seen in agglutination of two or all three chalazal nuclei of the embryo sac.

The Drusa-type embryo sac development is characteristic of genera *Anthemis, Crucianella, Majanthemum* and *Melampyrum*, is found in *Bowlesia, Drusa, Helichrysum* and *Tanacetum*, and along with other types occurs in *Erigeron, Erythronium, Rubia, Smilacina, Tamarix, Tulipa, Ulmis* and others.

Fritillaria-type Embryo Sac Development—is tetrasporic, bipolar, undergoing two mitotic divisions (Plate 66). Firstly described in *Fritillaria persica* (Bambachioni, 1928a). The meiosis results in forming of the coenocyte with various disposition of the four megasporial nuclei (4-nucleate embryo sac). In the interphase of the divisions II-III the nuclei are located according to the scheme 1+3. By the beginning of the division III three chalazal nuclei come close to each other, which differs greatly this from the Drusa-type. All the four megasporial nuclei enter the prophase of the division III simultaneously, and further on, the figures of the division of three lower nuclei unite in one. As a result there forms the secondary 4-nucleate embryo sac with two haploid nuclei at the micropylar end and two triploid nuclei at the chalazal end. After the division IV there forms the embryo sac, which consists of the egg cell apparatus (the egg cell is, usually, not differentiated), central cell with two polar nuclei and antipodals. The nuclei of the egg apparatus cells and the upper polar nucleus are diploid, while the antipodal nuclei and lower polar nucleus are triploid.

Quite often there can be observed the different degree of the chalazal end depression, which is characterized by anomalous mitotic divisions with the prophase falling out or transition from the early prophase or interphase into the metaphase with chromatine agglutination. The degree of the depression grows during its earlier display and as far as approaching to the chalazal end. In case of strong depression (*Tulipa*) the antipodals are reduced to 2-1, and the lower polar nucleus is inactivated. In extremis, in Clintonia-form and *Tulipa maximowiczii* the lower polar nucleus and antipodals are absent.

Distortions in the embryo sac ontogenesis occur quite rarely and are, usually, observed during the polarisation of the megaspore nuclei and chalazal nuclei fusion. The Fertillaria-type of embryo sac development is characteristic for most of genera of the Liliodeae subfamily, for the genera of *Gaillardia, Heckeria, Rudbeckia* (the sexual forms), *Piper, Ratibida*; it can be found in *Leptachis, Myricaria, Platystemon,* and along with other, mainly, tetrasporic types—in *Armeria, Cornus, Erythronium, Euphorbia, Limonium, Statica, Tamarix, Tulipa.*

Plumbagella-type Embryo Sac Development—tetrasporic, bipolar, undergoing one mitotic division (Plate 67). This type of embryo sac development was first discovered in *Plumbagella micranta* (Fagerlind, 1939a; Boyes, 1939) and later in *Gagea capusii* and some *Tamarix* species. After meiosis in the mother cell of the megaspores, 4 nuclei form and move to the two poles: one nucleus into the micropylar part, the other three into the chalazal part. The three nuclei in the chalazal end then fuse, forming a triploid nucleus (Carano-Bambacioni phenomenon); thus the embryo sac becomes 2-nucleate with one haploid nucleus situated in its micropylar part and one triploid nucleus in the chalazal part. The next division leads to formation of 4 nuclei: two haploid nuclei in the micropylar part and two triploid in the chalazal part of the embryo sac. At the stage of the secondary 4-nucleate embryo sac, cellularization and formation of its elements take place. At the micropylar end the haploid egg cell forms and at the chalazal end the only triploid antipodal. The two nuclei (upper haploid and lower triploid) become polar nuclei of the central cell. The mature, ready for fertilization embryo sac of the Plumbagella-type consists of an egg cell, central cell with tetraploid nucleus (the result of fusion of the haploid and triploid polar nuclei) and one antipodal, usually displaying depression symptoms. Synergids are absent.

Dahlgren (1916) considered the Plumbagella-type embryo sac development the most advanced among the other tetrasporic types, assuming the fact that this type is characterized by formation of only 4 nuclei. Boyes (1939) established that it is a reduced variant of the Fritillaria-type embryo sac development.

Adoxa-type Embryo Sac Development—tetrasporic, bipolar, undergoing one mitotic division (Romanov, 1944) (Plate 68). First described in *Adoxa moschatellina* (Jonsson, 1879-1880). Meiosis results in formation of the coenocyte with 4 megasporial nuclei. In the interphase of divisions II-III, the nuclei dispose according to the scheme 2 + 2. In *Tulipa* species, the chalazal nuclei can be rather smaller and less normal in shape than the micropylar ones. The 4-nucleate embryo sac of the Adoxa-type can thus be surprisingly similar to the secondary 4-nucleate embryo sac of the Fritillaria-type.

Division III usually passes synchronously. In some cases delay has been noted of the lowest chalazal nucleus and one of the micropylar nuclei and, as an exception, delay of the upper micropylar and anomalous division of the lowest of the chalazal nuclei. Consequent to division III, an embryo sac with a nuclei distribution 4 + 4 is formed. The formed embryo sac consists of an egg cell apparatus, three antipodals and central cell with two polar nuclei.

The characteristic depression of the chalazal end is manifested by the nuclei of two antipodals being smaller and more chromatinic than the antipodal nucleus, which is a sister with respect to the lower polar nucleus. In extremis, the antipodals are reduced to two or degenerate before fertilization.

To reliably define the Adoxa-type it is necessary to observe all developmental stages. Considerable variability during distribution of the megasporial nuclei—from 4 + 0 to 0 + 4—has been noted as a deviation in type.

The Adoxa-type embryo sac development is characteristic of genera *Adoxa* and *Sambucus*, has also been discovered in *Hydnora, Nolina, Phyllocactus, Rhytidophyllum* and *Sanicula*, and along with other types can be found in almost 20 other genera, including *Amaryllis, Erigeron, Limonium, Nerine, Tamarix, Tulipa, Solanum* and *Ulmis*.

Tulipa tetraphylla-type Embryo Sac Development—tetrasporic, bipolar, undergoing one mitotic division (Romanov, 1944) (Plate 69). First described in *Tulipa tetraphylla* and *T. ostrovskiana* (Romanov, 1938). The single hypodermal archesporial cell immediately becomes a megasporocyte. The nucleus is often located closer to the micropylar end. In meiosis I, the achromatinic spindle is most often oriented obliquely with respect to the long axis of the megasporocyte and, rarely, parallel. In the 2-nucleate coenocyte, the lower nucleus is often situated a considerable distance away from the chalazal end. Meiosis II passes strictly synchronously. After divisions I and II, cell plates form between sister nuclei.

In the interphase of divisions II-III, the nuclei rearrange according to the scheme 3 + 1, after which the central vacuole forms. Division III usually passes synchronously. Sometimes delay of the chalazal nucleus is noticed and, as an exception, its anomalous division. Consequent to division III an embryo sac with nuclei distribution 6 + 2 is formed.

In the mature embryo sac, the egg apparatus consists of five morphologically similar cells, of which one or two are situated higher than the others and are probably synergids. The upper polar nucleus is located near the egg cell apparatus. In the chalazal part there are only the antipodal and inactivated lower polar nucleus, which is smaller than the upper one. The antipodal nucleus often degenerates soon after formation but sometimes remains because more chromatinic than the lower polar one. Deviations in development are often noted, especially at the stage of polarization of the megasporial nuclei.

Along with the Adoxa-type, Tulipa tetraphylla-type embryo sac development is characteristic of *Tulipa kolpakovskiana, T. ostrovskiana* and *T. tetraphylla* (Pechenitsyn, 1989). It was also found in *Ulmis pumila* (Walker, 1950).

Eriostemones-type Embryo Sac Development—tetrasporic, unipolar, undergoing one mitotic division (Romanov, 1944; Fagerlind, 1944a) (Plate 70). First described in *Tulipa celsiana* and *T. silvestris* (Liliaceae) (Bambacioni-Mezzetti, 1931). The single archesporial cell is laid in the subepidermal layer and immediately becomes a megasporocyte.

The characteristic peculiarity of development is early cytoplasm vacuolization, already observed at the early prophase of meiosis I. In meiosis I, the achromatinic spindle is usually oriented transversely with respect to the long axis of the megasporocyte. In the 2-nucleate coenocyte, both nuclei are located at the micropylar end. Thus nuclear distribution starts as early as meiosis I (division) and is totally completed in the interphase of divisions II-III. Nuclear divisions II and III are synchronous. In telophase II, the initial and secondary phragmoplasts between daughter cells are formed. The formed embryo sac is 8-cellular: seven cells of the egg apparatus and the central cell with one polar nucleus disposed in the upper part and characterized by larger size. Later, the central cell nucleus moves down and is situated closer to the chalazal end.

The egg apparatus cells are usually disposed in two rows. The upper row consists of two cells whose nuclei differ from the rest in smaller size and apparently being more chromatinic. The five cells of the second row do not differ morphologically from each other. One functions as an egg cell. The lower polar nucleus and antipodals are absent.

The Eriostemones-type embryo sac development is uniquely constant and characteristic only of *Tulipa* species from subgenus *Eriostemones* (Liliaceae).

Penaea-type Embryo Sac Development—tetrasporic, tetrapolar, undergoing two mitotic divisions (Romanov, 1944; Fagerlind, 1944a) (Plate 71). First described for representatives of *Sarcocolla, Penaea* and *Brachysiphon* (Penaeaceae) (Stephens, 1909) and species *Euphorbia procera* (Euphorbiaceae) (Modilewski, 1909).

Meiosis results in formation of a tetrasporic coenocyte with rhombic disposition of nuclei: 1 + 1 + 1 + 1. Then two mitotic divisions occur which pass synchronously in the peripheral cytoplasm, forming 16 nuclei disposed four each at the four poles. Consequent to cellularization, 3-cellular groups form at each embryo sac pole and four free polar nuclei come together in the centre of the embryo sac and then fuse. Equal in terms of egg apparatus cells, differentiation of all 3-cellular groups is the characteristic peculiarity of further embryo sac development. The egg cell in the mature embryo sac is large, with the nucleus in the apical part and the vacuole in the basal; the synergids have a flat base, filiform apparatus, vacuole in the apical part and nucleus in the basal.

The lateral and chalazal groups look exactly the same but sometimes their cells are smaller.

The mature embryo sac is spherical, the central cell penetrated with cytoplasmic bands, and at its centre polar nuclear fusion before and during fertilization takes place.

Such structure of the embryo sac allows interpreting it as a complex of four unipolar embryo sacs developing from each megasporial nucleus (Stenar, 1937) and to speak of the absence of real antipodals (Tobe and Raven, 1984). However, afterwards only the micropylar group of cells function as the egg apparatus: the pollen tube enters one of the synergids and an embryo develops from the fertilized egg cell. Disposition of the chalazal group, connection of these cells with the hypostase (Biebersteiniaceae) or the band of nucellus cells linked with the conductive bundle of the ovule (Penaeaceae), make it possible to assume that these cells maintain the function of the antipodals which, however, are quite ephemeral. The synergid-like cells of the lateral groups are also ephemeral while the egg cell-like ones and the intact synergid persist for a long time, up to the early stages of endospermogenesis; moreover, their walls become greatly thickened (Biebersteinia).

Variability is observed in the embryo sac structure: group disposition is not always symmetrical and the number of cells and free nuclei in the embryo sac is not always the same. This is especially characteristic of *Acalypha indica* (Maheshwari and Johri, 1941; Kapil, 1960), which allows singling out in this type the Acalypha indica variation, characterized by formation of four 2-cellular groups at the embryo sac poles and eight free nuclei in the central cell.

Penaea-type embryo sac development is characteristic of all the genera of family Penaeaceae studied, as well as of *Malpighia* (Malpighiaceae), *Acalypha* and *Euphorbia procera* (Euphorbiaceae) and species of family Biebersteiniaceae.

PLUMBAGO-TYPE EMBRYO SAC DEVELOPMENT—tetrasporic, tetrapolar, undergoing one mitotic division (Plate 72). First described in *Plumbago capensis* (Haupt, 1934). In the mother cell of the megaspores, after two divisions of meiosis passing without cytokinesis, four nuclei form which are located one each at the four corners of the rhomb. Consequent to the following mitotic nuclear division, an 8-nucleate embryo sac forms with two nuclei at each of its four poles. Cytokinesis now takes place in the embryo sac. One nucleus of the micropylar couple becomes the nucleus of the egg cell; synergids are absent. Four nuclei of each couple, acting as polar nuclei, fuse and form the tetraploid nucleus of the central cell. The destiny of the remaining three nuclei differs: they either degenerate or become the nuclei of cells which probably function as antipodals.

The characteristic peculiarity of species with a Plumbago-type embryo sac development lies in the tendency towards degeneration of those cells and nuclei that do not participate immediately in the fertilization act. In extremis, it leads to formation by the time of fertilization of unipolar 2-nucleate embryo sacs consisting only of the egg cell and central cell instead of the typical tetrapolar 8-nucleate embryo sac of the Plumbago-type. In *Plumbago capensis* and *P. zeylanica* (Cass, 1972a, b; Russell, 1980) a filiform apparatus was discovered in the egg cell. According to the authors who described this phenomenon, the egg cell in such cases additionally maintains the function of a synergid. It was found that the pollen tube does not penetrate such an egg cell but simply enters its filiform apparatus and then grows into the thickness of the egg cell wall and discharges its content together with sperm and vegetative nucleus between the egg cell and the central cell.

Ultrastructural research on the tetrasporic embryo sacs of a number of species (Plyushch, 1992) points to their characteristic features of local cytoplasm autolysis, presence of an atypical egg apparatus structure and passing of karyogamy slower than in a monosporic embryo sac.

Variations exist in the Plumbago-type embryo sac development. For example, a unipolar 4-cellular embryo sac is often found (*Plumbago europaeum*). Cases are known of atypical nuclear disposition during embryo sac formation and delay or absence of nuclear divisions in some species of genus *Plumbago* (Haupt, 1934; D'Amato, 1940b).

Plumbago-type embryo sac development is characteristic of certain representatives of genus *Plumbago: P. capensis, P. zeylanica, P. scandens, P. coccinea* (Dahlgren, 1937; Fagerlind, 1939a), as well as two other genera of family Plumbaginaceae: *Ceratostigma* (*C. plumbaginioides* and *C. willmottiuna*) (D'Amato, 1940b) and *Vogelia* (Mathur and Khan, 1941).

Peperomia-type Embryo Sac Development—tetrasporic, unipolar, undergoing two mitotic divisions (Plate 73). First described in *Peperomia pellucida* (Piperaceae) (Campbell, 1899a, b, 1901). Specific embryo sac development of the Peperomia-type was demonstrated by Brown (1908), Johnson (1914) and a number of other researchers.

The megasporocyte is pyriform with a rather widened chalazal part where the nucleus is situated. Both meiotic divisions pass without cytokinesis, due to which the megasporocyte cell becomes the initial cell of the embryo sac, and all the four megaspore nuclei participate in its formation. Usually one nucleus is situated in the micropylar part, the other three in the chalazal. At this stage growth of the coenocyte cell and formation of the large vacuole in its centre are discernible. Then, after two mitotic divisions 16 nuclei form which are disposed by groups in the peripheral layer of the cytoplasm, among which four nuclei have a fixed position near the micropylar pole of the embryo sac, while the rest are located in its chalazal part.

During cellularization the spindles of mitotic division II are oriented tangentially in the embryo sac (Smirnov, 1982, 1983); each 4-nucleate group there can potentially form two cells and two free nuclei, i.e., in total eight cells along the embryo sac wall and an 8-nucleate central cell. However, such exact nuclear disposition is rarely found. Peperomia-type embryo sac development is characterized by high variability in number of central cell nuclei and lateral cells.

The mature embryo sac of Peperomia-type development is of characteristic globular shape, sometimes with a small outgrowth at the micropylar end where the egg apparatus is situated; the chalazal pole of the embryo sac is not morphologically prominent. The question of the egg apparatus structure is under discussion. It has been suggested that genus *Peperomia* typically has a 2-cellular egg apparatus, consisting of the egg cell and one synergid (Johnson, 1914; Maheshwari, 1954; Romanov, 1971a, 1981; Smirnov, 1982, 1983). From our point of view, the egg apparatus described by Johnson (1914) as 2-nucleate is, in fact, a zygote and intact synergid, as is evident from Plate 73. There is also an opinion that the egg apparatus of *Peperomia* is represented by the egg cell only (Poddubnaya-Arnoldi, 1976). For some *Peperomia* species both 2-cellular and 3-cellular egg apparatus have been noted (Fagerlind, 1939b; Murty, 1959). In a number of species (*Peperomia blanda, P. obtusifolia*), the egg apparatus always consists of three weakly differentiated cells without the cellulosic wall (Nikiticheva, 1981b; Nikiticheva *et al.*, 1981), which has been confirmed by ultrastructural studies (Plyushch, 1982, 1992).

The lateral cells, which vary in number (0-8), are located randomly along the embryo sac wall. Most researchers consider them homologues of the antipodals participating in transport of nutrients from the nucellus to the embryo sac. They remain in the fertilized embryo sac at the first stages of endosperm development and sometimes divide into two to four cells. The central cell is not large. The number of polar nuclei in it varies from 7 to 13. Their fusion usually starts only after contact with sperm. The initial endosperm nucleus is polyploid and occupies a considerable part of the cell.

Ultrastructural research on the *Peperomia blanda* embryo sac has revealed low density of ribosome distribution, extremely weak ER and Golgi apparatus in all its cells, which indicates low activity.

The question of the origin and evolutionary level of the Peperomia-type embryo sac remains debatable. Some authors (Campbell, 1901; Fagerlind, 1944a; Smirnov, 1982) homologize the *Peperomia* embryo sac and gametophyte of genus *Gnetum*. Others (Rutgers, 1923) see it as a totality of four 'simple' embryo sacs. However, most researchers support the secondary nature of the Peperomia-type embryo sac reformed during formation and its origin from the monosporic Polygonum-type (Coulter, 1908; Zomanov, 1945, 1971a; Gerassimova-Navashina, 1954; Poddubnaya-Arnoldi, 1976; Kordyum, 1978; Teryokhin, 1984, 1991; Plyushch, 1992). Nevertheless, it is hard to believe that the Peperomia-type embryo sac whose development follows no strict regulations originated from the Polygonum-type with its stable, well-ordered formational pattern (Gvaladze, 1973). Furthermore, weak differentiation of the egg apparatus and disposition of the egg cell immediately under the micropyle are considered less perfect in terms of adaptation to fertilization than in Polygonum-type embryo sacs (Plyushch, 1992). Nor should one exclude the possibility that the embryo sac of the type described is independent in origin from the other types of embryo sacs (Gerassimova-Navashina, 1958). The peculiarities of the Peperomia-type embryo sac discovered in genus *Peperomia* are evidently partly determined by the nature of the interrelations between all the structures formed during ovule and seed development (Gerassimova-Navashina, 1958; Nikiticheva *et al.*, 1981; Smirnov, 1982) and, in particular, by the nucellus (weak tissue differentiation of the nucellus persisting as trophic tissue in the mature seed, and absence of the conductive band in its chalazal part).

Questionable Types of Embryo Sac Development

Schisandra-type Embryo Sac Development (see above: Oenothera-type Embryo Sac Development)

Gunnera-type Embryo Sac Development—tetrasporic, bipolar, undergoing two mitotic divisions (Plate 74). The embryo sac development of *Gunnera macrophylla* (Ernst, 1908; Samuels, 1912) and *G. chilensis* (Modilewski, 1908) reveals certain peculiarities. Meiosis passes in the megasporocyte without cytokinesis and nuclei in the 4-nucleate coenocyte are disposed at the corners of the tetrad. Subsequently the embryo sac enlarges and nuclei entering mitosis are redistributed. At the 8-nucleate stage 2 nuclei are located at the micropylar pole and 6 nuclei at the chalazal pole; at the 16-nucleate stage 4 and 12 nuclei are divided by a large central vacuole. Cellularization in the embryo sac is completed with formation of a 3-cellular egg apparatus situated at the micropylar pole, a 6-cellular antipodal apparatus at the chalazal pole and a central cell containing 7 nuclei. Both simultaneous fusion of all the nuclei or successive fusion of the 6 chalazal nuclei, followed by their fusion with the micropylar polar nucleus have been observed (Romanov, 1944; Modilewski, 1953).

Some studies have singled out the Gunnera-subtype within limits of the Penaea-type (Mezzetti-Bambacioni, 1929) or the Gunnera-variation of the Peperomia-type (Schnarf, 1936). Considering such criteria as number of spores, number of mitotic divisions and nuclei, this variation can be assigned to the types described but differs in nuclear disposition in the female gametophyte and organization of the mature embryo sac. Hence the Gunnera-type embryo sac was isolated as an independent one. In current embryo sac classifications, however, the Gunnera-type is not treated as an independent type (Romanov, 1971b, 1981; Teryokhin, 1991).

Chrysanthemum-type Embryo Sac Development—tetrasporic, bipolar, undergoing two mitotic divisions (Plate 75). First described in *Chrysanthemum cinerariaefolium* from family Asteraceae (Martinoli, 1939). Meiosis results in formation of tetrasporic coenocyte in which the nuclei are disposed according to the scheme 1 + 2 + 1. In contrast to the Fritillaria-type, it is not the three chalazal nuclei which fuse, but only the two central nuclei. After the first mitosis six nuclei are formed: two haploid nuclei each at the micropylar and chalazal poles and two diploid nuclei in the centre of the developing embryo sac. After the next division the embryo sac is 12-nucleate. Cellularization is completed with formation of an 11-cellular embryo sac: 3-cellular egg apparatus (nuclei haploid), seven antipodals (four lower with haploid nuclei and three upper, diploid nuclei) and the central cell with two nuclei (haploid and diploid). The embryo sac of this species may develop differently, however, without fusion of the central nuclei in the tetrasporic coenocyte and with formation of 16 haploid nuclei (result of two mitoses), which conform to the Pyrethrum-form of the Drusa-type (Romanov, 1971b).

In *Gagea stipitata* from family Liliaceae (Shamrov 1990c), the embryo sac develops according to the Chrysanthemum-type (with distortions). The embryo sac arises through two mitotic divisions; however, only derivatives of the upper megasporial nucleus undergo a second divison, the lower megasporial nucleus often not dividing. The mature embryo sac is 6-7-cellular: the egg cell, two synergids (nuclei haploid), two to three antipodals (one with diploid, the other(s) with haploid nucleus) and 2-nucleate central cell (with haploid and diploid nuclei).

The status of Chrysanthemum-type embryo sac development is debatable. It is either singled out as a special type whose position is not clear (Maheshwari, 1950) or considered a type close to the Fritillaria-type (Martinoli, 1939, 1967; Fagerlind, 1944a; Battaglia, 1959). Some authors place it as a subtype in the Fritillaria-type (Mezzetti-Bambacioni, 1949), others as a special form in the Drusa-type, where it is intermediate between the Drusa- and Fritillaria-types (Romanov, 1981).

Evolution of Types of Embryo Sac Development (Fig. 9)

An embryo sac type is defined according to the three main modes of development which control the number of megaspores, number of postmeiotic divisions and the peculiarities of nuclear behaviour (definitive distribution, secondary redistribution, fusion of chalazal nuclei) (Romanov, 1944, 1971b, 1981; Gerassimova-Navashina, 1954a, b). In discussing the evolutional problems of the Angiospermae embryo sac most researchers do not go beyond establishing the initial type of embryo sac development. Starting with Schnarf (1931), it is almost universally recognized that the monosporic Polygonum-type is the most primitive and initial, although some authors have considered the 16-nucleate tetrasporic embryo sac such (Schurhoff, 1926; Fagerlind, 1944a; Modilewski, 1953; Sundara Rajan, 1976; Smirnov, 1982).

With reference to postmeiotic divisions, a decrease in their number is considered progressive mainly by analogy with the evolutional direction from the Gymnospermae female gametophyte to the Angiospermae embryo sac (Palm, 1915; Romanov, 1957, 1981; Kordyum, 1971; Battaglia, 1989). Romanov was the strongest proponent of this concept; he considered the decrease in number of mitotic divisions a major evolutional trend. However, the adaptational significance of shortening of the Angiospermae embryo sac development is not clear. Furthermore, it has been shown (Pechenitsyn, 1989) that mitotic divisions are even passed by in certain non-viable coenocytes, indicating considerable self-sufficiency in mode of development. Apparently, in the course of evolution of the Angiospermae embryo sac this mode of development exhibited some ambiguity.

The most debatable question is that of the definitive distribution of nuclei and their further behaviour. Discussing the reasons for nuclear distribution some

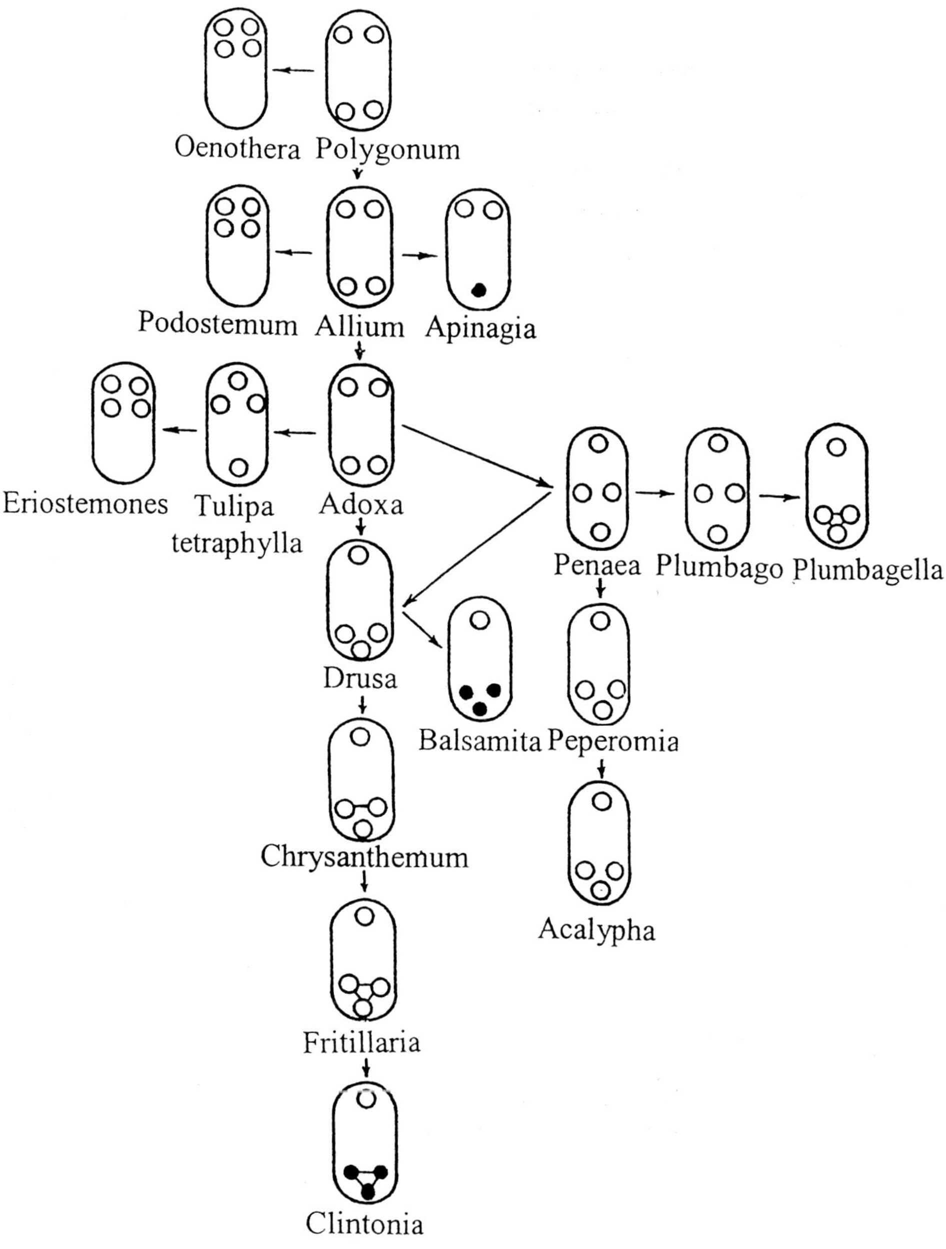

Fig. 9: Evolution of types of embryo sac development.

researchers (Maheshwari, 1948; Chiarugi, 1950; Romanov, 1965a, b, 1971b; Smirnov, 1982) evolved the notion of passive nuclear movement, while others took the opposite view, viz. active movement (Gerassimova-Navashina, 1954a, b, 1971; Teryokhin *et al.*, 1971; Pechenitsyn, 1989). The opinion of Teryokhin and coauthors (1971), supported by us, that nuclei are disposed in places where the most intensive metabolic processes occur, has also been confirmed by results of ultrastructural studies on the embryo sac (Plyushch and Bannikova, 1989).

Differences in the definitive distribution of nuclei and their further behaviour during evolution have hardly been discussed. Romanov's opinion (1965a, b) that the depression of the embryo sac chalazal pole was delineated in earlier phases of ontogenesis[6] seems rational. Battaglia's scheme (1989), according to which embryo sac evolution of Angiospermae proceeded in the direction of decrease in number of nuclei at the micropylar end during cell formation, appears quite artificial to us because it definitely excludes the Eriostemones- and Tulipa tetraphylla-types of embryo sac development which consequently have a 7- and 5-cellular egg apparatus.

To solve the question of evolutional trends of Angiospermae embryo sacs, it is imperative that comparative studies be carried out on representatives of different taxa (families, genera and species) characterized by possession of several types of embryo sac development, as for example in *Erythronium, Euphorbia, Limonium, Tamarix, Tulipa, Ulmus, Delosperma cooperi* and others. Such examples allow one to suppose that the variety of developmental types of megagametophyte in each particular taxon is not random but represents stages or steps in the entire evolutionary process.

Let us examine this concept on the example of genus *Tulipa*. Romanov concluded the priority of the Fritillaria-type embryo sac development in *Tulipa*. In his opinion, the other known types of embryo sac development in this taxon (Adoxa, Drusa, Tulipa tetraphylla, Eriostemones) originated from the Fritillaria-type during evolution of the genus.

More detailed examination of *Tulipa* species allowed us to propose another scheme of interrelation of different developmental types of tetrasporic embryo sacs (Pechenitsyn, 1989). We believed that if in following most embryologists by taking the monosporic Polygonum-type embryo sac development as primitive and initial, it would be rational to consider the peculiarities inherent in this type likewise primitive, namely isobipolarity, absence of chalazal pole depression and, probably, its late delineation, and fusion of the polar nuclei or at least their contact before fertilization.

On the assumption of the latter, we define the Adoxa-type, possessing a number of peculiarities similar to the initial ones of the Polygonum-type, as the initial type of embryo sac development in genus *Tulipa*. Such peculiarities as isobipolarity, late delineation of the chalazal pole depression, and drawing together of the polar nuclei before fertilization. As a distinct atavism, mention may be made of the phenomenon of temporary formation of cell plates after both meiotic divisions.

Divergent evolution in embryo sacs of tulips led to unipolarity in some species according to the following scheme: Adoxa-type—Tulipa tetraphylla-type—Eriostemones-type; in other species heterobipolarity has taken place with gradual increase in delineation of the chalazal pole depression according to the scheme: Adoxa-type–Drusa-type–Fritillaria-type. Increase in the chalazal pole depression of the embryo

[6]Based on a number of works (Teryokhin and Yakovlev, 1968; Teryokhin and Kamelina, 1969, 1972; Teryokhin, 1977; Teryokhin, 1991), it becomes clear that the depression of the chalazal pole of an embryo sac can be considered the starting stage of reduction of chalazal structures, which is fully evidenced in such taxa as Orchidaceae, Pyroloideae-Monotropoideae, Balanophoraceae and others.

sac was accompanied by its earlier delineation, which finally led to association of three chalazal nuclei during division III so characteristic of Fritillaria-type development.

The main determinative factors in evolution of the embryo sac in tulips were aridization and increased continental climate of the Ancient Mediterranean—the centre of tulip origin. Eventually species of subgenus *Eriostemones* became characterized by microthermoty and the Eriostemones-type of embryo sac development, and species of subgenus *Leiostemones* by mesothermoty and mainly the Fritillaria-type. The former spread much farther, occupying the northern parts of the generic area and regions with temperate climate, developing in ecological niches promising the earliest inflorescence. Doubtless, the settlement process was also determined by the fact that embryo sac development of the Fritillaria-type and especially of the Eriostemones-type, is considerably less influenced by temperature variations compared to the initial Adoxa-type.

Thus the most evolutionally advanced among the *Tulipa* tetrasporic embryo sacs are the unipolar Eriostemones-type and Fritillaria-type, characterized by heterobipolarity, the highest degree of chalazal pole depression accompanied by fusion of three chalazal nuclei, and also absence of drawing together of polar nuclei before fertilization.

Considering the increase in number of spores, changing of isopolarity towards unipolarity or heteropolarity which, in turn, is accompanied by growth of the chalazal pole depression, as the main directions in embryo sac evolution, the following scheme of their development types is proposed.

Regardless of number of megaspores participating in embryo sac formation, the isobipolar type is the initial one for embryo sac development: among the monosporic embryo sacs—Polygonum-type, among bisporic—Allium-type and among tetrasporic—Adoxa-type. The interrelation between mono- (Oenothera-type) and bisporic (Apinagia-form) embryo sacs is evident: development proceeded in the direction of unipolarity formation.

Connections between tetrasporic embryo sacs are much more complicated though both directions of evolution are clearly seen here also, namely formation of monopolarity (Adoxa-type—Eriostemones-type) and heterobipolarity with increase in chalazal pole depression (Adoxa-type—Clintonia-form). The Chrysanthemum cinerariaefolium-form characterized by fusion of not all chalazal nuclei is considered by us to be a stage of evolution from the Drusa-type to the Fritillaria-type, and the Clintonia-form a continuation of the Fritillaria-type evolution accompanied by further increase in chalazal pole depression.

While the branches of Adoxa-type—Eriostemones-type and Adoxa-type—Clintonia-form look quite natural, this is not so evident for the group of tetrapolar embryo sacs. Some idea of the possible evolutional pathways can be derived from the fact that the Penaea-type is present along with the Adoxa and other tetrasporic types in *Limonium* and *Delosperma cooperi*. This gives grounds for supposing that the Penaea-type embryo sac development originated from the Adoxa-type and is an intermediate with reference to distribution of megasporial nuclei 2 + 2 and 1 + 3. With prolongation of the embryo sac and increase in chalazal pole depression, the Penaea-type could transform into the Drusa-type and with the spherical shape remaining, into the Peperomia-type. In the formation of a 1-cellular egg apparatus, the following direction can be observed: Penaea-type–Plumbago-type–Plumbagella type, which is characterized like the branch Adoxa-type—Clintonia-form, by formation of heterobipolarity and increase in chalazal pole depression up to nuclear fusion. Precisely due to the presence of a

1-cellular egg apparatus, we derive the Plumbagella-type from the Plumbago-type, and do not consider it a reduced form of Fritillaria-type development. The position of the Balsamita-type is not absolutely clear. However, it may be supposed that this type originated from the Drusa-type.

The proposed scheme of interrelation of different types of embryo sac development is favoured by the results of analysis of distribution among families, counting two and more various types of development (total families more than 50). According to the analysis given in the 5-volume book 'Comparative Embryology of Flowering Plants' (1981-1990), the most widespread are development types of the branch Polygonum-type—Fritillaria-type, which are disposed in a strictly decrescent sequence: the Polygonum-type is found in 51 families, Allium-type 41, Adoxa-type 18, Drusa-type 13 and Fritillaria-type 8. Penaea- and Peperomia-types are found much more rarely, in 5 and 4 families respectively. The remaining types of development are characteristic of 1-3 families.

Determinative factors in the evolution of types of embryo sac development are climate change and increase in chalazal pole depression observable at earlier stages of development. The latter reflects the process of 'complete reduction of the chalazal complex of cells as a vestige of the trophic tissue of the female prothalis' (Smirnov, 1982) and is clearly seen in the sequences Allium–Adoxa, Adoxa–Clintonia, Adoxa–Plumbagella.[7]

The second direction of evolution—unipolarity formation—led to formation of Oenothera-, Podostemum- and Eriostemones-types of embryo sac development. Thus an increase in chalazal pole depression and unipolarity formation led to exclusion of the lower polar nucleus from the process of double fertilization. The adaptational significance of this phenomenon lies in hastening embryo development since in this case the zygote starts to divide much earlier than in species with triple fusion (Gvaladze, 1976; Pechenitsyn, 1989), which may certainly be considered a progressive feature (Smirnov, 1982).

Hierarchic Classification of Embryo Sacs

At this point it should be noted that the term 'embryo sac' covers the period of female gametophyte development commencing with its initiation up to completion of primary cytokinesis. The period of so-called 'specialization' (maturation) of the embryo sac cells is likewise included.

Classification of natural embryo sacs is based on differences in their morphogenetic origin, mode of formation and type of structure. Such classification is necessary for an appropriate look into female gametophyte evolution and is additionally used in taxonomic research, systematics, phylogeny and related applied disciplines, for instance studies of apomixis.

Embryo sac classification was first undertaken by Palm (1915). The most famous of the classifications known today are those by Schnarf (1936), Romanov (1944, 1971b), Maheshwari (1950), Johri (1963), Rutishauser (1969), and Ambegaokar (cited by Willemse and van Went, 1984). Romanov (1971b) and Maheshwari (1950) proposed

---

[7]At the same time a number of works (Strasburger, 1900; Smirnov, 1982; Teryokhin, 1991) show that there is no general reduction in female gametophyte in flowering plants since development of the gametophyte continues after fertilization. Only selective reduction takes place.

three principal criteria for distinguishing the types of embryo sacs, namely: (1) number of megaspores participating in sac formation; (2) number of free nuclear divisions contributing to sac formation following completion of meiosis; and (3) behaviour of the nuclei, determinative cellular set-up and polarity pattern. The derivative (secondary) increase in either nuclei or cells is discarded, itself sometimes happening at a later stage compared to primary cytokinesis.

In our opinion, the proposed criteria do not fullly represent some particulars and certain instances of complexity in embryo sac formation. These criteria require introduction of some adjustments and additional details. In some cases, for instance in embryo sac description procedure, it is irrelevant to speak of a clearly defined number of megaspores forming the embryo sac. In fact, it is only relevant to speak of a particular number of nuclei produced by meiotic divisions and engaged in its initiation (see Balsamita-variation below). The notion of *megaspore* construed as a distinct specialized cell type loses its direct meaning when applied to angiosperms in this context. But in light of the fact that meiosis results in formation of a haploid nuclei gametophyte, one can apply concepts and terms employed in descriptions of bi- and tetrasporic embryo sacs, keeping the aforesaid comment in mind, however.

Since the aforementioned criteria are not sufficiently precise, almost no two independent classifications comprise the same *type inventory* of embryo sacs. Moreover, the morphogenetic and evolutional disparity of the above criteria has inevitably to lead to a hierarchic classification of embryo sacs. An attempt at such a classification is presented in Table 3.

The proposed classification is grounded on the following three levels of hierarchy: types, subtypes and variations. It is evident already at the current stage that many variations may include several forms of development or organization; however, development of a detailed classification is still some distance in the future. Considering that each hierarchic level has its own defining criteria, we have to discuss the grounds sufficient for such definition. The considered types, subtypes and variations, being the major components of the suggested classification, also require a few brief comments. Mono-, bi- and tetrasporic female gametophytes are not limited to angiospesm only; they are also encountered in gymnosperms. Despite the apparent relative uniformity of the intial (free-nuclear) developmental stages in gymnosperms, bisporic female gametophytes are found in *Juniperus* and tetrasporic ones in *Gnetum* and *Welwitschia*. These facts possess evolutionary implications of high importance. They signify that following passage to seed reproduction certain functions of the megaspore (for instance, spreading or survival in adverse conditions) were lost, and the gametophyte function became 'extended' its application to all stages of megasporogenesis: megaspore mother cell, dyad cells and megaspores per se. Concomitantly, the functions of meiosis, as the means to transfer nuclei into the haploid state, had inevitably to be retained as related to sexual reproduction functions and were in many cases included in the initial stages of female gemetophyte development. Here it needs to be emphasized that the passage to regular uniparental reproduction, which requires retention of the diploid state of nuclei during expression of genes responsible for stimulating the gametophyte development (in case of inherited gametophyte apomixis), itself requires dramatically different genetic adaptations and ways of maintaining the diploid state of nuclei in the gametophyte.

It is quite evident that such changes possess a meaning broader and more important in terms of evolution compared, for example, to either the number of fluctuations in

Table 3. Hierarchic classification of embryo sacs. Cruved arrow indicates reorganization

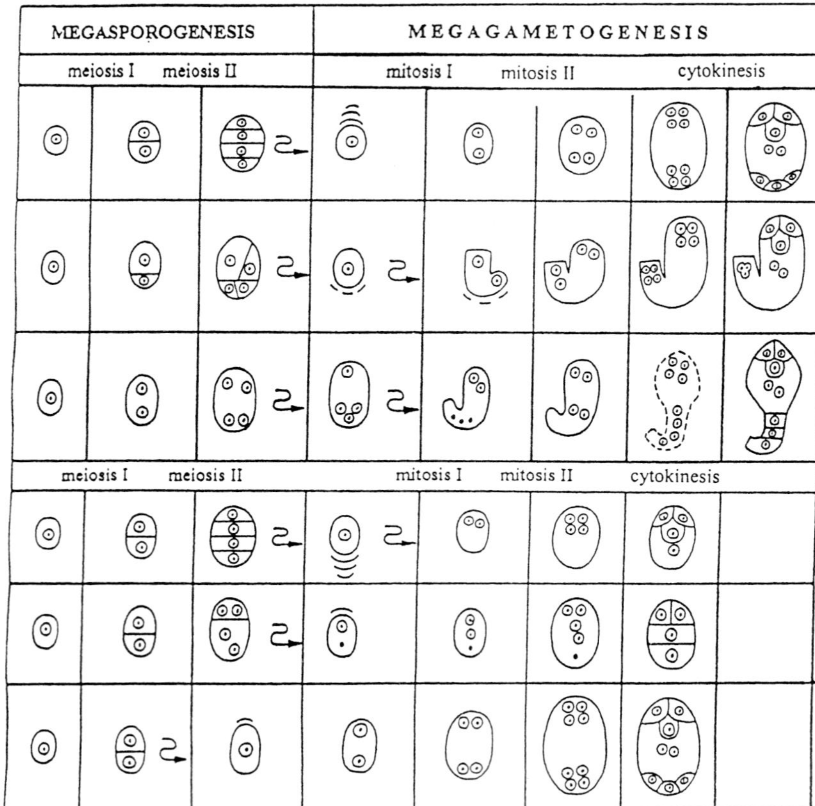

CLASSIFICATION		
TYPE	SUBTYPE	VARIATION
MONOSPORIC	Polygonum 5 divisions (2) + 3	Polygonum 8 (3 + 2 + 3), bipolar; depressed forms
		Balanophora 8 (3 + 2 + 3 or 1 tripl.) bipolar, with polarity inversion, depressed forms
		Balsamita 8 (3 + 2 +3), bipolar; with substitution and polarity inversion
	Oenothera 4 divisions (2) + 2	Oenɔthera 4 (3 + 1), unipolar
		Apinagia I (= Podostemum 4 (2 + 1 + 1), bipolar; with substitution
BISPORIC	Allium 4 divisions (2) + 2	Allium 8 (3 + 2 + 3) bipolar

Table 3. (*Contd.*)

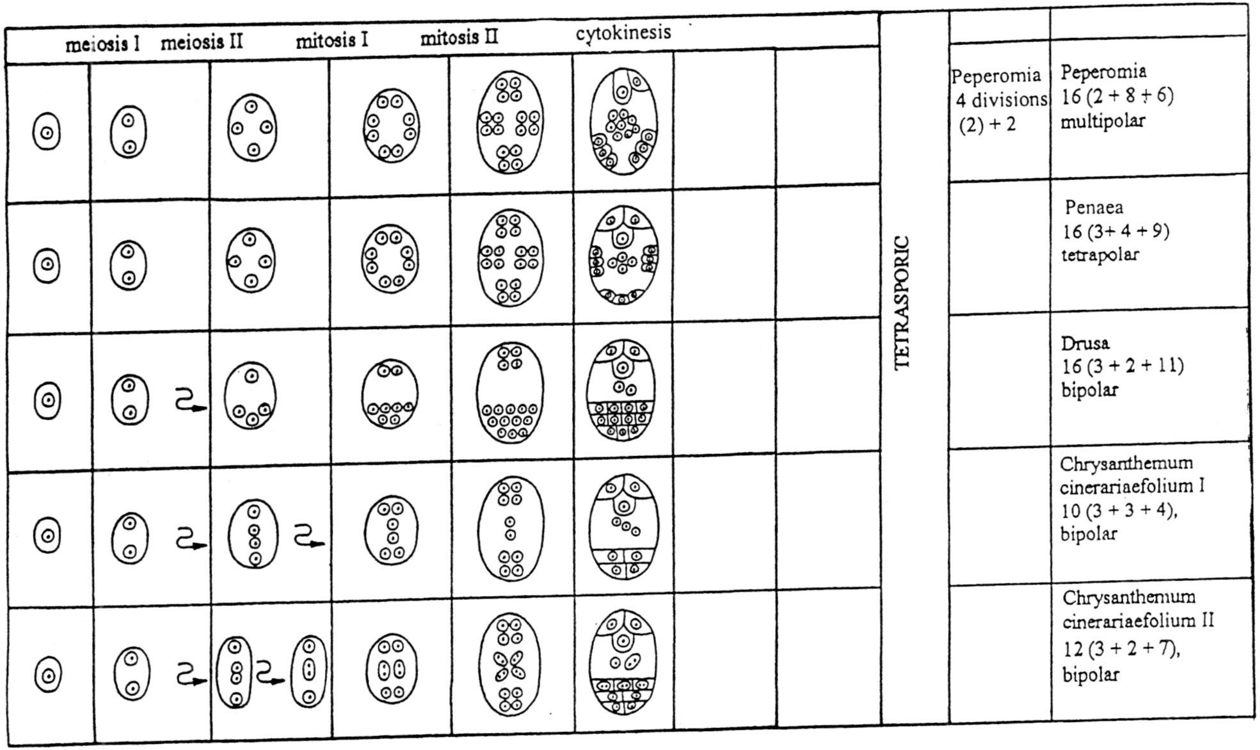

Table 3. (*Contd.*)

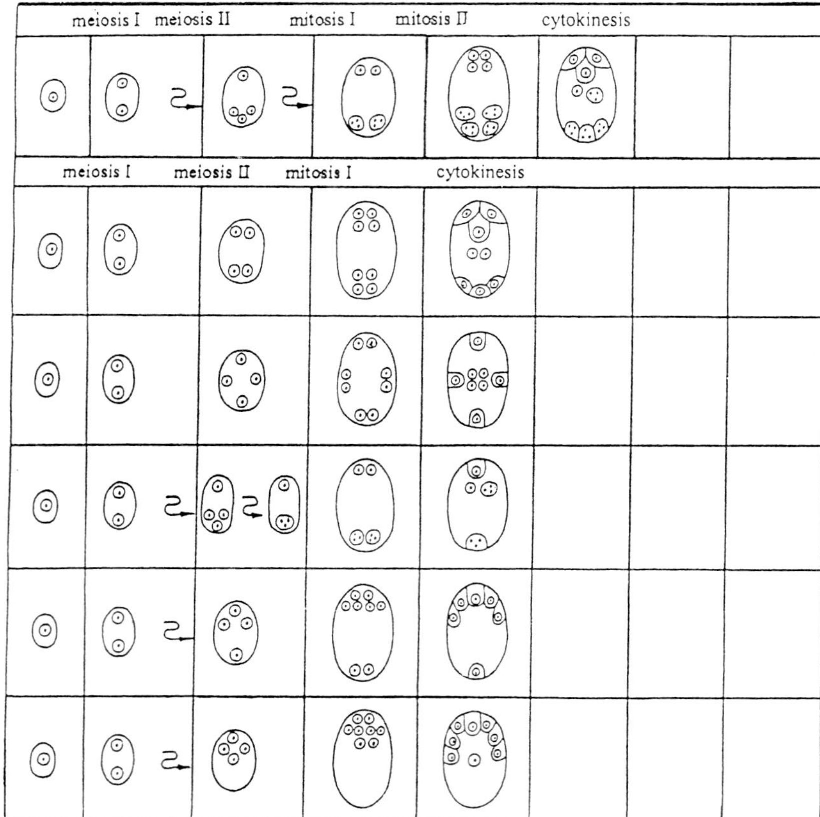

TETRASPORIC		Fritillaria 8 (3 + 2 + 3) bipolar, Bambachioni's effect
	Adoxa 3 divisions (2) + 1	Adoxa 8 (3 + 2 + 3), bipolar
		Plumbago 8 (1 + 4 + 3) tetrapolar depressed forms
		Plumbagella 4 (1 + 2 + 1) bipolar
		Tulipa tetraphylla 8 (5 + 2 + 1) bipolar
		Eriostemones 8 (7 + 1) unipolar

divisions of free nuclei over the embryo sac development process or migration of nuclei prior to cytokinesis. Shifting attention from cells to nuclei when determining the type of embryo sac is, as we shall see, of vital importance for their classification. However, subdivision of embryo sacs into three types (mono-, bi- and tetrasporic) still prevails (see Table 3).

'Endosporic' development of embryo sacs within the framework of coenocyte structure (of basically a polynucleate cell) to a large degree presupposes the nature of regularities and peculiarities of aberrances in their development. For instance, development within the framework of a coenocyte structure allows for nuclei migration, etc.

In determining the subtypes demonstrative of the varieties of embryo sac development, we ventured to refrain from the second criterion—number of nuclear divisions (instances of mitosis) which follow meiosis completion (Romanov and Maheshwari). This criterion applies only to monosporic embryo sacs, whereas development of bi- and tetrasporic formations of the same kind includes meiotic divisions also. It is more appropriate to count the number of divisions starting with the initial cell of the embryo sac, distinguishing if necessary between meiotic and mitotic divisions.

Apart from that, in the process of embryo sac development there often occur morphogenetic transformations unrelated to nuclear divisions but significant in the subsequent steps of development. Such transformations primarily include: the Carano-Bambacioni phenomenon (fusion of nuclei resulting in ploidy increase), replacement effect (wherein degenerative nuclei are eliminated as they develop and get replaced with those from their other clone) and, lastly, polarity inversion in embryo sac (wherein various accretions therein cause one pole to be replaced by the other). The listed transformations assist in determining the hierarchic class (level) of an embryo sac and its position in the classification. However, the number of nuclear divisions participating in embryo sac development still holds as the main subtype definition criterion.

At equal numbers of nuclear divisions a matured embryo sac may contain a different number of either nuclei or cells. This is provoked by a number of various factors, such as, for instance, the Carano-Bambacioni phenomenon, different nuclear fusion type, rhythm fluctuations of nuclear divisions or nature of cytokinesis. Additionally, the regular nuclear shifts and groupings create an original, taxon-settled structure of the completely formed embryo sac. We employed precisely these criteria to classify embryo sac variations. Any growth deviations brought about by other causes may be regarded as particular forms or teratologically induced.

The layout and manner of development of embryo sac are reflected in variations. We currently distinguish 17 such variations. A brief comment on each, with reference also to the types and subtypes to which it belongs, is given below.

Monosporic type. The nucleus responsible for initiation of embryo sac development under this type is contained in one of the four megaspores. Development itself runs within the framework of the coenomegaspore.

Polygonum-subtype, characterized by three mitotic divisions. For correct comparison with various subtypes of the other types one would need a reference point common to all. As such, only the mononuclear mother cell of the megaspores qualifies. Thus we conditionally add two meiotic divisions to the number of mitotic ones, the latter not being part of the Polygonum-subtype embryo sac development, but forming a real morphogenetic reserve depleted in the process of evolution when other types and the majority of subtypes and variations evolved. Polygonum-subtype comprises three variations (see Table 3).

Polygonum-variation. The first illustrated description of this variation was offered by Strasburger (1879a) based on the example of *Polygonum divaricatum*. This variation is formed by three consecutive mitotic divisions of the nuclei commencing with that of the viable megaspore. The three remaining megaspores perish. In cases of multicelled archesporium (Casuarinaceae, Rosaceae, Loranthaceae, etc), competition occurs not between the individual megaspores within a single tetrad, but among the viable (usually chalazal) megaspores of different tetrads. The isobipolar embryo sac, usually 8-nucleate, has the following characteristic nuclei distribution: 3 + 2 + 3 (with egg apparatus, two synergids, central 2-nucleate cell, three antipodals). The most variable is the chalazal region. Due to nuclear divisions missed and no corresponding formation of cells, the number contained in the emerging embryo sacs of *Polygonum*-variation is often incomplete. This is characteristic of many taxa in Orchidaceae, as well as of mycoparasite representatives of Ericaceae, e.g. *Hypopithys monotropa* (Teryokhin, 1963).

Balanophora-variation. The first detailed and well-illustrated description of this variation was offered by Fagerlind (1945) based on the example of *Balanophora elongata* and *Langsdorffia hypogaea*. The embryo sac is formed from the micropylar megaspore of the tetrad. At the 2-nucleate gametophyte stage the chalazal end produces an excrescence which the lower nucleus passes into. The excrescence actively grows in the micropylar direction on which pole the egg cell apparatus is formed following the third mitosis; the two polar nuclei are located at a lower level. The behaviour of the chalazal nuclei is highly unstable. It may be the case that four nuclei are formed on the chalazal end of the embryo sac; however, seemingly, antipodes are never formed. Nuclear divisions on the embryo sac poles in *Balanophora thwaitesii* are essentially asynchronous. Often only two or even one antipodal nucleus appears, which is located within the central cell. Nuclear fusions also occur quite frequently (Teryokhin and Yakovlev, 1967). All the aforesaid peculiarities of embryo sac development in *Balanophora* suggest that its development should be defined as an independent variation. Development of polar inversion of the embryo sac in *Balanophora*, an ancient and highly specialized parasite, is possibly conditioned by its development directly in the tissue of the miniature pistil. The ovule, evidently of anatropous type, was fully eliminated in the course of evolution. Embryo sac inversion serves as a means of orientation necessary to ensure effective fertilization.

Balsamita-variation. This type of development was first described by Fagerlind (1939c) based on the example of *Balsamita vulgaris*. Maheshwari (1950) placed the *Balsamita* example among the plants outside classification, whereas Romanov (1981) distinguished a separate tetrasporic Balsamita-type. The peculiarity of *Balsamita* embryo sac development is represented by two specific highly original processes. The initial embryo sac cell is a mother megaspore cell which feature apparently suggested to Romanov that the *Balsamita* embryo sac should be placed among tetrasporic gametophytes. As shown in Table 3 four megaspore nuclei are distributed in the coenomegasporocyte by 1 + 3 ratio. Only the micropylar nucleus is able to remain viable, while the other three perish. Thus, only one of four megasporocyte nuclei participates in embryo sac development (based on the substitution effect). Meiotic divisions are excluded from its development, which is grounded on three mitotic divisions of one of the megasporocyte nuclei. Apart from that, in the process of embryo sac development polarity inversion occurs similar to that described above under *Balanophora*. The evolutionary background for this inversion remains uncertain. However, we deem there are sufficient grounds for classifying the development of *Balsamita* as a specific Polygonum-subtype variation of monosporic development.

Oenothera-subtype. As opposed to Polygonum, results from two instances of mitosis.

Oenothera-variation. Geerts (1908) established that in *Oenothera lamarckiana* the embryo sac is usually formed from a micropylar tetrad megaspore and completes in two mitotic stages. Four nuclei emerge in this fashion form the egg cell, two synergids and only one polar nucleus. Thus, the *Oenothera* embryo sac is a unipolar formation. It is unclear though why there is no antipodal apparatus. It was supposed until recently that this variation is a characteristic property of Onagraceae taxa only; however, Oenothera-variation of embryo sac was then discovered in several taxa of other families.

Apinagia (= Podostemum)-variation. Maheshwari (1950) considered the extremely original embryo sac development in Podostemaceae to be a diversion from the bisporic Allium-type of embryo sac. Romanov (1981) grouped the developments of several Podostemaceae representatives in a special type, conventionally labelled Podostemon. The most important contribution to interpretation of the embryo sac development types in representatives of this highly original family of aquatic plants was made by Japanese and Indian researchers (Mukkada, 1962a, b; Arekal and Nagendran, 1975a, b) as well as by the famous Italian embryologist-analyst Emilio Battaglia (1971, 1987). Battaglia meticulously analyzed all available data and concluded that all Podostemaceae representatives possess peculiar embryo sac development features, the type of which was first described based on the example of *Apinagia*. The differences concern behaviour of the nuclei excluded from embryo sac development (the additional divisions, fast or prolonged die-off period).

Under this type it is the chalazal dyad cell which develops into the embryo sac; during this process two nuclei form in a each of the dyad cells. The micropylar dyad cell soon perishes completely, and of the two nuclei of the chalazal dyad cell only one micropylar nucleus participates in embryo sac development. The chalazal nucleus then gradually degenerates. Alternatively, this process starts following the chalazal nucleus division, in which case one is actually able to observe the deteriorating nuclei in the formed embryo sac. Thus, it is the replacement effect that manifests in Apinagia-variation embryo sac development. Resulting from two instances of mitosis of the micropylar nucleus of the chalazal dyad cell, which layout enables one to relate this kind of development to the monosporic type, a 4-nucleate embryo sac develops which later evolves into a 4-celled one.

Bisporic type. The nucleus responsible for initiation of embryo sac development under this type is that of the chalazal dyad cell. Second meiotic division is included here in embryo sac development.

Allium-subtype, characterized by three subsequent nuclear divisions, the first of which is the second meiotic division.

Allium-variation. This variation was first described by Strasburger (1879a) based on the example of *Allium fistulosum*. The embryo sac evolves from the chalazal dyad cell resulting from three subsequent divisions (one meiotic and two mitotic). The reorganization process is that of the micropylar dyad cell degenerating. The nuclear and subsequently cellular distribution of Allium-variation is identical to that of

Polygonum: 7 cells, 8 nuclei (3 + 2 + 3) and isobipolar embryo sac. Allium-variation is sporadically present in taxa of several families of various systematic areas of flowering plants.

Tetrasporic type. The nucleus responsible for initiation of embryo sac development under this type is that of the megasporocyte. The developmental process is conditioned by participation of all four megaspore nuclei and both meiotic divisions.

Peperomia-subtype, characterized by four subsequent nuclear divisions. Variations exist in number of nuclei and cells in the embryo sacs, as well as in modes of their distribution.

Peperomia-variation. This variation was described on the example of *Peperomia pellucida* (Campbell, 1899a, b; Johnson, 1900). In the process of development each of the four coenomegasporocyte nuclei divides two times forming 16 nuclei, which get distributed more or less evenly in the peripheral cytoplasm layer of the embryo sac. Thus, one may define its organization as multipolar, the distinguishing feature being the presence of only one synergid. Major development deviations occur in respect to number of nuclei participating in formation of the initial central cell nucleus of the embryo sac. Emrbyo sacs of this variation have been identified in several species of *Gunnera* (Halorrhagidaceae).

Penaea-variation. Direct transformation-free development under this variation is highly similar to the Peperomia-variation. The principal differences include clear-cut tetrapolar organization of the embryo sac and differentiation of two synergids. This variation was first identified in three genera: *Penaea, Brachysiphon* and *Sarcocolla* (Stephens, 1909). It was later also discovered in some genera of Malpigiaceae and Euphorbiaceae. It is sporadically present in a number of other genera.

Drusa-variation. An embryo sac of 16 nuclei was first identified in *Drusa oppositifolia* (Håkansson, 1923). Under this type of development following completion of meiosis in the coenomegasporocyte, three megaspore nuclei descend into its basal end. Subsequent to two instances of mitosis, this distribution transmutes into 2 + 6 and then 4 + 12. The micropylar nuclei form the egg apparatus and the upper polar nucleus, while the 12 lower nuclei form 11 antipodes and the lower polar nucleus. The overall layout of the embryo sac is 3 + 2 + 11 and heterobipolar in character. This variation is sporadically encountered in some other plants, for instance is genera *Crucianella, Tanacetum* and *Ulmus,* and of particular interest in *Chrysanthemum parthenium*. This brings us closer to describing the next two variations.

Martinoli (1939) described two rather unique embryo sac formation types within the same species, namely *Chrysanthemum cinerariaefolia*. Maheshwari (1950) averred that these defeat attempts at regular classification. But like Ambegaokar (cited by Willemse and van Went, 1984) we reckon it is possible to classify them as two original variations.

Chrysanthemum cinerariaefolia variation-I. Following meiosis completion in coenomegasporocyte the nuclei of four megaspores shift to form one vertical row in which the two extreme nuclei are located at the poles and the two middle ones drawn together in the megasporocyte centre. The central nuclei neither divide nor perish and subsequently act as the polar nuclei of the embryo sac. The micropylar nucleus divides two times and its four derivatives later form the egg apparatus and the upper polar nucleus (the third one!). The chalazal nucleus divides twice too giving rise four antipodes.

Chrysanthemum cinerariaefolia variation-II. Of the four formed coenomegasporocyte nuclei, two central ones merge to create one diploid nucleus. All coenomegasporocyte nuclei then successively and synchronously divide. The micropylar quadruple of haploid nuclei forms the egg apparatus and the upper polar nucleus. The chalazal quadruple of haploid nuclei forms four antipodal cells. Of the central quadruple of diploid nuclei, one functions as the lower polar nucleus and the remaining three form additional antipodal cells. The number of haploid nuclear divisions at the chalazal end is unstable as one or two may drop out.

Thus, in this development variation we encounter two transformations: two nuclei fusions effect a diploid nucleus formation, and partial replacement of haploid antipodes with diploid cells.

Fritillaria-variation. Bambacioni (1928a, b) demonstrated that in some species of genera *Fritillaria* and *Lilium* formation of our megaspore nuclei in the coenomegasporocyte is conducive to the secondary four-nucleate stage. At 1 + 3 megaspore nuclei distribution the chalazal nuclei are located in mutual proximity. At the next division three spindles of the chalazal nuclei merge and resulting from subsequent division form two triploid nuclei (Carano-Bambacioni effect). The micropylar nucleus retains its haploid nature and divides. Resulting from the next division there appear four haploid nuclei (on micropylar pole) and four triploid ones. The outcome of this is that the formed embryo sac, following the polar nuclei fusion, consists of haploid cells of the egg apparatus, tetraploid central cell nucleus and three triploid antipodes. This development type was later identified in several members of Liliaceae (Lilioidea), as well as in representatives of some other families, such as *Piper, Rudbeckia, Erithronium, Cardiocrinum, Tamarix, Cornus,* several species of *Tulipa,* etc.

Adoxa-subtype. This subtype is characterized by three consecutive divisions of the megasporocyte nuclei.

Adoxa-variation. This variation was first described in *Adoxa moschatellina* (Jonsson, 1879-1880). It is formed by one mitosis of the four megasporocyte nuclei from which emerges a 7-cell, 8 -nucleate (3 + 2 + 3), isobipolar embryo sac. The variation was later identified in *Adoxa* and *Sambucus,* as well as in some species of *Erithronium, Tulipa* and *Ulmus.* In *Erithronium americana* and some species of *Ulmus,* embryo sac development may from time to time follow either the Adoxa- or Fritillaria-variation. It is possible that the latter is a derivative of the former.

Plumbago-variation. This variation was first described in *Plumbago capensis* (Haupt, 1934). Its peculiarity lies in the nuclear distribution in the coenomegasporocyte following the first mitosis. A solitary synergid-free egg-cell appears in the formed embryo sac. The central cell has four polar nuclei. The remaining three nuclei form egg cell-like structures which usually perish and are not present in the mature embryo sac. This variation is specifically limited to three genera of Plumbaginaceae.

Plumbagella-variation. First described exclusively in *Plumbagella micrantha* (Fagerlind, 1939a). Four nuclei in the megasporocyte are distributed by 1 + 3 formula. Resulting a secondary 2-nucleate stage is formed from the fusion of three chalazal nuclei. A rather unique embryo sac structure (1 + 2 + 1) is thus created. It includes a single egg cell, a tetraploid central cell nucleus and one triploid antipode.

Tulipa tetraphylla-variation. This variation was first described by Romanov (1971b) using the plant of the same name as the example. The distribution ratio of megaspore nuclei in the coenomegasporocyte is 3 + 1. The egg apparatus with one egg cell and four synergids is formed following one mitosis. There are two polar nuclei and one antipode.

Eriostemones-variation. This variation was first described by Romanov (1971b) based on the example of the Eriostemones section of genus *Tulipa*. Development is similar to that of *Tulipa tetraphylla*. However, all 8 nuclei appear in the micropylar end of the embryo sac where the egg apparatus is formed from the egg cell and 6 synergids. One nucleus acts as the upper polar one. There are no antipodes.

The megasporogenesis and megagametogenesis processes analyzed above allow certain conclusions to be drawn.

Some complicated repeated adaptive transformations occurred in the process of female gametophyte evolutional development in various divisions of seed plants, mostly angiosperms. Those transformations exerted various influence on the seed itself and varied among taxa.

Differences in embryo sac formation, their final organization uniformly being 8 (3 + 2 + 3), demonstrate the fact that differences in female gametophyte development are of significant importance for the adaptive evolutional transformations of the ovule. Regretfully, this topic has yet to be properly explored by embryologists.

Manifold morphogenetic processes participate in formation of embryo sacs in various types, subtypes and variations. Respectively, it does not suffice to define female gametophyte development based on only a few rigidly defined criteria.

Lastly, bearing in mind the genetic aspects of the problem, while deciding how to properly classify this or that variation, it would be more appropriate to be guided by the extent of participation of the megaspore nuclei, dyad nuclei and coenomegasporocyte in embryo sac development.

Evolution of Female Gametophyte

Following discovery of fertilization (syngamy) in phanerogams (flowering plants) by Strasburger (1884), Le Monnier (1887) put forward the hypothesis that polar nuclei fusion in the central embryo sac cell may likewise be an instance of fertilization comparable to that identified in egg cells. He suggested that the endosperm of phanerogams be considered the second embryo transformed during the evolutionary process into a nutrient tissue for the developing embryo originating from the zygote. For Nawaschin (1898) who discovered the *'triple' fusion effect*—fusion of the two polar nuclei or the central cell nucleus derivative thereof with the other sperm cell nucleus—it was logical not only to subscribe to this opinion, but also to designate the functions of both sperm cells collectively as *'double fertilization'* (Nawaschin, 1900). One can still encounter the 'second embryo' endosperm theory (Favre-Duchartre, 1984) but with the important amendment that the antipodes are included in the somatic gametophyte tissue.

Most researchers acknowledging the 'double fertilization' phenomenon consider the embryo sac (female gametophyte) of angiosperms a neotenous formation, inasmuch as it is understood as a consequence of the overall female gametophyte reduction in gymnosperms (Romanov, 1944, 1971b; Battaglia, 1951; Gerassimova-Navashina, 1958; and others). Thus, considering that female gametophyte development in phanerogams results in formation of sexual structures (as is the case with gymnosperms) and unable to reconcile themselves with the 'second embryo' notion of endosperm, many researchers hold that the endosperm is a new formation separated from the female gametophyte (Schnarf, 1942; Swamy, 1974). This outlook rests on two principal points:

(1) the endosperm in phanerogams is viewed as a fertilization product and (2) the endosperm in angiosperms is 'normally' a triploid formation, as opposed to monoploid endosperm (virtually, a prothallium) and other embryo sac structures.

On the other hand, even Strasburger (1900), after thorough analysis of the problem, supposed that the 'triple fusion' was not proper fertilization, but only a stimulus to endosperm development. Following Maheshwari (1950), Battaglia (1951), reasonably rejecting both variants of archegonial theory put forward by Porsch (1907) and Schurhoff (1919, 1928), considered the 'triple fusion' to be a stimulus to mitotic activity of the central cell. However, he still supported the idea of overall endosperm reduction in the process of evolutional transition to angiospermous type and held the view that the endosperm in phanerogams is a secondary endosperm, while antipodes are construed as a residual feature of the primary endosperm homologous to that of gymnosperms.

Fagerlind (1941b) formulated the principal highlights of the 'Gnetum' theory very precisely: (1) the polar nuclei of the central cell of the embryo sac in angiosperms correspond to the last remnants of the free nuclei in *Gnetum* and to the early stages of gametophyte development in other gymnosperms; (2) the central vacuole of the embryo sac in phanerogams in homologous to that in *Gnetum* and other gymnosperms; (3) the egg cell is homologous to the fertile peripheral cells in *Gnetum* or the retarded archegonia, while the antipodes correspond to the 'lower' trophic tissue of the *Gnetum* gametophyte; (4) the endosperm in phanerogams is a 'retarded' gametophyte tissue whose further development is stimulated by its mergence with a male gamete. None of these conclusions signifies that phanerogams developed from *Gnetum*. They do, however, mark the direction of changes in female gametophyte evolution over the time of transition to angiospermous type.

It seems surprising that despite all the above described efforts, most embryologists support the concept of 'double fertilization' and the idea of the endosperm as a new outside-gametophyte formation. Quite possible there is a pressing need to occasionally recall the ideas and the arguments already expressed and to adduce new facts and additional arguments if the assumptions of the republic of science are to correspond to the true results of the evolutionary process. Battaglia (1951) and later Gerassimova-Navashina (1958) developed Fagerlind's ideas propounded in Fagerlind (1941b) concerning egg cell substitutions in gymnosperms archegonia (with the exception of *Gnetum* and *Welwitschia*) by their predecessors, i.e., cells formed in the micropylar gametophyte zone in angiosperms at early stages of their development. We see Battaglia's mistake (Battagila, 1951) in that while having noted the archegonial structure reduction, he still considered it a part of the overall female gametophyte reduction.

In summary, it is possible to single out certain aberrations as expressed by most embryologists nowadays: (1) notion of overall female gametophyte reduction in seed plants related to their transition to the angiospermous type; (2) conception of the second sperm cell nucleus mergence with the somatic central cell nucleus (nuclei) as a 'double fertilization' act; (3) conception of the endosperm in phanerogams as a 'new formation', not homologous to the prothallium (endosperm) of gymnosperms or to a part thereof.

Now we shall critically scrutinize primarily the last of the above-listed aberrations, which in turn will enable elucidation of the first two. It was Smirnov (1982) who first doubted the fact that the gymno- to angiospermous type evolutional transition was accompanied by overall reduction of the female gametophyte, construed by Smirnov as including the endosperm because the latter is a formation substantially homologous to the prothallium in gymnosperms. He held the opinion that attempts to consider the endosperm in phanerogams as a 'second embryo' or as a 'new formation' dramatically alter the composition (organization) of the interacting structural elements in a seed in the process of generation successions, which, firstly, is unlikely to be the case and, secondly, is harldy necessary. He also put forward a hypothesis of from-single-to-double-phase female gametophyte development in the process of transition to the angiospermous type. The core of his assumption is that not exactly effective egg cell fertilization taking place in gymnosperm archegonia formed at the final development stage of the female gametophyte (and the seed in general) in phanerogams is shifted to the early stages of gametophyte development, becoming a factor conducive to development of only fertile seeds.

As expressed by Teryokhin (1991), this was occasioned by selective reduction of the sexual (archegonial) gametophyte structures and substitution thereof by ontogenetic predecessors.

In this respect it would be interesting to consider the 'compositional' changes in homologies with respect to the angiospermous type. It is sufficiently obvious that gymnosperms have a uniform areal zonality of the female gametophyte related to its polarization (so-called 'polarized zonality'). It would be beneficial to briefly discuss this phenomenon based on the least complicated *'Welwitschia'* and *'Gnetum'* models. Sexual structures (egg cells) are formed in the micropylar gametophyte area, the central area being occupied by a reservoir trophic tissue referred to in Pearson (1929) as the *trophophyte,* the chalazal area being a structural and functional complex of the haustoria. Such a set-up is quite rational considering the ways the nutrients feed in from the mother plant on the one hand, and the sperm cell passage to the female gametophyte, on the other. This has been confirmed by all works on seed development in gymnosperms known to us and in reality observed by us on the example of seed development in *Thuja occidentalis* and *Taxus baccata.*

Such organization in gymnosperms proceeds gradually over the time of gametophyte development. In angiosperms with a 'normal' embryo sac type, a similar composition sets in speedily (momentarily) and synchronously in the process of primary cell formation. Here we construe the egg apparatus (egg cell and synergids) as a part of the sexual micropylar complex, the central cell as an initial element homologous to the trophophyte, and the antipodes on the chalazal gametophyte pole as the haustorial structures. During the period of further development in the chalazal part of the trophophyte structure a new haustorial complex is often formed replacing the ephemeral antipodes. Occasional deviations, for instance formation of the endospermal haustoria in the micropylar trophophyte area (as observed in figworts), or haustorial activity of the synergids, do not interfere with the overall principles of organization.

Arguments favouring the trophophyte (endosperm) in phanerogams as a new formation, prompted by the facts of its ploidy and 'triple fusion', do not appear convincing. Battaglia (1951) considered the issue of the secondary nucleus ploidy in the central cell and concluded that it did not alter the concept of this structure of female gametophyte as somatic, not sexual. In fact, when looking at embryo sacs with unreduced (diploid) cells in certain forms of apomixis, we do not deny their homology to haploid embryo sacs of the same type. So why then should this position be changed in relation to endosperm? Despite the fact that there have been no general observations of the overall endosperm and respectively gametophyte reduction in angiosperms, in certain taxa of angiosperm parasites specializing in the most prolific production of seeds in each ovary, endosperm reduction is present, which in extreme cases sometimes develops into complete endosperm elimination. Several cases are briefly studied below to illustrate and support the above-mentioned reasonings in relation to the morphology of the endosperm in phanerogams.

In pulverulent seeds of representatives of subfamily Pyroloideae of Ericaceae with reduced embryos, the cell endosperm is also underdeveloped, the degree of its development related to the degree of reduction of the embryos. This process is often accompanied of the chalazal nuclei reduction of the embryo sac (Teryokhin, 1977). Instead of three antipodal cells, two or sometimes even one cell with pycno-nuclei is formed. Thus, because of the underdeveloped endosperm (trophophyte) the haustorial structures responsible for it are likewise reduced. We observed a similar situation in *Balanophora thwaitesii*, also (Teryokhin and Yakovlev, 1967). As a consequence the chalazal pycno-nuclei often merge and share one cell with the polar nuclei, this cell functioning as initial (central) in the underdeveloped trophophyte.

In highly specialized Orchidaceae microparasites the extent of this process reaches considerably further. Owing to their specific biology these plants do not require an endosperm following ripening of the seed (Teryokhin, 1977). Consequent to the most considerable embryo reduction, some extreme cases of endosperm reduction are noted here. For instance, in *Neottia nidus-avis* only four free nuclei are formed in the central cell, whereas the predecessors of the antipodal nuclei and cells (including the chalazal polar nucleus) pycnotize already at the 4-nucleate embryo sac stage and disappear by the time the first or second division of the 'fertilized' micropylar nucleus occurs (Teryokhin and Kamelina, 1969, 1972). A similar situation is observed in several other specialized orchids, e.g. in genus *Epipactis* and several others.

In *Habenaria linearifolia* the reduction process reaches even further compared to that described above (Teryokhin and Kamelina, 1972). Although three antipodal cells and both antipodal nuclei are formed here, following the mergence of the central cell nucleus with that of the sperm cell the resultant nucleus does not divide but pycnotizes soon upon having considerably expanded in size. Thus the stimulating role of the second sperm cell is lost in the central cell developmental process.

Let us now compare the two example. In *Neottia*, following the upper polar nucleus mergence with the second sperm cell the diploid nuclear endosperm develops weakly, and in *Habenaria* the central cell nucleus does not divide in the same situation. Evidently, the second sperm cell only performs a signalling (trigger) function, and not the mitosis-stimulating function (as suggested by Battaglia, 1951). It is also probable that the ploidy level of the central cell exerts no decisive impact on its development. It should also be noted that only the extreme cases of suppressed trophophyte development worked through in the process of evolution have been described above.

We do not touch upon gametophyte apomixis phenomena in this article (the importance of those phenomena for female gametophyte evolution was dwelled upon somewhat by Battaglia in 1951). However, we shall look at two more highly demonstrative examples having direct bearing on the topic under discussion. In *Ophrys insectifera* (Savina, 1972) and possibly *Arundia speciosa,* in two genera, of *Phajus* (Nawaschin, 1900) and in *Trichosma suavis* (Baranov, 1971), reduction of the chalazal and even the central section of the embryo sac is so serve that the sperm cell upon reaching the central cell interior (where the chalazal nuclei may also be present) does not merge with the polar nuclei. It remains isolated until the overall degradation of the central cell nuclei is completed. In this state, however, the central cell (a superminitrophophyte) is capable of providing the nutrients for development of the reduced embryo. It is particularly in this case that the initial endosperm contains haploid nuclei which do not merge, whereat the entire structure functions as a haploid miniprothallium, or, more prescisely, as somatic prothallium tissue, following formation of the sexual structures. The overall dynamic 'composition' (polarized zonality) is retained here.

The heretofore adduced examples suffice, in our opinion, to clarify homology issues regarding female gametophyte structure in angio- and gymnosperms. They confirm the assumptions expressed by Fagerlind (1941b) with an amendment that the central cell of the embryo sac is homologous to the central area (trophophyte) of gymnosperms, as are the antipodes to the basal haustorial area in their gametophytes.

We shall adduce yet another example wherein the disbalance between the evolutionary origin of a structure and its chaining function is demonstrative of the unity in gametophyte composition in gymno- and angiosperms. In Apinagia (= Podostemum)-type embryo sac development in family Podostemaceae (Battaglia, 1971, 1987), the same trend towards reduction of chalazal structures of the embryo sac is encountered as was observed in the above examples. Adaptation to very special life conditions in fast-streaming waters with abrupt changes of seasons of rain and drought resulted in the following changes. The embryo sac is formed on the basis of four micropylar nuclei. Underneath the base of the developing embryo sac there forms a special perisperm represented by a free-nucleate structure (Arekal and Nagendran, 1975a, b). Two synergids usually form on the micropylar pole with the egg cell underneath and another cell in the base homologous to the central one but having the antipodal (haustorial) function (Teryokhin, 1991). Fertilization is of the unitary type wherein only the sperm cell functions whose nucleus merges with that of the egg cell (Mukkada, 1969; Battaglia 1971). The impossibility of the second sperm cell of the embryo sac taking part in fertilization is implicitly confirmed by changes in both number and distribution of the embryo sac structures where in some cases one synergid, the egg cell and two basal cells are formed. Sometimes even the general distribution of cells in the embryo sac is disturbed. It remains unclear, however, whether this is the normal run of things or an anomally in development.

This example, in our opinion, highlights once again the similarity in female gametophyte organizations of phanerogams and gymnosperms. Such examples, which considerably narrow the gap between angio- and gymnosperms, within the domain of seed reproduction, will evidently grow in number. On the other hand, it is not likely that we shall witness even one example capable of proving the 'embryo' nature of the endosperm or its special role as a 'new formation'.

It is surprising how easily embryologists recognized the 'double fertilization' concept proposed by Nawaschin (1898) and that several generations of embryologists have mostly failed to analyze it critically. The arguments regarding homology of the basic female gametophyte structures in phanerogams and gymnosperms supplied by Fagerlind (1941b), Battaglia (1951), Smirnov (1982), and the present article appear to be quite convincing. It is probably because of this property that no one has so far ventured to dispute them. It should be noted that penetration into more and more 'intimate' (subtle) structures of female gametophyte using constantly improving state-of-the-art technology cannot apparently answer the questions heretofore discussed. The answers to these questions will be found in another research domain, namely, in comparative and ecological morphology, which nowadays, given the current infatuation with 'experimental' embryology, unfortunately enjoys less and less attention.

The data supplied in the previous chapter demonstrate the principal correspondence in organization (layout) of female gametophyte in angio- and gymnosperms (i.e., 'compositional' similarity, as worded by Smirnov, or 'polarized zonality', as we termed it above, which is probably a more precise definition). As rightly noted by Smirnov (1982), the system of ontogeny wherein the endosperm in phanerogams is taken to be a second embryo (to which conclusion one inevitably arrives if working within the framework of any variety of the archegonial hypothesis) demands a fundamental readjustment of the historically moulded relationships between the female gametophyte, the embryo and the maternal sporophyte. The requirement here is that the natural metabolic and morphogenetic continuous interrelation between the embryo and the prothallium be substituted with that between the embryo and another *endosperm-resembling* structure of the additional embryo wherein the latter will acquire all properties and characteristics of a porthallium. In this case, by the way, it will be impossible to establish the origin of phanerogams either from gymnosperms or from any other known group of higher or lower plants. At this point we would like to highlight that since we consider the homology of the endosperm in phanerogams and the trophophyte (the central section of the endosperm) in gymnosperms firmly established, that the central embryo sac cell is a somatic and not a sexual structure is now beyond dispute. Therefore fertilization in phanerogams, as in all other taxa of the plant world, is a single phenomenon consisting of sperm cell and egg cell nuclei fusion, i.e., it is a single (individual) conjugation of made and female sexual cells. We showed above that in certain specialized taxa of parasitic and aquatic plants the singalling role ('trigger role' in Teryokhin's terminology; Teryokhin, 1991) of the second sperm cell nucleus in relation to the ones of the central somatic cell can be partially or, later, completely lost. To develop a more precise notion of the nature of the evolutionary processes pertaining to fertilization or, more specifically, to the functioning of the 'additional' sperm cells, it is evidently necessary to analyze behaviour of those cells in some taxa of gymnosperms.

According to the information available today, in most cases in gymnosperms the second sperm cell (or second nucleus thereof) does not participate in fertilization. For instance, in *Araucaria, Pinus* and *Podocarpus* the recipient vacuole takes in both the sperm cells, but only one takes part in fertilization while the other soon degrades. In Taxodiaceae and *Juniperus* the larger sperm cell is usually taken in by the egg cell, whereas the other sperm cell stays in the pollen tube and likewise degrades (Chamberlain, 1935; Kondratyaeva and Tikhova, 1982).

Meanwhile, in gymnosperms there are some cases of spontaneous or even regular participation of the second sperm cell in fertilization related processes. Favre-Duchartre (1984) divides such phenomena into two categories. Firstly, there are cases of interarchegonial fertilization when both sperm cells (or spermatozoids) which share one pollen tube are taken into the neighbouring archegonia. Such cases have been described in *Ginkgo* and *Callitris* (Lotsy, 1899; Favre-Duchartre, 1956).

Secondly, there is the second sperm cell mergence with non-sexual nuclei of the same archegonium. Such casees have been identified in *Thuja, Abies, Pseudotsuga,*

Ephedra and *Ginkgo* (Land, 1907; Hutchinson, 1915; Herzfeld, 1922, 1927; Khan, 1940, 1943; Mulay, 1941). Pearson (1929), Gerassimova-Navashina (1958) and Friedman (1990) see such cases as a prototype to 'double fertilization'. Sometimes such spontaneous phenomena acquire biological meaning in competition. For instance in Cycadaceae and *Ginkgo* both spermatozoids are taken into the egg cell or try to the taken therein (Lee, 1955; Swamy, 1973). In *Gnetum* both sperm cells are taken into the terminal part of the female gametophyte and move towards the closest sexual nucleus (Swamy, 1973). Sometimes the nuclei of both sperm cells merge with two neighbouring sexual nuclei (Lotsy, 1899). In such cases the sperm cells seem to continue the male gametophyte competition which they had begun earlier. For instance, in the pollen chamber of *Ginkgo* a number of pollen grains usually germinate, but the moment the spermatozoids are released from any of the pollen tubes the other pollen tubes start to degenerate (Lee, 1955). A similar situation with such competitive behaviour was observed in *Welwitschia* (Pearson, 1929).

Friedman (1990) drew the conclusion that 'double fertilization' in *Ephedra nevadensis* is a regular process because the second sperm cell nucleus merges with the abdominal channel nucleus (a sister egg cell nucleus) on a regular basis. In this particular case we indeed observe a phenomenon similar to fertilization in phanerogams, when the second sperm cell nucleus naturally merges with a non-sexual nucleus, in essence with a nucleus with the somatic structures of the sexual organ.

Our understanding is that in the process of evolutional development of the female gametophyte in gymnosperms in accordance with the *Welwitschia* model, transition from spontaneous to regular functioning of both sperm cells, i.e., to 'double' fertilization (based on the preceding competitive behaviour) is quite possible. Note that in *Welwitschia* where there is apparently no clear-cut distinction between somatic and sexual cells of the micropylar gametophyte zone, fertilization occurs at the cellular developmental stage and the nuclei in the primary cells, including sexual ones, are capable of fusion.

Thus in several 'model' taxa of gymnosperms one will find individually all basic constituents of 'double' fertilization in phanerogams: spontaneous fusion of the second sperm cell nucleus with the somatic one (in a number of taxa); mutual fusion of nuclei in one somatic cell in the area adjacent to the sexual cells location (in *Welwitschia*); and transition of the former phenomenon into a regular process (in *Ephedra canadensis*).

It was this combination of the above-described elements into a unified system in some heretofore unidentified taxon (or taxa) ancestral to gymnosperms that resulted in formation of the 'double fertilization' model known today. To make this possible there apparently must have been a combination with a development type of female gametophyte where concurrent with cell formation the sexual (gametes), transitional (synergids), trophic (central cell) and haustorial (antipodes) structures were differentiated. The closest to such sexual reproduction system composition in phanerogams is the *Welwitschia* model, which, however, does not fit the aforesaid case.

In phanerogams the case described above is most appropriately met in 7-cellular, 8-nucleate embryo sacs in Polygonum (monosporic), Allium (bisporic) or Adoxa (tetrasporic) variations. Considering the widest distribution of monosporic reproduction type in angiosperms, it is probably this fact, and not extreme antiquity, that would explain the wide distribution of Polygonum-subtype embryo sac in phanerogams.

Rather numerous deviations from normal developmental type in phanerogams obviously require broader and more detailed research with the objective of elucidating significance of such deviations as far as their concrete adaptive ontogenetic or ecological aspects are concerned. Apparently breaking free from the idolized 'double fertilization' and other 'germination' or 'new formation' hypotheses would allow us to move along in elucidating the evolutional meaning and purport of the fractionated female gametophyte development in angiosperms, of fertilization as the key link therein, and of the entire problem of sexual reproduction of seed plants.

TRANSPORT OF SUBSTANCES IN THE OVULE

Vascular Supply of the Ovule is provided through the terminal branches of the conductive bundles of the gynoecium, leading to the funiculus through the placenta and in most angiosperms ending in the chalaza (Plate 76).

The ends of the conductive bundles can branch out to some extent. The branches are mostly short and do not extend beyond the chalaza, as is the case with some representative of Euphorbiaceae (Landes, 1946), or else they may have extensions into the integuments or into the nucellus, which rarely occurs. The conductive bundle entry into the nucellus was described by Kershaw (1909) in *Myrica gale*. However, in the illustration adduced by him, one can see three rows of oblong cells in the central part of the nucellus, which can only be likened to procambial elements and differ from the differentiated vascular elements of the integumental conductive bundles at the same stage of ovule development. Unfortunately, the author provides no information about further life of the procambial nucellus strand. As described in Fagerlind (1947), the nucellar conductive bundles in *Anacolosa frutescens* and *Strombosia zeylanica* (Olacaceae) reach as far as the embryo sac base surrounding its basal part. In some representatives of Asclepiadaceae, Capparidaceae, Resedaceae, Casuarinaceae, Liliaceae, Thymelaeaceae and Olacaceae, so-called, nucellar tracheids have been identified, which are dispersed in the nucellus and not directly related to the conductive bundles (Johri and Kapil, 1953; Gupta and Rajeswari, 1977). According to Landes (1946), appearance of vascular elements in the nucellus is related to hypertrophied growth of the chalazal tissue. On the whole, vascular elements in the nucellus have been reported in genera of 17 families. Nevertheless the issue of whether there i vascular tissue in the nucellus cannot be considered fully clarified. Substantiated doubts have been expressed by certain authors as to whether singular tracheid-like cells encountered in the nucellus are in fact related to the conductive tissue system or are just the cells with reticular wall thickening (Bouman, 1984).

Differentiated conductive bundles have been found in the integuments of a large number of angiosperms. As reported by Bouman (1984), the conductive bundles in the outer integument have been identified in representatives of at least 80 families (Annonaceae, Cururbitaceae, Euphorbiaceae, Leguminosae, Myristicaceae, Rosaceae, Sterculiaceae, Vitaceae). The character of vascularization varies from a simple bundle in the antiraphal region to a dense network of minute bundles throughout the entire integument (Kuhn, 1928). Vascular bundles have been identified in the inner integument in representatives of Dipterocarpaceae, Elaeocarpaceae, Euphorbiaceae and Myristicaceae (Corner, 1976). Conductive bundles have been reported in both integuments for *Aleutrites*, *Dalechampia* and *Cremophyllum*.

The most developed conductive system in the ovule is found in representatives of some groups of palms (Arecaceae). Thus in *Hyphaene indica*, as reported by Mahabale and Chennaveeraiah (1957), a massive integument is permeated 2/3 its length by 16-17 conductive bundles. Concomitantly, on the example of palms the dense network of vascular bundles in the seed coats can be mostly traced not to the integuments but to the overgrown chalaza. *Hyphaene indica* apparently forms pachychalazal ovules (and seeds) in which a considerable part of the coats is formed from the overgrown chalaza with the entire system of bundles branched therein.

In Polymoniaceae, Diapensiaceae, Podostemonaceae, some genera of Solanaceae, many genera of Campanulaceae the conductive bundle does not reach the chalaza and sometimes even stays in the funiculus. Complete reduction of the conductive system is found in Pyrolaceae, Orchidaceae, Orobanchaceae, Crassulaceae, Loganiaceae and Loasaceae (Savchenko, 1973).

Concerning the overall structure and cytological composition of the conductive bundles the information available is quite meagre. In the funiculus and the raphal area a bundle is mostly collateral, in the integuments and the nucellus its structure can change to amphicribral or amphivasal. Conductive bundles usually consist of a few tracheids and even fewer elements of the phloem.

Considering the fact that in most ovules in angiosperms the conductive bundles end in chalaza, one is bound to draw the conclusion that the function of the specialized vascular tissue is largely limited to supply of water and plastic substances to the ovule. Redistribution of nutrients and metabolically active substances within the ovule itself is apparently carried out by other systems of transport (please see below), in which the chalazal zone in the ovule plays a special role.

The nature of the conductive system of the ovule can in some cases be of taxonomic significance. This is beyond dispute for palms (Uhl and More, 1971) and also a number of other families. Regarding the phylogenetic significance of the actual fact of conductive system elements being present in the ovules of angiosperms, particularly in the integuments and the nucellus, contradictory opinions have been expressed. It cannot be denied that of all the plant parts the ovules have the weakest vascularization and this feature can generally be considered evolutionally advanced.

Nutrient Supply and Transport in the Ovule are carried out in accordance with its two functions: embryo sac and seed coat development. Information on the composition of the nutrients and their pathway to the developing embryo sac and the seed coat is limited. However, some conclusions can be drawn based on the morphological and physiological parameters of differentiation of ovular tissues.

At the early stages the ovule is directly connected to the placenta. In the funiculus and the chalaza there evolves a tissue compounded of elongated cells differentiated into primary xylem and phloem elements. This provascular strand often terminates in the chalazal part of the nucellus. The degree of development of the vascular tissue varies: it may be absent in plants with minute ovules or well elaborated and branching towards both or one of the integuments. Differentiation of the vascular system in some taxa may be completed after fertilization in the process of seed development.

In the initial stage of ovular development when all its cells are meristematic, transport is effected both as symplastic (along the plasmodesmata) and apoplastic (along the young cell walls). During this period evidently no significant change in either supply or transport of nutrients occurs. As is well known, in the crassinucellate ovule the cells may divide longitudinally parallel to the main axis of the ovule. As a result, rows of cells are formed with the preferred plasmodesmal contact longitudinal in direction. The first signs of change in the system of transport are evidenced at the time of meiosis and isolation of meiotic cells by development of a callosic wall. The cytoplasm mass with organelles increases and a thick cell wall is formed consequent to an additional nutrient supply. During the time of linear tetrad development some nuclear cells start to elongate. The initial phase of functional megaspore development is accompanied by degeneration and disappearance of the callosic walls. According to De Boer-de Jeu (1978), the degrading megaspore products are reutilized as nutrients for the functional megaspore. The megaspore contains various substances in reserve: lipids, starch and, probably, proteins. After formation of the callose shell a temporary molecular sieve forms; some changes additionally occur in the apoplastic exchange as well as storage of carbohydrates which are essential for proper nutrition of the functional megaspore. In some ovules the hypostase begins to form at the dyad or tetrad stage. Diversity in both differentiation and layout of the hypostase is considerable. In some cases this diversity is manifested as booming activity of the group of cells located at the chalazal end of the functional megaspore. Cellular activity is determined by peroxidase, as well as acid and alkaline phosphatase reactions (Willemse and Bednara, 1979). The hypostase cell walls display strong fluorescence due to the presence of phenol compounds and react with lignin and suberin. This peculiarity demonstrates the changes in apoplastic transport towards and away from the developing megasporocyte. In this area nutrient intake through the hypostasic cells is effected mainly by the symplast; saturation of the cell walls with phenol compounds hampers apoplastic transport from the strongly vacuolated coenocyte and later from the embryo sac. In the opinion of Boesewinkel and Bouman (1984), the overall hypostasic function involving hormonal secretion or other cytoplasmic products can be conditioned solely by symplastic transport in this system.

Information on the origin and function of the epistase is rather meagre. That available highlights the presence of lignin- or suberin-like substances in the coatings and possibly obstruction of apoplastic transport in the micropyle.

Development of the functional megaspore into a 2-nucleate coenocyte begins with formation of vacuoles and later their fusion into one large central vacuole. This process denotes water absorption from the surrounding nucellus tissue. The substances reserved in the functional megaspore are consumed during the time of coenocyte development from the 2-nucleate to the 8-nucleate stage. Starch hydrolysis may cause osmotic pressure in the coenocyte. Using fluorochromes it was possible to demonstrate the absorption of substances mainly around the developing coenocyte and later around the embryo sac (Mogensen, 1981). Around the growing coenocyte, in particular in the micropylar area, nucellar cells die off, thereby providing not only space but also nutrients for the growing coenocyte and later the embryo sac. In some plants the entire nucellus in the top part and lateral sides of the embryo sac disappears. In such cases a special integumentary tissue responsible for nutrient supply to the embryo sac forms, namely the integumentary tapetum.

In a mature embryo sac, wall transformations can be seen in both the synergids and the central cell into distinctive transmissive structures with a labyrinthal coating. The antipodal cells in the embryo sac base may also contain such structures. The function of such compartments appears to be short-distance transport through the coating, which can intensify following fertilization (Johri and Ambegaokar, 1984).

Together with the 'transmissive' walls the tubular cell growth offshoots appear directed towards the surrounding tissues, which function like haustoria and are responsible for nutrient absorption. In some plants these haustoria are found already in the megaspore, in others the haustorium is formed in the upper part of the coenocyte or in the synergids. In the basal part there may also be haustoria originating from the retained antipodal cells. Antipodal cells are those participating in nutrient supply to the embryo sac. In the case of early degradation, their participation is highly ephemeral. Reserved substances such as starch or sometimes lipids may be present in the central embryo sac cell.

In the *Gasteria* ovule two rings of cells containing glucose are found: one coils around the upper part of the embryo sac and the other in the basal part near the chalaza. Soluble carbohydrates accumulate in these areas and in the upper part a starch deposit also. Starch accumulation additionally takes place in the central cell. Such zonality in distribution of some nutrients shows just how complex the system of nutrient supply to the embryo sac can be.

The nucellus tissue is surrounded by the inner and outer integuments. These tissues grow longitudinally due to cell divisions and later stretching. With final formation of the embryo sac, further development of the inner and outer integuments is temporarily suspended and intercellular spaces may appear therein. These spaces may be filled with gas, which is proof of its transport in these tissues.

Nutrient transport in the ovule undergoes changes following fertilization when the embryo and endosperm are formed.

Metabolite Translocation Pathways in Developing Ovules (Plate 77): Space-time coordination of the growth, differentiation and metabolic processes in the ovule, as well as in the entire plant on the whole, is ensured at different organizational levels by three types of transport (Kursanov, 1976; Lüttge and Higinbotham, 1984); (1) between the ovule tissues and the placenta—long-distance transport; (2) between cells of various tissues—medium-distance transport; and (3) between organelles of each cell—short-distance or intercellular transport.

Long-distance transport is effected by means of the phloem and the xylem. Trophic connection of the ovule and the placenta is effected through the funicular conductive bundle, which is most plants ends in the chalaza and more rarely extends into the integuments (e.g. in Campanulaceae, Magnoliaceae, Paeoniaceae). In a number of highly specialized taxa this connection is achieved either through medium-distance transport or through procambial cell strands (Buddlejaceae, Caprifoliaceae, Gentianaceae), or through the cell system of all funicular tissues (Gesneriaceae, Orchidaceae). In this case the plastic substances from the placental conductive bundles undergo symplastic transfer into the ovular structures. In the formed ovule short-distance metabolite transport is found primarily in the nucellus, chalaza and integumentary parenchyma, whose thin cell walls are permeated by numerous plasmodesmata. In the epidermis of the integuments plasmodesmata are found only on the radial cell walls and hence transport of substances is effected layer by layer (Mogensen, 1973; Wilms, 1980; Belyaeva, 1983). Short-distance transport in each cell of the ovule is effected by cellular organelles.

In the ovules, substances are transferred mostly from the receptacle, the ovary walls and the placenta. They emerge primarily in the chalaza, where the conductive bundle ends in many taxa, and then are transported into various reproductive and somatic ovule tissues. Metabolites can accumulate for a certain period of time in the nucellus and the integuments, and then be moved again to the chalaza (Buell, 1952; Coe, 1954; Mogensen, 1973; Bhandari *et al.*, 1980, 1985; Tilton and Lersten, 1981a, c). In ovules with an integumentary tapetum, the cells produce various physiologically active substances, which can be transferred directly into the chalaza. Apart from that, under the influence of hydrolytic enzymes (amylase, protease) secreted by the integumentary tapetum cells, the contiguous cells of the integumentary parenchyma may perish, following which the hydrolytic products also move to the chalaza (Zinger, 1958; Newcomb, 1973a, b; Van Went, 1989; Shamrov, 1990d).

Transport of substances from the chalaza to the developing sporogenous and gametophyte structures is effected by means of various specialized tissues, among which the hypostase plays a significant role (Lonay, 1922; Zinger, 1958; Batygina, 1974, 1984b; Malik and Vermani, 1975; Tilton, 1980; Shamrov, 1990d). Location of the hypostase at the base of the nucellus and the integuments and its contact with the conductive system of the ovule probably indicate its role as a *sui generis* 'distributor' of nutrients in the ovule, providing them to both the reproductive structures and the nucellus with integuments. In typical tenuinucellate ovules, the hypostase immediately adjoins the embryo sac. In crassinucellate and intermediate ovular types, transport of nutrients from the hypostase into the embryo sac is effected from either the central part of the podium (Ceratophyllaceae, Juncaceae—Shamrov, 1992a) or through the postament (Fagaceae, Grossulariaceae, some representatives of Orchidaceae—Mogensen, 1973; Shamrov and Nikiticheva, 1992). In ovules of some plants various structures (synergids, antipodes, endosperm, suspensor, etc.) form the haustoria, whose cells contain hypertrophied nuclei and dense cytoplasm. They absorb nutrients from the adjoining tissues and provide them to the developing embryo sac, and later the endosperm and embryo.

In the embryo sac cells wall outgrowths are formed, which expand the surface of the plasmalemma and play an important part in transmembrane substance transport, being responsible for the functions of absorption and secretion (Vijayaraghavan *et al.*, 1988a, b). The time of outgrowth appearance on the embryo sac walls correlates either with the time of depletion of substances reserved in the surrounding nucellus cells or the integument, or degradation of those cells (Schulz and Jensen, 1968a, 1969, 1977; Fougère-Rifot, 1978; Kudlicka *et al.*, 1981; Tilton *et al.*, 1984; Vijayaraghavan *et al.*, 1988a, b).

Information about the contents and quantities of transported substances is fragmentary and describes either the separate tissues or individual stages of the ovular development. Information about metabolite translocation pathways in developing ovules, despite being extremely meagre does demonstrate the specifics, conditioned by the peculiarities thereof (Malik and Vermani, 1975; Bhat, Vijayaraghavan, 1980; Kallarackal and Bhatnagar, 1981; Vasilyeva *et al.*, 1984). Hystochemical investigations have identified the predecessors of nucleic acids, proteins and polysaccharides in the placenta during the time of archesporial cell differentiation, these substances probably being fed into all primordial ovular tissues (Shamrov, 1990d, 1992a; Anisimova and Shamrov, 1992).

Until formation of strands in the procambial cells is completed (for instance, in the crassinucellate ovule of *Luzula pedemontana* at the stage of complete megasporocyte formation, and in the tenuinucellate ovule of *Gentiana cruciata* following completion of megasporogenesis) metabolites continue to be fed into all structures under formation and accumulate therein in large amounts. In the crassinucellate ovule at the stage of megasporogenesis both new and already reserved substances are used up as callose deposits in the megasporocyte and megaspores under formation. This is probably related to the growth processes and intensive cell divisions in locations where ovular structures are formed. However, with commencement of ovule development, when the podium and conductive bundle with elements of the phloem and the xylem are formed, metabolites accumulate once again in the somatic ovular tissues.

In the tenuinucellate ovule of *G. cruciata*, when formation of the embryo sac is initiated, the nucellus perishes and the procambial cell tissues together with the integumentary tapetum are formed, redistribution in metabolite translocation is effected. Substances are transported both through the procambial cells from the placenta and from the majority of the somatic ovular tissues into the developing embryo sac, which process is identifiable by increase in starch and proteins therein. In the crassinucellate ovule of *L. pedemontana* this peculiarity is detectable only after fertilization: upon initiation of the early globular embryo the substances begin to be fed mostly into the embryo and the endosperm.

It has been hypothesized that in the tenuinucellate ovule with a developing embryo sac the N-metabolites mostly arrive through the ovular epidermis, the integumental tapetum and the hypostase, and the C-metabolites mostly through the procambial strand cells and the hypostase (Shamrov, 1990d). Possibly, the transport of fluorochrome from the funiculus both into the chalazal and micropylar sections of the formed ovule in some plants is likewise demonstrative of the same pattern of metabolite translocation into the ovule (Mogensen 1981; Zhang and Zheng, 1988). In the crassinucellate ovule with a well-differentiated conductive system the metabolites are fed into the surrounding tissues from the placenta predominantly by the conductive bundle through the system of specialized tissues.

In accordance with the assumptions expressed by Batygina (1984b), in the ovary and the ovule there are several substance transport pathways which can change during the process of development and be effected by various structures. Based on studies carried out on *Nelumbo nucifera* and *Triticum aestivum* (Batygina, unpublished data) and analysis of available publications, at least three transportation pathways appear likely: apical, basal and lateral. At different development stages one of the pathways may prevail, whereas it may also be completely absent in some plants. Such a situation largely predetermines the progress of sporogenesis and embryo sac development type.

Transport of assimilators towards the recipient tissues (meristems, storage tissues) in a plant is provided by the system of attractant zones (Pénot, 1965, 1966; Kursanov, 1976). At early stages of development when the structures of the ovule are differentiated, such an attractant zone might possibly be the entire ovule, then the embryo sac and after fertilization the endosperm and embryo. The latter two may then act concurrently provided the endosperm is not consumed by the growing embryo.

Structural Aspects of Embryo sac Nutrition (Plate 78): The embryo sac and later the embryo developing therein act as a sink for metabolites, a fact confirmed by some structural and functional parameters, for instance the osmotic gradient increasing towards the embryo (Ryczkowski, 1962; Yeung and Brown, 1982).

The processes underway in the embryo sac require primary delivery of nutrients and protection from unfavourable physical and chemical factors in the medium, which is effected by means of special structural mechanisms. The embryo sac is not related directly to the conductive bundle of the ovule. In principle, all metabolites fed into the embryo sac from the conductive system must pass through the maternal sporophyte cells—the nucellus, the integument, etc. or—following fertilization—through the developing seed coat tissues and the endosperm (Murray, 1987).

The necessity of nutrition and protection of the embryo sac conditioned development of structures both common to and specifically individual for certain taxa. The structures typical for embryo sacs of most taxa comprise: (a) plasmodesmata; (b) isolation layers (cuticular, callosic, suberic); and (c) structures responsible for conductivity.

Although the functional megaspore can temporarily retain the connection with the nucellus, in the ripened ovule the plasmodesmata unite into one symplast exclusively one generation of cells, i.e., the gametophyte and the sporophyte cells (Carr, 1976). Between the gametophyte cells and the sporophyte tissues 'isolation' layers often evolve on the outer embryo sac wall or on the inner wall of the nucellus or the integument cells. Isolation layers appear prior to fertilization and may perish partly or completely after fertilization during the time of embryo development.

The sporophyte and the gametophyte are bound in apoplastic connection by means of structures responsible for provision of short-distance transport. Transmissive structures often develop in the most active areas of metabolism, expanding the surface and facilitating transport. They can emerge on the embryo sac, antipodes, synergids, suspensor or the surface endosperm layer walls and in other locations (Berger and Erdelská, 1973; Tilton *et al.*, 1984).

Prior to fertilization, the prevalent direction whereby nutrients are fed into the embryo sac is through its chalazal part, from the conductive bundle terminus through the nucellus and integument cells. Transport of substances in this direction is unobstructed by isolation layers but supported by transmissive structures. In a large number of plants, following fertilization the substances temporarily deposited in the seed and fruit coatings begin to be fed into the embryo sac. Most of these substances arrive in the coating tissues from the mother plant through the transmissive system, while a small portion may be the product of photosynthetic activity of the coating tissues themselves (Caley *et al.*, 1990). As a result of genetically programmed degradation, the content of the cell in the coatings is fed into the embryo sac. The nucellus also degrades, in most cases prior to fertilization. Thus, following fertilization, besides the chalazal pathway decomposition products of certain ovular tissues may be supplied to the embryo sac directly through its walls when the isolation layer is disturbed during growth. For instance, in the embryo sac of *Jasione montana*, following fertilization the isolation layer is destroyed in the submicropylar area marked by considerable expansion. In this area transmissive structures develop on the embryo sac wall, which connect the sac with the integumentary tapetum cells (Berger and Erdelská, 1973).

The substances supplied from the conductive bundle are temporarily stored in the coating tissues (Ryczkowski *et al.*, 1986). The temporary nutrient deposition system emerges not only for the 'sporophyte–gametophyte' relation but for the 'endosperm–embryo' relation as well; this happens exclusively in genera in which embryo nutrition at certain stages of embryogenesis depends on the endosperm. The predominant pathway of embryo nutrition effected through the suspensor at the initial stages of embryogenesis is related, firstly, to emergence of the new or expansion of the existing pathways of nutrient supply to the micropylar area of the embryo sac and, secondly, to development of special structures on the walls of the suspensor cells, the suspensor or the endosperm haustoria.

Haustoria are taxon-specific structures. As far as their appearance is concerned, they often resemble largely expanded cells with a polyploid nucleus and a characteristic set of organelles to perform secretory functions. Haustorial cells may have simple or branching outgrowths, which expand the surface and enhance the absorption of substances from the surrounding cells.

Among the specific structures also facilitating substance intake into the embryo sac are the hypostase in the chalazal part of the embryo sac and the epistase in the micropylar area (Subramanyam, 1960).

In some genera the number of antipodes increases as a result of multiple divisions. Embryo sacs with multiple antipodes are found for instance in Poaceae. It is assumed that the antipodal apparatus takes part in substance supply to the embryo sac through the chalazal part.

A similar function may be performed in the micropylar part by the intact synergid, which in some genera does not degrade after fertilization but continues to function (Maze and Lin, 1975).

Special structures controlling substance intake in the ovule, including the embryo sac, represent the cells and layers allegedly responsible for termination of water and substance supply into a mature seed. An example of such a structure is the pigment layer in the caryopsis of maize.

In angiosperms the ovule and (following fertilization) the seed ontogenetically represent a formation containing the tissues belonging the maternal sporophyte (nucellus, integuments), the gametophyte (embryo sac) and the filial sporophyte (embryo, endosperm). A prerequisite for this interaction is formation of special isolating and communicative structures providing the specifics of genetic programmes of separate cell generations on the one hand, and their metabolic connection on the other. Studies carried out on mutants with variable ovules have demonstrated the connection between development deviations of the embryo sac and the integuments (Robinson-Beers *et al.*, 1992).

Thus, isolation of the sporophyte and the gametophyte is provided by absence of symplastic connection and presence of isolation layers. The metabolic connection is rendered apoplastic by means of structures facilitating short-distance transport (for instance transmissive structures) from regions of either the conductive bundle terminus or the temporary nutrient depots.

Ovule and Seed Viewed from Reliability of Biological Systems: The 'ovary–ovule–embryo sac' (subsequently 'fruit–seed–embryo') system is a dynamic integrated system closely associated with development of the entire plant organism and characterized by the constantly changing state of its integrity. Each stage of its development is noted for certain morphogenetic and morphophysiological correlations responsible for conditioning of the normal flow of sporogenesis and megagametogenesis on the one hand, and determined by the genetics of the individual organism and the entire taxon on the other. Structural and functional changes of this complex reproductive system may influence different levels, such as those of the organism, organs, tissues, cells or molecules, and be manifested at different stages of structural development.

For a successful study of such a dynamic system with such functional connections as the ovary, ovule and seed, a systemic approach is expedient. Various aspects of the systemic approach in biology have been developed (Lishchitovich, 1974; Grodzinsky, 1977, 1983; Batygin, 1977; Urmantsev, 1979; Batygina, 1983, 1984a, b; Vasilyeva and Batygina, 1981; Batygina and Vasilyeva, 1983). When formulating a systemic approach to the problem of embryo sac differentiation, the following must be taken into account:

1. The ovule—a system comprising certain structural elements (nucellus, integument, embryo sac, etc.) between which there are strictly determined morphogenetic and morphophysiological correlations.
2. Differentiation of the ovule in the 'ovule—surrounding medium' system, the latter component usually being the mother plant tissues.
3. Organization of the ovule and the seed as viewed from the standpoint of reliability of biological systems.

In relation to the above the following research strategy has been developed:

1. **selection of** plant genera to act as **model objects, which must have contrasting embryological features, ecology and germination particulars, and which must occupy different positions in the system of flowering plants;**
2. study of the **dynamics** of ovule and seed structures **as well as their interconnection with the surrounding tissues of the ovary (and embryo); comparison of stages and rates of structural development with reference to processes of sporo-, gametophyte- and embryogenesis;**
3. **comparison of kinetics of morphogenetic and physiologobiochemical processes in the ovule and ovary (as well as in the seed and fruit);**
4. **comparison of differentiation regularities of the ovule and seed** in situ, in vivo and in vitro.

The suggested 'study block' allows definition of the morphogenetic and morphophysiological correlations conditioning spatial and temporal coordination of various structures and processes, as well as determination of critical periods (Vasilyeva *et al.*, 1987). Because the organization of the ovule and seed is adaptive by nature and changes while they develop and mature, special attention must be exercised in identifying the factors determining their constitution. For instance, the nature of ovular development at the early stages is possibly determined mainly by the nature and type of pollination and fertilization, and at the later stages by the type of embryo development, germination particulars, etc. (Batygina, 1984a, b).

One of the most crucial aspects in studying the organization of the ovule is definition of the biological meaning of the multitude of doubling structures (archesporial, sporogenous cells, megaspore tetrads, aposporic and diplosporic embryo sacs) **in one ovule and the roles they play in seed reproduction.** In this respect, it is of interest to study the ovule from the standpoint of the new trend in science—biological system reliability.

In studying the genesis of the ovule and seed, one should depart from the following premises. All cells are **totipotent, however, the time and degree of this property vary. In the ovule and the seed totipotency conditions continuity of morphogenesis and the multiple character of reproduction types** (sexual, asexual, apomixis) **and forms** (e.g. viviparity), **as well as the pathways of morphogenesis.** Totipotency may be accompanied by polyvariance (ambiguity of matrix processes) and heterogeneity of cells (Inge-Vechtomov, 1977).

In the process of evolution, the reproductive system of each taxon developed its own specifics of 'reserves' and 'failures'. This manifests in various nature and degree of oligomerization. Functioning of the doubling mechanisms is connected with 'competitiveness' and the system of 'failures' (manifestation of anomalies—aberrations, destructions, etc.) at different levels of biohierarchy. For instance, cell-by-cell selection, manifested as degradation of the larger part of sporogenous cells facilitates maintaining genetic and physiological purity of the egg cell and, in the long run, genetic purity of the progeny. Therefore, perishing of the sporogenous cells occurring at the lower level of the biohierarchic system, demonstrates that a high degree of biological system reliability has been retained and thus manifested. The system of 'hidden' reproduction reserves (changes in pollen sexualization[8], etc.) is present at the early stages of flower development as also at other stages of ontogeny. The taxon-specific character of those reserves is easily identified in a tissue culture ('cauliflower' effect, stamen and pollen locule formation on petals and ovary, etc.).

The character of the first stages of ovular development (apospory, diplospory, dormant meristem) mainly defines the diversity of the reserves (parthenogenesis, nucellar and integumentary embryoidogeny, etc.) and, respectively, heterogeneity of the seeds. These are the reserves allowing the 'failure' system to come into action at individual seed developmental stages, while ensuring there will not be complete failure in the seed reproduction system whereby the sexual process (meiosis and fertilization) might fail, resulting in the demise of the 'ovule-seed' system.

The above-described phenomena can be illustrated by comparative analysis of various taxonomic representatives, contrasted by the embryological features (ovule type, nucellus and embryo sac structures, etc.).

For example, the system of reserves in the crassinucellate ovule of peony (*Paeonia anomala, P. lactiflora*) (Plate 79) is manifested as a multitude of archesporial cells (4-5), megasporocytes (16-20), megaspore tetrads (5-7), developing embryo sacs (3-4), and in the developing seed also as nucellar embryoidogeny (Yakovlev and Yoffe, 1965; Batygina and Shamrov, 1993). In addition, an interesting phenomenon has been discovered in peony, designated as 'cloning'. The core of this phenomenon is that at the coenocyte-cellular stage of the sexual proembryo as many as 25-27 somatic embryos are deposited. Of this number resulting from subsequent competitive development, only one embryoid (not a sexual embryo!) usually remains (Batygina, 1988, 1989, 1993). Thus, the integrated system 'ovule-seed' in peony possesses various reserves at different biohierarchical levels, providing for reliability of reproduction.

Another type of reserves is found in fowl grass. In the crassinucellate ovule of fowl grass (*Poa pratensis*) the system of reserves is represented by a 'dormant meristem' manifested in various types of embryo sac formation (reduced, apo- and diplosporic), and subsequently as parthenogenesis, apogamy and various forms of nucellar embryoidogeny (Batygina and Mametyeva, 1979; Batygina, 1991). Nucellar embryoids in the fowl grass ovule differ in origin. Some probably form from the dormant meristem, others from cells of the proliferating nucellus (Batygina and Freiberg, 1979).

---

[8]Please see in Part II: Development of Embryo Sacs in the Anther.

In the maize (*Zea mays*) ovule, as in species of *Poa*, usually one archesporial cell is formed, and subsequently on embryo sac and one embryo (Maheshwari, 1950; Chebotaru, 1972; Korobova, 1982; Van Lammeren, 1987). In this case, the reliability system is manifested as metamerism of structures: multitude of cops, large number of ovules in a cop, etc. However, in experimental conditions the mutant of maize *lar-487* shows the reproduction system out of balance due to gene mutation in a recessive homozygous state. This is displayed as a multitude of doubling structures (archesporial and sporogenous cells, megaspore tetrads, reduced and diplosporic embryo sacs) (Batygina *et al.*, 1994; Sheridan, Golubovskaya, Batygina, Shamrov, in press).

Therefore, experimental data attest to possible reversal and emergence of a number of specific ancestral features in ovule layout, typical for taxa located at the base of the flowering plant system. Interestingly the majority of doubling structures are typical for crassinucellate ovules of a number of primitive taxa (Fagaceae, Casuarinaceae, Lauraceae, etc.).

The system of reserves in tropical orchids becomes apparent in formation of nucellar and integumental embryoids at the early stages of ovular development, as well as in apogamy and polytypic monozygotic polyembryoidogeny (Afzelius, 1928; Swamy, 1943, 1948a). Neither apospory nor diplospory is present and metamerism is expressed as a large number of tenuinucellate ovules. In boreal orchids the system of reserves is realized mainly in formation of a large number of protonutritives from one embryo during germination (Batygina and Vasilyeva, 1983; Batygina and Shevtsova, 1985) and also in viviparity (Kerner, 1901).

The high degree of totipotency of the ovule and seed cells in orchids, manifested in differentiation and reserve functions taking place at different times, performs something similar to a 'standby' function in relation to the specifically organized reduced tenuinucellate ovule in orchids. **The range of various forms and means of reproduction is realized in various pathways of morphogenesis: embryogenesis, embryoidogenesis and gemmorhizogenesis** (Batygina *et al.*, 1978; Batygina, 1987, 1989). They provide not only for reliability of reproduction systems, but also for the wide habitat of orchid distribution. A similar situation, as noted above, is found in *Poa pratenesis*, where various types of apomixis have been identified.

Specifics of ovule genesis and layout in flowering plants permit forecasting the time of differentiation and functioning of the doubling structures in case of non-availability or disruptions of the sexual process. The ovule and seed present themselves as a self-regulating system, providing the homeostasis and reproductive strategy of a certain genus.

Thus the reliability of reproduction systems (seed and vegetative) is taken care of by the system of reserves and failures manifested at different levels of biohierarchy. They all appear as links in a single chain in the reproduction system of flowering plants, providing for plasticity and tolerance (adaptation) of not only seed as an ontogenetic stage, but also of the plant itself and of the entire population.

One can succeed in managing various stages of ontogeny only if all the ongoing processes in the ovule and the seed described above are considered in terms of interdependence and their influence on the overall productivity of the population or species is determined.

BIBLIOGRAPHY

Abe, K. 1972a. Contribution to the embryology of the family Orchidaceae. VI. Development of the embryo sac in 15 species of orchids. Science Reports Tôhoku University. Sendai, Japan. Series IV, Biology, 36 (3): 135–178.

Abe, K. 1972b. Contribution to the embryology of the family Orchidaceae. VII. A comparative study of the orchid embryo sac. Science Reports Tôhoku University. Sendai, Japan. Series IV, Biology, 36 (3): 179–201.

Adatia, R.D. and G.L. Shah. 1952. A contribution to the life history of *Coriandrum sativum* L. J. Univ. Bombay, Ser. B. 20: 34–46.

Afzelius, K. 1928. Die Embryobildung bei *Nigritella nigra*. Svensk Botanisk Tidskrift, 22 (1–2): 82–91.

Agarwal, S. 1963. Morphological and embryological studies in the family Olacaceae—I. *Olax* L. Phytomorphology, 12 (2): 185–196.

Ahokas, H. 1978. Cytoplasmic male sterility in barley. 2. Physiology and anther cytology of msml. Hereditas, 89 (1): 7–21.

Albani, D., I. Altosaar, P.G. Arnison and S.F. Fabijanski. 1991. A gene showing sequence similarity to pectin esterase is specifically expressed in developing pollen of *Brassica napus*. Sequences in its 5′ flanking region are conserved in other pollen-specific promoters. Plant Molecular Biology, 16: 501–513.

Albani, D., L.S. Robert, P.A. Donaldson, I. Altosaar, P.G. Arnison and S.F. Fabijanski. 1990. Characterization of a pollen-specific gene family from *Brassica napus* which is activated during early microspore development. Plant Molecular Biology, 15: 605–622.

Albani, D., R. Sardana, L.S. Robert, I. Altosaar, P.G. Arnison and S.F. Fabijanski. 1992. A *Brassica napus* gene family which shows sequence similarity to ascorbate oxidase is expressed in developing pollen. Molecular characterization and analysis of promoter activity in transgenic tobacco plants. Plant Journal, 2: 331–342.

Albertini, L., A. Souvré and J.C. Audran. 1987. Le tapis de l'anthère et ses relations avec les microsporocytes et les grains de pollen. Revue Cytologie et Biologie Végétales. Botaniste, 10 (3–4): 211–242.

Alberts, B. and D. Brey, J. Luise, and M. Reff, K. Roberts, and J. Wotson 1987. Molecular biology of cell, 4. Mir, Moscow, Russia.

Alexandrov, V.G. 1954. Anatomy of Plants. Sov. Nauka, Moscow, Russia.

***Alexandrov, V.G., O.G. Alexandrova and M.S. Yakovlev. 1944.** Typical features of fruit morphology of maize (*Zea mays* L.) type grasses. Sov. Botanica, 6: 63–75.

Alexandrovskyi, E.S. and T.N. Naumova. 1985. Family Anacardiaceae, pp. 166–173. *In* T.B. Batygina and M.S. Yakovlev (eds.). Comparative Embryology of Flowering Plants. Brunelliaceae—Tremandraceae. Nauka, Leningrad, Russia.

Alimova, G.K. 1985. Family Staphylleaceae, pp. 176–178. *In* T. B. Batygina and M.S. Yakovlev (eds.). Comparative Embryology of Flowering Plants. Phytolacaceae—Thymelaeaceae. Nauka, Leningrad, Russia.

Alimova, G.K. 1987. Family Alangiaceae, pp. 15–18. *In* T. B. Batygina and M.S. Yakovlev (eds.). Comparative Embryology of Flowering Plants. Davidiaceae—Asteraceae. Nauka, Leningrad, Russia.

Alimova, G.K. 1990. Family Alstroemeriaceae, pp. 74–76. *In* T.B. Batygina and M.S. Yakovlev. (eds.). Comparative Embryology of Flowering Plants. Butomaceae—Lemnaceae. Nauka, Leningrad, Russia.

---

*Asterisked works in Russian—General Editor

Alimova, G.K. 1990. Family Haemodoraceae, pp. 134–138. *In:* T.B. Batygina and M.S. Yakovelv (eds.). Comparative Embryology of Flowering Plants. Butomaceae—Lemnaceae. Nauka, Leningrad, Russia, 5.

***Alimova, G.K. and M.S. Yakovlev. 1982.** Embryology of *Streptocarpus rexii* (Gesneriaceae). Botan. Zhurnal, 67 (4): 470–479.

Allen, G.S. 1946. Embryogeny and development of the apical meristems of *Pseudotsuga.* I. Fertilization and early embryogeny. Amer. J. Bot., 33(8): 666–677.

Allen, G.S. 1947. Embryogeny and development of the apical meristems of *Pseudotsuga.* II. Late embryogeny. Amer. J. Bot., 34(2): 73–80.

Allen, R.J. and D.M. Lonsdale. 1993. Molecular characterization of one of the maize polygalacturonase gene family members which are expressed during late pollen development. Plant Journal, 3: 261–271.

Alov, I.A., A.I. Braude and M.E. Aspiz. 1969. Principles of Cell Functional Morphology. Medicine, Moscow, Russia.

Andrews, H.N. 1961. Studies in Paleobotany. John Wiley and Sons, NY, USA.

***Anisimova, G.M. 1983a.** Development and structure of seed in *Orobanche cumana* and *O. alsatica* (Orobanchaceae). Botan. Zhurnal, 63 (4): 473–482.

Anisimova, G.M. 1983b. Family Begoniaceae, pp. 144–148. *In* T.B. Batygina and M.S. Yakovlev (eds.). Comparative Embryology of Flowering Plants. Phytolaccaceae—Thymelaeaceae. Nauka, Leningrad, Russia.

Anisimova, G.M. 1987. Family Araliaceae, pp. 30–34. *In* T.B. Batygina and M.S. Yakovlev (eds.). Comparative Embryology of Flowering Plants. Davidiaceae—Asteraceae. Nauka, Leningrad, Russia.

Anisimova, G.M. 1990. Family Juncaceae, pp. 183–187. *In* T.B. Batygina and M.S. Yakovlev (eds.). Comparative Embryology of Flowering Plants. Butomaceae—Lemnaceae. Nauka, Leningrad, Russia.

Anisimova, G.M. and I.I. Shamrov. 1992. Ovule development and structure peculiarities in *Luzula pedemontana* (Juncaceae), pp. 45–46. *In* T.B. Batygina (ed.). Proc. XI Intern. Symp. 'Embryology and Seed Reproduction', 3–7 July, 1990, Leningrad, Russia, Nauka, St.-Petersburg, Russia.

Anton, A.M. and A.E. Cocucci. 1984. The grass megagametophyte and its possible phylogenetic implications. Plant Syst. and Evol., 146 (1–2): 117–121.

Arbeloa, A. and M. Herrero. 1987. The significance of the obturator in the control of pollen tube entry into the ovary in peach (*Prunus persica*). Ann. Bot., 60 (6): 681–685.

Archibald, E.E. 1939. The development of the ovule and seed of jointed cactus (*Opuntia aurantiaca* Lindley). South Afr. J. Sci., 36 (2): 195–211.

Arekal, G.D. 1963. Embryological studies in Canadian representatives of the tribe Rhinantheae, Scrophulariaceae. Can. J. Bot., 41 (2): 267–303.

Arekal, G.D. and C.R. Nagendran. 1975a. Embryo sac of *Hydrobryopsis sessilis* (Podostemaccae)—origin, organization and significance. Botaniska Notiser, 128 (3): 332–338.

Arekal, G.D. and C.R. Nagendran. 1975b. Is there a Podostemum type of embryo sac in the genus *Farmeria*? Caryologia, 28 (2): 229–235.

Arekal, G.D. and C.R. Nagendran. 1976. A new type of embryo sac organization in angiosperms. Current Sci. (India), 45 (20): 717–719.

Arekal, G.D. and C.R. Nagendran. 1977a. The female gametophyte in two Indian genera of Tristichoideae (Podostemaceae)—a reinvestigation. Proc. Indian Acad. Sci. Section B, Biol. (Bangalore, India), 86: 287–294.

Arekal, G.D. and C.R. Nagendran. 1977b. Female gametophyte of *Zeylanidium* (Podostemaceae)—a clarification. Phytomorphology, 27: 123–129.

Ashley, T. 1972. Zygote shrinkage and subsequent development in some *Hibiscus* hybrids. Planta (Berlin), 108 (4): 303–317.

Asplund, E. 1920. Studien über die Entwicklungsgeschichte der Bluten einiger Valerianaceen. Kongliga Svenska Vetenskapsakademiens Handlingar, 61 (3): 3–66.

Audran, J.C. and M.T.M. Willemse. 1982. Wall development and its autofluorescence of sterile and fertile *Vicia faba* L. pollen. Protoplasma (Wien), 110 (2): 106–110.

Baillon, H.E. 1858. Etude général du groupe des Euphorbiacées. Paris, Comptes Rendus, XLII: 147–149.

Baillon, H.E. 1876. Sur l'origine du macis de la Muscade et des arilles en général. Adansonia, 11: 329–340.

Baillon, H.E. 1885. Recherches sur les ovaires acropilés Premier Memoire, pp. 59–67. *In* Bull. Congrès Intern. Botanique et d'horticulture. St.-Petersburg, Russia.

Baltz, R., C. Domon, D.T.N. Pillay and A. Steinmetz. 1992. Characterization of a pollen-specific cDNA from sunflower encoding a zinc finger protein. Plant Journal, 2: 713–721.

Bambacioni, V. 1928a. Ricerche sulla ecologia e sulla embryologia di *Fritillaria persica* L. Ann. Bot., 18 (1): 7–37.

Bambacioni, V. 1928b. Contributo alla embriologia di *Lilium candidum* L. C.R. Acad. Sciences, Lincei, 8 (6): 612–618.

Bambacioni-Mezzetti, V. 1931. Nuove ricerche sull'embriologia delle Giliaceae. Ann. Bot., 19: 365–382.

Banerjee, U.C. 1967. Ultrastructure of tapetal membranes in grasses. Grana Palynologica, 7 (1–2): 365–377.

***Bannikova, V.P. and T.A. Pluyshch. 1990.** Fusion of gamete nuclei and pollar nuclei in flowering plants. Cytology i Genetica, 24 (1): 62–64.

***Bannikova, V.P., O.A. Khvedynich, S.P. Shpilevaja et al. 1985.** Sexual cells and fertilization in angiosperms and algae. Nauk. Dumka, Kiev, Ukraine.

***Baranov, P.A. 1917.** Materials on the embryology of Orchidaceae. J. Rus. Bot. Soc., 2: 22–27.

Baranov, P.A. 1955. History of Plant Embryology. Izd. Akad. Nauk SSSR, Moscow-Leningrad, Russia.

Barnes, S.H. and S. Blackmore. 1987. Preliminary observations on the formation of the male germ unit in *Catananche caerulea* L. (Compositae: Lactuceae). Protoplasma, 138 (2–3): 187–189.

Barrat, D.H.P. and R.B. Flawell. 1975. Alterations in mitochondria associated with cytoplasmic and nuclear genes concerned with male sterility in maize. Theor. and Appl. Genet., 45 (7): 315–321.

Battaglia, E. 1947. Ricerche cariologiche e embriologiche sul genere *Rudbeckia*. 9. Le anomalie del gametofito femminie cellularizzato di *R. laciniata*. Nuovo Gior. Bot. Ital., 54: 377–405.

Battaglia, E. 1951. The male and female gametophytes of angiosperms—an interpretation. Phytomorphology, 1 (1–2): 87–116.

Battaglia, E. 1959. New types of embryo sac development in *Scilla autumnalis*. 9th Intern Bot. Congr. vol. 2, pp. 23–24.

Battaglia, E. 1971. The embryo sac of Podostemaceae—an interpretation. Caryologia, 24: 403–420.

Battaglia, E. 1986. Embryological questions: 7. Do new types of embryo sac occur in *Schisandra*. Ann. Bot., 44: 69–92.

Battaglia, E. 1987b. Embryological questions: 11. Has the debated case of Podostemaceae been resolved? Ann. Bot., 45: 37–64.

Battaglia, E. 1989. The evolution of the female gametophyte of angiosperms: an interpretative key. (Embryological questions: 14). Ann. Bot., 47: 7–144.

***Batygin, N.F. 1977.** The systems of higher plant reliability, pp. 136–143. *In* The Systems of Cell Reliability. Nauk. Dumka, Kiev, Ukraine.

***Batygina, T.B. 1961.** Transformation of male gametes during fertilization in wheat. Reports USSR Acad. Sciences, 137 (1): 220–223.

***Batygina, T.B. 1962a.** Fertilization process in wheat. Trans. Komarov Bot. Inst. USSR Acad. Sciences. Series VII, 5: 260–293.

***Batygina, T.B. 1962b.** Microsporogenesis and development of pollen grain in wheat. Reports USSR Acad. Sciences, 142 (5): 1205–1208.

Batygina, T.B. 1974a. Fertilization process of cereals, pp. 205–220. *In* H.F. Linskens (ed.). Proc. Intern. Symp. Fertilization Process in Higher Plants. Nijmegen, the Netherlands. North-Holland Publ. Co., Amsterdam. Oxford. American Elsevier Publ. Co., Inc., NY, USA.

Batygina, T.B. 1974b. Wheat Embryology. Kolos, Leningrad, Russia.

Batygina, T.B. 1981. Family Nelumbonaceae, pp. 116–120. *In* T.B. Batygina and M.S. Yakovlev (eds.). Comparative Embryology of Flowering Plants. Winteraceae—Juglandaceae. Nauka, Leningrad, Russia.

***Batygina, T.B. 1983a.** Anther as a model for study of morphogenetic potencies and morphogenetic pathways, pp. 9–10. *In* Abstracts All–Union Symp. 'Development of Male Generative Sphere of Plants (Morpho–physiological aspects)'. Sympheropol, Ukraine.

Batygina, T.B. 1983b. Family Portulacaceae, pp. 26–32. *In* T.B. Batygina and M.S. Yakovlev (eds.). Comparative Embryology of Flowering Plants. Phytolaccaceae—Thymelaeaceae. Nauka, Leningrad, Russia.

***Batygina, T.B. 1983c.** System approach to the problem of embryo differentiation in angiosperms, pp. 25–26. *In* Proc. VIII All-Union Conference on Plant Embryology 'Problems of Gametogenesis, Fertilization and Embryogenesis'. 'Phan', Tashkent, Uzbek Republic.

***Batygina, T.B. 1984a.** Ovary and ovule, a complex integrated system (concurrent development of structures and some aspects of substance transport in generative organs), pp. 6–8. *In* Materials All–Union Symp. 'Morpho-functional Aspects of Development of Female Generative Structures in Seed Plants'. Telavi, Georgia.

Batygina, T.B. 1984b. Necessity and possible ways of improvement of the exiting classifications of ovule types in angiosperms, pp. 18–19. *In* Abstracts I All-Union Conf. Plant Anatomy. Leningrad, Russia.

Batygina, T.B. 1985a. Family Onagraceae, pp. 104–110. *In* T.B. Batygina and M.S. Yakovlev (eds.). Comparative Embryology of Flowering Plants. Brunelliaceae—Tremandraceae. Nauka, Leningrad, Russia.

Batygina, T.B. 1985b. Family Tropaeolaceae, pp. 215–219. *In* T.B. Batygina and M.S. Yakovlev (eds.). Comparative Embryology of Flowering Plants. Brunelliaceae—Tremandraceae. Nauka, Leningrad, Russia.

Batygina, T.B. 1987a. Cereal grains (Atlas). Nauka, Leningrad, Russia.

Batygina, T.B. 1987b. New concept of asexual reproduction in flowering plants, p. 5. *In* Abstracts XIV Intern. Bot. Congr., West Berlin.

Batygina, T.B. 1988. Some aspects of reproductive biology: asexual reproduction in flowering plants and genetic heterogeneity of seeds, pp. 443–451. *In* Proc. Intern. Symp. 'Sexual Reproduction in Higher Plants'. Springer-Verlag, Sciena, Italy.

Batygina, T.B. 1989a. New approach to the system of reproduction in flowering plants. Apomixis Newsletter, 1: 52–55.

Batygina, T.B. 1989b. New approach to the system of reproduction in flowering plants. Phytomorphology, 39 (3–4): 311–325.

Batygina, T.B. 1989c. New concept of asexual reproduction in flowering plants, pp. 28–44. *In* J. Pare and M. Bugnicourt (eds.) Some Aspects and Actual Orientation in Plant Embryology, 'Dedicatory volume to Prof. A. Lebègue. Univ. Picardie, Amiens, France.

Batygina, T.B. 1991a. Nucellar embryoidogeny in *Poa pratensis* (Poaceae). Polish Botanical Studies, 2: 121–125.

Batygina, T.B. 1991b. New hypothesis about the initials and genesis of embryoids (somatic embryos) and position of embryoidogeny in the reproduction system. Apomixis Newsletter, 3: 19–24.

Batygina, T.B. 1992. Position of the phenomenon of embryoidogeny in the system of flowering plants reproduction, pp. 6–10. *In* T.B. Batygina (ed.). Proc. XI Intern. Symp. 'Embryology and Seed Reproduction.' 3–7 July 1990, Leningrad, Russia. Nauka, St.-Petersburg, Russia.

Batygina, T.B. 1993. Morphogenetic reserves of reproductive structures (totipotency and determination), p. 65. *In* Abstracts III Cong. RSPPh, St.-Petersburg, Russia.

***Batygina, T.B. and T.E. Freiberg. 1979.** Polyembryony in *Poa pratensis*. Bot. Zhurnal, 64 (6): 793–804.

Batygina, T.B. and T.B. Mametyeva. 1979. Embryology of genus *Poa* L., pp. 89–95. *In* M.S. Yakovlev (ed.). Actual Problems of Embryology of Angiosperms. Nauka, Leningrad, Russia.

***Batygina, T.B. and I.I. Shamrov. 1981a.** Embryology of Nymphaeales and Nelumbonales. I. Anther development. Bot. Zhurnal, 66 (12): 1696–1709.

Batygina, T.B. and I.I. Shamrov. 1981b. Family Ceratophyllaceae, pp. 110–115. *In* T.B. Batygina and M.S. Yakovlev (eds.). Comparative Embryology of Flowering Plants. Winteraceae—Juglandaceae. Nauka, Leningrad, Russia.

Batygina, T.B. and G.E. Kolesova. 1983. Family Euphorbiaceae, pp. 267–277. *In* T.B. Batygina and M.S. Yakovlev (eds.). Comparative Embryology of Flowering Plants. Phytolaccaceae—Thymelaeaceae. Nauka, Leningrad, Russia.

***Batygina, T.B. and V.E. Vasilyeva. 1983a.** Experience of system approach to the problem of embryo differentiation in angiosperms. Ontogenes, 14 (3): 304–311.

Batygina, T.B. and V.E. Vasilyeva. 1983b. System of reproduction of Orchidaceae (on example of *Dactylorhyza maculata* L. Soo.), pp. 27–33. *In* Proc. VII Intern. Symp. 'Fertilization and Embryogenesis in Ovulated Plants'. High Tatra (Račkova dolina), 14–17 June, 1982, Bratislava, Czechoslovakia.

Batygina, T.B. and G.E. Kolesova. 1985. Family Trapaceae, pp. 110–116. *In* T.B. Batygina and M.S. Yakovlev (eds.). Comparative Embryology of Flowering Plants. Brunelliaceae—Tremandraceae. Nauka, Leningrad, Russia.

***Batygina, T.B. and G.G. Shevtsova. 1985.** Morphogenesis in orchid ontogenesis (on the example of *Cymbidium hybridum,* Orchidaceae). Bot. Zhurnal, 70 (12): 1614–1621.

Batygina, T.B. and G.E. Kolesova. 1987. Family Menyanthaceae, pp. 145–153. *In* T.B. Batygina and M.S. Yakovlev (eds.). Comparative Embryology of Flowering Plants. Davidiaceae—Asteraceae. Nauka, Leningrad, Russia.

Batygina, T.B. and I.I. Shamrov 1993. Ovule and seed structure in *Poa, Paeonia* and orchids from the point of reliability, 96. *In* Abstracts XV Intern. Bot. Cong., Yokohama, Japan.

***Batygina, T.B., T.B. Mametyeva and V.E. Vasilyeva. 1978.** The problems of morphogenesis *in vivo* and *in vitro*. Bot. Zhurnal, 63 (1): 87–111.

***Batygina, T.B., I.I. Shamrov, and G.E. Kolesova. 1982.** Embryology of Nymphaeales and Nelumbonales. II. Development of female embryonic structures. Bot. Zhurnal, 67 (9): 1179–1195.

***Batygina, T.B., G.E. Kolesova and V.E. Vasilyeva. 1983.** Embryology of Nymphaeales and Nelumbonales. III. Embryogenesis of *Nelumbo nucifera* Gaertn. Bot. Zhurnal, 68 (3): 311--326.

Batygina, T.B., G.E. Kolesova and E.V. Dragunova. 1987. Family Rhamnaceae, pp. 94–98. *In* T.B. Batygina and M.S. Yakovlev (eds.) 1987. Comparative Embryology of Flowering Plants. Davidiaceae—Asteraceae. Nauka, Leningrad, Russia.

***Batygina, T.B. , E.S. Teryokhin, G.K. Alimova and M.S. Yakovlev. 1963.** Genesis of male sporangia in Gramineae and Ericaceae. Bot. Zhurnal, 48 (8): 1108–1120.

Batygina, T.B. and M.S. Yakovlev (eds.) 1981. Comparative Embryology of Flowering Plants. Winteraceae—Juglandaceae. Nauka, Leningrad, Russia, v.1.

Batygina, T.B. and M.S. Yakovlev (eds.) 1983. Comparative Embryology of Flowering Plants. Phytolaccaceae—Thymelaeaceae. Nauka, Leningrad, Russia, v.2.

Batygina, T.B. and M.S. Yakovlev (eds.) 1985. Comparative Embryology of Flowering Plants. Brunelliaceae—Tremandraceae. Nauka, Leningrad, Russia, v.3.

Batygina, T.B. and M.S. Yakovlev (eds.) 1987. Comparative Embryology of Flowering Plants. Davidiaceae—Asteraceae. Nauka, Leningrad, Russia, v.4.

Batygina, T.B. and M.S. Yakovlev (eds.) 1990. Comparative Embryology of Flowering Plants. Butomaceae—Lemnaceae. Nauka, Leningrad, Russia, v.5.

Beardsell, D.V., E.G. Williams and R.B. Knox. 1989. The structure and histochemistry of the nectary and anther secretory tissue of the flowers of *Thryptomene calucina* (Lindl.) Stapf (Myrtaceae). Australian J. Bot., 37: 65–80.

Bednara, J. and B. Rodkiewicz. 1974. Megasporocyte and megaspore ultrastructure in *Epilobium*. Bull. l'Acad. Polonaise des Sciences, Série des Sciences Biolog., 22 (12): 847–850.

Bednara, J. and B. Rodkiewicz. 1975. Cell polarization in female gametophyte development in *Epilobium* (Onagraceae). Ann. Med. Soc. Polish Acade. Science, 20 (2): 56–58.

Bednara, J., M. T.M. Willemse and A.A.M. van Lammeren. 1990. Organization of the actin cytoskeleton during megasporogenesis in *Gastria verrucosa* visualised with fluorescent labelled phaloidin. Acta Botanica Neerlandica, 39 (1): 43–48.

Bednara, J. , O. Erdelská, A. van Lammeren and M.T.M. Willemse. 1994. The embryo sac cytoskeleton of *Galanthuns nivalis* L. plant—before and after fertilization. *In* Abstracts 13th Intern. Cong. 'Sexual Plant Reproduction'. Frontiers in Sexual Plant Reproduction Research. Intern. Asso., Vienna, Austria.

Behnke, H. 1968. Zur Aufbau gitterartiger Membranstrukturen in Seibelementplasma von *Dioscorea.* Protoplasma, 66 (3): 287–310.

Bell, P. R. 1989. The alternation of generations. Advances in Botanical Research, 16: 55–93.

Belyaeva, N.S. 1973. On the difference within one pair of sperms in some angiosperms, pp. 17–18. *In* Materials All-Union Symp. 'Sexual Process and Embryogenesis of Plants', Moscow, Russia.

Belyaeva, N.S. 1983. Ultrastructure of chalazal end cells of ovule of *Delphinium* before fertilization, pp. 189–194. *In* Proc. VII Intern. Symp. 'Fertilization and Embryogenesis in Ovulated Plants'. High Tatra (Račkova dolina), 14–17 June, 1982, Bratislava, Czechoslovakia.

Belyaeva, L.E. and G.B. Rodionova. 1983. Family Brassicaceae, pp. 154–164. *In* T.B. Batygina and M.S. Yakovlev (eds.). Comparative Embryology of Flowering Plants. Phytolaccaceae—Thymelaeaceae. Nauka, Leningrad, Russia.

Benson, M. 1904. *Telangium scotti,* a new species of *Telangium* (Calymmatotheca) showing structure. Ann. Bot., 18 (79): 161–177.

Berg, R.J. 1962. Morphology and taxonomic position of *Medeola,* Liliaceae. Skrifter utgitt av det Norske videnskaps—akademii Oslo. 3: 1–55.

Berger, C. and O. Erdelská. 1973. Ultrastructural aspects of the embryo sac of *Jasione montana* L.: cell walls. Caryologia, 25 (Suppl.): 109–120.

Berkaloff, C., E. Casadevall, C. Largeau, P. Metzger, S. Peracca and J. Virilet. 1983. The resistant polymer of the walls of the hydrocarbon-rich alga *Botryococcus braunii.* Phytochemistry, 22 (2): 389–397.

Bhandari, N.N. 1970. Schisandraceae. *In:* Proc. Symposium on Angiosperms. Bull. Indian Nat. Science Acad., 41: 42–44.

Bhandari, N.N. 1971a. Embryology of the Magnoliales and comments on their relationships. J. Arnold Arboretum, 52 (1): 1–39.

Bhandari, N.N. 1971b. Embryology of the Magnoliales and comments on their relationships. J. Arnold Arboretum, 52 (2): 285–304.

Bhandari, N.N. 1984. Microsporangium, pp. 53–121. *In* B.M. Johri (ed.). Embryology of Angiosperms. Springer-Verlag, Berlin.

Bhandari, N.N. and R. Venkataraman. 1968. Embryology of *Drimys winteri.* J. Arnold Arboretum, 49 (4): 509–524.

Bhandari, N.N. and R. Khosla. 1982. Development and histochemistry of anther in Triticale cv. Tri–1. I. Some new aspects in early ontogeny. Phytomorphology, 32 (1): 18–27.

Bhandari, N.N. and Sharma, Madhu. 1988. Histochemical and ultrastructural studies during anther development in *Solanum nigrum* Linn. II. Distribution of organelles in the meiocytes during microsporogenesis. Phytomorphology, 38 (1): 1–8.

Bhandari, N.N. and P. Chitralekha. 1989. Cellularization of the female gametophytes in *Ranunculus sceleratus* Linn. Can. J. Bot., 67: 1325–1330.

Bhandari, N.N., P. Soman and M. Bhargava. 1980. Histochemical studies on the female gametophyte of *Argemone mexicana* L. Cytologia, 45 (1–2): 281–291.

Bhandari, N.N., A. Bhargava and A. Sachdeva. 1985. The mature ovule of *Papaver somniferum*—A histochemical study. Phytomorphology, 35 (1–2): 111–119.

Bhargava, H.R. 1932. Contribution to the morphology of *Boerhaavia repanda.* J. Indian Bot. Soc., 11 (4): 303–326.

Bhat, U. and M.R. Vijayaraghavan. 1980. Distribution of insoluble polysaccharides in *Linum usitatissimum*—zygote to seedling. Cytologia, 45 (1–2): 65–75.

Biddle, J.A. 1979. Anther and pollen development in garden pea and cultivated lentil. Can. J. Bot., 57 (18): 1883–1901.

Blackmore, S. and S.H. Burnes. 1988. Pollen ontogeny in *Catananthe coerulea* (Compositae: Lactucea). 1. Premeiotic phase to establishment of tetrads. Ann. Bot., 62 (2): 605–614.

Bleier, H. 1930. Untersuchungen über das Verhalten der verschiedenen Kernkomponenten bei der Reductionteilung von Bastarden Lierre, typ. J. Van In u Cie, Louvain A. Hystpruyst. Cellule, 15 (1): 83–144.

Bocquet, G. 1959. The campylotropous ovule. Phytomorphology, 9 (3): 222–227.

Boehm, K. 1931. Embryologische Untersuchungen an Zingiberaceae. Planta, 14 (2): 411–440.

Boesewinkel, F.D. 1981. Development of ovule and seed coat in the Rutales–Geraniales assembly. Ph.D. thesis, Acad. Proefschr., Univ. van Amsterdam.

Boesewinkel, F.D. 1984. Ovule and seed structure in Datiscaceae. Acta Botanica Neerlandica, 33 (4). p. 419–429.

Boesewinkel, F.D. and Bouman F. 1984. The seed structure, pp. 567–610. *In* B.M. Johri (ed.) 1984. Embryology of Angiosperms. Springer-Verlag, Berlin.

Boesewinkel, F.D. and F. Bouman. 1991. The development of bi- and unitegmic ovules and seeds in *Impatiens* (Balsaminaceae). Bot. Jahrb. für Systematik, 113: 87–104.

Bogdanov, Yu.F. and A.A. Strokov. 1976. Peculiarities of histone synthesis in meiosis and relation to chromosome structure and behaviour. Biologisches Zentralblatt, 95 (3): 335–342.

Bogdanov, Yu.F., A.A. Strokov and S.A. Reznickova. 1973. Histone synthesis during meiotic prophase in *Lilium*. Chromosoma, 43: 237–245.

***Bolkhovskikh, E.V. 1965.** Dynamics of nucleic acids during the formation of generative and vegetative cells of lily pollen grain. Cytology, 7 (6): 745–748.

Bor, J. and R.N. Kapil. 1976. Anatropy and ontogeny of the bitegmic ovule in *Chrozophora* A.H.L. Jussieu (Euphorbiaceae). Acta Botanica Neerlandica, 25 (6): 385–400.

Bouman, F. 1971. The application of integumentary studies to taxonomic and phylogenetic problems. Berichte der Deutschen Botanischen Gesellschaft, 74 (3–4): 169–177.

Bouman, F. 1974. Developmental studies of the ovule, integuments, and seed in some angiosperms. Univ. van Amsterdam. Naarden, Boek en Offsetdrukk. Los. 180 pp.

Bouman, F. 1984. The ovule, pp. 123–157. *In* B.M. Johri (ed.). Embryology of Angiosperms. Springer-Verlag, Berlin.

Boum, F. and J.I.M. Calis. 1977. Integumentary shifting—a third way to unitegmy. Berichte der Deutschen Botanischen Gesellschaft, 90 (1–2): 15–28.

Bouman, F. and S. Schrier. 1979. Ovule ontogeny and seed coat development in *Gentiana* with a discussion on the evolutionary origin of the single integument. Acta Botanica Neerlandica, 28 (6): 467–478.

Bouman, F. and A. Louis. 1989. Seed structure in *Voyria primuloides* Baker (Gentianaceae): taxonomic and ecological implication, pp. 261–270. *In* Dedicatory volume to Prof. A. Lebeque 'Some Aspects and Actual Orientations in Plant Embryology'. Univ Picardie, Amiens, France.

Boush, I.A. 1944. Handbook on systematics of higher plants. GUIZ, Moscow, Russia.

Boyes, J.W. 1939. Development of embryo sac of *Plumbagella micranta*. Amer. J. Botany, 26 (6): 539–547.

Boyes, J.W. and E. Battaglia. 1951. The tetrasporic embryo sacs of *Plumbago coccinea, P. scandens* and *Ceratostigma willmottianum*. Bot. Gazette, 112 (4): 485–489.

Brongniart, A. 1827. Mémoire sur la génération et le développement de l'embryon dans les végétaux phanérogames. Ann. Sciences Naturelles. Botanique, 12: 225–296.

Brooks, J. and G. Shaw. 1968. Chemical structure of the exine of pollen walls and a new function for carotenoids in nature. Nature, 219: 332–533.

Brooks, J. and G. Shaw. 1978. Sporopollenin: a review of its chemistry, palaeochemistry, and geochemistry. Grana, 17: 91–97.

Brown, R. 1818. On some remarkable deviations from the usual structure of seeds and fruits. Trans. Linn. Soc. London, 12: 143–151.

Brown, R. 1826a. Character and description of *Kingia*, pp. 534–565. *In* P. King (ed.). Voyages of Discovery. London.

Brown, R. 1826b. Sur la structure de l'ovule anténieurement à l'imprégnation dans les plantes phanérogames et sur la fleur femelle des Cycadées et des Coniferés. Ann. Sciences Naturelles. Botanique, 7: 211–244.

Brown, R. 1834. Vermischte botanische Schiften. Leonard Schrag, Nurnberg. Bd 5.

Brown, R.C. and H.L. Mogensen. 1972. Late ovule and early embryo development in *Quercus gambelii*. Amer. J. of Bot., 59 (4): 311–316.

Brown, R.C. and B.E. Lemmon. 1991. Pollen development in orchids, 1. Cytoskeleton and the control of division plane in irregular patterns of cytokinesis. Protoplasma, 136: 9–18.

Brown, S.M. and M.L. Crouch. 1990. Characterization of a gene family abundantly expressed in *Oenothera organensis* pollen that shows sequence similarity to polygalacturonase. Plant and Cell Physiology, 2: 263–274.

Brown, W.D. 1908. The nature of the embryo sac of *Peperomia*. Bot. Gazette, 46 (8): 445–460.

Bryan, J.H.D. 1951. DNA–protein relations during microsporogenesis of *Tradescantia*. Chromosoma, 4 (4): 369–392.

Budelier, K.A., A.G. Smith, and C.S. Gasser. 1990. Regulation of a stylar transmitting tissue-specific gene in wild-type and transgenic tomato and tobacco. Molecular and General Genetics, 224: 183–192.

Budell, B. 1964. Untersuchungen den Antherenentwicklung einiger Blutenpflanzen. Zeitschrift für Botanik, 52 (1): 1–28.

Buell, M.F. 1952. Developmental morphology in *Dianthus*. II. Starch accumulation in ovule and seed. Amer. J. Bot., 39 (7): 458–467.

Bugnicourt, M. 1969. Polyembryonie chez l'*Hypericum tetrapterum* Fries (=*Hypericum acutum* Moench.). Bull. Société botanique de France, 116 (9): 495–504.

Burton, G.W. and K.U. Ingold. 1984. *β*-carotene: an unusual type of lipid antioxidant. Science, 224: 569–573.

Bush, I.A. 1944. Handbook on Systematics of Higher Plants. GUIZ, Moscow, Russia.

Butenko, R.G. 1970. Totipotency of plant cell and tissue culture, pp. 84–92. *In* Culture of Separated Plant Organs, Tissues and Cells. Nauka, Moscow, Russia.

Buxbaum, F. 1953. The Flower, pp. 88–170. *In* E.B. Kuntz (ed.). Morphology of Cacti. Abbey Garden Press, Pasadena, California.

Caldwell, O.W. 1899. On the live history of *Lemna minor*. Bot. Gazette, 27 (1): 37–66.

Caley, C.Y., C.M. Duffus and B. Jeffcoat. 1990. Photosynthesis in the pericarp of developing wheat grains. J. Experim. Bot., 224: 303–307.

Camerarius, R.J. 1694. De sexu plantarum epistola, pp. 83–148. *In* Gmeliu, Joh. Georg. Sermo academicus de noxorum vegetabilium post creationem divinam exortu. Tabingae.

Camp, W.H. and M.M. Hubbard. 1963a. Vascular supply and structure of the ovule and aril in peony and aril in nutmeg. Amer. J. Botany, 50: 174–178.

Camp, W.H. and M.M. Hubbard. 1963b. On the origin of the ovule and cupule in lyginopterid pteridosperms. Amer. J. Bot., 50: 235–243.

Campbell, D.H. 1899a. A peculiar embryo sac in *Peperomia pellucida*. Annals of Botany, 13 (52): 626.

Campbell, D.H. 1899b. Die Entwicklung des Embryosackes von *Peperomia pellucida* Kunth. Berichte der Deutschen Botanischen Gesellschaft, 17 (10): 452–456.

Campbell, D.H. 1901. The embryo sac of *Peperomia*. Ann. Bot., 15 (57): 103–118.

Carniel, K. 1961. Das Antherentapetum von *Zea mays*. Österreichische Botanische Zeitschrift, 108 (1): 89–96.

Carniel, K. 1963. Das Antherentapetum. Österreichische Botanische Zeitschrift, 110 (2–3): 145–176.

Carpenter, J.L., S.E. Planse, D.P. Snustad and C. Silflow. 1992. Preferential expression of an α-tubulin gene of *Arabidopsis* in pollen. Plant Cell Physiol., 4: 557–571.

Carr, D.J. 1976. Plasmodesmata in growth and development. *In* D.E.S. Gunning and A.W. Robards (eds.) Intercellular Communication in Plants: Studies on Plasmodesmata. Springer, Berlin.

Cass, D.D. 1972a. Occurrence and development of a filiform apparatus in the egg of *Plumbago capensis*. Amer. J. Bot., 59 (3); 279–283.

Cass, D.D. 1972b. Ultrastructure of the egg of *Plumbago zeylanica*. Amer. J. Bot., 59 (6): 648.

Cass, D.D. 1981. Structural relationships among central cell and egg apparatus cells of barley as related to transmission of male gametes. Acta Societatis Botanicorum Poloniae, 50 (1–2): 177–180.

Cass, D.D. 1990. Studies of isolated embryo sacs and sperm cells of maize and winter wheat. Amer. J. Bot., 77 (Suppl.): 5–6.

Cass, D.D. and W.A. Jensen. 1970. Fertilization in barley. Amer. J. Bot., 57 (1): 62–70.

Cass, D.D. and J. Karas. 1974. Ultrastructural organization of the egg of *Plumbago zeylanica*. Protoplasma, 81 (1): 49–62.

Cass, D.D., D.J. Peteya and B.L. Robertson. 1985. Megagametophyte development in *Hordeum vulgare*. 1. Early megagametogenesis and nature of cell wall formation. Can. J. Bot., 63 (12): 2164–2171.

Cass, D.D., D.J. Peteya and B.L. Robertson. 1986. Megagametophyte development in *Hordeum vulgare*. 2. Late stages of wall development and morphological aspects of megagametophyte cell differentiation. Can. J. Bot., 64 (10): 2327–2336.

Chaban, I.A. 1975. Embryogenesis and endospermogenesis of *Reseda lutea* L. and *Chelidonium majus* L. from data on light and electronic microscopy. Ph.D. thesis, Komarov Botanical Institute, Leningrad, Russia.

Chaban, I.A. 1979. Electron microscopy investigation of endosperm in *Reseda lutea* L., pp. 58–69. *In* M.S. Yakovlev (ed.). Actual Questions on Embryology of Angiosperms. Nauka, Leningrad, Russia.

Chaban, I.A. 1983. Family Resedaceae, pp. 164–172. *In* T.B. Batygina and M.S. Yakovlev (eds.). Comparative Embryology of Flowering Plants. *Phytolaccaceae—Thymelaeaceae*. Nauka, Leningrad, Russia.

***Chaban, I.A. and M.S. Yakovlev. 1974.** Embryology of *Reseda lutea* L. I. Megasporogenesis and embryo sac development. Bot. Zhurnal, 59 (1): 24–37.

Chamberlain, C.J. 1895. Contributions to the embryology of *Aster novaeangliae*. Bot. Gazette, 20: 205–212.

Chamberlain, C.J. 1935. Gymnosperms. Structure and Evolution. Univ. Chicago Press, Chicago, USA.

Chandra, S. and S.P. Bhatnagar. 1975. Pollination and nucellus in *Abelmoschus esculentus*. Biologia Plantarum (Praha), 17 (2): 154–1 55.

Charzynska, M., F. Ciampolini and M. Cresti. 1988. Generative cell division and sperm cell formation in barley. Sexual Plant Reproduction, 1: 240–247.

Charzynska, M., M. Murgia and M. Cresti. 1989. Ultrastructure of the vegetative cell of *Brassica napus* pollen with particular reference to microbodies. Protoplasma, 152 (2/3): 22–28.

Chaubal, R. and B.J. Reger. 1994. Secretion of calcium by pearl millet synergids, p. 23. *In* Abstracts 13th Intern. Cong. Sexual Plant Reproduction 'Frontiers in Sexual Plant Reproduction Research', 10–14 July, 1994, University of Vienna, Vienna, Austria.

Chaudhry, Bharti. 1992. Histochemical and ultrastructural aspects of anther, pollen and seed development in *Prosopis juliblora* (Swartz) DC. with a note on insect infestation of seeds. Ph.D. thesis, Univ. Delhi, Delhi.

Chaudhry, Bharti and M.R. Vijayaraghavan. 1992. Structure and function of the anther gland in *Prosopis juliblora* (Leguminosae, Mimosoideae): A histochemical analysis. Phyton, 32: 1–7.

***Chebotaru, A.A. 1960.** On the megasporogenesis and embryo sac development in maize (megasporogenesis—a base of early onthogenesis), pp. 49–74. *In* Trans. Moldavian branch Biol. Inst. USSR Acad. Sciences, 1. Shtiintsa, Kishinev, Moldova.

Chebotaru, A.A. 1968. Electron microscopy investigation, ontogeny and the question of new formation of organelles, p. 11. *In* Abstracts XII Intern. Cong. Genetics. Tokyo, Japan.

***Chebotaru, A.A. 1969a.** Additional observations of pollen grain ultrastructure. Reports USSR Acad. Sciences, 184 (2): 449–451.

Chebotaru, A.A. 1969b. The development and ultrastructure of grass plastids, pp. 122–145. *In* Chloroplasts and Mitochondria. Nauka, Moscow, Russia.

***Chebotaru, A.A. 1969c.** The ultrastructure of male and female gametophytes, pollen tube and the question on double fertilization in plants. Reports Moldavian SSR Acad. Sciences, 3: 87–90.

***Chebotaru, A.A. 1970.** Cytoembryological and electronic microscopy investigation of maize (*Zea mays* L.). Ph.D. thesis, Kishinev, Moldova.

***Chebotaru, A.A. 1972.** Embryology of Maize. Shtiintsa, Kishinev, Moldova.

Chebotaru, A.A. 1978. Some characteristics of vegetative, generative and sperm cell cytoplasm of maize (*Zea mays* ssp. *sackharata* Sturt.) Bull. Société botanique de France. Actualities botaniques, France, Reims, (1–2): 155–160.

***Chebotaru, 1983.** On the morphofunctional status of higher plants. *In* Abstracts All-Union Symp. 'The Development of Male Generative Sphere of Plants (Morphophysiological aspects)', Sympheropol, Ukraine.

Chebotaru, A.A. 1985. The morphofunctional status of higher plant gametogenesis. *In* Proc. Intern. Symp. Sexual Reproduction. Wageningen, Netherlands.

Chen, Z.-K., F.-H. Wang and F. Zhou. 1988. On the origin, development and ultrastructure of the orbicules and pollenkitt in the tapetum of *Anemarrhena asphodeloides* (Liliaceae). Grana, 27: 273–282.

Chiarugi, A. 1927. Il gametofito femmineo delle Angiospermae nei suovo varitipi di construcione e di sviluppo. Nuovo Giornale Botanico Italiano, N. S. 34 (5): 1–133.

Chiarugi, A. 1933. Lo sviluppo del gametofito femmineo della *Weddelina squamulosa*. Comptes Rendus de l'Academie de Sciences Lincei, Roma. 17: 1095–1100.

Chiarugi, A. 1950. La poliploidia della genrazione aploide femminile delle fanerogame. Caryologia, 3 (2): 149–155.

Chiarugi, A. and E. Franchini. 1930. Apomissia in *Ochna serulata* Walp. Nuovo Giornale Botanico Italiano, 37 (1): 1–250.

Chikkannaiah, P.S. 1962. Embryological studies in Commelinaceae, pp. 23–36. *In* Symp. Council Scientific and Industrial Research. 'Plant Embryology'. New Delhi.

Christensen, J.E., H.T. Horner and N.R. Lersten. 1972. Pollen wall and tapetal orbicular wall development in *Sorghum bicolor*. Amer. J. Bot., 59 (1): 43–58.

Chu, H. and Z. Hu. 1981. Male gametophyte development in monosomics of wheat. Genome, 32 (1): 155–164.

Clauhs, R.P. and P. Grun. 1977. Changes in plastid and mitochondrion content during maturation of generative cells of *Solanum* (Solanaceae). Amer. J. Bot., 64 (4): 377–383.

Claussen, P. 1927. Über das Verhalten des Antherentapetum bei einiger Monocotylen und Ranales. Botanisches Archiv Königsberg (Kaliningrad, Russian SFSR) & Dahlem bei Berlin, 18: 1–27.

Cleveland, L.R. 1947. The origin and evolution of meiosis. Science, 105 (2724): 287–289.

Clifford, S.C. and S.J. Orvens. 1990. The stigma, style and ovarian transmitting tract in the Oncidiinae (Orchidaceae): morphology, developmental anatomy and histochemistry. Bot. Gazette, 151 (4): 440–451.

Cocucci, A.E. 1981. Aspectos ultrastructurales de la fertilization en angiospermas. Kurtziana, 14: 41–62.

Cocucci, A.E. and W.A. Jensen 1969. Orchid embryology: the mature megagametophyte of *Epidendrum scutella*. Kurtziana, 5: 23–28.

Coe, G.E. 1954. Distribution of carbon 14 in ovules of *Zephyranthes drummondii*. Bot. Gazette, 115 (4): 342–346.

Cooper, D.C. 1935. Macrosporogenesis and development of the embryo sac of *Lilium henryi*. Bot. Gazette, 97 (1): 346–355.

Cooper, D.C. 1941. Macrosporogenesis and the development of the seed of *Phryma leptostachya*. Amer. J. Bot., 28 (9): 755–761.

Cooper, D.C. 1952. The transfer of desoxyribose nucleic acid from the tapetum to the microsporocytes at the onset of meiosis. Amer. Naturalist, 86: 219–230.

Cooper, D.C. and R.A. Brink. 1945. Seed collapse following matings between diploid and tetraploid races *Lycopersicon pimpinellifolium*. Genetics, 30 (4): 376–401.

Copeland, H.F. 1966. The morphology and embryology of *Euonymus japonica*. Phytomorphology, 16 (3): 326–333.

Cori, P. 1977. Wall ingrowths in the embryo sac of Euphorbia helioscopia. Israel Journal of Botany, 26(4): 202–208.

Corner, E.J.H. 1949a. The Annonaceous seed and its four integuments. New Phytologist, 48 (3): 333–364.

Corner, E.J.H. 1949b. The Durian theory of the origin of the modern tree. Ann. Bot. New series, 13 (52): 367–414.

Corner, E.J.H. 1953. The Durian theory extended. I. Phytomorphology, 3 (4): 465–476.

Corner, E.J.H. 1954a. The Durian theory extended. II. The arillate fruit and the compound leaf. Phytomorphology, 4 (1–2): 152–165.

Corner, E.J.H. 1954b. The Durian theory extended III. Pachycauly and megaspermy conclusion. Phytomorphology, 4 (3–4): 263–274.

Corner, E.J.H. 1976. The Seeds of Dicotyledons. Vols. 1, 2. Cambridge Univ. Press. Cambridge.

Coulter, J.M. 1898. Contribution to the life history of *Ranunculus*. Bot. Gazette, 25: 73–88.

Coulter, J.M. 1908. Relation of megaspores to embryo sacs in Angiosperms. Bot. Gazette, 45: 361–366.

Coulter, J.M. 1909. Evolutionary tendencies among Gymnosperms. Bot. Gazette, Vol. 41.

Coulter, J.M. and C.J. Chamberlain. 1903. Morphology of Angiosperms. Univ. Chikago Press, Chicago, Illinois, USA.

Cousin, M.T. 1979. Tapetum and pollen grain of *Vinca rosea*, Apocynaceae. Grana, 18 (2): 115–128.

Cresti, M., S.A. Lancelle and P.K. Hepler. 1987. Structure of the generative cell wall complex after freeze substitution in pollen tubes of *Nicotiana* and *Impatiens*. J. Cell Science, 88: 373–378.

Cresti, M., J.L. van Went, M.T.M. Willemse and E. Pacini. 1976. Fibrous masses and cell and nucleus movement in the pollen tube of *Petunia hybrida*. Acta Botanica Neerlandica, 25 (5): 381–383.

Crété, P. 1937. Etude sur la strophiole du *Chelidonium majus* L. Bull. Société botanique de France, 84 (3–4): 196–199.

Cronquist, A. 1981. An Integrated System of Classification of Flowering Plants. Columbia Univ. Press, NY, USA.

D'Amato, F. 1940. Contribution all'embriologia delle Plumbaginaceae. Nuovo Giornale Botanico Italiano, N.S. 47 (2): 349–382.

D'Amato, F. 1984. Role of polyploidy in reproductive organs and tissues, pp. 519–566. *In* B.M. Johri (ed.). Embryology of Angiosperms. Springer-Verlag, Berlin.

Dahlgren, K.V.O. 1916. Zytologische und embryologische Studien über die Reichen Primulales und Plumbaginales. Kongliga Svenska Vetenskapsakademiens Handlingar, 56 (4): 1–80.

Dahlgren, K.V.O. 1927. Die Morphologie des Nuzellus mit besonderer Berücksichtigung der deckzellosen Typen. Jahrbücher für Wissensschaftliche Botanik, 67 (2): 374–426.

Dahlgren, K.V.O. 1928. Hakenformige Leistenbildungen bet Synergiden. Berichte der Deutschen Botanischen Gesellschaft, 46: 434—443.

Dahlgren, K.V.O. 1937. Die Entwicklung des Embryosacken bei *Plumbago zeylanica*. Botaniska Notiser, 6: 487–498.

Dahlgren, K.V.O. 1939. Endosperm und Embryobildung bei *Zostera marina*. Botaniska Notiser, (4): 607–615.

Dahlgren, K.V.O. 1940. Postamentbildungen in den Embryosacken der Angiospermen. Botaniska Notiser, (4): 347–369.

Dahlgren, R. 1975. The distribution of characters within an angiosperm system. Some embryological characters. Botaniska Notiser, 128: 181–197.

Dahlgren, R. 1980. A revised system of classification of the angiosperms. Bot. J. Linn. Soc. 80 (2): 91–124.

Danilova, M.F. and M.E. Kirpichnikov. 1985. Glossary of terms, pp. 8–32. *In* A.L. Takhtajan (ed.) 1985. Comparative Anatomy of Seeds. Monocotyledons. Nauka, Leningrad, Russia, Vol. 1.

Davis, G.L. 1966. Systematic Embryology of Angiosperms. John Wiley and Sons, Inc. NY, USA.

Davis, G.L. 1968. Apomixis and abnormal anther development in *Calotis lappulacea* Benth. (Compositae). Austral. J. Bot., 16 (1): 1–17.

De Boer-de Jeu, M.J. 1978. Megasporogenesis. A comparative study of the ultrastructural aspects of megasporogenesis in *Lilium, Allium* and *Impatiens*. Mededeelingen van de Landbouwhogeschool te Wageningen, 16: 128.

De Candolle, A.P. 1827. Organodraphie végétale, ou discription valsonnée des organes des plantes pour servir de suite et de développement à la théorie élémentaive de la botanique et d'introduction à la physiologie végétale et à la discription des familles. Deterville, Paris, France, Vols. 1–2.

De Fossard, R.A. 1969. Development and histochemistry of the endothecium in the anthers of in vitro grown *Chenopodium rubrum* L. Bot. Gazette, 130 (1): 10–22.

De Mason, D.A. and K.N.C. Sekhar. 1988. The breeding system in the date (*Phoenix dactylifera* L.) and its recognition by early cultivators. Advances in Economic Botany, 6: 20–35.

De Mol, W.E. 1923. Duplication of generative nuclei by means of physiological stimuli and its significance. Genetica, 5: 225–272.

Deschamps, R. 1969. Premiers stages du developpement de l'embryon et de l'albumen du Lin: etude au microscope electronique. Revue de Cytologie et de Biologie Végétales, 32 (3–4): 379–390.

Diboll, A.G. and D.A. Larson. 1966. An electron microscopic study of mature megagametophyte in *Zea mays*. Amer. J. Bot., 53 (4): 391–402.

Dickinson, H.G. 1970. The fine structure of a peritapetal membrane investing the microsporangium of *Pinus banksiana*. New Phytologist, 69 (4): 1065–1068.

Dickinson, H.G. 1976. The deposition of acetolysis resistant polymers during the formation of pollen. Pollenand Spores, 18 (3): 321–334.

Dickinson, H.G. 1982. The development of pollen. Revue de Cytologie et de Biologie Végétales, Botanique, 5: 5–19.

Dickinson, H.G. and D. Lewis. 1973. The formation of the tryphine coating the pollen grains of *Raphanus* and its properties relating to the self-incompatibility system. Proc. Roy. Soc. London. Series B. Biological Sciences. 184: 149–165.

Dickinson, H.G. and J. Heslop-Harrison. 1977. Ribosomes, membranes and organelles during meiosis in angiosperms. Phil. Trans. Roy. Soc. London, Series B, 277: 327–342.

Dickinson, H.G. and F.L. Li. 1988. Organelle behaviour during higher plant gametogenesis. Proc. Soc. Exper. Bio. Med. 5: 131–148.

Dong, J. and Hong-yuan Yang. 1989. Ultrastructure of embryo sac in *Oryza sativa* L. Acta Botanica Sinica, 31 (2): 81–88.

Dormer, K.J. 1962. The fibrous layer in the anthers of Compositae. New Phytologist, 61 (2): 150–153.

Dottori, N. 1991. Anatomia reproductiva en Ulmaceae sensu lato. III. Esporangius, esparogenesis y gametogenesis de *Phyllostylon rhamnoides* y *Celtis tala*. Kurtziana, 21: 81–110.

Drahowzal, G. 1936. Ein Beitrag zur Morphologie und Entwicklungsgeschichte der Pollenkorner. Österreichische Botanische Zeitschrift, 85 (4): 241–269.

Dressler, R.L. 1993. Phylogeny and Classification of the Orchid Family. Dioscorides Press, Portland, OR, USA.

Dumas, C., R.B. Knox and T. Gaube. 1985. The spatial association of the sperm cells and vegetative nucleus in the pollen grain of *Brassica*. Protoplasma, 124 (3): 168–174.

Dumas, C., R.B. Knox, C.A. MsConchie and S.D. Russel. 1984. Emerging physiological concepts in fertilization. What's New in Plant Physiology, 15: 17–20.

Dunbar, A. 1973. Pollen development in *Eleocharis palustris* group (Cyperaceae). I. Ultrastructure and ontogeny. Bot. Notiser, 126 (2): 197–254.

Duvick, D.N. 1959. The use of cytoplasmic male sterility in hybrid production. Econ. Bot., 13 (3): 167–195.

Dzevaltovskyi, A. K. 1983. Family Cucurbitaceae, pp. 127–138. *In* T.B. Batygina and M.S. Yakovlev (eds.). Comparative Embryology of Flowering Plants. *Phytolaccaceae—Thymelaeaceae.* Nauka, Leningrad, Russia.

Eames, A.J. 1961. Morphology of the Angiosperms. McGraw-Hill, NY, USA.

Earle, T.T. 1938. Embryology of certain Ranales. Bot. Gazette, 100 (2): 257–275.

Echlin P. 1971a. Production of sporopollenin by the tapetum, pp. 220–247. *In* Proc. Symp. 'Sporopollenin', London, 1970. London–NY.

Echlin, P. 1971b. The role of the tapetum during microsporogenesis of angiosperms, pp. 52–53. *In* J. Heslop-Harrison (ed.). Pollen: Development and Physiology. Butterworths, London.

Echlin, P. and H. Godwin. 1968. The ultrastructure and ontogeny of pollen in *Helleborus foetidus* L. II. Pollen grain development through the callose special wall stage. J. Cell Science, 3: 161–174.

Edwardson, J.R. 1970. Cytoplasmic male sterility. Bot. Review, 36 (4): 341–420.

El-Ghazaly, G. 1989. Pollen and orbicule morphology of some *Euphorbia* species. Grana, 28: 234–259.

El-Ghazaly, G. and W. Jensen. 1985. Studies of the development of wheat (*Triticum aestivum*) pollen. III. Formation of microchannels in the exine. Pollen et Spores, 27 (1): 5–14.

El-Ghazaly, G. and W. Jensen. 1986a. Studies of the development of wheat (*Triticum aestivum*) pollen. I. Formation of the pollen wall and Ubisch bodies. Grana, 25: 1–29.

El-Ghazaly, G. and W. Jensen. 1986b. Studies of the development of wheat (*Triticum aestivum*) pollen: formation of the pollen aperture. Can. J. Bot., 64 (2): 3141–3154.

Ellis Raymon, J. and Y. Chaffey Nigel. 1987. Structural differentiation of the nucellar epidermis in the caryopsis of rice (*Oryza sativa*). Ann. Bot., 60 (6): 671–675.

***Enaleeva, I.Kh., V.S. Tyrnov, and S.S. Khokhlov. 1972.** Separation of angiosperm embryo sacs by tissue maceration. Cytology i Genetica, 6 (5): 439–441.

Engler, A. and K. Prantl. 1926. Die natürlichen Pflanzenfamilien. 2-Aufl. Leipzig, 13.

Engell, K. 1987. Embryology and taxonomic position of *Retzia capensis* (Retziaceae). Nord. J. Bot., 7 (2): 117–124.

Engler, A. and K. Prantl. 1889. Die natürlichen Pflanzenfamilien. XII. T. 3. Abt. 1. Leipzig.

Erdelcká, O. 1973. Nucleolarvakuolen in den Zellen des Embryosackes von *Jasione montana* L. Protoplasma, 76: 123–127.

Erdelcká, O. 1983. Microcinematographical investigation of the female gametophyte, ferilization and endosperm development, pp. 49–54. *In* Proc. 7th Intern. Symp. 'Fertilization and Embryogenesis in Ovulated Plants', High Tatra (Račkova dolina), June 14–17, 1982, Bratislava, Czechoslovakia.

Erdtman, G. 1969. Handbook of Palynology. Munksgaard, Copenhagen.

Erdtman, G., B. Berglung, and J. Praglovgki. 1961. An introduction to a Scandinavian pollen flora. Grana Palynologica, 2 (3): 3–92.

***Ermakov, I.P. and O.V. Goryaeva. 1982.** Changes in the structure of ovule cells in *Lycopersicon esculentum* Mill., accompanying embryo sac activation before fertilization. Vestnik MGU, Biological Series: 1–11.

Ernst, A. 1908. Zur Phylogenie des Embryosackes der Angiospermen. Berichte der Deutschen Botanischen Gesellschaft, 26a: 419–438.

Esau, K. 1977. Anatomy of Seed Plants. J. Wiley and Sons, NY, USA.

Esau, K. and I.D. Charvat. 1975. An ultrastructural study of acid phosphatase localization in cells of *Phaseolus vulgaris* phloem by the use of the azo dyo method. Tissue and Cell, 7 (4): 619–630.

Esau, K. and L.L Hoefert. 1980. Endoplasmic reticulum and its relation to microtubulus in sieve elements of sugarbeet and spinach. J. Ultrastructure Res., 71 (3): 249–257.

Esser, K. 1963. Bildung und Abbau von Kallose in den Samenanlagen der *Petunia hybrida*. Zeitschrift für Botanik, 51 (1): 32–51.

Evert, R.F. 1984. Comparative structure of phloem, pp. 145–234. *In* R.A. White and W.C. Dickinson (eds.). Contemporary Problems on Plant Anatomy. Orlando, FL. USA.

Eymé, J. 1965. Recherhes sur la constitution cytoplasmique de l'archespore et du sac embryonnaire de *Lilium candidum* L. Botaniste, 48(1–6): 99–155.

Eymé, J. 1967. Nouvelles observation sur l'infrastructure de tissus nectarignes floraux. Botaniste, 50 (1–6): 169–183.

Faegri, K. and J. Iversen. 1975. Textbook of pollen analaysis . 3rd edition. Hafner, New York, USA.

Fagerlind, F. 1937. Embryologische, zytologische und Bestaubungs experimentelle Studien in der Familie Rubiaceae nebst Bemerkungen über einige Polyploiditatsprobleme. Acta Horti Bergiani, 11 (9): 195–470.

Fagerlind, F. 1939a. Der Embryosack von *Plumbagella* und *Plumbago*. Arkiv för Botanik, Ser.B, 29 (1): 1–8.

Fagerlind, F. 1939b. Die Entwicklung des Embryosackes bei *Peperomia pellucida* Kunth. Arkiv för Botanik, 29A (17): 1–15.

Fagerlind, F. 1939c. Drei Beispiele des Fritillaria-typus. Svensk Botanisk Tidskrift Utgifven af Svenska Botaniska Föreningen, 33 (2): 188–204.

Fagerlind, F. 1941a. Der Bau der Samenanlage und des Macrogametophyten bei *Quisqualis indica*. Bot. Notiser, 2: 217–222.

Fagerlind, F. 1941b. Bau und Entwicklung der Gnetum-Gametophyten. Kongliga Svenska Vetenskapsakademiens Handlingar, Ser. 3. 19: 1–55.

Fagerlind, F. 1944a. Der tetrasporische Angiospermen-Embryosack und dessen Bedeutung für das Verstandnis der Entwicklungsmechanik und Phylogenie des Embryosackes. Arkiv för Botanik, 31A (11): 1–71.

Fagerlind, F. 1944b. Die Samenbildung und die Zytologie bei agamospermischen und sexuellen Arten von *Elatostema* und einigen nahestehenden Gattungen nebst Beleuchtung einiger damit zusammenhängender Probleme. Kongliga Svenska Vetenskapsakademiens Handlingar, 21 (4): 1–130.

Fagerlind, F. 1945. Blüte und Blütenstand der Gattung *Balanophora*. Bot. Notiser, 4: 330–350.

Fagerlind, F. 1947. Gynoeceum—morphologische und embryologische Studien in der Familie Olacaceae. Bot. Notiser, 3: 207–230.

Fagerlind, F. 1961. The initiation and early development of the sporangium in vascular plants. Svensk Botanisk Tidskrift Utgifven af Svenska Botaniska Föreningen, 55 (2): 299–312.

Fahn, A. 1979. Secretory Tissues in Plants. Academic Press, London.

Favre-Duchartre, M. 1963. A propos des nucellus des plantes à ovules. Annales Scientifiques de lUniversité de Reims et de l'ARERS, 1–2.

Favre–Duchartre, M. 1984. Homology and Phylogeny, pp. 679–734. *In* B.M. Johri (ed.). Embryology of Angiosperms. Springer-Verlag, Berlin.

Fedorov, A.A. and Z.T. Artjushenko. 1975. Atlas on the Descriptive Morphology of Higher Plants. Leningrad, Russia.

Feudenstein, J.V. 1991. A systematic study of endothecial thickenings in the Orchidaceae. Amer. J. Bot., 78 (6): 766–781.

Figdor, W. 1897. Uber *Cotylanthera* Bl. Annales du Jardin Botanique de Buitenzorg, 14: 213–240.

Finn, W.W. 1928. Spermazellen bei *Vinca minor* und *V. herbacea*. Berichte der Deutschen Botanischen Gesellschaft, 46 (4): 235–236.

Fischer, A. 1880. Zur Kenntnis Embryosakentwicklung einiger Angiospermen. Jenaische Zeitschrift für Naturwissenschaft herausgegeben von der Medicinisch-naturwissenschaftlichen Gesellschaft zu Jena, 14: 9–132.

Flawell, R. 1974. A model for the mechanism of cytoplasmic male sterility in plants with special reference to maize. Plant Science Letters, 3: 259–263.

Fleurat-Lessard, G. 1986. Specialized formation of endoplasmic reticulum in parenchyma cells of *Mimosa pudica* L. Protoplasma, 130 (1): 1–4.

Folsom, M.W. and D.D. Cass. 1988. The characteristics and fate of the soybean inner nucellus. Acta Botanica Neerlandica, 37 (3): 387–393.

Folsom, M.W. and D.D. Cass. 1989. Embryo sac development in soybean: ultrastructure of megasporangenesis and early megagametogenesis. Can. J. Bot., 67: 2841 –2849.

Folsom, M.W. and D.D. Cass. 1990. Embryo sac development in soybean: cellularization and egg apparatus expansion. Can. J. Bot., 68: 2135–2147.

Ford, J.H. 1971. Ultrastructural and chemical studies of pollen wall development in the Epacridaceae. *In* J. Brooks, P. R. Grant, M. Muir, P. van Gijzel, G. Shaw (eds.). Sporopollenin. Acad. Press, London.

Foster, A.S. 1938. Structure and growth of the shoot apex in *Ginkgo biloba*. Bull. Torrey Botanical Club, 65: 531–556.

Fougère-Rifot, M. 1975. L'édification de l'appareil filiforme et l'évolution cytoplasmique des synergides du sac embryonnaire d'*Aquilegia vulgaris*. Comptes Rendus de l'Academie de Sciences, 280D (21): 2445–2447.

Fougère-Rifot, M. 1978. La cellule centrale du sac embryonnaire d'*Aquilegia vulgaris* L.: des noyaux polaires aux noyaux d'albumen leiteux. Bull. Société botanique de France. Actualities botaniques, 125 (1–2): 207–213.

Fougère-Rifot, M. 1981. Intercellular relations and wall structures in mature embryo sacs of four species of angiosperms. Acta Societatis Botanicorum Poloniae, 50 (1–2): 131–138.

Fougère-Rifot, M. 1987a. L'évolution du sac embryonnaire des Ranunculacées: les synergides d'*Aquilegia vulgaris* L. et l'apparation d'une sécrétion tardive. Revue de Cytologie et de Biologie Végétales—le Botaniste., 10 (1): 33–46.

Fougère-Rifot, M. 1987b. Macrosporo-et macrogametogeneses, fécondation, embryogénèse et albuminogénèse. Bull. Société botanique de France. Actualities botaniques, 134 (1): 113–168.

Freiberg, T.E. 1981. Family Winteraceae, pp. 26–30. *In* T.B. Batygina and M.S. Yakovlev (eds.). Comparative Embryology of Flowering Plants. Winteraceae—Juglandaceae. Nauka, Leningrad, Russia.

Freiberg, T.E. 1985. Family Parnassiaceae, pp. 43–48. *In* T.B. Batygina and M.S. Yakovlev (eds.). Comparative Embryology of Flowering Plants. Brunelliaceae—Tremandraceae. Nauka, Leningrad, Russia.

Freiberg, T.E. 1987a. Family Bignoniaceae, pp. 259–264. *In* T.B. Batygina and M.S. Yakovlev (eds.). Comparative Embryology of Flowering Plants. Davidiaceae—Asteraceae. Nauka, Leningrad, Russia.

Freiberg, T.E. 1987b. Family Verbenaceae, pp. 219–225. *In* T.B. Batygina and M.S. Yakovlev (eds.). Comparative Embryology of Flowering Plants. Davidiaceae—Asteraceae. Nauka. Leningrad, Russia.

Freiberg, T.E. 1990. Family Commelinaceae, pp. 198–204. *In* T.B. Batygina and M.S. Yakovlev (eds.). Comparative Embryology of Flowering Plants. Butomaceae—Lemnaceae. Nauka, Leningrad, Russia.

Freiberg, T.E. and O.P. Kamelina. 1981. Family Magnoliaceae, pp. 36–41. *In* T.B. Batygina and M.S. Yakovlev (eds.). Comparative Embryology of Flowering Plants. Winteraceae—Juglandaceae. Nauka, Leningrad, Russia.

French, I.C. 1985a. Patterns of endothecial wall thickenings in Araceae: subfamilies Pothoidaceae and Monsteroidaceae. Amer. J. Bot., 72: 472–486.

French, I.C. 1985b. Patterns of endothecial wall thickenings in Araceae: subfamilies Calloidaceae, Lasioidaceae and Philodendroidaceae. Bot. Gazette, 146: 521–533.

French, I.C. 1986. Patterns of endothecial thickenings in Araceae: subfamilies Colocasiodaceae, Aroidaceae and Pistioidaceae. Bot. Gazette, 147: 166–179.

Friedman, W.E. 1990. Double fertilization in Ephedra, a nonflowering seed plant: its bearing on the origin of Angiosperms. Science, 247: 951–954.

Fuchs, T.C.E. 1938. Beiträge zur Embryologie der Thymelaeaceae. Osterreichische Botanische Zeitschrift, 87 (1): 1–41.

Gabara B.A. 1974. Possible role for the endoplasmic reticulum in exine formation. Grana, 14 (1): 16–22.

***Gabarayeva, N.I. 1986.** Development of exine in *Mihelia fuscata* (Magnoliaceae) in connection with the changing of cytoplasmic organelles in microspores and tapetum. Bot. Zhurnal, 71 (3): 311–322.

***Gabarayeva, N.I. 1987.** Ultrastructure and development of endexine lamelles in *Manglietia tenuipes* (Magnoliaceae) in connection with the question of the existence of endexine lamelles in the primitive angiosperms. Bot. Zhurnal, 72 (10): 1310–1317.

***Gabarayeva, N.I. 1990.** Synthesis, place of sporopollenin precursors in the developing pollen grains of the family Magnoliaceae representatives. Bot. Zhurnal, 75 (6): 783–791.

Gabarayeva, N.I. 1991a. Patterns of development in primitive angiosperm pollen. Pollen and Spores: Pattern of diversification, 44 (4): 257–268.

***Gabarayeva, N.I. 1991b.** Ultrastructure and development of exine and orbicules in *Magnolia delavayi* (Magnoliaceae) at the tetrad and in the beginning of posttetrad periods. Bot. Zhurnal, 76 (1): 10–19.

***Gachechiladze, M.I. 1981.** Embryology of some species of the family Onagraceae. Ph.D. thesis, Tbilisi, Georgia.

Gaertner, J. 1788–1807. De fructibus et seminibus plantarum, Vols. 1–3, Stuttgart, Leipzig.

Galil J., Zeroni M. 1969. On the organization of the pollinium in *Asclepias curassavica*. Bot. Gazette, 130(1): 1–4.

Gates, R.R. 1911. Pollen formation in *Oenothera gigas*. Ann. Bot., 25: 909–940.

Geerts, G.M. 1908. Beitrag zur Kenntnis der zytologischen Entwicklung von *Oenothera lamarckiana*. Berichte der Deutschen Botanischen Gesellschaft, 26a: 608–614.

Geitler, L. 1935. Beobachtungen über die erste Teilung in Pollenkorn der Angiospermen. Planta, 24 (3): 361–386.

Geitler, L. 1941. Embryosäcke aus Pollenkornern bei Ornithogalum. Berichte der Deutschen Botanischen Gesellschaft, 59 (9): 419–423.

Gerasimoff, I. 1890. Einige Bemerkungen über die Function des Zellkerns. Moscow Soc. Natl. Bull., 4: 548–554.

Gerassimova, E.N. 1933. Fertilization in *Crepis capillaris*. Cellule, 42 (1): 103–148.

***Gerassimova-Navashina, H.N. 1951a.** Pollen grain, gametes and sexual process in flowering plants. Trans. Komarov Bot. Inst. USSR Acad. Sciences, Series VII, 2: 294–355.

Gerassimova-Navashina, H.N. 1951b. Pollen grain, gametes and sexual process in flowering plants, pp. 28–33. *In* Reports of Delegate Conf. All-Union Bot. Soc., Moscow-Leningrad, Russia, 4.

***Gerassimova-Navashina, H.N. 1952.** Cytological and embryological understanding of the pollination process. Trans. Komarov Bot. Inst. USSR Acad. Sciences, Series VIII, 3: 165–211.

***Gerassimova-Navashina, H.N. 1954a.** Embryo sac development and the question of origin of flowering plants. Reports USSR Acad. Sciences, 95 (4): 877–881.

***Gerassimova-Navashina, H.N. 1954b.** Embryo sac development, double fertilization and the question of origin of flowering plants. Bot. Zhurnal, 39 (5): 655–681.

***Gerassimova-Navashina, H.N. 1957.** Fertilization as an ontogenetic process. Bot. Zhurnal, 42 (11): 1654–1673.

***Gerassimova-Navashina, H.N. 1958.** Gametophyte and main features of development and function of reproductive elements in flowering plants. Prob. Bot., 3: 125–167.

***Gerassimova-Navashina, H.N. 1962.** Cytological data on the stimulus to development of embryo sac cells (of plants). Trans. Komarov Bot. Inst. USSR Acad. Sciences, Series VII, 5: 238–249.

***Gerassimova-Navashina, H.N. 1971.** Double fertilization in flowering plants and some theoretical aspects, pp. 113–152. *In* Problems of Embryology. Naukova Dumka, Kiev, Ukraine.

***Gerassimova-Navashina, H.N. and T.B. Batygina. 1958.** The process of fertilization in *Scilla sibirica* Andr. Bot. Zhurnal, 43 (7): 959–988.

Gerassimova-Navashina, H.N. and V.A. Gulyaev. 1973. Some data on ultrastructure of embryo sac cells after pollination in *Crepis capillaris* (L.) Wallr. Bull. Main Botanical Garden USSR Acad. Sciences, 89: 14–20.

Gifford, E.M. and G.E.I. Corson. 1971. The shoot apex in seed plants. Bot. Review, 37 (2): 143–229.

Glushchenko, G.I. 1998. Cytoembryological investigation of *Allium cepa* L. Ph.D. thesis, Moscow, Russia.

Godineau, J.-C. 1971. Ultrastructure du sac embryonnaire de *Crepis tectorum* L.: Etat après la cellularisation et la fusion des noyaux polaires. Ann. Scientifiques l'Univ. Reims et l'ARERS, 9 (1): 78–88.

Goebel, K. 1880. Beiträge zur vergleichenden Entwicklungsgeschichte der Sporangien. Bot. Zeitung, 38 (32): 545–553.

Goebel, K. 1898–1901. Organographie der Pflanzen. G. Fisher, Jena.

Goebel, K. 1905. Organography of Plants, Oxford Univ. Press, London, vol. 2.

Goebel, K. 1923. Organographie der Pflanzen, insbesondere der Archegoniaten und Samenpflanzen, vol. 3, IX, pp. 1209–1789. G. Fischer, Jena.

Goebel, K. 1932. Organographie der Pflanzen, insbesondere der Archegoniaten und Samenpflanzen, vol. 3, pp. 1380–1445. G. Fischer, Jena.

Goebel, K. 1933. Organographie der Pflanzen, insbesondere der Archegoniaten und Samenpflanzen, 10: 1379–2078. G. Fischer, Jena.

Goldberg, A. 1986. Classification, evolution and phylogeny of the families of dicotyledons. Smithsonian Contrib. Bot., 58: 1–205.

Goldshmidt, R. 1923. Mechanism and Physiology of Sex Determination. Gosizdatelstvo, Moscow, Russia.

Golenkin, M.I. 1927. Winners in the battle for survival. Investigation of causes and conditions of earth conquest by flowering plants in the Cretaceous period, pp. 1–101. *In* Trans. Botanical Institute, scientific institute association attached to the physics-mathematical faculty of the 1st Moscow University. Moscow, Russia.

***Golubeva, E.A. 1983.** The formation of acetolysostable membranes of pollen grains and anther tapetum cells in *Beta vulgaris.* Trans. Applied Botany, Genetics and Selection, All-Union Institute of Plant Industry, 74: 56–64.

***Golubeva, E.A. 1989.** Ultrastructure and function of anther wall middle layers in sugar beet. Trans. Applied Botany, Genetics and Selection. All-Union Institute of Plant Industry, 124: 78–82.

Golubovskaya, I.N. 1988. Chromosomes of maize, pp. 7–9. *In* Genetics of Cultivated Plants: Maize, Rice, Millet, Oat. Agropromizdat, Leningrad, Russia.

Gorska-Brylass, A. 1967. Transitory callose envelope surrounding the generative cell in pollen grains. Acta Societatis Botanicorum Poloniae, 36 (2): 419–422.

Gorska-Brylass, A., E. Bednarska and I. Swierzowicz. 1986. At which morphological maturation stage of pollen grain does DNA replication take place? Acta Universitatis Nicolai Copernice. Biologia. 28: 13–19.

Göthe, J. W. 1790. Versuch die Metamorphose der Pflanzen zu erklaren. Corl Wilhelm Ettinger, Gotha.

Grevtsova, N.A. 1987. Family *Apiaceae,* pp. 35–44. *In* T.B. Batygina and M.S. Yakovlev (eds.). Comparative Embryology of Flowering Plants. *Davidiaceae—Asteraceae.* Nauka, Leningrad, Russia.

***Grevtsova, N.A. and A.N. Sladkov. 1983.** Possibility of term unification at the description of microsporangium wall development. Biologicheskye Nauki, 3: 83–87.

Grevtsova, N.A. and Kh.Kh. Djalilova. 1992. Microsporangium wall structure in *Lupinus* L., pp. 191–192. *In* Proc. XI Inter. Symp. 'Embryology and Seed Reproduction', 1990, Leningrad, St.-Petersburg, Russia.

Grew, N. 1672. The anatomy of the vegetable begun. With a general account of vegetation founded thereon. Printed for Spencer Hickman, London.

*Grodzinsky, D.M. 1977. The systems of plant organism reliability, pp. 17–29. *In* D.M. Grodzinsky (ed.). The Systems of Cell Reliability. Naukova Dumka, Kiev, Ukraine.
*Grodzinsky, D.M. 1983. The Plant System Reliability. Naukova Dumka, Kiev, Ukraine.
Grootjen, C. J. 1983. Development of ovule and seed in Marantaceae. Acta Botanica Neerlandica, 32 (1–2): 69–86.
Grootjen, C.J. and F. Bouman. 1981. Development of ovule and seed in *Costus cuspidatus* (Zingiberaceae) with special reference to the operculum. Bot. J. Linn. Soc. 83 (1): 27–39.
Grootjen, C.J. and F. Bouman. 1988. Seed structure in Cannaceae: taxonomic and ecological implications. Ann. Botany, 61 (3): 363–371.
***Grossheim, A.A. 1945.** Question of graphic representation of flowering plants systems. Sov. Botanika, 13 (3): 3–27.
Grove, S.N. and C.E. Bracker. 1978. Protoplasmic changes during zoospore encystment and cyst germination in *Pythium aphanidermatum*. Experimental Mycology, 2 (1): 51–98.
Guèrin, P. 1916. Reliquiae Trebianae. Recherches sur la structure anatomique de l'ovule et de la graine des Thymeleacées. Ann. Jardin botanique de Buitenzorg, Série 2, 14: 3–35.
Guignard, L. 1881. Sur l'origine du sac embryonnaire et le role des antipodes. Bulletin de la Société botanique de France, 28: 197–201.
Guignard, L. 1882. Recherches sur le sac embryonnaire des phanerogames angiosperms. Ann. Sciences Naturelles. Botanique, 13 (6): 136–199.
Guignard, L. 1889. Etudes sur le phenomenes morphologiques de la fécondation. Bull. Société botanique de France, 36: 100–146.
Guignard, L. 1891. Nouvelles etudes sur la fécondation. Ann. Sciences Naturelles. Botanique, 14 (7): 163–296.
Guignard, L. 1893. Recherches sur le développement de la graine et en particulier de tégument seminal. J. Botany, British and Foreign, 16: 301–368.
Guignard, L. 1917. L'ovule chez les Apocynacées et Asclepiadacées. Memories de l'Académie Royale des Sciences de l'institut de France, 55: 1–34.
Guilford, W.J. , D.M. Schneider, J. Labovitz and S.J. Opella. 1988. High resolution solid state ^{13}C NMR spectroscopy of sporopollenins from different plant taxa. Plant Physiology, 86(1): 134–136.
Guinet, Ph. and A.Le Thomas. 1973. Interpretation de la repartition dissymetrique des couches de l'exine dans les pollens composes: consequences relatives a la notion d'aperture. Co. Re. Acade. Sciences, D. 276: 1545–1548.
Gunning, B.E.S. and J.S. Pate. 1974. Transfer cells, pp. 441–480. *In* A.W. Robards (ed.) Dynamic Aspects of Plant Ultrastructure. McGraw-Hill Book Co., London.
Gunning, B.E.S. and A.R. Hardham. 1982. Microtubules. Ann. Rev. Plant Physiology, 33: 651–698.
Gupta, S.C. and V.M. Rajeswari. 1977. Nucellar tracheids in ovules of *Luffa cylindrica*. Ann. Botany, 41 (175): 1005–1007.
Gurudeva, M.R. and R.K. Somashekar. 1985. Development of embryo sac in *Paracautleya bhatii* R. M. Smith., Zingiberaceae. Current Science (India), 54 (19): 1009–1010.
Guttenberg, H. 1960. Grundzuge der Histogenese höherer Pflanzen. I. Die Angiospermen. Borntraeger, Berlin-Nikolassee.
***Gvaladze, G.E. 1960.** Embryology of *Allium schoenoprasum* L. Reports Georgian SSR Acad. Sciences, 25 (3): 327–334.
***Gvaladze, G.E. 1966.** Investigation of polar nuclei in angiosperms. Reports Georgian SSR Acad. Sciences, 68 (3).
***Gvaladze, G.E. 1973.** Comparative embryological investigation of embryo sacs of *Polygonum, Oenothera, Allium, Fritillaria* types in some plants. Ph.D thesis, Tbilisi, Georgia.
***Gvaladze, G.E. 1976.** Chalazal polar nucleus of embryo sac central cell in Angiosperms. Metzniereba, Tbilisi, Georgia.
***Gvilava, M.N. 1987.** Embryo sac development in the species of family Onagraceae. Ph.D. thesis, Tbilisi, Georgia.

Gylarov, M.S. (ed.) 1989. Biological Encyclopaedic Glossary. Publ. 2. Sovetskaya Entziclopedia, Moscow, Russia.

Haan, H.R.M. de. 1920. Contribution to the knowledge of the morphological value and phylogeny of the ovule and its integuments. Recueil des Travaux Botaniques Neerlandais, 17: 219–324.

Haberlandt, G. 1921. Wundhormone als Erreger von Zellteilungen. Beiträge zur Allgemeinen Botanik, 2: 1–53.

Haberlandt, G. 1922. Die Vorstufen und Ursachen der Adventivembryonie. Sitzungsberichte für Preusse Acad. Wissenschaften, 25: 386–406.

Haberlandt, G. 1923. Zur Embryologie von *Allium odorum*. Berichte der Deutschen Botanischen Gesellschaft, 41: 174–179.

Haberlandt, G. 1925. Zur Embryologie und Cytologie von *Allium odorum*. Berichte der Deutschen Botanischen Gesellschaft, 43: 559–564.

Hagemann, R. 1981. Unequal plastid distribution during the development of the male gametophyte of angiosperms. Acta Societatis Botanicorum Poloniae, 50 (1–2): 321–327.

Hagemann, R. and M.B. Schröder. 1989. The cytological basis of the plastid inheritance in angiosperms. Protoplasma, 152 (1): 57–64.

Håkansson, A. 1921. Beiträge zur Entwicklungsgeschichte der Taccaceen. Botaniska Notiser, pp. 189–268.

Håkansson, A. 1923. Studien über die Entwicklungsgeschtchte der Umbelliferen. Acta Univ. Lundensis, Adv. 2. 1 (7): 1–120.

Hamilton, D.A., D.M. Bashe, J.P. Stison and J.P. Mascarenhas. 1989. Characterization of pollen specific genomic clone from maize. Sexual Plant Reproduction, 2 (4): 208–212.

Harling, G. 1950. Embryological studies in the Compositae. I. Anthemideae—Anthemedinae. Acta Horti Bergiani, 15: 135–168.

Harling, G. 1951. Embryological studies in the Compositae. II. Anthemideae—Chrysantheminae. Acta Horti Bergiani, 16 (1): 1–56.

Hartig, Th. 1858. Entwicklungsgeschichte des Pflanzenkeims, dessen Stoffbildung und Stoffwandlung während der Vorgange des Reifens und des Keimens für Pflanzenbau. Forstnersche Buchhandl, Leipzig.

Haupt, A.W. 1934. Ovule and embryo sac of *Plumbago capensis*. Bot. Gazette, 95 (3): 649–659.

Hayashi, H. 1960. On the microsporogenesis and pollen morphology in the family Magnoliaceae. Scientific Reports of the Tôhoku University, Series 4, 29 (1–2): 27–33.

Hayashi, Y. 1963. The embryology of the Magnoliaceae (sens. lat.) II. Megasporogenesis, female gametophyte and embryogeny of *Schisandra repanda* Radlkofer and *Kadsura japonica* Dunal. Science Reports Tôhoku Imperial Univ. Ser. 4 (Biology), 29 (3–4): 403–411.

Hayashi, Y. 1965. The comparative embryology of the Magnoliaceae (s. l.) in relation to the systematic consideration of the family. Science Reports Tôhoku Imperial Uni. Ser. 4 (Biology), 31: 29–44.

Hayward, M.D. and M.A.P.P. Manthriratna. 1972. Pollen development and variation in the genus *Lolium*. 1. Pollen size and tapetal relationships in male-fertiles and male-steriles. Zeitschrift für Pflanzenzüchtung, 67: 131–144.

Hegelmaier, F. 1868. Die Lemnaceen. Eine monographische Untersuchung. W. Engelmann, Lepzig.

Hemsley, A.R, W.G. Chaloner, A.C. Scott and P.J. Barric. 1992. Composition of sporopollenin in fossil and living spore systematics, p. 62. *In* Abst. 8th Intern. Palynological Cong., Aix-en-Provence.

Henfrey, A. 1853. The phenomenon of rejuvenescence in nature ... by Dr. A. Braun; transl. by A. Henfrey. Royal Society, London.

Herich, R. and A. Lux. 1985. Lytic activity of Ubisch bodies (orbicules). Cytologia, 50: 563–569.

Hertwig, O. 1894. Zeit und Streitfragen der Biologie. Heft 1. Praformation oder Epigenese? Grundzuge einer Entwicklungs-theorie der Organismen. G. Fischer, Jena.

Herzfeld, S. 1922. *Ephedra campylopoda* Mey. Morphologie der weiblischen Blute und Befruchtungsvorgang. Denkschriften der Königlichen Acad. Wissenschaften zu München, 98: 243–268.

Herzfeld, S. 1927. Beiträge zur Kenntnis von *Ginkgo*. Jahrb. Wissensschaftliche Botanik, 66: 814–862.

Heslop-Harrison, J. 1962. Origin of exine. Nature, 195: 1069–1071.

Heslop-Harrison, J. 1963. An ultrastructural study of pollen wall ontogeny in *Silene pendula*. Grana Palynologica, 4 (1): 7–24.

Heslop-Harrison, J. 1964. Cell walls, cell membranes, and protoplasmic connections during meiosis and pollen development, pp. 39–47. *In* H.F. Linskens (ed.). Pollen Physiology and Fertilization. North-Holland Publ. Co., Amsterdam.

Heslop-Harrison, J. 1966a. Cytoplasmic connections between angiosperm meiocytes. Ann. Botany, 30 (118): 221–230.

Heslop-Harrison, J. 1966b. Cytoplasmic continuities during spore formation in flowering plants. Endeavour, 25 (95): 65–72.

Heslop-Harrison, J. 1968a. Anther carotenoids and the synthesis of sporopollenin. Nature, 220 (5167): 605.

Heslop-Harrison, J. 1968b. Wall development within the microspore tetrad of *Lilium longiflorum*. Can. J. Bot., 46 (10): 1185–1192.

Heslop-Harrison, J. 1969. An acetolysis-resistant membrane investing tapetum and sporogenous tissue in the anthers of certain Compositae. Can. J. Bot., 47: 541–542.

Heslop-Harrison, J. 1979a. Aspects of the structure, cytochemistry and germination of the pollen of rye (*Secale cereale* L.). Ann. Botany, Supplement 1, 44: 1–47.

Heslop-Harrison, J. 1979b. Pollen walls as adaptive systems. Ann. Missouri Botanical Garden, 66 (4): 813–829.

Heslop-Harrison, J. 1987. Pollen germination and pollen tube growth, pp. 1—78. *In* K.L. Giles and J. Prakash (eds.) International Review of Cytology. Pollen: Cytology and Development. London.

Heslop-Harrison, J. and A. Mackenzie. 1967. Autoradiography of soluble (^{14}C) thymidine derivatives during meiosis and microsporogenesis in *Lilium* anthers. J. Cell Science, 2: 387–400.

Heslop-Harrison, J. and H.G. Dickinson. 1969. Time relationships of sporopollenin synthesis associated with tapetum and microspore in *Lilium*. Planta, 84 (3): 199–214.

Heslop-Harrison, J. and Y. Heslop-Harrison. 1982. The growth of the grass pollen tube. 1. Characteristics of the polysaccharide particles ("P-particles") associated with apical growth. Protoplasma, 112 (1/2): 71–80.

Heslop-Harrison, J. and Y. Heslop-Harrison. 1989. Actomyosin and movement of the vegetative nucleus of the angiosperm pollen tube: association with the actin cytoskeleton. J. Cell Science, 93: 229–308.

Heslop-Harrison, J., J.S. Heslop-Harrison and Y. Heslop-Harrison. 1986. The compartment of the vegetative nucleus and generative cell in the pollen and pollen tubes of *Helleborus foetidus* L. Ann. Botany, 58 (1): 1–12.

Hesse, M. 1984. An exine architecture model for viscin threads. Grana, 23: 69–75.

Hesse, M. 1986. Orbicules and the ectexine are homologous sporopollenin concretions in Spermatophyta. Plant Systematics and Evolution, 153 (1/2): 37–48.

Hesse, M. and M.W. Hess. 1993. Recent trends in tapetum research. A cytological and methodological review. Plant Systematics and Evolution, 7: 127–145.

Heusser, K. 1915. Die Entwicklung der generativen Organe von *Himanloglossum hircinum*. Beihefte zum Botanischen Centralblatt, 1 (32): 218–277.

Hoare, G. 1934. Gametogenesis and fertilization in *Scilla nonscripta*. Cellule, 42: 269–292.

Hodge, R, W. Paul, J. Draper and R. Scott. 1992. Cold-plaque screening: a simple technique for the isolation of low abundance, differentially expressed transcripts from conventional cDNA libraries. Plant J., 2: 257–260.

Hoefert, L.L 1969. Ultrastructure of *Beta* pollen. I. Cytoplasmic constituents. Amer. J. Bot., 56 (4): 363–368.

Hoeppener, E. and O. Renner. 1929. Genetische und zytologische Oenothereen-Studien. Botanische Abhandlungen, 1(15): 1–66.

Hofmeister, W. 1847. Untersuchungen des Vorganges bei der Befruchtung der Oenothereen. Botanische Zeitschrift, 5: 785–792.

Hofmeister, W. 1849. Die Entstehung des Embryo der Phanerogamen. Eine Reihe microscopischer Untersuchungen von Wilhelm Hofmeister. Verlag von Fr. Hofmeister, Leipzig.

Hofmeister, W. 1851. Vergleichende Untersuchungen der Keimung, Entfaltung und Fruchtbildung der höherer Kryptogamen und Samenbildung der Coniferen. Verlag von Friedrich Hofmeister, Leipzig.

Hofmeister, W. 1852–1857. Beiträge zur Kenntnis der Gefasskryptogamen, 1–2. Abhandlungen der Mathematisch-physischen Klasse der Sächsischen Academie der Wissenschaften, vols. 4, 5.

Hofmeister, W. 1859. Neue Beiträge zur Kenntnis der Embryobildung der Phanerogamen. II. Monocotyledonen. Druck von Breitkopf und Härtel, Leipzig, 2: 629–760.

Hofmeister, W. 1861. Über die durch die Schwerkrott bestimmer Richtunger von Pflazentheilen. Annales Scientifiques Naturelles (Botanique). 15: 179–219.

Hofmeister, W. 1867. Die Lehre von der Pflanzenzelle. Verlag von Wilhelm Engelmann, Leipzig. 12, 404 p.

Hogan, D.J. 1987. Microtubule patterns during meiosis in two higher plant species. Protoplasma, 138: 126–136.

Holmgren, I. 1913. Zur Entwicklungsgeschichte von *Butomus umbellata*. Svensk Botanisk Tidskrift Utgifven af Svenska Botaniska Föreningen, 7: 58–77.

Holmgren, I. 1915. Die Entwicklung des Ernbryosackes bet *Anthemis tinctoria*. Svensk Botanisk Tidskrift Utgifven af Svenska Botaniska Föreningen, 9: 171–183.

Horner, H.T. 1977. A comparative light- and electronmicroscopic study of microsporogenesis in male-fertile and cytoplasmic male-sterile sunflower (*Helianthus annuus*). Amer. J. Bot., 64 (56): 745–759.

Hrich, R. 1983. Study of specific programmes of cytodifferentiation and degradation of tapetal cells. Acta Facultatis Rerum Naturalium Universitatis Comeniana. Physiologie Plantarum. 19: 1–7.

Hu, S.Y. 1990. Male germ unit and sperm heteromorphism: the current status. Acta Botanica Sinica, 32 (2): 230–240.

Huang, B.Q. and S.D. Russell. 1989. Isolation of fixed and viable eggs, central cells and embryo sacs from ovules of *Plumbago zeylanica*. Plant Physiology, 90: 9–12.

Humphrey, J.E. 1896. The development of the seed in the Scitamineae. Ann. Botany, 10 (37): 1–40.

Hutchinson, A.N. 1915. Fertilization in *Abies*. Bot. Gazette, 60: 457–472.

Hyde, B.B. 1970. Mucilage-producing cells in the seed coat of *Plantago ovata*: developmental fine structure. Amer. J. Bot., 57 (210): 1197–1206.

Ilyina, G.M. 1981. Family Papaveraceae, pp. 142–150. *In* T.B. Batygina and M.S. Yakovlev (eds.). Comparative Embryology of Flowering Plants. Winteraceae—Juglandaceae. Nauka, Leningrad, Russia.

Ilyina, G.M. 1990. Family Lemnaceae, pp. 279–288. *In* T.B. Batygina and M.S. Yakovlev (eds.). Comparative Embryology of Flowering Plants. Butomaceae—Lemnaceae. Nauka, Leningrad, Russia.

***Inge-Vechtomov, S.G. 1977.** The difference in matrix processes as a factor of adaptation, pp. 75–85. *In* D.M. Grodzynsky. The Systems of Cell Reliability. Naukova Dumka, Kiev, Ukraine.

Inge-Vechtomov, S.G. 1989. Genetics with the Bases of Selection. Vysshaya shkola, Moscow, Russia, 592 pp.

Ishioka, K. and S. Sato. 1990. Polyploidy of the nuclear DNA contents of the antipodal cells of *Hemerocallis middendorfii* var. *esculenta*. Scientific Reports Hirosaki Univ., 37 (2): 128–131.

Israel, H.W. and J. Sagawa. 1964. Post-pollination ovule development in *Dendrobium* orchids. II. Fine structure of the nucellar and archesporial phases. Caryologia, 17 (2): 301–306.

Ivanovskaya, E.V. 1983. Cytoembryological Investigation of Plant Cell Differentiation. Izdatelstvo Moskovskogo Gosudarstvennogo Universiteta, Moscow, Russia.

Izhar S. and R. Frankel. 1971. Mechanism of male sterility in *Petunia*: the relationship between pH, callose activity in the anthers and the breakdown of the microsporogenesis. Theor. Appl. Gen., 41: 104–108.

Jalouzot, M.-F. 1971. Aspects ultrastructuraux de la mégasporogénèse d'*Oenothera lamarkiana* en rapport avec les depots callosiques observes. Ann. Scientifiques l'Univ. Reims et l'ARERS, 9 (1): 36–46.

Jalouzot, M.-F. 1978. Differentiation des eléménts de la tetrad femelle chez *Oenothera erythrosepala*. Bull. Société botanique de France. Actualities botaniques, 12 5(1–2): 167–170.

Jalouzot, R. 1969. Differentiation nucleaire et cytoplasmique du grain de pollen de *Lilium candidum*. Experim. Cell Research, 55 (1): 1–8.

Jensen, W.A. 1963 (1964). Cell development during plant embryogenesis. Meristems and differentiation. Brookhaven Symposia in Biology. 16: 179–202.

Jensen, W.A. 1965a. The composition and ultrastructure of the nucellus in cotton. J. Ultrastructure Research, 13 (1–2): 112–128.

Jensen, W.A. 1965b. The ultrastructure and composition of egg and central cell of cotton. Amer. J. Bot., 52 (8): 781–797.

Jensen, W.A. 1965c. The ultrastructure and histochemistry of the synergids of cotton. Amer. J. Bot., 52 (3): 238—256.

Jensen, W.A. 1968. Cotton embryogenesis: the zygote. Planta, 79 (4): 346–366.

Jensen, W.A. 1969. Cotton embryogenesis: pollen tube development in the nucellus. Can. J. Bot., 47 (3): 383–386.

Jensen, W.A. 1972. The embryo sac and fertilization in Angiosperms. Univ. Hawaii, Harold L. Lyon Arboretum lecture. 3: 2–31.

Jensen, W.A. 1973. Fertilization of flowering plants. Bioscience, 23 (1): 21–27.

Jensen, W.A. and D.B. Fisher. 1968a. Cotton embryogenesis: the entrance and discharge of the pollen tube in the embryo sac. Planta, 78 (2): 158–183.

Jensen, W.A. and D.B. Fisher. 1968b. Cotton embryogenesis: the sperm. Protoplasma, 65 (3): 277–286.

Jensen, W.A., M. Ashton and L.R. Heckard. 1974. Ultrastructural studies of the pollen of subtribe Castilleiinae, family Scrophulariaceae. Bot. Gazette, 135: 210–218.

Johansen, D.A. 1928. The hypostase and seed sterility in the Onagraceae. Madroño, 1 (10): 165–167.

Johansen, D.A. 1931. Studies on the morphology of the Onagraceae. 6. *Anogra pallida*. Amer. J. Bot., 18 (10): 854–864.

Johnson, D.S. 1900. On the endosperm and embryo of *Peperomia pellucida*. Bot. Gazette, 30 (1): 1–11.

Johnson, D.S. 1914. Studies of the development of the Piperaceae. 2. The structure and seed development of *Peperomia hispidula*. Amer. J. Bot., 1 (7): 323–339; (8): 357–397.

Johnson, L.E.B., R.D. Wilcoxson and F.J. Frosheirser. 1975. Transfer cells in tissues of the reproductive system of alfalfa. Can. J. Bot., 53: 952–956.

Johri, B.M. 1935. Studies in the family Alismaceae III. *Sagittaria quayanensis* H.B.K. and *S. latifolia* Willd. Proc. Indian Acad. Science. Section B, Biological Sciences (Bangalore, India), 2 (1): 33–48.

Johri, B.M. 1936. Studies in the family Alismaceae. IV *Alisma plantago* L., *A. plantago-aquatica* L. and *Sagittaria graminea* Mich. Proc. Indian Acad. Science. Section B, Biological Sciences (Bangalore, India), 4 (2): 128–138.

Johri, B.M. 1963. Female gametophyte, pp. 69–103. *In* P. Maheshwari (ed.) Recent Advances in the Embryology of Angiosperms. Intern. Soc. Plant Morphologists, Univ. Delhi, Delhi, India.

Johri, B.M. (ed.) 1984. Embryology of Angiosperms. Springer-Verlag, Berlin.

Johri, B.M. 1992. Nutrition of the anther, ovule and seed, pp. 11–12. *In* T.B. Batygina (ed.). Proc XI Intern. Symp. 'Embryology and Seed Reproduction'. 3–7 July, 1990, Leningrad, USSR, Nauka, St.-Petersburg, Russia.

Johri, B.M. and R.N. Kapil. 1953. Contribution to the morphology and life history of *Acalypha indica* L. Phytomorphology, 3 (3): 137–151.

Johri, B.M. and C.B. Seghal. 1963. Growth of *Anethum graveolens* L., pp. 245–256. *In* P. Maheshwari and N.S. Ranga Swamy (eds.) Plant Tissue and Organ Culture. A Symposium.

Johri, B.M. and S. Agarwal. 1965. Morphology and embryological studies in the family Santalacaceae. VIII. *Quinchamalium chinense* Lam. Phytomorphology, 15 (4): 360–372.

Johri, B.M. and K.B. Ambegaokar. 1984. Embryology: then and now, p. 28. *In* B. M. Johri (ed.) Embryology of Angiosperms. Springer-Verlag, Berlin.

Johri, B.M., K. Ambegaokar and P.S. Srivastava. 1992. Comparative embryology of Angiosperms, vols. 1, 2. Springer-Verlag, Berlin.

Jones, D. 1950. The interrelation of plasmogenes and chromosomes in pollen production in maize. Genetics, 35 (5): 101–104.

Jones, D., H. Stinson and U. Khoo. 1957. Pollen restoring genes. Connecticut Agri. Experi. St. Bull., 610: 1–43.

Jonsson, B. 1879–1880. Om embryosackens utveckling hos Angiosperma. Lunds Universitets arsberatt., 16: 1–86.

Joshi, A.C. 1938. Systematic distribution of the Fritillaria-type of embryo sac and the mono- or polyphyletic origin of Angiosperms. Chronica Botanica, 4 (6): 507–508.

Joshi, A.C. 1939. Morphology of *Tinospora cordifolia* with some observation on the origin of the single integument, nature of synergidae, and affinities of the Menispermaceae. Amer. J. Bot., 26(26): 433–439.

Joshi, A.C. and J. Venkateswarlu. 1935. Embryological studies in the Lythraceae. I. *Lawsonia inermis*. Proc. Indian Acad. Science. Section B, Biological Sciences (Bangalore, India), 2 (4): 481–493.

Joshi, A.C. and J.V. Pantulu. 1941. A morphological and cytological study of *Polianthes tuberosa*. J. Indian Bot. Soc., 20: 37–71.

Joshi, P.C. 1960. Morphological and embryological studies in the family Santalacaceae. V. *Osynis wightiana* Wall. Phytomorphology, 10 (3): 239–248.

Kajale, L.B. 1937. Embryology of *Achyranthes aspera* Linn. Proc. Indian Acad. Science, 5 (5): 195–205.

Kajale, L.B. 1938. Embryo and seed development in the Nyctaginaceae. 1. Studies in the genus *Boerhaavia*. J. Indian Bot. Soc., 17 (4): 243–254.

Kallarackal, I. and S.P. Bhatnagar. 1981. Cytochemical studies on the integument and integumentary tapetum in *Linaria bipartita* (Vent.) Will. Caryologia, 34 (4): 447–455.

Kamalova, G.V., G.Ya. Zhukova and V.A. Rumi. 1983. Family Malvaceae, pp. 260–266. *In* T.B. Batygina and M.S. Yakovlev (eds.) Comparative Embryology of Flowering Plants. Phytolaccaceae—Thymelaeaceae. Nauka, Leningrad, Russia.

Kamelina, O.P. 1978. Tapetum in Dipsacaceae and classification of tapetum types in angiosperms, pp. 78–79. *In* Proc. VI Delegate Cong. All-Union Bot. Soc., Leningrad, Russia.

Kamelina, O.P. 1980. Comparative Embryology of Families Dipsacaceae and Morinaceae. Nauka, Leningrad, Russia.

Kamelina, O.P. 1981a. Anther, pp. 18–21. *In* T.B. Batygina and M.S. Yakovlev (eds.). Comparative Embryology of Flowering Plants. Winteraceae—Juglandaceae. Nauka, Leningrad, Russia.

Kamelina, O.P. 1981b. Family Calycanthaceae, pp. 69–74. *In* T.B. Batygina and M.S. Yakovlev (eds.). Comparative Embryology of Flowering Plants. Winteraceae—Juglandaceae. Nauka, Leningrad, Russia.

Kamelina, O. P. 1981c. Family Daphniphyllaceae, pp. 172–176. *In* T.B. Batygina and M.S. Yakovlev (eds.). Comparative Embryology of Flowering Plants. Winteraceae— Juglandaceae. Nauka, Leningrad, Russia.

*__Kamelina, O. P. 1981d.__ Embryology of poorly investigated taxa. 1. Some data on the embryology of the family Eupomatiaceae. Bot. Zhurnal, 66 (6): 854–859.

Kamelina, O.P. 1983a. Family Datiscaceae, pp. 139–144. *In* T.B. Batygina and M.S. Yakovlev (eds.). Comparative Embryology of Flowering Plants. Phytolaccaceae—Thymelaeaceae. Nauka, Leningrad, Russia.

Kamelina, O.P. 1983b. Family Phytolaccaceae, pp. 7–13. *In* T.B. Batygina and M.S. Yakovlev (eds.). Comparative Embryology of Flowering Plants. Phytolaccaceae—Thymelaeaceae. Nauka, Leningrad, Russia.

***Kamelina, O.P. 1984.** On the embryology of genus *Escallonia* L. (Escalloniaceae). Bot. Zhurnal, 69 (10): 1304–1316.

Kamelina, O.P. 1985a. Family Cneoraceae, pp. 138–141. *In* T.B. Batygina and M.S. Yakovlev (eds.). Comparative Embryology of Flowering Plants. Brunelliaceae—Tremandraceae. Nauka, Leningrad, Russia.

Kamelina, O.P. 1985b. Family Coriariaceae, pp. 173–175. *In* T.B. Batygina and M.S. Yakovlev (eds.). Comparative Embryology of Flowering Plants. Brunelliaceae—Tremandraceae. Nauka, Leningrad, Russia.

Kamelina, O.P. 1985c. Family Nitrariaceae, pp. 152–156. *In* T.B. Batygina and M.S. Yakovlev (eds.). Comparative Embryology of Flowering Plants. Brunelliaceae—Tremandraceae. Nauka, Leningrad, Russia.

Kamelina, O.P. 1985d. Family Sonneratiaceae, pp. 88–93. *In* T.B. Batygina and M.S. Yakovlev. (eds.). Comparative Embryology of Flowering Plants. Brunelliaceae—Tremandraceae. Nauka, Leningrad, Russia.

Kamelina, O.P. 1985e. Family Zygophyllaceae, pp. 138–141. *In* T.B. Batygina and M.S. Yakovlev (eds.). Comparative Embryology of Flowering Plants. Brunelliaceae—Tremandraceae. Nauka, Leningrad, Russia.

Kamelina, O.P. 1987a. Family Dipsacaceae, pp. 184–192. *In* T.B. Batygina and M.S. Yakovlev. (eds.). Comparative Embryology of Flowering Plants. Davidiaceae—Asteraceae. Nauka, Leningrad, Russia.

Kamelina, O.P. 1987b. Family Morinaceae, pp. 177–184. *In* T.B. Batygina and M.S. Yakovlev (eds.). Comparative Embryology of Flowering Plants. Davidiaceae—Asteraceae. Nauka, Leningrad, Russia.

***Kamelina, O.P. 1988a.** Embryology of genus *Brexia* in connection with its systematics. Bot. Zhurnal, 73 (3): 355–366.

Kamelina, O.P. 1988b. The types of tapetum in Angiosperms. Ann. Scientifiques l'Univ. Reims et l'ARERS, 23: 71–73.

Kamelina, O.P. 1990a. Family Marantaceae, pp. 247–254. *In* T.B. Batygina and M.S. Yakovlev. (eds.). Comparative Embryology of Flowering Plants. Butomaceae—Lemnaceae. Nauka, Leningrad, Russia.

***Kamelina, O.P. 1990b.** The male and female embryonic structure development in family Marantaceae. Bot. Zhurnal, 75 (4): 480–493.

***Kamelina, O.P. 1991.** Comparative embryological analysis as a method of phylogenetic systematics of flowering plants. Postdoctoral thesis, Tashkent, Uzbekistan.

Kamelina, O.P. 1992. Distribution of embryological features in the orders of Magnoliophyta classes, pp. 242–243 *In* T.B. Batygina (ed.). Proc. XI Intern. Symp. 'Embryology and Seed Reproduction', 3–7 July, 1990, Leningrad, USSR, Nauka, St.-Petersburg, Russia.

Kamelina, O.P. and A.K. Dzevaltovsky. 1987. Family *Lamiaceae*, pp. 225-236. *In* T.B. Batygina and M.S. Yakovlev. (eds.). Comparative Embryology of Flowering Plants. *Davidiaceae—Asteraceae.* Nauka, Leningrad, Russia, 4.

***Kamelina, O.P. and M.S. Yakovlev. 1976.** Anther development and microgametogenesis in representatives of families Dipsacaceae and Morinaceae. Bot. Zhurnal, 61 (7): 932–945.

***Kamelina, O.P. and N.G. Tuchina. 1982.** Embryology of poorly investigated taxa. II. Anther and pollen grain development in *Nandina domestica* Thunb. (Berberidaceae). Bot. Zhurnal, 67 (11): 1459–1468.

***Kamelina, O.P. and G.V. Yakovleva. 1982.** Megasporogenesis and embryo sac development in *Dianthus caryophyllus* L. (Caryophyllaceae). Bot. Zhurnal, 67 (1): 80–87.

*Kamelina, O.P. and O.B. Proscurina. 1985. Embryology of *Stegnosperma halimifolium* (Stegnospermataceae). Distribution of embryological features in the order *Caryophyllales*. Bot. Zhurnal, 70 (6): 721–730.

Kamelina, O.P. and O.B. Proskurina. 1987. Family Elaeagnaceae, pp. 105–110. *In* T.B. Batygina and M.S. Yakovlev (eds.). Comparative Embryology of Flowering Plants. Davidiaceae—Asteraceae. Nauka, Leningrad, Russia.

Kamelina, O.P. and G.Ya. Zhukova. 1987. Family Valerianaceae, pp. 171–177. *In* T.B. Batygina and M.S. Yakovlev (eds.). Comparative Embryology of Flowering Plants. Davidiaceae—Asteraceae. Nauka, Leningrad, Russia.

*Kamelina, O.P. and N.A. Zhinkina. 1989. Embryology of *Ostrowskia magnifica* (Campanulaceae). The male embryonic structure development. Bot. Zhurnal, 74 (9): 1293–1301.

Kamelina, O.P. and E.S. Teryokhin 1990. Family Zannichelliaceae, pp. 44–50. *In* T.B. Batygina and M.S. Yakovlev (eds.). Comparative Embryology of Flowering Plants. *Butomaceae—Lemnaceae*. Nauka, Leningrad, Russia.

Kamelina, O.P., N.G. Tuchina and M.A. Vishnyakova. 1987. Family Caprifoliaceae, pp. 159–168. *In* T.B. Batygina and M.S. Yakovlev (eds.). Comparative Embryology of Flowering Plants. Davidiaceae—Asteraceae. Nauka, Leningrad, Russia.

*Kamelina, O.P., Sh. Khorolsuren and O.B. Proskurina. 1990. Embryology of representatives of Mongolian People's Republic flora—*Incarvillea potaninii* (Bignoniaceae) and *Sphaerophysa salsula* (Fabaceae). Bot. Zhurnal, 75 (9): 1248–1261.

Kapil, R.N. 1960. Embryology of *Acalypha* Linn. Phytomorphology, 10 (2): 174–184.

Kapil, R.N. 1962. Some recent examples of the value of embryology in relation to taxonomy. Bulletin of the Botanical Survey of India, 4 (1–4): 57–66.

Kapil, R.N. and N. Ahluwalia. 1963. Embryology of *Peganum harmala* Linn. Phytomorphology, 13 (2): 127–140.

Kapil, R.N. and I.K. Vasil 1963. Ovule, pp. 41–67. P. Maheshwari (ed.). Recent Advances in the Embryology of Angiosperms. Intern. Soc. Plant Mophologists, Univ. Delhi, Delhi, India.

Kapil, R.N. and N.N. Bhandari. 1964. Morphology and embryology of *Magnolia* Dill. ex Linn. Proc. Natl Inst. Sciences India, 30 (5): 245–262.

Kapil, R.N. and S. Jalan. 1964. *Schisandra* Michaux—its embryology and systematic position. Bot. Notiser, 117 (3): 285–306.

Kapil, R.N. and S.C. Tiwari. 1978a. The integumentary tapetum. Bot. Review, 44 (4): 457–490.

Kapil, R.N. and S.C. Tiwari. 1978b. Plant embryological investigations and fluorescence microscopy; an assessment of integration. Intern. Rev. Cytology, 53: 291–331.

Kapil, R.N., J. Bor and F. Bouman. 1980. Seed appendages in Angiosperms. Bot. Jahrb. für Systematik, Pflanzengeshichte und Pflanzengeographie, 101 (4): 555–574.

Kapil, R.N. and A.K. Bhatnagar. 1981. Ultrastructure and biology of female gametophyte in flowering plants. Intern. Rev. Cytology, 69 (3/4): 291–341.

*Kapinos, G.E. and Ph.V. Kagramanova. 1966. Flowering biology and cytoembryology of *Leontice* L. Reports Azerbaijan SSR Acad. Sciences. Series, Biological Sciences, 1: 10–16.

Kaul, M.L.H. 1988. Male Sterility in Higher Plants. Springer-Verlag, Berlin.

Keijzer, C.J., M.C. Reinders and H.B. Leferinkten Kloster. 1988. A micromanipulation method for artificial fertilization in *Torenia*, pp. 119–124. *In* M. Cresti, P. Gori and E. Pacini (eds.). Sexual Reproduction in Higher Plants. Springer-Verlag, Berlin.

Kenrick, J. and R.B. Knox. 1979. Pollen development and cytochemistry in some Australian species of Acacia. Australian J. Bot., 27: 41 3–427.

Kerner, A. 1901. The Life of Plants. St-Petersburg, Russia.

Kershaw, E.M. 1909. The structure and development of the ovule of *Myrica gale*. Ann. Botany, 23 (91): 353–362.

Khan, R. 1940. A note on double fertilization in *Ephedra foliata*. Current Science. (India), 9: 323–324.

Khan, R. 1943. Contributions to the morphology of *Ephedra foliata* Boiss. 11. Fertilization and Embryogeny. Proc. Indian Acad. Science, 13: 357–375.

Khanna, P. 1967. Morphological and embryological studies in Nymphaeaceae. III. *Victoria cruziana* D'orb and *Nymphaea stellata* Willg. Bot. Magazine. Tokyo, 80 (950): 309–312.

Khvedynich, O.A. and S.A. Petrik. 1986. The dynamics of RNA accumulation in pollen grains. Botanicheskyi Zhurnal, 71 (1): 47–51 (in Russian).

Khvostova, V.V. and Ju.F. Bogdanov (eds.). 1975. Cytology and Genetics of Meiosis. Nauka, Moscow, Russia.

Kihara, H. and T. Hori. 1966. The behavior of nuclei in germinating pollen grain of wheat, rice and maize. Der Züchter. Zeitschrift für theoretische und angewandte Genetik, 36 (4): 145–150.

Knoth, R., M. Wriacher and J. Vetter. 1980. Phytoferritin—accumulating plastids in the male generative cell of *Pelargonium hortorum* Bailey. Zeitschrift für Pflanzenphysiologie, 98: 365–370.

Knox, R.B. 1984. The pollen grain, pp. 197–271. *In*. B.M. Johri (ed.). Embryology of Angiosperms, vol. 1. Springer-Verlag, Berlin.

Knox, R.B. and J. Heslop-Harrison 1970. Pollen wall proteins: localization and enzymatic activity. J. Cell Science, 6: 1–27.

*__Koltsov, N.K. 1935.__ The gene role in the physiology of development. Biol. Zhurnal, IV (5): 753–774.

Komar, G.A. 1965a. Structure of arillus of some representatives of family Grossulariaceae, pp. 114–130. *In* M.S Yakovlev (ed.). Flower Morphology and Reproductive Process in Angiosperms. Nauka, Moscow-Leningrad, Russia.

Komar, G.A. 1965b. Arils, their nature, structure and function. Bot. Zhurnal, 50 (5): 715–724.

Komar, G.A. 1973. Arils, pp. 55–64. *In* M.I. Savchenko (ed.). Morphology of the Ovule in Angiosperms. Nauka, Leningrad, Russia.

*__Komar, G.A. 1982.__ Ovule type of some species of subfamily Asphodeloideae (Liliaceae). Bot. Zhurnal, 67 (7): 800–805.

Kondratyaeva, V.P. and M.A. Tikhova. 1982. Female gametophyte. Fertilization, pp. 72–97. *In* A.A. Yazenko-Khmeleisky (ed.). Reproductive Structures of Gymnosperms. Nauka, Leningrad, Russia.

Konycheva, V.I. and R.U. Kadyrova. 1983. Family Chenopodiaceae, pp. 49–52. *In* T.B. Batygina and M.S. Yakovlev (eds.). Comparative Embryology of Flowering Plants. Phytolaccaceae—Thymelaeaceae. Nauka, Leningrad, Russia.

Korchagina, I.A. 1965. Ovule types in the birch family and possible trend of their evolution, pp. 159–160. *In* Proc. Moscow Interhigher School Conference on Plant Morphology, Moscow, Russia.

*__Korchagina, I.A. 1969.__ Attempt at classification of the ovule morphological types in angiosperms. Bot. Zhurnal, 54 (10): 1513–1530.

Korchagina, I.A. 1981. Family Betulaceae, pp. 201–210. *In* T.B. Batygina and M.S. Yakovlev (eds.). Comparative Embryology of Flowering Plants. Winteraceae—Juglandaceae. Nauka, Leningrad, Russia.

*__Kordyum, E.L. 1967.__ Cytoembryology of the Umbellatae Family. Naukova Dumka, Kiev, Ukraine.

*__Kordyum, E.L. 1968.__ Peculiarities of early ontogenesis of the ovule with different archesporium type in the number of representatives of angiosperms. Cytology i Genetica, 2 (5): 415–428.

*__Kordyum, E.L. 1971.__ Significance of embryology for the solution of questions on systematics and phylogeny of angiosperms, pp. 196–216. *In* V.P. Zosimovich (ed.). Problems of Embryology. Naukova Dumka, Kiev, Ukraine.

*__Kordyum, E.L. 1978.__ Evolutionary Cytoembryology of Flowering Plants. Naukova Dumka, Kiev, Ukraine.

*__Korobova, S.N. 1971.__ Comparative investigation of staminate and pistillate flowers of maize, pp. 86–87. *In* Proc. V All-Union Conf. Plant Embryology. Kishinev, Moldova.

Korobova, S.N. 1982. Formation of female gametophyte, fertilization, development of embryo and endosperm in maize, pp. 151–176. *In* GE. Shmaraev and O.N. Korovina (eds.) Cultural Flora of the USSR. Kolos, Moscow, Russia.

Korobova, S.N. 1983. Family Plumbaginaceae, pp. 59–64. *In* T.B. Batygina and M.S. Yakovlev (eds.). Comparative Embryology of Flowering Plants. Phytolaccaceae—Thymelaeaceae. Nauka, Leningrad, Russia.

Korobova, S.N. 1985. Family Saxifragaceae, pp. 26–29. *In* T.B. Batygina and M.S. Yakovlev (eds.). Comparative Embryology of Flowering Plants. Brunelliaceae—Tremandraceae. Nauka, Leningrad, Russia.

Korobova, S.N. and N.A. Zhinkina. 1987a. Family Campanulaceae, pp. 300–308. *In* T.B. Batygina and M.S. Yakovlev. (eds.). Comparative Embryology of Flowering Plants. Davidiaceae—Asteraceae. Nauka, Leningrad, Russia.

Korobova, S.N. and N.A. Zhinkina. 1990. Family Araceae, pp. 275-279. *In* T.B. Batygina and M.S. Yakovlev. (eds.). Comparative Embryology of Flowering Plants. *Butomaceae—Lemnaceae.* Nauka, Leningrad.

Kosmath, L. 1927. Studien über das Antherentapetum. Österreichische Botanische Zeitschrift, 76 (3): 235–241.

Kozo-Polyanskyi, B.M. 1922. Introduction to Phylogenetic Systematics of Higher Plants. Priroda i Kultura, Voronezh, Russia.

Kranz, E., J. Bautor and H. Lorz. 1991. In-vitro fertilization of single, isolated gametes of maize mediated by electrofusion. Sexual Plant Reproduction, 4: 12–16.

Kress, W.J. and D.E. Stone. 1982. Nature of the sporoderm in monocotyledons, with special reference to the pollen grain of *Canna* and *Heliconia*. Grana, 21: 129–148.

Krishna Iyengar, C.V. 1941. Development of the embryo sac and endosperm-haustoria in *Torenia cordifolia* Roxb. and *T. hirsuta* Benth. Proc. Natl. Inst. Sciences (India), 7: 61–71.

Kronestedt-Robards, E.C. and J.R. Rowley. 1989. Pollen grain development and tapetal changes in *Strelitzia reginae* (Strelitziaceae). Amer. J. Bot., 76 (6): 856–870.

Kudlicka, K., M. Kuras, and B. Rodkiewicz. 1981. Embryo sac development in *Stellaria media*. Acta Societatis Botanicorum Poloniae, 50 (1–2): 143–146.

Kühn, G. 1928. Beiträge zur Kenntnis der intraseminalen Leitbundel bei den Angiospermen. Bot. Jahrb. für Systematik, Pflanzengeshichte und Pflanzengeographie, 61: 325–385.

Kuprianova, L.A. and L.A. Aleshina. 1967. Palynological Terminology of Angiosperms. Nauka, Leningrad, Russia.

Kuran, H. 1972. Callose localization in the walls of megasporocytes and megaspores in the course of development of monospore embryo sacs. Acta Societatis Botanicorum Poloniae, 41 (4): 519–534.

***Kurbanov, E.A. 1981.** Comparative ultrastructural investigation of zyziphora and thyme microsporogenesis. Reports Azerbaijan SSR Acad. Sciences, Series Biological Sciences, 4: 6–11.

Kursanov, A.L. 1976. Transport of Assimilates in Plants. Nauka, Moscow, Russia.

Kursanov, L.I., F.I. Krasheninnikov, N.A. Komarnitskyi and A.L. Kursanov,. 1940. Handbook on Botany for the Higher Pedagogical Educational Establishments and Universities, vol. 1. Gosudarstvennoe Uchebno-Pedagogicheskoe Izdatelstvo Narkomprosa RSFSR, Moscow, Russia.

La Cour, L.F. 1949. Nuclear differentiation in the pollen grain. Heredity, 3 (3): 319–337.

La Fountain, Jr. J.R. and K.L. La Fountain. 1973. Comparison of density of nuclear pores on vegetative and generative nuclei in pollen of *Tradescantia*. Experim. Cell Research, 78: 472–476.

La Rue, C.D. 1954. The extraordinary funiculus of *Acacia confusa* Merril. Rhodora, 56 (670): 229–231.

Lakshmanan, K.K. 1968. Preliminary note on the embryology of *Diplanthera univervis* Aschers. Current Science. (India), 37 (18): 534–535.

Lakshmanan, K.K. and S. Poornima. 1988. Microsporogenesis in *Rhizophora lamarckii* Montr. Current Science. (India), vol. 57, no. 19.

Land, W.J. 1907. Fertilization and embryogeny in *Ephedra trigurcata*. Botanical Gazette, 44: 237-292.

Landes, M. 1946. Seed development in *Acalypha rhomboidea* and some other Euphorbiaceae. Amer. J. Bot., 33 (6): 562–568.

Largeau, C., K. Casadevall, A. Kadouri and P. Metzger. 1984. Formation of *Botryococcus*-derived kerogens—Comparative study of immature torbanites and of the extant alga *Botryococcus braunii*. Org. Geochem., 6: 327–332.

Larson, D.A. 1965. Fine-structural changes in the cytoplasm of germinating pollen. Amer. J. Bot., 52 (2): 139–154.

Larson, D.A. and C.W. Lewis. 1963. Pollen wall development in *Parkinsonia aculeata*. Grana Palynologica, 3 (3): 21–27.

Laser, K.D. and N.R. Lersten. 1972. Anatomy and cytology of microsporogenesis in cytoplasmic male sterile angiosperms. Bot. Review, 38 (3): 325–454.

Lee, C.L. 1955. Fertilization in *Ginkgo biloba*. Bot. Gazette, 117: 79–100.

***Levitskyi, G.A. 1928.** Experimentally causing chromosome motion from one cell to another. Zhurnal Russkogo Botanicheskogo Obshchestva, 13 (1-2): 19–25.

Lindley, J. 1836. A Natural Systems of Botany. Longman, Rees, Orme, Brown, Green & Longman, London.

Lindley, J. 1853. The Vegetative Kingdom. Bradbury, Evans, London.

***Lishchitovich, L.I. 1974.** Introduction to the conception of ontogenetic chain, pp. 23–32. *In* D.M. Grodzinsky (ed.). The System Approach to Plant Biology, Kiev, Ukraine.

***Lodkina, M.M. 1971.** Pollen sac formation in two species of duckweed, p. 107. *In* Proc. V All-Union Conf. Plant Embryology, Kishinev, Moldova.

Lombardo, G. and F.M. Gerola. 1968. Cytoplasmic inheritance and ultrastructure of the male generative cell of higher plants. Planta, 82: 105–110.

Lonay, H. 1922. Contribution à l'étude des rélations entre la structure des differentes parties de l'ovule et la nutrition générale de celui-ci avant et aprés la fécondation. Bull. Classe des Sciences Acad. Royale Belgique, Ser. 5. 8 (1): 24–45.

Long, A.G. 1966. Some lower carboniferous fructification from Berwickshire, together with a theoretical account of the evolution of ovules, cupules and carpels. Trans. Royal Society Edinburgh, 66: 345–375.

Lotsy, J.P. 1899. Contribution to the life-history of the genus *Gnetum*. Ann. Jardin Botanique Buitenzorg, Leiden. Ser. 11, 1: 2–11.

Lüttge, U. and N. Higinbotham. 1977. Transport in Plants. New York etc.: Springer.

***Magakyan, Yu. A., N.U. Nadjapyan and N.A. Koltukhcheva. 1987.** The content of total proteins in cells as a regulated factor of their proliferation. Cytology. 29 (4): 465–471.

Magnus, W. 1913. Die atypische Embryosackentwicklung der Podostemaceen. Flora, 105 (5): 275–336.

Mahabale, T.S. and M.S. Chennaveeraiah. 1957. Studies on *Hyphaene indica* Becc. I. Morphology. Phytomorphology, 7 (2): 184–194.

Maheshwari, D.H. and K. Lakshminarayana. 1977. Embryological studies in Gentianaceae. J. Indian Bot. Soc., 56 (2): 182–188.

Maheshwari, P. 1937. A critical review of the types of embryo sacs in angiosperms. New Phytologist, 36 (5): 359–416.

Maheshwari, P. 1947. Tetranucleate embryo sacs in angiosperms. Lloydia, 10 (1): 1–18.

Maheshwari, P. 1948. The angiosperm embryo sac. Botanical Review, 14 (1): 1–56.

Maheshwari, P. 1950. An Introduction to the Embryology of Angiosperms. McGraw-Hill Book Company, NY, USA.

Maheshwari, P. and B.M. Johri. 1941. The embryo sac of *Acalypha indica* L. Beihefte zum Botanischen Centralblatt, 61A: 125–136.

Maheshwari, P. and A.R. Srinivasan 1944. A contribution to the embryology of *Rudbeckia bicolor* Nutt. New Phytologist, 43: 135–142.

Maheshwari, P. and R.N. Kapil. 1963. Fifty Years of Science in India. Progress of Botany. Indian Science Congress Association. Calcutta.

Maheshwari, S.C. 1954. The embryology of *Wolffia*. Phytomorphology, 4 (4): 355–365.

Maheshwari, S.C. 1956. The endosperm and embryo of *Lemna* and systematic position of the Lemnaceae. Phytomorphology, 6 (1): 51–55.

Malik, C.P. and S.Vermani. 1975. Physiology of sexual reproduction. I. A histochemical study of the embryo sac development in *Zephyranthes rosea* and *Lagenaria vulgaris*. Acta Histochemica et Cytochemica, 53: 244–280.

Malpighi, M. 1675. Anatome plantarum. J. Martyn, London.

Mametyeva, T.B. and O.P. Kamelina. 1981. Family Myristicaceae, pp. 47–50. *In* T.B. Batygina and M.S. Yakovlev (eds.) Comparative Embryology of Flowering Plants. Winteraceae—Juglandaceae. Nauka, Leningrad, Russia.

Mandrik, V.Yu. and Yu.Yu, Petrus. 1985. Family Rosaceae, pp. 55–64. *In* T.B. Batygina and M.S. Yakovlev (eds.) Comparative Embryology of Flowering Plants. Brunelliaceae—Tremandraceae. Nauka, Leningrad, Russia.

Manning, J.C. and H.P. Linder. 1990. Cladistic analysis of patterns of endothecial thickening in the Poales (Restionales). Amer. J. Bot., 77 (2): 196–210.

Mansfield, S.G., L.G. Briarty and S. Erni. 1991. Early embryogenesis in *Arabidopsis thaliana*. I. The mature embryo sac. Can. J. Bot., 69 (3): 447–460.

Mariami, C., M. De Beuckeller, J. Truettner, J. Leemons and R. Goldberg. 1990. Induction of male sterility in plants by a chimeric ribonuclease gene. Nature, 347 (6295): 737–741.

Marquardt, H., O.M. Barth and U. von. Rahden 1968. Zytophotometrische und elektronenmikroskopische Beobachtungen über die Tapetumzellen in den Antheren von *Paeonia tenuifolia*. Protoplasma, 65 (4): 407–421.

Martens, P. 1957. Bee nucellaire micropyly et paroi megasporale chez *Encephalartos poggei* Asch. Cellule, 58 (2): 151–168.

Martinoli, G.O. 1939. Contributo all'embriologia delle Asteracee. I–III. Nuovo Giornale Botanico Italiano, 46 (2): 259–298.

Martinoli, G.O. 1948. Embriologia della *Peperomia maculosa* Hook. Nuovo Giornale Botanico Italiano, 55: 235–250.

Martinoli, G.O. 1967. Position and significance of the *Pyrethrum cinerariaefolium* (*Chrysanthemum cinerariaefolium*) type among the tetrasporic types of female gametophyte. Phytomorphology, 17 (1–4): 207–210.

Maruyama, K., H. Gay and B.P. Kaufmann. 1965. Nature of the wall between generative and vegetative nuclei in the pollen grain of *Tradescantia paludosa*. Amer. J. Bot., 52 (6): 605–610.

Masand, P. 1963. Embryology of *Zygophyllum fabago* L. Phytomorphology, 13 (3): 293–302.

Mascarenhas, J.P. 1975. The biochemistry of angiosperm pollen development. Bot. Review, 41 (3): 259–314.

Mascarenhas, J.P. 1989. The male gametophyte of flowering plants. Plant Cell, 1: 657–664.

Mascarenhas, J.P. 1990. Gene activity during pollen development. Ann. Rev. Plant Physiology, Plant Molecular Biology, 41: 317–338.

Mathur, K.L. and R. Khan. 1941. The development of the embryo sac in *Vogelia indica* Lamk. Proc. Indian Acad. Science. Section B, Biological Sciences (Bangalore, India), 13 (6): 360–368.

Mauritzon, J. 1933. Studien über die Embryologie der Familien Crassulaceae und Saxifragaceae. Ph.D. thesis, Botanischen Inst. Lund., H. Ohlssons Buchdruckerei, Gleerupska Univ., Bokhandeln, Lund, Sweden.

Mauritzon, J. 1936. Samenbau und Embryologie einer Scitamineen. Acta Universitatis Lundensis, 31 (9): 1—31.

Maze, J. and S.C. Lin, 1975. A study of the mature megagametophyte of *Stipa elmeri*. Can. J. Bot., 53 (24): 2958–2977.

Medina, F.J., M.C. Risueno and M.J. Rodriguez-Garcia. 1981. Evolution of female meiosis in *Pisum sativum*. Planta, 151: 215–228.

***Medvedeva, G. and V. Bazavluk. 1951.** Question of the sterility of branchy wheat. Reports USSR Acad. Sciences, 77 (6): 1099–1102.

Meeuse, A.D. 1963. From ovule to ovary: A contribution to the phylogeny of the megasporangium. Acta Biotheoretics, 16 (2): 127–182.

Meeuse, A.D. 1964. The bitegmic spermatophytic ovule and the cupule—a reappraisal of the so-called pseudo-monomerous ovary. Acta Botanica Neerlandica, 13: 97–112.

Meeuse, A.D. and F. Bouman. 1974. The inner integument—its probable origin and homology. Acta Botanica Neerlandica, 23: 237–249.

Mepham, R.H. and G.R. Lane. 1969. Nucleopores and polyribosome formation. Nature, 221: 288–289.

Mepham, R.H. and G.R. Lane. 1970. Observation on the fine structure of developing microspores of *Tradescantia bracteata*. Protoplasma, 70: 1–20.

Messere, A. 1931. Ricerche embriologiche e cariologiche sopra i generi *Allium* e *Nothoscordon*. Nuovo Giornale Botanico Italiano, 38 (3): 409–441.

Mezzetti-Bambacioni, V. 1949. Sull'embriologia della *Fritillaria messanensis* Raf. e sull'opportunita di distinguere una serie *Fritillaria* net tipo *Euphorbia dulcus*. Ann. Botanica. 23: 116–138.

Mignotte, J., J. Vallade and F. Bugnon. 1989. Apex caulinaires végétatifs à proméristème supère ou infère: Quelques variantes principales chez les plantes Vasculaires. Bull. Société botanique de France. Letters Botaniques, 136 (3): 213–224.

Mirbel, C.F.B. 1829. Nouvelles recherches sur la structure et les développements de l'ovule végétale. Ann. Sciences Naturelles. Botanique, 17: 302–318.

Mirbel, S.F.B. 1830. Additions aux nouvelles recherches sur la structure et les développements de ovule. Mem. Acad. Roy. Sciences de l'institut de France, 9: 629–684.

Modilewski, J.S. 1908. Zur Embryobildung von *Gunnera chilensis*. Berichte der Deutschen Botanischen Gesellschaft, 26: 550–556.

Modilewski, J.S. 1909. Zur Embryobildung von einiger Onagraceae. Berichte der Deutschen Botanischen Gesellschaft, 27: 287–291.

Modilewski, J.S. 1925. Research of polyembryony in *Allium odorum* L. Vestnik Kievskogo Botanicheskogo Sada, 2: 9–20 (in Ukrainian).

***Modilewski, Ya.S. 1928a.** Further investigation on embryology and cytology of *Allium* species. Vestnik Kievskogo Botanicheskogo Sada, 7/8: 57–64.

***Modilewski, Ya.S. 1928b.** Female gametophyte in angiosperms. Ukrainskyi Botanicheskyi Zhurnal, 5: 5–40.

***Modilewski, Ya.S. 1953.** Embryology of Angiosperms. Kiev, Ukraine.

Mogensen, H.L. 1972. Fine structure and composition of the egg apparatus before and after fertilization in *Quercus gambelii*: the functional ovule. Amer. J. Bot., 59 (9): 931–941.

Mogensen, H.L. 1973. Some histochemical, ultrastructural, and nutritional aspects of the ovule of *Quercus cambelii*. Amer. J. Bot., 60 (1): 48–54.

Mogensen, H.L. 1978. Synergids of *Proboscidea lousianica* (Martineaceae) before fertilization. Phytomorphology, 28 (1): 114–122.

Mogensen, H.L. 1981. Translocation of uranin within the living ovules of selected species. Amer. J. Bot., 68 (2): 195–199.

Mogensen, H.L. 1982. Double fertilization in barley and the cytological explanation for haploid embryo formation, embryoless caryopses and ovule abortion. Carlsberg Research Communications, 47: 313–354.

Mogensen, H.L. 1984. Quantitative observations on the pattern of synergid degeneration in barley. Amer. J. Bot., 71: 1448–1451.

Mogensen, H.L. 1986. Juxtaposition of the generative cell and vegetative nucleus in the mature pollen grain of amaryllis (*Hippeastrum vitatum*). Protoplasma, 134: 67–72.

Mogensen, H.L. 1988. Exclusion of male mitochondria and plastids during syngamy in barley as a basis for maternal inheritance. Proc. Natl. Acad. Sciences USA, 85: 2594–2597.

Mogensen, H.L. 1990a. Fertilization and early emhryogenesis, pp. 76–99. *In* G.P. Chapman (ed.). Reproductive Versatility in the Grasses. Cambridge Univ. Press, Cambridge.

Mogensen, H.L. 1990b. Ultrastructure of male sex cells and fertilization in flowering plants. Amer. J. Bot., 77 (6) Suppl. 6.

Mogensen, H.L. 1992. The male germ unit: concept, composition, and significance, pp. 129–147. *In* S.D Russell and C. Dumas (eds.). Sexual Reproduction in Flowering Plants. Intern. Review of Cytology, Acad. Press, San Diego, CA, USA.

Mogensen, H.L. and H.K. Suthar. 1979. Ultrastructure of the egg apparatus of *Nicotiana tabacum* (Solanaceae) before and after fertilization. Bot. Gazette, 140 (2): 168–179.

Mogensen, H.L. and M.L. Rusche. 1985. Quantitative ultrastructural analysis of barley sperm. I. Occurrence and mechanism of cytoplasm and organelle reduction and the question of sperm dimorphism. Protoplasma, 128 (1): 1–13.

Mogensen, H.L. and A.R. Ladyman. 1989. A structural study on the mode of action of CHA (Chemical Hybridizing Agent) in wheat. Sexual Plant Reproduction, 2: 173–183.

Monnier, G. 1887. Sur la valeur morphological de l'albumen chez les Angiosperms. Journal de Botanique, (Paris). 1: 140–142.

Moscatelli, A., A. Tiezzi, R. Vigniani, G. Cai, A. Bartalesi and M. Cresti. 1988. Presence of kinesin in tobacco pollen tube, pp. 205–209. *In* M. Cresti, O. Gori and E. Pacini (eds.). Sexual Reproduction in Higher Plants. Springer, Heidelberg: Berlin.

Moss, G.J. and J. Heslop-Harrison. 1967. A cytochemical study of DNA, RNA and protein in the developing maize anther. II. Observations. Ann. Botany. N. S., 31 (123): 555–572.

***Movsesyan, S.N. 1968.** Comparative cytoembryological investigation of some *Rudbekia* species, 138–140. *In* Proc. All-Union Symp. Plant Embryology, Kiev, Ukraine.

Mücke, M. 1908. Über den Bau und die Entwicklung der Früchte und über die Herkunft von *Acorus calamus* L. Bot. Zeitung, 66 (1): 1–23.

Mukkada, A.J. 1962a. Morphological and embryological studies on some Indian Podostemaceae: Ph.D. thesis, Univ. Delhi, Delhi. 72 pp.

Mukkada, A.J. 1962b. Some observations on the embryology of *Dicraea stylosa* Wight. *In* Plant Embryology. New Dehli, pp. 139–145.

Mukkada, A.J. 1969. Some aspects of the morphology, embryology and biology of *Ternida zeylanica* (Carduer) Tulasne. New Phytologist, 68: 1145–1153.

Mulay, B.N. 1941. A study of the pistillate plants of *Ephedra foliata* Boiss. found at Drigh Road near Karashi in Sind. Proc. Indian Science Cong. Assoc., 28: 158.

Müntzing, A. 1967. Genetics. Basic and Applied (2nd ed.). LTs Förlag, Stockholm, Sweden.

Murray, D.R. 1987. Nutritive role of seed—coats in developing legume seeds. Amer. J. Bot., 74: 1122–1137.

Murthy, S.K.N. 1935. A reply to the post-script of the paper on "Studies in the family Alismataceae. 1. *Limnophyton obtusifolium* Mig", by B. M. Johri. J. Cl. Bot. Soc. 14 (4): 351.

Murty, Y.S. 1959. Studies in the order Piperales. III. A contribution to the study of floral morphology of some species of *Peperomia*. J. Indian Bot. Soc., 38 (1): 120—139.

Nägeli, K. 1842. Zur Entwicklungsgeschichte des Pollens bei den Phanerogamen. Zürich.

Nair, N.C. and N. P. Sukumaran. 1960. Floral morphology and embryology of *Brucea amarissima*. Bot. Gazette, 121: 175–185.

Naithani, S.P. 1937. Chromosome studies in *Hyacinthus orientalis* L. III. Reversal of sexual state in the anthers of *H. orientalis* L. var. 'Yellow Hammer'. Ann. Botany, 1: 369–377.

Nakamura, S. and H. Miki-Hirosige. 1985. Fine structural study on the formation of the generative cell wall and intine-3 layer in a growing pollen grain of *Lilium longiflorum*. Amer. J. Bot., 72 (3): 365–375.

Narayana, H.S. and K. Jain 1962. A contribution to the embryology of *Limeum indicum*. Lloydia, 25 (2): 100–108.

***Naumova, T.N. 1970.** Polyembryomy in *Euonymus macroptera* Rupr. and *E. planipes* (Koehne) Koehne. Bot. Zhurnal, 55 (9): 1270–1280.

Naumova, T.N. 1983a. Family Cactaceae, pp. 19–26. *In* T.B. Batygina and M.S. Yakovlev (eds.). Comparative Embryology of Flowering Plants. Phytolaccaceae—Thymelaeaceae. Nauka, Leningrad, Russia.

Naumova, T.N. 1983b. Family Orchidaceae, pp. 78–83. *In* T.B. Batygina and M.S. Yakovlev (eds.). Comparative Embryology of Flowering Plants. Phytolaccaceae—Thymelaeaceae. Nauka, Leningrad, Russia.

Naumova, T.N. 1987a. Family Asclepiadaceae, pp. 132–137. *In* T.B. Batygina and M.S. Yakovlev (eds.). Comparative Embryology of Flowering Plants. Davidiaceae—Asteraceae. Nauka, Leningrad, Russia.

Naumova, T.N. 1987b. Family Celastraceae, pp. 49–54. *In* T.B. Batygina and M.S. Yakovlev (eds.). Comparative Embryology of Flowering Plants. Davidiaceae—Asteraceae. Nauka, Leningrad, Russia.

Naumova, T.N. 1990. Family Trilliaceae, pp. 151–159. *In* T.B. Batygina and M.S. Yakovlev (eds.). Comparative Embryology of Flowering Plants. Butomaceae—Lemnaceae. Nauka, Leningrad, Russia.

***Navashin, M.S. 1947.** Nuclei of meristematic tissue arising during regeneration in plants. Reports scientific works in 1945. Department Biological Sciences USSR Acad. Sciences, pp. 262–263.

Navashin, M.S. and L.M. Makushenko. 1956. First stages of regeneration in connection with the question of plant cell genesis, p. 356. *In* Trans. Conf. Embryology 'The Problems of Modern Embryology'. Leningrad, Russia.

Nawaschin, S.G. 1894. On the ordinary birch (*Betula alba* L.) and the morphological significance of chalazogamy. V. Demakov, St.-Petersburg, Russia.

Nawaschin, S.G. 1898. Resultate einer Revision der Befruchtungesvorgänge bei *Lilium martagon* und *Fritillaria tenella*. Bull. Acad. Sciences, St.-Petersburg, 9: 377–382.

***Nawaschin, S.G. 1900.** Fertilization in Compositae and Orchids (preliminary report). Reports Imperial Acad. Sciences, 13 (3): 335–340.

Nawaschin, S.G. 1911. Details on the formation of male sexual nuclei in *Lilium martagon*. Reports Kiev Soc. Nature Experientists, 21 (4): 119–151.

Nawaschin, S.G. 1927. The experience of structural representation of sexual cells characteristics, pp. 94–114. *In* 1927. Dedicatory volume to I.P. Borodin, Leningrad, Russia.

Nemec, B. 1898. Über den Pollen der petaloiden Antheren von *Hyacinthus orientalis* L. Rozpravy Československé Akademie Věd a Uměení (Prague), 7: 17–23.

Netolitzky, F. 1926. Die Anatomie den Angiospermen-Samen. Verlag von Gebr. Borntraeger, Berlin.

Newcomb, W. 1973a. The development of the embryo sac of sunflower *Helianthus annuus* before fertilization. Can. J. Bot., 51 (5): 863–878.

Newcomb, W. 1973b. The development of the embryo sac of sunflower *Helianthus annuus* after fertilization. Can. J. Bot., 51 (5): 879–890.

Newcomb, W. and T.A. Steeves. 1971. *Helianthus annuus* embryogenesis: embryo sac wall projection before and after fertilization. Bot. Gazette, 132 (4): 367–371.

Newcomb, W. and L.C. Fowke. 1973. The fine structure of the change from the free-nuclear to cellular condition in the endospenm of chickweed *Stellaria media*. Bot. Gazette, 134(3): 236—241.

Niezgoda, C.J., S.M. Feuer and L.I. Nevling. 1983. Pollen ultrastructure of tribe Ingeae (Mimosoideae: Leguminosae). Amer. J. Bot., 70 (5): 650–667.

***Nikiticheva, Z.I. 1979a.** Interconnections in the development of endosperm and ovule tissues in Scrophulariaceae and Orobanchaceae species. Bot. Zhurnal, 64 (11): 1567–1580.

Nikiticheva, Z.I. 1979b. The development of embryo, endosperm and seed in parasitic members of the family Scrophulariaceae, pp. 130–157. *In* M.S. Yakovlev (ed.). Actual Questions on the Embryology of Angiosperms. Nauka, Leningrad, Russia.

Nikiticheva, Z.I. 1981a. Embryological features of some Piperales. Acta Societatis Botanicorum Poloniae, 50 (1–2): 329–332.

Nikiticheva, Z.I. 1981b. Family Piperaceae, pp. 84–90. *In* T.B. Batygina and M.S. Yakovlev (eds.). Comparative Embryology of Flowering Plants. Winteraceae—Juglandaceae. Nauka, Leningrad, Russia.

Nikiticheva, Z.I. 1981c. Family Saururaceae, pp. 81–84. *In* T.B. Batygina and M.S. Yakovlev (eds.). Comparative Embryology of Flowering Plants. Winteraceae—Juglandaceae. Nauka, Leningrad, Russia.

Nikiticheva, Z.I. 1983a. Family *Ebenaceae*, pp. 229-233. *In* T.B. Batygina and M.S. Yakovlev. (eds.). Comparative embryology of flowering plants. Phytolaccaceae—Thymelaeaceae, Nauka, Leningrad, Russia, 2.

Nikiticheva, Z.I. 1983b. Family Styraceae, pp. 221-225. *In* T.B. Batygina and M.S. Yakovlev. (eds.). Comparative embryology of flowering plants. Phytolaccaceae—Thymelaeaceae. Nauka, Leningrad, Russia, 2.

Nikiticheva, Z.I. 1985a. Family Crassulaceae, pp. 29–34. *In* T.B. Batygina and M.S. Yakovlev (eds.). Comparative Embryology of Flowering Plants. Brunelliaceae—Tremandraceae. Nauka, Leningrad, Russia.

Nikiticheva, Z.I. 1985b. Family Haloragaceae, pp. 116–120. *In* T.B. Batygina and M.S. Yakovlev (eds.). Comparative Embryology of Flowering Plants. Brunelliaceae—Tremandraceae. Nauka, Leningrad, Russia.

Nikiticheva, Z.I. 1985c. Family Linaceae, pp. 196–201. *In* T.B. Batygina and M.S. Yakovlev (eds.). Comparative Embryology of Flowering Plants. Brunelliaceae—Tremandraceae. Nauka, Leningrad, Russia.

Nikiticheva, Z.I. 1985d. Family Polygalaceae, pp. 225–230. *In* T.B. Batygina and M.S. Yakovlev (eds.). Comparative Embryology of Flowering Plants. Brunelliaceae—Tremandraceae. Nauka, Leningrad, Russia.

Nikiticheva, Z.I. 1987. Family Scrophulariaceae, pp. 251–259. *In* T.B. Batygina and M.S. Yakovlev (eds.). Comparative Embryology of Flowering Plants Daridiaceae—Asteraceae. Nauka, Leningrad, Russia.

Nikiticheva, Z.I. and M.S. Yakovlev. 1985. Family Rhizophoraceae, pp. 120–125. *In* T.B. Batygina and M.S. Yakovlev (eds.). Comparative Embryology of Flowering Plants. Brunelliaceae—Tremandraceae. Nauka, Leningrad, Russia.

Nikiticheva, Z.I. and O.B. Proskurina. 1987. Family Convolvulaceae, pp. 196–200. *In* T.B. Batygina and M.S. Yakovlev (eds.). Comparative Embryology of Flowering Plants. Davidiaceae—Ateraceae. Nauka, Leningrad, Russia.

Nikiticheva, Z.I. and O.B. Proskurina. 1990. Family Juncaginaceae, pp. 28–34. *In* T.B. Batygina and M.S. Yakovlev (eds.). Comparative Embryology of Flowering Plants. Butomaceae—Lemnaceae. Nauka, Leningrad, Russia.

***Nikiticheva, Z.I. and O.B. Proskurina. 1992.** Embryology of *Scheuchzeria palustris* (Scheuchzeriaceae). Bot. Zhurnal, 77 (1): 3–18.

Nikiticheva, Z.I., M.S. Yakovlev and T.A. Pluyshch. 1981. The development of ovule, embryo sac, embryo and endosperm in some species of genus *Peperomia* (Piperaceae). Bot. Zhurnal, 66 (10): 1388–1397.

Niogret, M.-F., and M. Dubald, P. Mandaron and R. Mache. 1991. Characterization of pollen polygalacturonase encoded by several cDNA clones in maize. Plant Mol. Biol., 17: 1155–1164.

Noher de Halac, J. 1980. Fine structure of nucellar cells during development of the embryo sac in *Oenothera biennis* L. Ann. Botany, 45 (5): 515–521.

Norstog, K. 1974. Nucellus during early embryogeny in barley: fine structure. Botanical Gazette, 135 (2): 97–103.

***Ogorodnikova, V.F. 1976a.** The ultrastructure of the tapetal membrane in some cultivated plants. Trans. Applied Botany, Genetics and Selection, 58: 124–129.

***Ogorodnikova, V.F. 1976b.** The ultrastructure of anther tapetum membranes in *Triticum activum* L., *Secale cereale* L. and *Avena sativa* L., pp. 380–382. *In* Proc. X All-Union Conf. Electron Microscopy, Tashkent, Uzbekistan.

***Ogorodnikova, V.F. 1986.** Genesis and ultrastructure of sporopollenin membrane of tapetum cells in grasses. Bot. Zhurnal, 71 (10): 1366–1371.

***Ogorodnikova, V.F. 1990a.** Dynamics of the ultrastructure of tapetum cells organelles in *Triticum aestivum* (Poaceae). Bot. Zhurnal, 75 (2): 186–192.

Ogorodnikova, V.F. 1990b. Sporopollenin membrane in tapetum of the anthers, p. 116. *In* T.B. Batygina (ed.). Abstracts XI Intern. Symp. 'Embryology and Seed Reproduction'. 3–7 July, 1990, Leningrad, USSR. Nauka, St.-Petersburg, Russia.

Ogur, M., R.O. Erickson and G.V. Rosen. 1951. Nucleic acids in relation to cell division in *Lilium longiflorum*. Experim Cell Research, 2 (1): 73–89.

Oksiyuk, P.F. 1927. Embryology of sugar-beet. Vestnik Kievskogo Botanichnogo Sadu, 5/6: 145–164 (in Ukrainian).

Oksiyuk, P.F. 1937. Comparative cytological and embryological investigation of the family Resedaceae (*Reseda* and *Astrocarpus*). Zhur. Inst. botaniki Acad. Nauk Ukrainskoi SSR, 12 (20): 1–40.

Ono, T. 1928. Embryologische Studien an einiger Pontederiaceen. Science Reports Tôhoku Imperial Univ. Ser. 4 (Biology), 3: 405–415.

***Orlova, I.N. 1971.** Cytomixis and its consequences in the microsporogenesis of tetraploid *Triticale*. Genetica, VII (12).

***Oryol, L.I. 1968.** Cytoembryological research of cytoplasmic male sterility of onion. Trans. Applied Botany, Genetics and Selection, 40 (1): 163–176.

***Ostapenko, E.K. and O.A. Khvedynich. 1987.** Cytochemical investigation of embryo sacs in *Galanthus nivalis* (Amaryllidaceae). Bot. Zhurnal, 72 (3): 291–298.

Osterwalder, A. 1898. Beiträge zur Embryologie von *Aconitum napellus* L. Flora, 85 (3): 254–292.

Pacini, E. 1990. Tapetum and microspore function, pp. 213–237. *In* S. Blackmore and R.B. Knox (eds.). Microspores: Evolution and Ontogeny. Acad. Press, London.

Pacini, E. and M. Cresti. 1978. Ultrastructural characteristics of tapetum and microspore mother cells in *Lycopersicum peruvianum* during meiotic prophase. Bull. Soc. botanique de France. Actualities botaniques, 125 (1–2): 121–128.

Pacini, E. and G. Casadoro. 1981. Tapetum plastids of *Olea europea*. Protoplasma, 106 (3–4): 289–297.

Pacini, E. and G.G. Franchi. 1983. Pollen grain development in *Smilax aspera* L. and possible functions of the loculus, pp. 183–190. *In* D. L. Mulcahy E. Ottaviano (eds.). Biology and Implications for Plant Breeding. NY, USA.

Pacini, E. and B.E. Juniper. 1983. The ullrastructure of the formation and development of the amoeboid tapetum in *Arum italicum* Miller. Protoplasma, 117: 116–129.

Pacini, E. and C.J. Keijzer. 1989. Ontogeny of intruding non-periplasmodial tapetum in wild chicory (*Cichorium intybus*: Compositae). Plant Systematics and Evolution, 167: 149–164.

Pacini, E. and G.G. Franchi. 1991. Diversification and evolution of the tapetum, pp. 301–316. *In* S. Blackmore and S.H. Barnes (eds.). Pollen and Spores. Oxford, spec. vol. 44.

Pacini, E., G.G. Franchi and M. Hesse. 1985. The tapetum: its form, function and possible phylogeny in Embryophyta. Plant Systematics and Evolution, 149: 155–185.

Paliwall, R.L. 1950. Life of *Coriandrum sativum* L. vol. 3, p. 47. *In* Proc. 37th Indian Science Congress (Poona).

Palm, B. 1915. Studien über Konstruktionstypen und Entwicklungswege des Embryosackes der Angiospermen. Diss. Stockholm.

Panchaksharappa, M.G. 1962. Embryological studies in some members of the Zingiberaceae, pp. 224–238. *In* Plant Embryology. A Symposium. New Dehli, India.

Panchaksharappa, M.G. 1966. Embryological studies in some members of the Zingiberaceae. II. *Elettaria cardamonum, Hitchenia caulina* and *Zingiber macrastachyum*. Phytomorphology, 16 (4): 412–417.

Panchakscharappa, M.G. and J. Syamasundar. 1975. A cytological study of ovule development in *Dipcadi montanum* Dalz. Cytologia, 40 (2): 141–149.

Pearson, H.H. 1929. Gnetales. NY, USA.

Pechenitsyn, V.P. 1989. Morphology and embryology of *Tulipa* species. Postdoctoral thesis, Leningrad, Russia.

Pèchoutre, F. 1902. Contributionà l'étude du développement de l'ovule et de la graine des Rosacées. Ann. Sciences Naturelles. Botanique, Sr. 8. 16 (1–2): 1–158.

Peddada, L. and J.P. Mascarenhas. 1975. The synthesis of 5 S ribosomal RNA during pollen development. Develop. Growth. Diff. 17 (1): 1–8.

Pénot, M. 1965. Etude du rôte des appels moléculaires dans la circulation liberiénne. Theses Fac. Sci. Univ. Strasbourg. 1–49.

Pénot, M. 1966. Etude de la migration orientée du ^{14}C dans le liber de *Tradescantia viridis*. C.R. l'Acad. Sciences, Paris, 266: 874–877.

Periasamy, K. 1962a. Studies on seeds with ruminate endosperm. II. Development of rumination in the Vitaceae. Proc. Indian Acad. Sciences. Section B, Biological Sciences (Bangalore, India), 56 (1): 13–26.

Periasamy, K. 1962b. The ruminate endosperm: development and types of rumination, 62–74. *In* Plant Embryology. A Symposium. New Dehli, India.

Periasamy, N. and B.G.L. Swamy. 1961. Studies in the Annonaceae. II. The development of ovule and seed in *Cananga odorata* and *Milmiusa wightiana*. J. Indian Bot. Soc., 40 (2): 206–216.

Periasamy, N. and B.G.L. Swamy. 1966. Morphology of anther tapetum of Angiosperms. Current Science. (India), 35 (17): 427–431.

Peterson, P.A., R.B. Flawell and D.H.P. Barrat. 1975. Altered mitochondrial membrane activities associated with cytoplasmically inherited disease sensitivity in maize. Theoretical and Applied Genetics, 45: 309–314.

Peterson, R.L., M.G. Scott and S.L. Miller. 1979. Some aspects of carpel structure in *Caltha palustris* L. (Ranunculaceae). Amer. J. Bot., 66: 334–342.

Petrova, T.F. 1977. Cytoembryology of Liliaceae. Subfamily Lilioideae. Nauka, Moscow, Russia.

***Petrova, T.F., O.B. Solovyanova, and Yu.S. Chentsova. 1985.** The ultrastructure of gigantic chromosomes in antipodal cells of barley. Cytology, 27 (5): 499–503.

***Petrovskaya, T.P. 1955.** The dynamics of nucleic acids in the reproductive processes of orchids *Calanthe veitchii*. Bull. Main Bot. Garden USSR Acad. Sciences, 22: 67–74.

Pfeiffer, A. 1891. Die Arillargebilde der Pflanzensamen. Bot. Jahrb. für Systematik, Pflanzengeshichte und Pflanzengeographie, vol. 13.

Pierson, E.S., H.M.P. Kengen and J. Derksen. 1989. Microtubules and actin filaments colocalize in pollen tubes of *Nicotiana tabacum* L. and *Lilium longiflorum* Thunb. Protoplasma, 150: 75–77.

Pipkin, J.L. and D.A. Larson. 1973. Changing patterns of nucleic acids, basic and acidic proteins in generative and vegetative nuclei during pollen germination and pollen tube growth in *Hippeastrum belladonna*. Experim. Cell Research, 79 (1): 28–42.

Planchon, J.E. 1845. Développement et caractères des vrais et des faux arilles. Ann. Sciences Naturelles. Botanique, 3: 275–312.

Plaut, W. S. 1953. DNA synthesis in microsporocytes in *Lilium henry*. Hereditas, 39: 438–444.

***Plisko, M.A. 1971.** Electron microscopy investigation of megagametogenesis peculiarities in *Calendula officinalis* L. Bot. Zhurnal, 56 (5): 582–598.

***Plisko, M.A. 1977.** The ultrastructure of sinergids in *Calendula officinalis* L. at the initial developmental stage (before the entrance of pollen tube in embryo sac). Bot. Zhurnal, 62 (2): 161–172.

***Plisko, M.A. 1980.** The ultrastructure of nucellus in *Alcea rosea* (Malvaceae). Bot. Zhurnal, 65 (6): 775–786.

***Plisko, M.A. 1982.** Some peculiarities of degeneration of nucellus cells. Bot. Zhurnal, 67 (11): 1469–1478.

***Plyushch, T.A. 1982.** The ultrastructure of embryo sac in *Peperomia blanda* (Piperaceae). Ukrainskyi Botanicheskyi Zhurnal, 39 (4): 88–91.

Plyushch, T.A. 1983. Fine structure of ovular reproductive elements in *Peperomia blanda* (Piperaceae). Abstracts of Ph.D. Thesis. Leningrad, Russia.

***Plyushch, T.A. 1988a.** Question of the function of sinergids in angiosperms. Bot. Zhurnal, 73 (8): 1140–1146.

Plyushch, T.A. 1988b. The walls of egg apparatus cells in angiosperms, p. 60. *In* Proc. VI All-Union Symp. 'Ultrastructure of Plants' .Kiev, Ukraine.

***Plyushch, T.A. 1989.** The ultrastructure of Allium type embryo sac in *Scilla sibirica* (Liliaceae) and *Nicotiana rustica* (Solanaceae). Bot. Zhurnal, 74 (11): 1598–1602.

***Plyushch, T.A. 1991.** Question of the egg apparatus ultrastructural angiosperms. Bot. Zhurnal, 76 (8): 1049–1055.

***Plyushch, T.A. 1992.** The Ultrastructure of Embryo Sac in Angiosperms. Naukova Dumka, Kiev, Ukraine.

*Plyushch, T.A. and V.P. Bannikova. 1989.** The ultrastructure of embryo sac and fertilization in *Adoxa moschatellina* (Adoxaceae). Bot. Zhurnal, 74 (1): 59–65.

***Poddubnaya-Arnoldi, V.A. 1930.** Embryological method in systematics of flowering plants. Reports Assoc. Physical-Mathematical Scientific Research Institute MGU, 3 (1–A).

***Poddubnaya-Arnoldi, V.A. 1951.** The significance of embryological investigations of higher plants for systematics. Uspekhi sovremennoi biologii, 32 (3): 352–392.

***Poddubnaya-Arnoldi, V.A. 1958.** The investigation of fertilization process in some flowering plants on living material. Bot. Zhurnal, 43 (2): 178–193.

***Poddubnaya-Arnoldi, V.A. 1959.** The investigation of embryonical processes in some orchids on living material. Trans. Main Botanical Garden, 6: 49–89.

Poddubnaya-Arnoldi [Poddubnaja-Arnoldi], V.A. 1960. Study of the fertilization in living material of some angiosperms. Phytomorphology, 10 (2): 185–198.

Poddubnaya-Arnoldi, V.A. 1964. General Embryology of Angiosperms. Nauka, Moscow, Russia.

Poddubnaya-Arnoldi, V.A. 1976. Cytoembryology of the Angiosperms: Principles and Perspectives. Nauka, Moscow, Russia.

Poddubnaya-Arnoldi, V.A. 1982. Characteristics of Flowering Plant Families According to Cytoembryological Features. Nauka, Moscow, Russia.

***Polunina, N.N. 1959.** The flowering biology and embryology of *Eucalyptus*. Trans. Main Botanical Garden, 6: 191–210.

***Polyakova, T.F. 1961.** Question of the role of vegetative nucleus in male gametophyte development. Cytology, 3 (3): 254–265.

***Polyakova, T.F. 1963.** Research on the vegetative nucleus in the pollen grain of *Echinops sphaerocephalus*. Cytology, 5 (1): 52–60.

***Polyakova, T.F. and S.E. Podstavek. 1960.** Behaviour of the vegetative nucleus in plants with binuclear and three-nuclear types of pollen grains. Reports USSR Acad. Sciences, 133 (6): 1433–1436.

Ponzi, R. and P. Pizzolongo. 1976. *Cytinus hypocistis* L. embryogenesis: ultrastructural aspects of megasporogenesis and megagametogenesis. J. Submicroscopic Cytology, 8 (4): 327–336.

Popham, R.A. and A.P. Chan. 1952. Zonation in the vegetative stem tip of *Chrysanthemum morifolium* Bailey. Amer. J. Bot., 37 (6): 476–482.

Porsch, O. 1907. Versuch einer phylogenetischen Erklärung des Embryosackes und der dopelten Befruchtung der Angiospermen. Jena.

Porter, E.K., J.M. Bird and H.G. Dickinson. 1982. Nucleic acid synthesis in microsporocytes of *Lilium* sv. *cinabar*: events in the nucleus. J. Cell Science, 57 (3): 229–246.

Prahl, A.-K., M. Rittscher and R. Wiermann. 1986. New aspects of sporopollenin biosynthesis, pp. 313–318. *In* D. H. Malcahy (ed.). Biotechnology and Ecology of Pollen. Springer-Verlag, NY, USA.

Prahl, A.-K., H. Springstubbe, K. Grumbach and R. Wiermann. 1985. Studies on sporopollenin biosynthesis: the effect of inhibitors of carotenoid biosynthesis on sporopollenin accumulation. Zeitschrift für Naturforschung, 40: 621–626.

Proskurina, O.B. 1990. The embryology of family Elaeagnaceae. Ph.D. thesis, Komarov Botanical Institute, Leningrad, Russia.

Puri, V. 1951. The role of floral anatomy in the solution of morphological problems. Bot. Review, 17: 471–553.

Purkinje, J.E. 1830. De cellulis antherarum fibrosis nec non de granorum pollinarium formis commentatio phytotomica. Bratislaviae.

Rafnar, T., I.J. Griffith, M. Kuo, and J.F. Bond, B.L. Rogers and D.G. Klapper. 1991. Cloning of Amb a I (antigen E), the major allergen family of short ragweed pollen. J. Biological Chemistry, 266: 1229–1236.

Raikov, I.B. 1975. The problem of origin and evolution of meiosis, pp. 344–371. *In* V.V. Khvostova, and Yu.F. Bogdanov (eds.).Cytology and Genetics of Meiosis. Nauka, Moscow, Russia.

Raj, B. 1970. Morphological and embryological studies in the family Santalaceae—*Buckleya lanceolata* Sieb. et Zucc. Beiträge zur Biologie der Pflanzen, 46 (2): 193–207.

Raj, B. and G. El-Ghazaly. 1987. Morphology and taxonomic application of orbicules (Ubisch bodies) in Chloranthaceae. Pollen Spores, 29: 151–166.

Ram, M. 1959a. Morphological and embryological studies in the family Santalaceae. III. *Leptomeria* R. Br. Phytomorphology, 9 (1): 20–33.

Ram, M. 1959b. Occurrence of embryo sac-like structures in the microsporangia of *Leptomeria billardierii* R. Br. Nature, 184: 914–915.

Randolph, L.F. 1936. Developmental morphology of the caryopsis in maize. J. Agricultural Research, 53 (12): 881–916.

Rangaswamy N.S. and Janaki Subramanyan. 1992. Computer graphics to infer microspore tetrad configurations, pp. 451–452. *In* T.B. Batygina (ed.). Proc. XI Intern. Symp. 'Embryology and Seed Reproduction'. 3–7 July, 1990, Leningrad, USSR. Nauka, St.-Petersburg, Russia.

Rao, D. 1966. A contribution to the floral anatomy and embryology of Humiriaceae, Linaceae and Erythroxylaceae. Ph.D. thesis, Univ. Osmania, India.

Rappaport, J., S. Satina and A.F. Blakeslee. 1950. Extracts of ovular tumors and their inhibition of embryo growth in *Datura*. Ameri. J. Bot., 37 (8): 586–595.

Rasch, E. and J.W. Woodard. 1959. Basic proteins of plant nuclei during normal and pathological cell growth. J. Biophysical and Biochemical Cytology, 8 (3): 263–276.

Razi, B.A. 1949. Embryological studies of two members of the Podostemaceae. Bot. Gazette, 111 (1): 211–218.

Rees-Leonard, O.L. 1935. Macrosporogenesis and development of the macrogametophyte of *Solanum tuberosum*. Bot. Gazette, 96: 734–750.

Reznickova, S.A. 1975. The regularity of cell differentiation during microsporo- and gametogenesis (on the example of *Lilium candidum* L.). Ph.D thesis, Moscow, Russia.

Reznickova, S.A. 1979. Biochemical and ultrastructural features of meiotic cells. Nucleus (India), 22: 50–66.

Reznickova, S.A. 1984. Cytology and Physiology of the Developing Anther. Nauka, Moscow, Russia.

Reznickova, S.A. and V.O. Ostroverkhov. 1970. Cytophotometric investigation of DNA contents in microsporogenesis process in *Lilium candidum* L. Reports USSR Acad. Sciences, 194 (3): 695–697.

Reznickova, S.A. and M.T.M. Willemse. 1980. Formation of pollen in the anther of *Lilium*. II. The function of the surrounding tissues in the formation of pollen and pollen wall. Acta Botanica Neerlandica, 29 (2/3): 141–156.

***Reznickova, S.A., A.M. Bugara and A.I. Erokhina. 1979.** Cytochemical investigation of DNA and histones in the anther tissues of lily in microsporo- and gametogenesis. Ontogenes, 10 (3): 277–284.

Richard, L.C.M. 1808. Demonstrations botaniques ou analyse du fruit considere en générale. Paris, 111 pp.

Rifot, M. 1973. Evolution structural du pole chalazien du sac embryonnaire d'*Aquilegia vulgaris* en liaison avec son activit trophique. C.R. l'Acad. Sciences, Paris, 277 (14): 1313–1316.

Roberts M.R., R. Hodge, J.H.E. Ross, A. Sorensen, D.J. Murphy, J. Draper and R. Scott. 1993. Characterization of a new class of oleosins suggest a male gametophyte-specific lipid storage pathway. Plant Journal, 3: 629–636.

Robinson-Beers, K., R.E. Pruit and Ch. Gasser. 1992. Ovule development in wild-type *Arabidopsis* and two female sterile mutants. Plant Cell, 4: 1237–1249.

***Rodionova, G.B. 1972.** The development of ovule and integuments in *Heliophila amplexicaulis* L. Bot. Zhurnal, 57 (9): 1153–1161.

Rodionova G.B. 1983. Family Capparaceae, pp. 150–154. *In* T.B. Batygina and M.S. Yakovlev (eds.). Comparative Embryology of Flowering Plants. Phytolaccaceae—Thymelaeaceae. Nauka, Leningrad, Russia.

Rodionova, G.B. 1992. Family Brassicaceae, pp. 191–201. *In* A.L. Takhtajan (ed.). Comparative Anatomy of Seeds. Dicotyledones. Dilleniidae. Nauka, Leningrad, Russia, 4.

Rodkiewicz, B. 1970. Callose in cell walls during megasporogenesis in angiosperms. Planta, 93: 39–47.

Rodkiewicz, B. 1978. Przebudowa cytoplasmy isolacia i polaryzacja w melocytach podczas megasporogenezy u angiospermae [Cytoplasm restructuring isolation and polarization in meiocytes during megasporogenesis in angiosperms]. Postepy Biologii Komorki, 5: 135–154 (in Polish).

Rodkiewicz, B. and E. Mikulska. 1964. Electron microscope observations of cytoplasmic changes in developing megasporocyte of *Lilium candidum*. Flora, 154, Abt. A, (2): 383–387.

Rodkiewicz, B. and E. Mikulska. 1965a. Electron microscope observations of endoplasmic reticulum in *Lilium* megasporocytes. Flora, 155, Abt. A, (3): 341–346.

Rodkiewicz, B. and E. Mikulska. 1965b. The development of cytoplasmic structures in the embryo sac of *Lilium candidum* as observed with the electron microscope. Planta, 67 (3): 297–304.

Rodkiewicz, B. and J. Bednara. 1976. Cell wall ingrowths and callose distribution in megasporogenesis in some Orchidaceae. Phytomorphology, 26 (3): 276–281.

Rodkiewicz, B. and R. Sniezko. 1978. Differentiation of functional megaspore in Onagraceae. Bull. Soc. botanique de France, 125 (1–2): 163–166.

Rodkiewicz, B. and J. Bednara. 1986. Development and ultrastructure of female gametophyte in angiosperms. Postepy Biologii Komorki, 13: 125–144. (Polish with English summary).

Rodkiewicz, B. and E. Duda. 1988. Aggregations of organelles in meiotic cells of higher plants. Acta Societatis Botanicorum Poloniae, 52: 637–659.

***Rodkiewicz, B., J. Bednara and R. Sneizko. 1976.** Polarization of cells in megasporogenesis of Onagraceae. Bot. Zhurnal, 61 (7): 905–909.

Rodriguez-Garcia, M.I. and I.D. Alche. 1992. Contribution to knowledge of the special callose wall during microsporogenesis in *Olea europaea* L., pp. 463–464. *In* T.B. Batygina (ed.). Proc. XI Intern. Symp. 'Embryology and Seed Reproduction'. 3–7 July, 1990, Leningrad, USSR. Nauka, St.-Petersburg, Russia.

Roeckel, P., A. Chaboud, and E. Matthys-Rochon, S.D. Russel and C. Dumas. 1990. Sperm cell structure, development and organization, pp. 281–307. *In* Microspores: Evolution and Ontogeny. Acad. Press, San Diego, CA, USA.

Rogers B.L., J.P. Morgenstern, I.J. Griffith, X.B. Yu, C.M. Counsell, A. W. Brauer, T.P. King, R.D. Graman and M-C. Kuo. 1991. Complete sequence of the allergen Am a II. J. Immunology, 147: 2547–2552.

Rogers, H.J., A. Harvey and D.M. Lonsdale. 1992. Isolation and characterization of a tobacco gene with homology to pectate lyase which is specifically expressed during microsporogenests. Plant Molecular Biology, 20: 403–502.

Roland, F. 1971. Characterization and extraction of the polysaccharides of the intine and the generative cell wall in the pollen grains of some Ranunculaceae. Grana, 11: 101–106.

Roland-Heydacker, F. 1979. Aspect ultrastructuraux de l'ontogenie du pollen et du tapis chez *Mahonia aquifolium* Nutt. Berberidaceae. Pollen et Spores, 31 (3): 259–278.

***Romanov, I.D. 1936.** Embryo sac development in the genus *Gossypium*, pp. 17–35. *In* The Questions on Cytology, Embryology and Anatomy of Cotton. Tashkent, Uzbekistan.

***Romanov, I.D. 1938.** A new form of the Adoxa type embryo sac in *Tulipa tetraphylla* and *T. ostrovskiana*. Reports USSR Acad. Sciences, 19 (1–2): 113–115.

***Romanov, I.D. 1939.** Two new forms of embryo sac in the genus *Tulipa*. Reports USSR Acad. Sciences, 22: 140–142.

***Romanov, I.D. 1944.** The evolution of angiosperm embryo sacs. Postdoctoral thesis, Tashkent, Uzbekistan.

***Romanov, I.D. 1945.** Embryo sac evolution in the Tulipeae group of the family Liliaceae. Byull. sredneasiatskogo gosudarstvennogo universiteta, 23: 118–119.

***Romanov, I.D. 1957.** The embryo sac in the genus *Tulipa*. Reports USSR Acad. Sciences, 115 (5): 1025–1027.

***Romanov, I.D. 1961.** Origin of the special structure of endosperm nuclei in *Gagea*. Reports USSR Acad. Sciences, 141 (4): 984–986.

*Romanov, I.D. 1965a.** Analytical experience of some peculiar features in *Fritillaria* type embryo sac. Bot. Zhurnal, 50 (9): 1276–1287.

***Romanov, I.D. 1965b.** Anomalous mitoses and cytoplasmic gradient in the embryo sacs of some plants. II. Species of *Tulipa*. Cytology, 7 (1): 23–32.

***Romanov, I.D. 1966.** The specific features of pollen development in grasses. Reports USSR Acad. Sciences, 169 (2): 456–459.

***Romanov, I.D. 1970a.** The peculiar features of pollen development in grasses and their significance for some genetic investigation. Genetica, 6 (10): 11–25.

***Romanov, I.D. 1970b.** The specific features of pollen development in grasses and their significance for some genetic investigation. Genetica, 6 (6): 11–23.

***Romanov, I.D. 1970c.** Wheat pollen development according to the observation on living material. Bull. Vavilov All-Union Plant Industry Institute, 15: 38–45.

***Romanov, I.D. 1971a.** The nuclei motion during embryo sac development in *Solanum viraultorum* (Bitt.) Card. et Haukes. Bot. Zhurnal, 55 (11): 1563–1568.

***Romanov, I.D. 1971b.** The types of embryo sac development in flowering plants, pp. 72–112. *In* V.P. Zosimovich (ed.). Problems of Embryology. Naukova Dumka, Kiev, Ukraine.

Romanov, I.D. 1981. Embryo sac, pp. 11–16. *In* T.B. Batygina and M.S. Yakovlev (eds.) Comparative Embryology of Flowering Plants. Winteraceae—Juglandaceae. Nauka, Leningrad, Russia.

***Romanov, I.D. and I.N. Orlova. 1971.** Cytomixis and its consequences in microsporocytes Triticale. Genetica, 7 (12): 5–13.

Rosanoff, S. 1865. Zur Kenntnis des Baues und der Entwicklungsgeschichte des Pollens der Mimoseae. Jahrb. für Wissen Schaftliche Botanik, 4: 441–450.

Roth, I. 1957. Die Histogenese der Integumente von *Capsella bursa-pastoris* und ihre morphologische Bedeutung. Flora, 145 (1–2): 212–235.

Rougier, M., N. Inoud, and C. Said, S. Russell and C. Dumas. 1991. Male gametophyte development and formation of the male germ unit in *Populus deltoides* following compatible pollination. Protoplasma, 162 (2–3): 140–150.

Rowley, J.R. 1959. The fine structure of pollen wall in the Commelinaceae. Grana Palynologica, 2 (1): 3–31.

Rowley, J.R. 1990. The fundamental structure of the pollen exine. Plant Systematics and Evolution, Suppl. 5. P. 13–29.

Rowley, J.R. and A.O. Dahl. 1977. Pollen development in *Artemisia vulgaris* with special reference to glucocalyx material. Pollen et Spores, 19: 169–284.

Rowley, J.R. and B. Prijanto. 1977. Selective destribution of the exine of pollen grains. Geophytology, 7 (1): 1–23.

Rowley, J.R. and B. Walles. 1987. Origin and structure of Ubisch bodies in *Pinus sylvestris*. Acta Societatis Botanicorum Poloniae, 56 (2): 215–227.

Rowley, J.R. and G. El-Ghazaly. 1992. Lipid in wall and cytoplasm of *Solidago* pollen. Grana, 31: 273—283.

Rowley, J.R., A.O. Dahl and J. S. Rowley. 1981a. Substructure in exines of *Artemisia vulgaris* (Asteraceae). Rev. Palaeobotany and Palynology, 35: 1–38.

Rowley, J.R., A.O. Dahl, S. Sengupta and J.S. Rowley. 1981b. A model of exine substructure based on dissection of pollen and spore exines. Palynology, 5: 107–152.

Rowley, J.R., N.I. Gabarayeva and B. Walles. 1992. Tapetal cell modifications around microspores of Nymphaea, pp. 467–468. *In* T.B. Batygina (ed.). Proc XI Intern. Symp. 'Embryology and Seed Reproduction'. 3–7 July, 1990, Leningrad, USSR. Nauka, St.-Petersburg, Russia.

Russell, S.D. 1979. Fine structure of megagametophyte development in *Zea mays*. Can. J. Bot., 57 (10): 1093–1110.

Russell, S.D. 1980. Participation of male cytoplasm during gamete fusion in Angiosperm, *Plumbago zeylanica*. Science, 210: 200–201.

Russell, S.D. 1983. Fertilization in *Plumbago zeylanica*: gametic fusion and fate of the male cytoplasm. Amer. J. Bot., 70 (3): 416–434.

Russell, S.D. 1984. Ultrastructure of the sperm of *Plumbago zeylanica*. II. Quantitative cytology and three-dimensional organization. Planta, 162 (5): 385–391.

Russell, S.D. 1991. Isolation and characterization of sperm cells in flowering plants. Ann. Rev. Plant Physiology, Plant Mole. Biol., 42: 189–204.

Russell, S.D. 1993. The egg cell: development and role in fertilization and early embryogenesis. Plant Cell, 5 (10): 1349–1359.

Russell, S.D. and T. W. Mislan. 1986. Microtubule-organelle associations during generative cell polarization in *Plumbago zeylanica*, pp. 280–281. *In* Proc. 44th Ann. Meeting of Electron Microscopy Society of America.

Rutgers, F.L. 1923a. The female gametophyte of Angiosperms. Ann. Jardin Botanique de Buitenzorg, 33 (1): 1–5.

Rutgers, F.L. 1923b. Embryo sac and embryo of *Moringa oleifera*. Ann. Jardin Botanique de Buitenzorg, 33 (1): 6–66.

Rutishauser, A. 1969. Embryologie und Fortpflanzungsbiologie der Angiospermen. Springer-Verlag, Berlin.

Ryczkowski, M. 1962. Changes in the osmotic value of the sap from embryos, the central vacuole and cellular endosperm during development of the ovules. Bull. l'Acad. Polonaise des Sciences, Série des Sciences Biologuigues, pp. 375–380.

Ryczkowski, M., A. Kowalska and W. Reczynski. 1986. Element concentrations in the sap surrounding the developing embryo of *Aesculus hybrida*. J. Plant Physiology, 122: 367–472.

Sachs, J. 1877. Über die Anordnung der Zellen in jungsten Pflanzentheile. Verhandlungen der medizinisch-physiologische Gesellschaft, Wizburg, 5 (11): 1–26.

Sachs, J. 1882. Vorlesungen über Pflanzenphysiologie. Leipzig.

Sahni, B. 1925. The ontogeny of vascular plants and the theory of recapitulation. J. Indian Botanical Society, 4.

Sampson, F.B. 1969. Cytokinesis in pollen mother cells of angiosperms, with emphasis on *Laurelia novazelandiae* (Monimiaceae). Cytologia, 34: 511–634.

Samuels, J.A. 1912. Etudes sur le développement du sac embryonnaire du *Gunnera macrophylla* Bl. Archiv der Zellforschung, 8 (1): 52–120.

Sanger, J.M. and W.T. Jackson. 1971. Fine structure study of pollen development in *Haemanthus katherinae* Baker. J. Cell Science, 8: 303–315.

Sassen, M.M. 1964. Fine structure of *Petunia* pollen grain and pollen tube. Acta Botanica Neerlandica, 13: 175–181.

Satarova, T.N. 1990. Family Asparagaceae, pp. 114–124. *In* T.B. Batygina and M.S. Yakovlev (eds.). Comparative Embryology of Flowering Plants. Butomaceae—Lemnaceae. Nauka, Leningrad, Russia.

Satina, S., J. Rappaport and A.F. Blakeslee 1950. Ovular tumors connected with incompatible crosses in *Datura*. Amer. J. Bot., 37: 576–586.

Sato, Joshoniko. 1988. Pollen development of *Aucuba japonica* Thunb. Science Reports Yokohama National Univ. Sec. 2. Biological and Geological Sciences, no. 35.

Sato, S. 1990. The nuclear DNA contents in the synergids and antipodal cells of the mature embryo sac of *Allium tuberosum*. Science Reports Hirosaki Univ. 37 (2): 124–127.

***Savchenko, M.I. 1959.** Abnormalities in ovule development of some flowering plants. Bot. Zhurnal, 44 (6): 786–804.

Savchenko, M.I. 1960. Abnormalities in the angiosperm ovule structure. Reports of the USSR Academy of Sciences, 130 (5): 1154–1157 (in Russian).

Savchenko, M.I. 1973. Morphology of the Flowering Plant Ovule. Nauka, Leningrad, Russia.

***Savchenko, M.I. and L.R. Petrova. 1963.** Morphology of the ovule in barley *Hordeum vulgare* L. and some peculiar features of its development. Bot. Zhurnal, 48 (11): 1623–1638.

Savchenko, M.I. and G.A. Komar. 1965. Morphology of the ovule in some Monocotyledons, pp. 74–113. *In* M.S. Yakovlev (ed.). Flower Morphology and Reproductive Process in Angiosperms, Moscow-Leningrad, Russia.

***Savina, G.I. 1972.** Fertilization process in *Ophrys insectifera*. Bot. Zhurnal, 57 (3): 389–392.

Savina, G.I. 1983. Family Sterculiaceae, pp. 248–251. *In* T.B. Batygina and M.S. Yakovlev (eds.). Comparative Embryology of Flowering Plants. Phytolaccaceae—Thymelaeaceae. Nauka, Leningrad, Russia.

Savina, G.I. 1987. Family Proteaceae, pp. 113–115. *In* T.B. Batygina and M.S. Yakovlev (eds.) 1987. Comparative Embryology of Flowering Plants. Davidiaceae—Asteraceae. Nauka, Leningrad, Russia.

Savina, G.I. and G.Ya. Zhukova. 1983. Ultrastructural aspects of the embryo sac development in *Epipactis atrorubens* (Hoffm.) Schult. *In* Proc. VII Intern. Cytoembryological Symp. 'Fertilization and Embryogenesis in Ovulated Plants'. High Tatra (Račkova dolina), 14–17 June, 1982. Bratislava, Czechoslovakia.

Schacht, H. 1857. Ueber Pflanzen-Befruchtung. Jahrb. für Wissenschaftliche Botanik, 1: 193–232.

Schaffner, J.H. 1896. The embryo sac of *Alisma plantago*. Bot. Gazette, 21: 123–132.

Scheridan, W.F. 1973. Nonaqueous isolation of nuclei from lily pollen and examination of their histones. Zeitschrift für Pflanzenphysiologie, 68 (4): 450–459.

Scheridan, W.F. and H. Stern. 1967. Histone of meiosis. Experim. Cell Research, 45 (3): 323–335.

Schell, R. and M. Wolter. 1986. On the presence of elastoviscin in all subfamilies of the Orchidaceae and the homology to pollenkitt. Nordic J. Bot., 6 (3): 321–324.

Schlag, M. and M. Hesse. 1992. The formation of the generative cell in *Polystachia pubescens* (Orchidaceae). Sexual Plant Reproduction, 5: 131–137.

Schleiden, M.J. 1837. Einige Blicke auf die Entwicklungsgeschichte des vegetablischen Organismus bei den Phanerogamen. Archiv für Naturgeschichte, (Wiegmann). Jg. III. 1: 289–320.

***Schmalhauzen, I.I. 1940.** Inconstancy and changing of adaptive norms in evolution process. Zhurnal Obshchei Biologii, 1(A).

Schmidt, A. 1924. Histologische Studien an phanerogamen Vegetationspunkten. Botanisches Archiv Königsberg (Kaliningrad, Russian SFSR) & Dahlem bei Berlin, 8 (1–2): 345–404.

Schmitz, F. 1872. Die Bluten-Entwicklung der Piperaceen. Botanishe Abhandlungen, 2 (1): 1–74.

Schnarf, K. 1923. Kleine Beiträge zur Entwicklungsgeschichte der Angiospermen. VI. Über das Verhalten des Antherentapetums einiger Pflanzen. Österr. Botanische Zeitung, 72 : 242–245.

Schnarf, K. 1929. Embryologie der Angiospermen. Archegoniaten. Gebr. Bornträger, Berlin.

Schnarf, K. 1931. Vergleichende Embryologie der Angiospermen. Gebr. Bornträger, Berlin.

Schnarf, K. 1936. Contemporary understanding of embryo sac development among Angiosperms. Bot. Review, 2 (12): 565–585.

Schnarf, R. 1942. Archegonium und Archegontheorie. Biologia Generales. Vienna & Leipzig, 16 (1–3): 198–224.

Sheridan, W.F. 1973. Nonqueous isolation of nuclei from lily pollen and an examination of their histones. Zeitschrift für Pflanzenphysiologie, 68: 450–459.

Schröder, M.-B. and H. Oldenburg. 1990. Ultrastructural studies on plastids of generative and vegetative cells in Liliaceae. 7. Plastid distribution during generative cell development in *Tulbaghia violacea* Harv. Flora, 184: 131–136.

Schulz, R. and W.A. Jensen. 1968a. *Capsella* embryogenesis: the synergids before and after fertilization. Amer. J. Bot., 55 (5): 542–552.

Schulz, R. and W.A. Jensen. 1968b. *Capsella* embryogenesis: the egg, zygote and young embryo. Amer. J. Bot., 55 (7): 807–819.

Schulz, R. and W.A. Jensen. 1969. *Capsella* embryogenesis: the suspensor and the basal cell. Protoplasma, 67 (2–3): 139–163.

Schulz, R. and W.A. Jensen. 1971. *Capsella* embryogenesis: the chalazal proliferating tissue. J. Cell Science, 8 (2): 139–163.

Schulz, R. and W.A. Jensen. 1973. *Capsella* embryogenesis: the central cell. J. Cell Science, 12 (3): 741–763.

Schulz, R. and W.A. Jensen. 1977. Cotton embryogenesis: the early development of the free nuclear endosperm. Amer. J. Bot., 64 (4): 384–394.

Schulz, R. and W.A. Jensen. 1981. Pre-fertilization ovule development in *Capsella*: ultrastructure and ultracytochemical localization of acid phosphatase in the meiocyte. Protoplasma, 104: 27–45.

Schulz, R. and W.A. Jensen. 1986. Pre-fertilization ovule development in *Capsella*; the dyad, tetrad, develozping megaspore, and two-nucleate megagametophyte. Can. J. Bot., 64: 875–884.

Schulze Osthoff, K. and J. Wiermann. 1987. Phenols as integrated compounds of sporopollenin from *Pinus* pollen. J. Plant Physiology, 131: 5–15.

Schürhoff, P.N. 1919. Zur Phylogenie des Angiospermen—Embryosackes. Berichte der Deutschen Botanischen Gesellschaft, 37: 160—168.

Schürhoff, P.N. 1926. Die Zytologie der Blutenpflanzen. Enke Verlag, Stuttgart.

Schürhoff, P.N. 1928. Über Entwicklung des Eiapparates der Angiospermen. Berichte der Deutschen Botanischen Gesellschaft, 46: 560—572.

Schweiger, J. 1905. Beiträge zur Kenntnis der Samenentwicklung der Euphorbiaceen. Flora, 94: 339—379.

Schwere, S. 1896. Zur Entwicklungsgeschichte der Frucht von *Taraxacum officinale* Web. Ein Beitrage zur Embryologie der Compositen. Flora, 82 (1): 32—66.

Sears, B.B. 1980. Elimination of plastids during spermatogenesis and fertilization in the plant kingdom. Plasmid, 4: 233—255.

Senyaninova-Korchagina, M.V. 1959. On the Triumph of Flowering Plants from the Position of Heliogeophysics. Izdatelstvo LGU, Leningrad, Russia.

Severtsov, A.N. 1939. Morphological Patterns of Evolution. Izdatelstvo Akademii Nauk SSSR, Moscow-Leningrad, Russia, 601 pp.

Shah, Rajesh. 1992. Light and electron microscopic studies on development of polyads and seeds with special reference to seed infestation and destruction in *Albizzia lebbek* (L.) Benth. Ph.D. thesis, Univ. of Delhi, Delhi, India.

*__Shamrov, I.I. 1981.__ Peculiar features of anther development in *Ceratophyllum demersum* and *C. pentacanthum* (Ceratophyllaceae). Bot. Zhurnal, 66 (10): 1464–1473.

*__Shamrov, I.I. 1988.__ Ovule development and structural characteristics of embryo sac in some members of the Gentianaceae family. Bot. Zhurnal, 73 (2): 213–222.

Shamrov, I.I. 1990a. Family Alliaceae, pp. 76–83. *In* T.B. Batygina and M.S. Yakovlev (eds.). Comparative Embryology of Flowering Plants. Butomaceae—Lemnaceae. Nauka, Leningrad, Russia.

Shamrov, I.I. 1990b. Family Hemerocallidaceae, pp. 83–87. *In* T.B. Batygina and M.S. Yakovlev (eds.). Comparative Embryology of Flowering Plants. Butomaceae—Lemnaceae. Nauka, Leningrad, Russia.

Shamrov, I.I. 1990c. Family Liliaceae, pp. 63–74. *In* T.B. Batygina and M.S. Yakovlev (eds.). Comparative Embryology of Flowering Plants. Butomaceae—Lemnaceae. Nauka, Leningrad, Russia.

*__Shamrov, I.I. 1990d.__ The ovule of *Gentiana cruciata* (Gentianaceae): structural–functional aspects. Bot. Zhurnal, 75 (10): 1363–1379.

Shamrov, I.I. 1991. The ovule of *Swertia iberica* (Gentianaceae): structural and functional aspects. Phytomorphology, 41 (3–4): 213–229.

Shamrov, I.I. 1992a. Structural and functional aspects of ovule development in some flowering plants, pp. 501–502. *In* T.B. Batygina (ed.). Proc. XI Intern. Symp 'Embryology and Seed Reproduction'. 3–7 July, 1990, Leningrad, USSR. Nauka, St.-Petersburg, Russia.

Shamrov, I.I. 1992b. The ovule morphogenesis in *Gagea stipitata* (Liliaceae), p. 85. *In* Abstracts XII Intern. Congr. Sexual Plant Reproduction. Columbus, Ohio, USA.

Shamrov, I.I. 1995. Ovule of flowering plants. Principles of organization and typification. Postdoctoral thesis, Komarov Bot. Inst. St.-Petersburg, Russia.

Shamrov, I.I. and T.B. Batygina. 1988. Family Ceratophyllaceae, pp. 153–156. *In* A.L. Takhtajan (ed.). Comparative Anatomy of Seeds. Dicotyledons. Magnoliidae—Ranunculidae. Nauka, Leningrad, Russia, vol. 2.

*Shamrov, I.I. and A.N. Winter. 1991. Ovule development in representatives of the genera *Nymphaea* and *Victoria* (Nymphaeaceae). Bot. Zhurnal, 76 (8): 1072–1083.

*Shamrov, I.I. and Z.I. Nikiticheva. 1992. The ovule and seed morphogenesis in *Gymnadenia conopsea* (Orchidaceae): structural and histochemical investigation. Bot. Zhurnal, 77: 45–60.

*Shamrov, I.I. and G.M. Anisimova. 1993a. The development of ovule and seed in *Juncus filiformis* (Juncaceae). Bot. Zhurnal, 78: 18–33.

*Shamrov, I.I. and G.M. Anisimova. 1993b. The ovule morphogenesis in *Luzula pedemontana* (Juncaceae): structural-histochemical investigation. Bot. Zhurnal, 78: 47–59.

Shamrov, I.I., G.M. Anisimova, G. Lakshmi Sita and T.B. Batygina. 1992. The peculiarity of ovule development and structure in *Santalum album* (Santalaceae), p. 62. *In* Abstracts XII Inter. Cong. Sexual Plant Reproduction. Columbus, Ohio, USA.

Shevchenko, S.V. 1990. Family Cannaceae, pp. 245–247. *In* T.B. Batygina and M.S. Yakovlev (eds.). Comparative Embryology of Flowering Plants. Butomaceae—Lemnaceae. Nauka, Leningrad, Russia.

Shevchenko, S.V. and O.P. Kamelina. 1987. Family Davidiaceae, pp. 7–12. *In* T.B. Batygina and M.S. Yakovlev (eds.). Comparative Embryology of Flowering Plants. Davidiaceae—Asteraceae. Nauka, Leningrad, Russia.

Shi, L., H.L. Mogensen, T. Zhu and S.E. Smith. 1991. Dynamics of nuclear pore density and distribution patterns within developing pollen: implications for a functional relationship between the vegetative nucleus and the generative cell. J. Cell Sci., 99: 115–120.

*Shishkin, I.A. 1984. Individual development and natural selection. Ontogenes, 15 (2): 4–18.

Shivanna, K.R. and B.M. Johri. 1985. The Angiosperm Pollen: Structure and Function. Wiley Eastern, New Delhi.

Shkutina, F.M. and V.V. Khvostova. 1966. Cytological analysis of wheat-rye amphiploids, p. 261. *In* Experimental Polyploidy in Plant Selection. Moscow, Russia.

*Shneider, T.M. 1988. The peculiarities of meiosis in remove hybrids of wheat received with the participance of mutant ph. Cytology i genetica, 22 (3): 18–22.

Shreve, F. 1906. The development and anatomy of *Sarracenia purpurea*. Bot. Gazette, 42: 107–126.

Sidorova, N.V. 1985. The ultrastructure features of isolated embryo sac of tobacco. Reports Ukrainian SSR Acad. Sciences, "B", 12: 63–66.

Silvanovich, A., J. Astwood, L. Zhang, F. Kisil, A. Sehon, S. Mohapatra and R. Hill. 1991. Nucleotide sequence analysis of the cDNA coding for *Poa* p IX isoallergens of Kentucky bluegrass pollen. J. Biological Chemistry, 266: 1204–1210.

*Simonenko, V.K. 1982. The development of anther and microspores in the fertile and CMS-lines of sunflower. Cytology i Genetica, 16 (5): 34–41.

Singh, B. 1961. Studies in the family Malpighiaceae. III. Development and structure of seed and fruit of *Malpighia glabra* Linn. Horticultural Advance, 5: 145–155.

Singh, M.B., T. Hough, P. Theerakulpisut, A. Avjioglu, S. Davies, P. M. Smith, P. Taylor, R.J. Simpson, L.D. Ward, J. McCluskey, R. Ruy and R.B. Knox. 1991. Isolation of cDNA encoding a newly identified major allergenic protein of rye-grass pollen: intracellular tatring to the amyloplast. Proc. Natl. Acad. Sciences USA, 88: 1384–1488.

Singh, R.P. 1968. Structure and development of seeds in Euphorbiaceae: *Melanthesa rhamnoides* Wt. Beiträge zur Biologie der Pflanzen, 45 (1): 127–133.

Singh, S. and B. Walles. 1992. The ovarian transmitting tissue in *Lilium regale*. Intern. J. Plant Science, 153 (2): 205–211.

Skvarla, J.J. and J.A. Larson. 1966. Fine structural studies of *Zea mays* pollen. I. Cell membranes and exine ontogeny. Ame. J. Bot., 53: 1112–1125.

*Sladkov, A.N. 1962. Polar axis of spores and nuclei arrangment in meiosis and different types of tetrads in ferns. Reports USSR Acad. Sciences, 146 (1): 225–229.

*Sladkov, A.N. 1970. The types of tetrads, hexads and spores. Vestnik MGU, Series 6, Biology, Soil Science, 2: 82–91.

*Sladkov, A.N. 1989. Microspore tetrad types of flowering plants. Bull. Moscow Society of Nature Experience, Biological department, 94 (6): 48–52.

*Sladkov, A.N. and N.A. Grevtsova. 1988. Formation of the angiosperms microsporangium wall. Bull. Moscow Society of Nature Experience, Biological department, 93 (4): 69–74.

*Sladkov, A.N. and N.A. Grevtsova. 1989. Formation of the angiosperm megasporangium wall. Bull. Moscow Society of Nature Experience, Biological department, 94 (3): 75–79.

Sladkov, A.N. and N.A. Grevtsova. 1991. Introduction to the Embryology of Angiosperms. Izdatelstvo MGU, Moscow, Russia.

Smirnov, A.G. 1982. Female gametophyte of angiosperms and its evolution. Izdatelstvo Kazanskogo Universiteta, Kazan, Russia.

*Smirnov, A.G. 1983. Some results of embryological research in *Peperomia langsdorffii*, pp. 124–125. *In* Proc. VIII All-Union Conf. Plant Embryology 'Problems of Gametogenesis, Fertilization and Embryogenesis'. 'Phan', Tashkent, Uzbek Republic.

*Smirnova, V.A. 1954. Movement of nuclei of some Monocotyledons with mosaics. Bull. Moscow Society of Nature Experience, Biological department, 59 (5): 71–76.

Smith, F.H. 1943. Megagametophyte of *Clintonia*. Bot. Gazette, 105 (2): 263–267.

Smith, S.E. 1989. Biparental inheritance of organelles and its implications in crop improvement. Plant Breeding Reviews, 9: 361–393.

Sniezko, R. 1976. Distribution of polysaccharides during megasporogenesis in Onagraceae. Ann. Univ. M. Curi-Sklodowska. 31: 183–191.

Sniezko, R. and C. Harte. 1984. Polarity and competition between megaspores in the ovule of *Oenothera* hybrids. Plant Systematics and Evolution, 144 (1): 83–97.

*Sokolov, I.D. 1968. Embryo sac development in bulb onion (*Allium cepa* L.). Reports Turkmen SSR Acad. Sciences, Series Biological Sciences, 1: 87–89.

Sokolovskaya, T.B. 1981. Family Ranunculaceae, pp. 130–138. *In* T.B. Batygina and M.S. Yakovlev (eds.). Comparative Embryology of Flowering Plants. Winteraceae—Juglandaceae. Nauka, Leningrad, Russia.

Sokolovskaya, T.B. 1983. Family Elatinaceae, pp. 96–98. *In* T.B. Batygina and M.S. Yakovlev (eds.). Comparative Embryology of Flowering Plants. Phytolaccaceae—Thymelaeaceae. Nauka, Leningrad, Russia.

*Solntseva, M.P. 1971. On the behaviour of sexual cell nuclei and the formation of mosaic embryo in semigamy in *Rudbeckia laciniata* L. Reports USSR Acad. Sciences, 197 (1): 234–237.

Solntseva, M.P. 1983. Family Polygonaceae, pp. 53–58. *In* T.B. Batygina and M.S. Yakovlev (eds.). Comparative Embryology of Flowering Plants. Phytolaccaceae—Thymelaeaceae. Nauka, Leningrad, Russia.

Solntseva, M.P. 1985. Family Melastomataceae, pp. 94–96. *In* T.B. Batygina and M.S. Yakovlev (eds.). Comparative Embryology of Flowering Plants. Brunelliaceae—Tremandraceae. Nauka, Leningrad, Russia.

Solntzeva, M.P. 1987. Family Asteraceae, pp. 318–331. *In* T.B. Batygina and M.S. Yakovlev (eds.). Comparative Embryology of Flowering Plants. Davidiaceae—Asteraceae. Nauka, Leningrad, Russia.

Soodan A.S. and B.A Wafai. 1987. Spontaneous occurrence of cytomixis during microsporogenesis in almond *Prunus amygdalus* Batsch. and peach *P. persica* Batsch. Cytologia, 52 (2): 361–364.

Southworth, D. 1990a. Exine biochemistry, pp. 193–212. *In* S. Blackmore and R.B. Knox (eds.). Microspores: Evolution and Ontogeny. Acad. Press, London.

Southworth, D. 1990b. Approaches to the study of flowering plant sperm cells. Amer. J. Bot., 77 (6): 7–8.

Southworth, D., M.B. Singh, T. Hough, I.J. Smart, P. Taylor and R.B. Knox. 1988. Antibodies to pollen exines. Planta, 176: 482–487.

Souvre, A., L. Albertini and J.C. Aurdan. 1987. Le grain du pollen des angiosperms. Bull. Soc. botanique de France, 137: 87–112.

Sreedevi, P. and A.N. Namboodiri. 1982. The germination of pollinium and the organization of germ furrow in some members of Asclepiadaceae. Can. J. Bot., 60 (2): 166–172.

*Starshova, N.P. and M.P. Solntseva. 1973. On the characteristics of embryo sac in *Philodendron amurense* Rupr. Bot. Zhurnal, 58 (11): 1644–1654.

Stebbins, G.L. 1974. Flowering Plants. Evolution above the Species Level. Belknap Press, Harvard Univ. Press, Cambridge.

***Steffen, K. 1951.** Zur Kenntnis des Befruchtungsvorganges bei *Impatiens glandulifera* Lindl. Planta, 39(3): 175–244.

Steffen, K. 1955. Kern- und Nucleolenwachstum bei endomitotischer Polyploidisierung. Ein Beitrag zur karyologischen Anatomie von *Pedicularis palustris* L.Planta, 45: 379–394.

Steffensen, D.M. 1966. Synthesis of ribosomal RNA during growth and division in *Lilium*. Experim. Cell Research, 44 (1): 1–12.

Stenar, H. 1935. Embryologische Beobachungen über *Scheuchzeria palustris* L. Bot. Notiser, 1: 78–86.

Stenar, H. 1937. Zur Embryosackentwicklung einiger Malpighiaceen. Bot. Notiser, 110–118.

Stephens, E.L. 1909. The embryo sac and embryo of certain Penaeaceae. Ann. Botany, 23 (91): 363–378.

Sterling, C. 1963. The affinities of *Prinsepa* (Rosaceae). Amer. J. Bot., 50 (7): 693–699.

Stevens Valda, A.M. and B.G. Murray. 1981. Studies on heteromorphic selfincompatibility systems: the cytochemistry and ultrastructure of the tapetum of *Primula obconica*. J. Cell Science, 50: 419–431.

Stieglitz, H. 1977. Role of beta-1,3 glucan in postmeiotic microspore release. Developmental Biology, 57: 87–97.

Stinson, J.R. A.J. Eisenberg, R.P. Willing, M.E. Pe, D.D. Hanson and J.P. Mascarenhas. 1987. Genes expressed in the male gametophyte of flowering plants and their isolation. Plant Physiology, 83: 442–447.

Stopes, M. 1905. On the double nature of the cycadean integument. Ann. Botany, 19: 561–566.

Stosser, R. 1986. Über die Anziehung der Pollenschlauche durch die Samenanlagen bei Fruchten. Angewandte Botanik, 60 (5–6): 421–426.

Stow, I. 1930. Experimental studies on the formation of the embryo sac-like pollen in the anthers of *Hyacinthus orientalis*. Cytologia, 1: 417–439.

Stow, I. 1934. On the female tendencies of the embryo sac-like giant pollen grains of *Hyacinthus orientalis*. Cytologia, 5: 88–108.

Stozharova, I.A. 1973. Role of synergids in the fertilization process of *Allium nutans* L., pp. 221–222. *In* N.V. Zizin (ed.) Sexual Process and Embryogenesis in Plants. Moscow, Russia.

***Stozharova, I.A. and V.A. Poddubnaja-Arnoldi. 1977.** Role of synergids in the fertilization, embryo- and endospermogenesis of *Allium nutans* L. Bull. Main Botanical Garden USSR Acad. Sciences, 105: 70–78.

Strasburger, E. 1872. Die Coniferen und die Gnetaceen. Fischer, Jena.

Strasburger, E. 1877. Über Befruchtung und Zellteilung. Fischer, Jena.

Strasburger, E. 1879a. Die Angiospermen und die Gymnospermen. Fischer, Jena.

Strasburger, E. 1879b. Neue Beobachtungen über Zellbildung und Zellteilung. Bot. Zeitung, 37: 265–279, 28–288.

Strasburger, E. 1884. Neue Untersuchungen über den Befruchtungsvorgang bei den Phanerogamen als Grundlage für eine Theorie der Zeugung. Fischer, Jena.

Strasburger, E. 1900. Einige Bemerkungen zur Frage nach der doppelten Befruchtung bei Angiospermen. Bot. Zeitung, 58,Abt. 2, (19–20): 293–316.

Strokov, A.A. 1977. The investigation of histones in microsporogenesis of lily. Ph.D. thesis, Moscow, Russia.

Subramanyam, K. 1960. Nutritional mechanism of the seed, 1. Nutritional mechanism of the embryo sac. Madras University, 30B (1): 29–44.

Subramanyam, K. and H.C. Govindu. 1948. The development of the female gametophyte in *Oenothera*. J. Indian Botanical Society, 27 (1): 25–30.

Sumner, M.J. 1989. Development of the megagametophyte central cell of *Brassica campestris* (Pap.). Amer. J. Bot., 76 (6, Suppl.): 60.

Sundara Rajan, S. 1976. Is Polygonum type embryo sac really primitive? Acta Botanica Indica, 4: 89–92.

Swamy, B.G.L. 1943. Embryology of Orchidaceae. Current Science. (India), 12: 13–17.

Swamy, B.G.L. 1948a. Agamospermy in *Spiranthes cernua*. Lloydia, 11 (3): 149–162.

Swamy, B.G.L. 1948b. A contribution to the embryology of the Marcgraviaceae. Amer. J. Bot., 35 (9): 628–633.

Swamy, B.G.L. 1964. Macrogametophytic ontogeny in *Schisandra chinensis*. J. Indian Botanical Society, 43 (3): 391–396.

Swamy, B.G.L. 1973. Contribution to the monograph of *Gnetum*. 1. Fertilization and proembryo. Phytomorphology, 23: 48–53.

Swamy, B.G.L. 1974. On the presumed ancestry of the embryo sac. Phytomorphology, 24: 102–106.

Swamy, B.G.L. and K. Periasamy. 1955. Contribution to the embryology of *Acrotrema arnottianum*. Phytomorphology, 5 (2 3): 301–314.

Swift, H. 1950. The constancy of deoxyribose nucleic acid in plant nuclei. Proc. Natl. Acad. Sciences, USA, 36 (5): 643–654.

Szujkó-Lacza, J. 1978. Hypostase, embryonic sac and endosperm in *Anethum graveolens* L., and in various families. Acta Biologica Academiae Scientiarum Hungaricae, 29 (3): 255–271.

Tackholm, G. and E. Soderberg. 1918. Neue Beispiele der simultanen und sukzessiven Wandbildung in der Pollenmutterzellen. Svensk Botanisk Tidskrift Utgifven af Svenska Botaniska Föreningen, 12 (2): 189–201.

***Taimla, E.A., I.P. Ermakov, M.I. Tukeeva and L.V. Karpova. 1984.** Electrophoresic analysis of the male gametophyte nucleus histones in *Tradescantia paludosa* Anders et Woodson. Vestnik Moskovskogo Gosudarstvennogo Universiteta, Series 16, Biology, 3: 47–53.

Takahashi, M. and N. Sohma. 1979. Pollen wall formation and tapetutn in *Disporum smilacinum* A. Iray (Liliaceae). Science Reports Tôhoku Imperial Univ., Ser.4 (Biology), 4: 37.

Takats, S.T. 1962. An attempt to detect utilization of DNA breakdown products from the tapetum for DNA synthesis in microspores of *Lilium longiflorum*. Amer. J. Bot., 49 (7): 748–758.

Takhtajan, A.L. 1948. Morphological Evolution of Angiosperms. Moscow Society of Nature Experience, Moscow, Russia.

Takhtajan, A.L. 1964. Foundation of Evolutionary Morphology of Angiosperms. Nauka, Moscow-Leningrad, Russia.

Takhtajan, A.L. 1980a. Outline of the classification of flowering plants (Magnoliophyta). Bot. Review, 46 (3): 225–359.

Takhtajan, A.L. 1980b. The ovule, pp. 49–53. *In* A.A. Fedorov (ed.). The Life of Plants, 5(1) Prosveshchenie, Moscow, Russia.

Takhtajan, A. L. 1980c. Stamens, pp. 28–30. *In* A.A. Fedorov (ed.). The Life of Plants, 5(1) Prosveshchenie, Moscow, Russia.

Tarasenko, LV. 1990. Ultrastructural analysis of the generative and sperm cells in the pollen of *Matthiola bicornis* L., p. 168. *In* T.B. Batygina (ed.). Abstracts XI Intern. Symp. 'Embryology and Seed Reproduction', 3–7 July, 1990, Leningrad, USSR. Nauka, St.-Petersburg, Russia.

***Tarasenko, L.V., V.P. Bannikova and K.M. Sytnik. 1988.** Interconnection of cells in the pollen grains of *Cecale* and *Mattiola*. Reports Ukrainian SSR Acad. Sciences, "B", 3: 81–84.

Tarkowska, J. 1973. The nature of cytomixis. Caryologia, 25 (Suppl.): 151–157.

Tarkowska, J., M. Wazynska and A. Jablonska. 1987. Translocation on nuclei and reorientation of mitotic apparatus in the ontogeny of stomata in *Tradescantia virginiana* L. Acta Societatis Botanicorum Poloniae, 56 (4): 599–610.

Tatintseva, S.S. 1973. Spermiogenesis of some angiosperms, pp. 228–229. *In* Proc. All-Union Symposium 'Sexual Process and Embryogenesis of Plants'. Moscow, Russia.

***Tatintseva, S.S. 1975.** The formation and development of sperms in some angiosperms. I. Poaceae (Gramineae). II. Asteraceae (Compositae). Bot. Zhurnal, 60 (12): 1707–1720.

*Tatintseva, S.S. 1988. The formation and development of sperms in some angiosperms. III. Boraginaceae. IV. Najadaceae. Reports Turkmenian SSR Acad. Sciences, Series Biological Sciences, 1: 3–9.

Taylor, J.H. 1953. Autoradiographic detection of incorporation of P^{32} into chromosomes during meiosis and mitosis. Experim. Cell Research, 4: 164–173.

Taylor, J.H. 1958. Incorporation of phosphorus-32 into nucleic acids and proteins during microgametogenesis of *Tulbaghia*. Amer. J. Bot., 45 (2): 123–131.

Taylor, J.H. 1959. Autoradiographic studies of nucleic acids and proteins during meiosis in *Lilium longiflorum*. Amer. J. Bot., 46 (7): 477–484.

Telepova, M.N. 1985. The ultrastructure of the phloem cells in the root of *Hordeum vulgare* L. Ph.D. thesis, Komarov Bot. Inst., Leningrad, Russia.

***Teryokhin, E.S. 1963.** The development of ovule and female gametophyte in Pyroleae and Monotropeae. Bot. Zhurnal, 48 (3): 406–414.

Teryokhin, E.S. 1977. Parasitic Flowering Plants. The Evolution of Ontogenesis and the Mode of Life. Nauka, Leningradskoe otdelenie, Leningrad, Russia.

***Teryokhin, E.S. 1984.** Some evolutionary features of female seed plants, p. 58. *In* Materials of All-Union Symp. 'Morphofunctional Aspects of Development of Female Generative Structures in Seed Plants'. Telavi, Georgia.

Teryokhin E.S. 1985. Family Droseraceae, pp. 48–51. *In* T.B. Batygina and M.S. Yakovlev (eds.). Comparative Embryology of Flowering Plants. Brunelliaceae—Tremandraceae. Nauka, Leningrad, Russia.

Teryokhin E.S. 1987a. Family Balanophoraceae, pp. 89–93. *In* T.B. Batygina and M.S. Yakovlev (eds.). Comparative Embryology of Flowering Plants. Davidiaceae—Asteraceae. Nauka, Leningrad, Russia.

Teryokhin E.S. 1987b. Family Orobanchaceae, pp. 268–272. *In* T.B. Batygina and M.S. Yakovlev (eds.). Comparative Embryology of Flowering Plants. Davidiaceae—Asteraceae. Nauka, Leningrad, Russia.

Teryokhin E.S. 1987c. Family Viscaceae, pp. 82–86. *In* T.B. Batygina and M.S. Yakovlev (eds.). Comparative Embryology of Flowering Plants. Davidiaceae—Asteraceae. Nauka, Leningrad, Russia.

Teryokhin E.S. 1990a. Family Najadaceae, pp. 54–56. *In* T.B. Batygina and M.S. Yakovlev (eds.). Comparative Embryology of Flowering Plants. Butomaceae—Lemnaceae. Nauka, Leningrad, Russia.

Teryokhin E.S. 1990b. Family Zosteraceae, pp. 52–54. *In* T.B. Batygina and M.S. Yakovlev (eds.). Comparative Embryology of Flowering Plants. Butomaceae—Lemnaceae. Nauka, Leningrad, Russia.

***Teryokhin, E.S. 1991.** The problems of seed plant ontogenesis. Trans. Komarov Bot. Inst. USSR Acad. Sciences. Morphology and Anatomy of Plants. 2: 68.

***Teryokhin, E.S. and M.S. Yakovlev. 1967.** Embryology of Balanophoraceae. Botanicheskyi Zhurnal, 52 (6): 745–758.

***Teryokhin, E.S. and O.P. Kamelina. 1969.** Endosperm of Orchidaceae (question of reduction). Bot. Zhurnal, 54 (5): 657–669.

***Teryokhin, E.S. and O.P. Kamelina. 1972.** Endosperm of *Habernaria linearifolia* Maxim. and *Neottia nidus-avis* (L.) Rich. (question of evolution of endosperm in orchids). Bot. Zhurnal, 57 (11): 1373–1383.

Teryokhin E.S. and G.M. Anisimova. 1981. Family Rafflesiaceae, pp. 96–100. *In* T.B. Batygina and M.S. Yakovlev (eds.). Comparative Embryology of Flowering Plants. Winteraceae—Juglandaceae. Nauka, Leningrad, Russia.

Teryokhin E.S. and G.V. Shibakina. 1985. Family Zosteraceae, pp. 59–62. *In* A.L. Takhtajan (ed.). Comparative Anatomy of Seeds. Monocotyledons. Nauka, Leningrad, Russia, vol. 1.

***Teryokhin, E.S., Z.I. Nikiticheva and O.P. Kamelina. 1971.** Movement of nuclei in endosperm, embryo and in megasporogenesis of some flowering plants, pp. 176–178. *In* Proc. V All-Union Conf. Plant Embryology, Kishinev, Moldova.

*Teryokhin, E.S., M.S. Yakovlev and Z.I. Nikiticheva. 1975. The development of microsporangium, pollen grains, ovule and embryo sac in *Cynomorium songaricum* Rupr. (Cynomoriaceae). Bot. Zhurnal, 60 (2): 153–162.

*Teryokhin, E.S., T.B. Batygina and I.I. Shamrov. 1993. The classification of microsporangium type wall in angiosperms. Terminology and Conceptions. Bot. Zhurnal, 78 (6): 16–24.

Theunis, C.H., E.S. Piterson and M. Cresti. 1991. Isolation of male and female gametes in higher plants. Sexual Plant Reproduction, 4 (3): 145–154.

Thomson, W.W. and J.M. Watley. 1980. Development of nongreen plastids. Ann. Rev. Plant Physiology, 31: 375–394.

Tiagi, B. 1965. Development of the seed and fruit of *Melampyrum nemorosum* L. and *M. arvense* L. Can. J. Bot., 43 (12): 1511–1521.

Tiezzi, A., A. Moscatelli, F. Ciampolini, C. Milanesi, M. Murgia and M. Cresti. 1988. The cytoskeletal apparatus of the generative cell in several angiosperm species, pp. 215–220. *In* M. Cresti, P. Cori and E. Pacini (eds.). Sexual Reproduction in Higher Plants. Springer, Heidelberg, Berlin.

Tilton, V.R. 1980. Hypostase development in *Ornithogalum caudatum* (Liliaceae) and notes on other types of modifications in the chalaza of angiosperms ovules. Can. J. Bot., 58 (19): 2059–2066.

Tilton, V.R. 1981. Ovule development in *Ornithogalum caudatum* (Liliaceae) with a review of selected papers on angiosperm reproduction. IV. Egg apparatus structure and function. New Phytologist, 88 (3): 505–531.

Tilton, V.R. and H.L. Mogensen. 1979. Ultrastructural aspects of the ovule of *Agave parryi* before fertilization. Phytomorphology, 29 (3–4): 338–350.

Tilton, V.R. and H.T. Horner. 1980. Stygma, style and obturator of *Ornithogalum caudatum* (Liliaceae) and their function in reproductive process. Amer. J. Bot., 67 (7): 1113–1131.

Tilton, V.R. and N.R. Lersten 1981a. Ovule development in *Ornithogalum caudatum* (Liliaceae) with a review of the selected papers on angiosperms reproduction. I. Integuments, funiculus and vascular tissue. New Phytologist, 88 (3): 439–457.

Tilton, V.R. and N.R. Lersten 1981b. Ovule development in *Ornithogalum caudatum* (Liliaceae) with a review of selected papers on angiosperm reproduction. III. Nucellus and megagametophyte. New Phytologist, 88 (3): 477–504.

Tilton, V.R., L.W. Wilcox, R.G. Palmer and M.C. Albertsen. 1984. Stigma, style and obturator of soybean, *Glycine max* (L.) Merr. (Leguminoseae) and their function in the reproductive process. Amer. J. Bot., 71 (4): 676–686.

Tischler, G. 1915. Die Periplasmodienbildung in den Antheren der Comubrigen Monocotylen. Jahrb. für Wissen Schaftliche Botanik, 55: 52–90.

Titova, G.E. 1988. The embryology of some water plants (*Nelumbo nucifera* Gaertn., *Cabomba caroliniana* A. Gray, *Trapa astrachanica* (Fler.) Winter). Ph.D. thesis, Komarov Bot. Inst., Leningrad, Russia.

Titova, G.E. 1990. The development of the female generative structures in *Cabomba caroliniana* A. Gray (Cabombaceae), p. 210. *In* T.B. Batygina (ed.). Abstracts XI Intern. Symp. 'Embryology and Seed Reproduction', 3–7 July, 1990, Leningrad, USSR. Nauka, St.-Petersburg, Russia.

Tiwari, S.C. 1988. The cytoskeleton during pollen development in *Tradescantia verbiniana*: A study employing chemical fixation, freeze substitution, immunofluorescence and colchicine administration. Can. J. Bot., 67: 1244–1253.

Tiwari, S.C. and B.E.S. Gunning. 1986a. Development of tapetum and microspores in *Canna* L.: an example of an invasive but non-syncytial tapetum. Ann. Botany (USA), 57 (4): 557–563.

Tiwari, S.C. and B.E.S. Gunning. 1986b. Development and cell surface of a non-syncytial invasive tapetum in *Canna*: ultrasrtuctural, freeze substitution, cytochemical and immunofluorescence study. Protoplasma, 134: 1–16.

Tobe, H. and P.H. Raven. 1984. The embryology and relationships of Penaeaceae (Myrtales). Plant Systematics and Evolution, 146 (3–4): 181–195.

Traas, J.A., S. Burgain and R. Dumas de Vaulx. 1989. The organization of the cytoskeleton during meiosis in eggplant (*Solanum melongena* L.) microtubules and F-actine are both necessary for coordinated meiotic division. J. Cell Science, 92: 541–550.

Treub, M. and J. Mellink. 1880. Notice sur le dévelopement du sac embryonnaire dans quelques angiosperms. Arch. Neerl. Sci. 15: 452–457.

Treviranus, G.R. 1805. Biologie oder Phylosophie der lebenden Natur. Johann Friedrich Rower, Gottingen, vol. 3.

Tschermak-Woess, E. 1956. Notizen über die Riesenkemne und "Riesenchromosomen" in den Antipoden von *Aconitum*. Chromosoma, 8 (2): 113–134.

Tuchina, N.G. 1985. The development of the male and female embryonic structures in *Sambucus coreana* (Nakai) Kom. et Alis. and *Sambucus racemosa* L., pp. 96–109. *In* Trans. 7 Young Scientists Conf. Komarov Bot. Inst. USSR Acad. Sciences, Leningrad, Russia.

Turala, N. 1966. Strukturen endopolyploider nerne in Bereich der Samenanlage einiger Monocotylen. Österreichische Botanische Zeitschrift, 113: 529–541.

Turpin, P. 1806. Memoire sur l'organe par lequel le fluide fécondant peut s'introduire dans l'ovule des végétaux. Ann. Muséum Nationale d'Histore Naturelle, 7: 199–211.

Twell, D.R. Wing, J. Yamaguchi and S. McCormick. 1989. Isolation and expression of an anther-specific gene from tomato. Molecular and General Genetics, 247: 240–245.

Tzvelev, N.N. 1981. The family Asclepiadaceae, pp. 362–365. *In* A.A. Fedorov (ed.). The Life of Plants, 5 (2), Prosveshchenie, Moscow, Russia.

Uhl, N.W. and H.E. Moore. 1971. The palm gynoecium. Amer. J. Bot., 58 (10): 945–992.

***Urmantsev, Yu.A. 1979.** A system approach to the problem of plant stability (example of investigation of the dependence of pigment content in haricot bean leaves on the simultaneous effect on it of drought and salting). Physiology Rastenyi, 26 (4): 762–777.

***Ustinova, E.I. 1944.** Comparative embryological investigation of normal and viviparous species of onion. Bot. Zhurnal, 29 (5): 232–239.

Van Aelst, A.C. and J.L. van Went. 1991. The ultrastructure of mature *Papaver dubium* L. Pollen grains. Acta Botanica Neerlandica, 40 (4): 319–328.

Van der Pijl, L. 1955. Sarcotesta, aril, pulpa and the evolution of the Angiosperm fruit. Proc. Koniklijke Nederlandse Akad. van Wetenschappen. Series C 58 (2–3): 154–161.

Van der Pijl, L. 1957. On the arilloides of *Nephelium, Euphorbia, Litchi* and *Aesculus* and the seed of Sapindaceae. Generale Acta Botanica Neerlandica, 6 (5): 618–641.

Van der Pijl, L. 1969. Principles of Dispersal in Higher Plants. Springer-Verlag, Berlin.

Van der Pluijm, J.E. 1964. An electron microscopic investigation of the filiform apparatus in the embryo sac of *Torenia fourniori*, pp. 8–16. *In* H.F. Linskens (ed.). Pollen Physiology and Fertilization. North-Holland Publ. Co., Amsterdam, The Netherlands.

Van Lammeren, A.M., J. Bednara and M.T.M. Willemse. 1989. Organization of the active cytoskeleton during pollen development in *Gasteria verrucosa* (Mill.) H. Duval visualized with rhodamin-phalloidin. Planta, 178 (4): 531–539.

Van Lammeren, A.M., C.J. Keijer, M.T.M. Willemse and H. Kieft. 1985. Structure and function of the microtubular cytoskeleton during pollen development in *Gasteria verrucosa* (Mill.) H. Duval. Planta, 185 (1): 1-11.

Van Miegroet, Fr. and M. Dujardin, 1992. Cytologie et histologie de la reproduction chez le *Nymphaea heudelotii*. Can. J. Bot., 70 (10): 1991–1996.

Vannereau, A. 1978. Etude ultrastructurale du complete micropylaire du sac embryonnaire adulte de *Plantago lanceolata* L. Bull. Soc. botanique de France, 125 (1–2): 201–205.

Van Tieghem, Ph. 1898. Structure des quelques ovules et parti qu'on peut pour ameliorer la classification. J. Bot. France, 12: 197–220.

Van Tieghem, Ph. 1901a. Epideepharide genre nouveau Luxembourgiacées. J. Bot. France, 15, (12): 389.

Van Tieghem, Ph. 1901b. L'hypostase, sa structure et son rôle constans, sa position et sa forme variables. Bull. Mus. Natn. Hist. Nat., 7 (8): 412–418.

Van Tieghem, Ph. 1903. Sur l'hypostase. Ann. Sci. Nat. Botanique. Ser. 8 (17): 347–362.

Van Tunen, A.J., L.A. Mur, O.S. Brouns, J.-D. Rienstra, R.E. Noes and J.N.M. Mol. 1990. Pollen- and anther-specific promoters from *Petunia:* tandem promoter regulation of the *chiA* gene. Plant Cell, 2: 393–401.

Van Went, J.L. 1970a. The ultrastructure of the egg and central cell of *Petunia.* Acta Botanica Neerlandica, 19 (3): 313–322.

Van Went, J.L. 1970b. The ultrasturcture of the synergids of *Petunia.* Acta Botanica Neerlandica, 19 (2): 121–132.

Van Went, J.L. 1989. Integumentary tapetum in *Impatiens walleriana* ovules, pp. 271–279. *In* J. Pare and M. Bugnicourt (eds.). 1989. Some Aspects and Actual Orientations in Plant Embryology. Dedicatory volume to Prof. A. Lebègue. Cedex, Amiens, France.

Van Went, J.L. and H.F. Linskens. 1967. Die Entwicklung des sogenannten "Fadenapparates" in Embryosak von *Petunia hybrida.* Gen. Breed. Res., 37 (1): 51–56.

Van Went, J.L. and M.T.M. Willemse. 1984. Fertilization, pp. 273–317. *In* B.M. Johri (ed.). Embryology of Angiosperms. Springer-Verlag, Berlin.

Van Went, J.L. and M. Cresti. 1988a. Pre-fertilization degeneration of both synergids in *Brassica campestris* ovules. Sex. Plant Reprod., 1: 208–216.

Van Went, J.L. and M. Cresti. 1988b. Cytokinesis in microspore mother cells of *Impatiens sultani.* Sex. Plant Reprod., 1: 228–233.

Vasilevskaya, V.K. and E.A. Kondratyeva-Melvyl. 1958. Some questions about vegetative shoot apex structure, pp. 288–298. *In* V.N. Sukachev (ed.). Problems of Botany. Moscow-Leningrad, Russia.

***Vassilyeva, V.E. and T.B. Batygina. 1981.** The cultivation in vitro of lotus embryos and ovules isolated at different stages of development. Physiology rastenyi, 28 (2): 319–327.

Vassilyeva, V.E., T.B. Batygina, and G.E. Titova. 1987. Morphophysiological correlations in the development of the reproductive structures of *Nelumbo nucifera* Gaertn. Phytomorphology, 37(4): 349-358.

***Vassilyeva, V.E., G.E. Kolesova, K.A. Chochiya and T.B. Batygina. 1984.** Morpho-physiological correlations in the development of lotus ovule, 11. *In* Materials All-Union Symp. 'Morphofunctional Aspects of Development of Female Generative Structures in Seed Plants', Telavi, Georgia.

Vassilyeva, V.E., G.E. Titova, J.P. Ermakov, N.M. Morozova and N. A. Chochiya. 1988. Time and space concordance of the growth, differentiation and metabolism in the developing seed of *Nelumbo nucifera* Gaertn, p. 487. *In* M. Cresti, P. Gori and E. Pacini (eds.). Proc. 10th Intern. Symp. Sexual Reproduction in Higher Plants. 30 May–4 June, 1988. Siena Univ., Siena, Italy. Springer-Verlag, Berlin.

Vassilyev, A.E. 1977. Functional Morphology of Plant Secretory Cells. Nauka, Leningrad, Russia.

***Vassilyev, A.E. and M.A. Plisko. 1971a.** Electron microscopy investigation of integumentary tapetum and the adjacent ovule tissues in *Calendula officinalis* L. Bot. Zhurnal, 56 (3): 335–346.

***Vassilyev, A.E. and M.A. Plisko. 1971b.** Ultrastructure of embryo sac in angiosperms, pp. 29–30. *In* Proc. V All-Union Conf. Plant Embryology. Kishinev, Moldova.

Vassilyev, A.E. and P.A. Lobkov. 1990. Filiform apparatus of synergids: structure and functions, p. 129. *In* T.B. Batygina (ed.). Abstracts XI Inter. Symp. 'Embryology and Seed Reproduction'. 3–7 July, 1990, Leningrad, USSR. Nauka, St.-Petersburg, Russia.

Vaughn, K.C. 1981. Organelle transmission in higher plants: organelle alteration vs. physical exclusion. J. Heredity, 72 (5): 335–337.

Vazart, B. 1958. Befruchtung und Kernversehmelzung. Protoplasmatologia, 7: 158.

Vazart, B. 1971. Premiere division haploide et formation de la cellule generatrice dans le pollen de tabac. Ann. Scientifiques l'Univ. Reims et l'ARERS, 9 (1): 179–187.

Vazart, B. and J. Vazart. 1965. Infrastructure de l'ovule du Lin, *Linum usitatissimum* L. I. L'assise jaquette ou endothelium. C. R. l'Acad. Sciences, Paris, 261 (11): 2927–2930.

Vazart, B. and J. Vazart. 1966. Infrastructure du sac embryonnaire du Lin (*Linum usitatissimum* L.). Rev. Cytologie et Biologie Végétales, 29 (3–4): 251–266.

Vazart, J. 1968. Ultrastructure de l'ovule du Lin, *Linum usitatissimum* L. Le complexe antipodial. C. R. l'Acad. Sciences, Paris. 266 (3): 211–213.

Vazart, J. 1969. Organization et ultrastructure du sac embryonnaire du Lin (*Linum usitatissimum* L.). Rev. Cytologie et Biologie Végétales, 3 2(3–4): 227–240.

Vazart, J. 1971. Degenerescence d'une synergide et penetration du tube pollinique dans le sac embryonnaire de *Linum usitatissimum* L. Ann. Scientifiques l'Univ. Reims et l'ARERS, 9 (1): 89–97.

Velenovsky, I. 1913. Vergleichende Morphologie der Pflanzen. Verlagsbuchhandlung von Fr. Řivnač Praga, 4.

Venkata Rao, C. 1953. Contributions to the embryology of Sterculiaceae. V. J. Indian Bot. Soc., 32 (4): 208–238.

Venkateswarlu, Prakasa Rao S. 1972. Embryological studies in some Combretaceae. Bot. Notiser, 125 (2): 161–179.

Ventura, M. 1938. Observazioni embriologiche su *Arechavaletaria uruguayensis* Speg. Ann. di Botanica (Roma), 21 (3): 527–533.

Venturelli, M. and F. Bouman. 1986. Embryology and seed development in *Mayaca fluviatilis* (Mayacaceae). Acta Botanica Neerlandica, 35 (4): 497–516.

*__Veselova, T.D. 1989.__ Embryological investigation of *Lychnis coronaria* (L.) Desv. (Caryophyllaceae). VINITI, 2003—B 89: 1–16.

*__Veselova, T.D. 1991.__ Morphological nature of obturator in Caryophyllaceae. Biological Sciences, 2: 93–103.

Vidajati, O. 1973. Cytoembryological studies on certain members of *Arctotideae* and *Calenduleae* (*Asteraceae*). Ph. D. thesis. Moscow, Russia.

Vijayalakshmi, A. and K.K. Lakshmanan. 1988. Histochemistry of meiocytes, microspores, pollen and tapetum in *Haworthia fasciata* Haw. Natl. Acad. Sciences, Letters, 11 (1).

Vijayaraghavan, M.R. and S. Ratnaparkhi. 1973. Dual origin and dimorphism of the anther tapetum in *Alectra thomsoni* Hook. Ann. Botany, 37: 355–359.

Vijayaraghavan, M.R. and S. Ratnaparkhi. 1977. Pollen-embryo sacs in *Heuchera micrantha* Dougl. Caryologia, 30 (1): 105–119.

Vijayaraghavan, M.R. and A.K Shukla. 1977. Absence of callose around the microspore tetrad and poorly developed exine in *Pergullaria daemia*. Ann. Botany, 41 (175): 923–926.

Vijayaraghavan, M.R. and Neeru Seth. 1991. Histochemical and ultrastructural studies on microspore mother cells during meiotic prophase in *Sesbania speciosa*. Phytomorphology, 41: 231–239.

Vijayaraghavan, M.R., W.A. Jensen and M. E. Ashton. 1972. Synergids of *Aquilegia formosa*—their histochemistry and ultrastructure. Phytomorphology, 22 (2): 144–159.

Vijayaraghavan, M.R., Misra Gayatri and Sujata Vefantam. 1988a. Wandlabyrinthe in the embryo sac of *Nigella damascena* Linn. Proc. Indian Acad. Science. Plant Sciences, 98 (4): 261–268.

Vijayaraghavan, M.R., Neeru Seth and Alka Jain. 1988b. Ultrastructure and histochemistry of angiosperm embryo sac—An overview. Proc. Indian Acad. Science. Section B, Biological Sciences (Bangalore, India), 54 (1): 93–110.

Vincent, P.L.D. and F.M. Getliffe. 1988. The endothecium in *Senecio* (Asteraceae). Bot. J. Linnean Society, 97: 63–71.

Vithanage, H.J.M.V. and R.B. Knox. 1980. Periodicity of pollen development and quantitative cytochemistry of exine and intine enzymes in the grasses *Lolium perenne* and *Phalaris tuberosa*. Ann. Botany, 45 (2): 131–142.

*__Vlasova, N.A. and L.E. Serebryakova. 1963.__ To the question about localization of nucleic acids concurrently with rates of development of embryo sac in cotton. Uzbekskyi Biologicheskyi Zhurnal 17 (1): 27–33.

Von Hanstein, J. 1868. Die Scheitelzellgruppe im Vegetationspunkt der Phanerogamen, pp. 109–134. *In* C. Georgi (ed.). Festschrifte der Niederrheinische Gesellschaft für Natur und Heilkunde zum 50 Jahre, Bonn.

Van Ubisch, M. 1927. Zur Entwicklungsgeschichte der Antheren. Planta, 3: 490–495.

Vorsobina, L.I. and M.P. Solntseva. 1990. Family Amaryllidaceae, pp. 87–96. *In* T.B. Batygina and M.S. Yakovlev (eds.). Comparative Embryology of Flowering Plants. Butomaceae—Lemnaceae. Nauka, Leningrad, Russia.

Wagner, V.T. and H.L. Mogensen. 1988. The male germ unit in the pollen and pollen tubes of *Petunia hybrida*: ultrastructural, quantitative and three-dimensional features. Protoplasma, 143: 101–110.

Walker, R.J. 1947. Megasporogenesis and embryo development in *Tropaeolum majus* L. Bull. Torrey Botanical Club, 74: 240–249.

Walker, R.J. 1950. Megasporogenesis and development of megagametophyte in *Ulmus*. Amer. J. Bot. 37 (1): 47–52.

Walles, B. 1992. Ultrastructure of the ovarian transmitting tissue in *Lilium regale,* pp. 607–608. *In* T.B. Batygina (ed.). Proc. XI Internl. Symp. 'Embryology and Seed Reproduction'. 3–7 July, 1990, Leningrad, USSR. Nauka, St-Petersburg, Russia.

Walles, R.I. and J.R. Rowley. 1979. Tapetal cells during pollen development in *Pinus sylvestris*. J. Ultrastructure Research, 1966, 1: 95.

Walton, J. 1952. L'évolution des téguments et de la protection du sporange. L'Année Biologique, Série 3, 28: 129–133.

Warming, E. 1873. Untersuchungen über pollenbildende Phyllome und Kaulome. Botanishe Abhandlungen Gebiet. Morphologie und Physiologie. 2 (2): 1–90.

Warming, E. 1874. Bemerkungen über das Eichen. Bot. Zeitung, 32: 465–470.

Warming, E. 1878. De l'ovule. Ann. Sciences Naturelles. Botanique. Ser. 6. 5: 175–266.

Warming, E. 1913. Observations sur la valeur systematique de l'ovule. Mindeskrift Yapetus Steenstrup, Kjbenavn, 24: 1–45.

Waterkeyn, L. 1962. Les parois microsporocytaires de nature callosique chez *Helleborus* et *Tradescantia.* Cellule, 62 (2): 223–225.

Waterkeyn, L. 1981. Cytochemical localization and function of the 3–linked callose in the developing cotton floor cell wall. Protoplasma, vol. 106.

Waterkeyn, L. and A. Bietait. 1969. On a possible function of the callosic special wall in *Jpomea purpurea* (L.) Roth. Grana, 10: 13–20.

Webb, M.C. and B.E.S. Gunning. 1990a. Embryo sac development in *Arabidopsis thaliana*. I. Megasporogenesis including the microtubular cytoskeleton. Sexual Plant Reproduction, 3: 244–256.

Weber, E. 1929. Entwicklungsgeschichte Untersuchungen über die Gattung *Allium*. Botanisches Archiv Königsberg (Kaliningrad, Russian SFSR) & Dahlem bei Berlin, 25: 1–44.

Wehling, N., Ch. Niester, J.J. Boon, M.T.M. Willemse and R. Wiermann. 1989. p-Coumaric acid—a monomer in the sporopollenin skeleton. Planta, 179: 376–380.

Weiling, F. 1965. Light- und elektronmicroskopische Beobachtungen zum Problem der Cytomixis sowie ihrer moglichen Beziehung zur Potocytose Untersuchungen bei *Cucurbita*-Arten und *Lycopersicum esculentum.* Planta, 67 (2): 182.

Weinstein, A.J. 1926. Cytological studies on *Phaseolus vulgaris*. Amer. J. Bot., 13: 248–263.

Went, F.A.F.C. 1908. The development of the ovule, embryo sac and egg in Podostemaceae. Recueil des Travaux Botaniques Néerlandais, 5: 1–16.

Went, F.A.F.C. 1910. Intersuchungen über Podostemaceen. *In* Verh. Kon. Akad. Wetensch. J. Muller, Amsterdam.

Westermaier, M. 1890. Zur Embryologie der Phanerogamen insbesondere über die sogenannten Antipoden. Nova Acta Leopold. 57 (1): 1–30.

Weterings, K., W. Rejnen, R. van Aarssen, A. Kortstee, J. Spijkers, M. van Herpen, J. Schrauwen and G. Wullems. 1992. Characterization of a pollen-specific cDNA from *Nicotiana tabacum* expressed during microgametogenesis and germination. Plant Molecular Biology, 18: 1101–1111.

Wettstein, R.R. 1911. Handbuch der Systematischen Botanik. Deuticke, Leipzig, Wien, vol. VIII.

Wiegand, K.M. 1900. Notes on the embryology of *Potamogeton*. Bot. Gazette, 30 (1): 25–47.

Wieliunska-Woitas, O. 1984. Embryology of some representatives of the family Onagraceae. Ph.D. thesis, Komarov Bot. Inst., Leningrad, Russia.

Wiermann, R. and S. Gubatz. 1992. Pollen wall and sporopollenin. Intern. Rev. Cytology, 140: 35–72.

Willemse, M.T.M. 1971. Morphological and quantitative changes in the population of cell organelles during microsporogenesis of *Pinus sylvestris* L.I. Morphological changes from zygotene until prometaphase I. Acta Botanica Neerlandica, 20 (3): 261–274.

Willemse, M.T.M. 1981. Polarity during megasporogenesis and megagametogenesis. Phytomorphology, 31 (1): 124–134.

Willemse, M.T.M. and M.A.W. Franssen-Verheijen. 1978. Cell organelles changes during megasporogenesis and megagametogenesis in *Gasteria venrucosa* (Mill). Haw. Bull. Soc. botanique de France. Actualities botaniques, 125: 187–191.

Willemse, M.T.M. and J. Bednara. 1979. Polarity during megasporogenesis in *Gasteria verrucosa.* Phytomorphology, 29 (1): 156–165.

Willemse, M.T.M. and S.A. Reznickova. 1980. Formation of pollen in the anther of *Lilium.* Development of the pollen wall. Acta Botanica Neerlandica, 29 (2–3): 127–140.

Willemse, M.T.M. and M.J. De Boer-De Jeu. 1981. Megasporogenesis and early megagametogenesis. Acta Societatis Botanicorum Poloniae, 50 (1–2): 111–120.

Willemse, M.T.M. and M.A.W. Franssen-Verheijen. 1983. Ovular development, pp. 177–181. *In* Proc. VII Intern. Symp. 'Fertilization and Embryogenesis in Ovulated Plants'. High Tatra (Račkova dolina), 14–17 June, 1982. Bratislava, Czechoslovakia.

Willemse, M.T.M. and J.L. van Went. 1984. The female gametophyte, pp. 159–196. *In* B.M. Johri (ed.). Embryology of Angiosperms. Springer-Verlag, Berlin.

Willemse, M.T.M. and A.A.M. van Lammeren. 1988. Structure and function of microtubular cytoskeleton during megasporogenesis and embryo sac development in *Gasteria verrucosa* (Mill.) H. Duval. Sexual Plant Reproduction, 1: 745–782.

Wilms, H.J. 1980. Development and composition of the spinach ovule. Acta Botanica Neerlandica, 29 (4): 243–260.

Wilms, H.J. 1981. Ultrastructure of the developing embryo sac of spinach. Acta Botanica Neerlandica, 30 (1–2): 75–99.

Wing R.A., J. Yamaguchi, S.K. Larabell, V.M. Yrsin and S. McCormic. 1989. Molecular and genetic characterization of two pollen-expressed genes that have sequence similarity to pectate lyases of the plant pathogen *Erwinia.* Plant Molecular Biology, 14: 17–28.

Winter, A.N. 1987. Embryology of the genus *Hydrastemma* (Barclaya) in connection with its systematics. Ph.D. thesis, Komarov Bot. Inst., Leningrad, Russia.

***Winter, A.N. and I.I. Shamrov. 1991a.** The development of ovule and embryo sac in *Nuphar lutea* (Nymphaeaceae). Bot. Zhurnal, 76 (3): 378–390.

***Winter, A.N. and I.I. Shamrov. 1991b.** Megasporogenesis and embryo sac development in members of genera *Nymphaea* and *Victoria* (Nymphaeaceae). Bot. Zhurnal, 76 (12): 1716–1728.

Wolter, M., C. Seuffert and R. Schill. 1988. The ontogeny of pollinia and elastoviscin in the anther of *Doritis pulcherrima* (Orchidaceae). Nordic J. Bot., 8 (1): 77–88.

Woodard, J.W. 1958. Intracellular amounts of nucleic acids and proteins during pollen grain growth in *Tradescantia.* J. Biophysical and Biochemical Cytology, 4 (4): 383–389.

Woodcock, C.L.F. and P.R. Bell. 1968. Features of the ultrastructure of the female gametophyte of *Myosurus minimus.* J. Ultrastructure Research, 22: 546–563.

Wooding, F.B. 1967. Endoplasmic reticulum aggregates of ordered structure. Planta, 76 (2): 205–208.

Worsdell, W.C. 1904. The structure and morphology of the ovule. An historical sketch. Ann. Bot., 18: 57–86.

Wunderlich, R. 1954. Über das Antherentapetum mit besondere Berucksichtigung seiner Kernzahl. Osterreichische Botanische Zeitschrift, 101 (1–2): 1–63.

Yakobson, L.Ya. and E.S. Teryokhin. 1983. Family Ericaceae, pp. 202–209. *In* T.B. Batygina and M.S. Yakovlev (eds.). Comparative Embryology of Flowering Plants. Phytolaccaceae—Thymelaeaceae. Nauka, Leningrad, Russia.

*Yakobson L.Ya., I.S. Litvinchik and V.T. Langenfeld. 1984. Male gametophyte development in *Viburnum* L. Bull. Main Botanical Garden USSR Acad. Sciences, 131: 105–108.

Yakovlev, M.S. 1981. The glossary of main terms, pp. 7–25. *In* T.B. Batygina and M.S. Yakovlev (eds.). Comparative Embryology of Flowering Plants. Winteraceae—Juglandaceae. Nauka, Leningrad, Russia.

Yakovlev, M.S. 1983. Family Paeoniaceae, pp. 70–77. *In* T.B. Batygina and M.S. Yakovlev (eds.). Comparative Embryology of Flowering Plants. Phytoloccaceae—Thymelaeaceae Nauka, Leningrad, Russia.

Yakovlev, M.S. and M.P. Solntseva. 1965a. Some questions about flower morphology and embryology of feather-grasses, pp. 61–73. *In* Flower Morphology and Reproductive Process in Flower Plants. Nauka (Trans. Komarov Bot. Inst. USSR Acad. Sciences). Moscow-Leningrad, Russia.

Yakovlev, M.S. and M.D. Yoffe. 1965b. Embryology of some representatives of genus *Paeonia*, pp. 140–176. *In* Flower Morphology and Reproductive Process in Flower Plants. Nauka, (Trans. Komarov Bot. Inst. USSR Acad. Sciences). Moscow-Leningrad, Russia.

Yamashita, T. 1976. Über die Pollenbildung bei *Halodule pinifolia* und *H. uninervis*. Beiträge zur Biologie der Pflanzen, 52 (1–3): 217–226.

Yamaura, A. 1933. Karyologische und embryologische Studien über einige Bambusa-Arten (vorlaufige Mitteilung). Bot. Magazine. Tokyo, 47: 551–555.

Yeung, E.C. and D.C.W. Brown. 1982. The osmotic environment of developing embryos of *Phaseolus vulgaris*. Zeitschrift für Pflanzenzüchtung, 106: 149–156.

***Yoffe, M.D. 1968.** Fertilization process in *Anemone nemorosa* L. and *A. ranunculoides* L. Bot. Zhurnal, 53 (7): 891–905.

Yoschida, O. 1962. Embryologische Studien über *Schisandra chinensis* Bailey. J. College Arts and Sciences, Chiba University. 3 (4): 459–462.

You, R. and W.A. Jensen. 1985. Ultrastructural observations of the mature megagametophyte and fertilization in wheat (*Triticum aestivum*). Can. J. Bot., 63 (2): 163–178.

Zavada, M.S. 1984. Pollen wall development of *Austrobaileya maculata*. Bot. Gazette, 145 (1): 11–21.

Zee, S.Y. and J.H.P. Siu. 1990. Studies on the ontogeny of the pollinium of a massulate orchid (*Perstylus spiranthes*). Rev. Palaeobotany and Palynology, 64 (1–4): 159–164.

Zenkteler, M. 1962. Microsporogenesis and tapetal development in normal and malesterile carrots (*Daucus carota*). Ann. Jardin Botanique, 49: 341–348.

Zetzsche, F., P. Kalt, J. Liecht and E. Ziegler. 1937. Zur Konstitution des Lycopodium-Sporonins, des Tasmanins und des Lange-Sporonins. XI. Mitteilung über die Membran der Sporen und Pollen. J. für Praktische Chemie, 148: 267–286.

Zhang, Zhen-jue and Guo-ji. Zheng. 1988. Translocation of uranin within the living ovules of *Vanilla*. Acta Botanica Sinica, 30 (5): 490–493.

Zhinkina, N.A. and O.P. Kamelina. 1985. Family Pittosporaceae, pp. 20–24. *In* T.B. Batygina and M.S. Yakovlev (eds.). Comparative Embryology of Flowering Plants. Brunelliaceae—Tremandraceae. Nauka, Leningrad, Russia.

Zhou, C. and H.Y. Yang. 1985. Observations on enzymatically isolated living and fixed embryo sacs in several angiosperm species. Planta, 165: 225–231.

Zhu, T., H.L. Mogensen and S.E. Smith. 1990a. Generative cell composition and its relation to male plastid inheritance patterns in *Medicago sativa*. Protoplasma, 158: 66–72.

Zhu, T., H.L. Mogensen and S.E. Smith. 1990b. Ultrastructural reconstruction and quantitative features of alfalfa sperms. Ame. J. Bot., Suppl. 77 (6): 42.

Zhu, T., H.L. Mogensen and S.F. Smith. 1991. Quantitative cytology of the alfalfa generative cell and its relation to male plastid inheritance patterns in three genotypes. Theoretical and Applied Genetics, 81: 21–26.

Zhu, T., H.L. Mogensen and S.E Smith. 1992. Heritable paternal cytoplasmic organelles in alfalfa sperm cells: ultrastructural reconstruction and quantitative cytology. European J. Cell Biology (submitted).

*Zhukova, G.Ya. 1983. Ultrastructural and functional peculiarities of flowering plant egg, pp. 30–31. *In* Proc. VIII All-Union Conf. Plant Embryology 'Problems of Gametogenesis, Fertilization and Embryogenesis'. 'Phan', Tashkent, Uzbek Republic.

*Zhukova, G.Ya. 1984. Ultrastructure of mature embryo sac cells in flowering plants (features of similarity and distinction), pp. 22–23. *In* Materials of All-Union Symp. 'Morpho-functional Aspects of Development of Female Generative Structures in Seed Plants'. Telavi, Georgia.

Zhukova, G.Ya. 1987. Family Polemoniaceae, pp. 206–209. *In* T.B. Batygina and M.S. Yakovlev (eds.). Comparative Embryology of Flowering Plants. Davidiaceae—Asteraceae. Nauka, Leningrad, Russia.

*Zhukova, G.Ya. 1993. Plastids of gametes and embryo in flowering plants. Bot. Zhurnal, 78 (4): 18–35.

*Zhukova, G.Ya. and T.B. Sokolovskaya. 1977. Ultratructure of antipodal cells of embryo sac in *Aconitum napellus* L. (Ranunculaceae) before fertilization. Bot. Zhurnal, 62 (11): 1600–1611.

*Zhukova, G.Ya. and T.B. Sokolovskaya. 1983. On some ultrastructural features of embryo sac in *Aconitum napellus* L., pp. 67–68. *In* Proc. VIII All-Union Conf. Plant Embryology 'Problems of Gametogenesis, Fertilization and Embryogenesis'. 'Phan', Tashkent, Uzbek Republic.

Zhukova, G.Ya. and V.A. Poddubnaya-Arnoldi. 1987. Family Solanaceae, pp. 241–248. *In* T.B. Batygina and M.S. Yakovlev (eds.). Comparative Embryology of Flowering Plants. Davidiaceae—Asteraceae. Nauka, Leningrad, Russia.

Zhukova, G.Ya. and N.G. Tuchina. 1990. Family Arecaceae, pp. 255–266. *In* T.B. Batygina and M.S. Yakovlev (eds.). Comparative Embryology of Flowering Plants. Butomaceae—Lemnaceae. Nauka, Leningrad, Russia.

Zhukovsky, P.M. 1949. Botany. Sovietskaya Nauka, Moscow, Russia.

Zinger, N.V. 1958. The Seed, Its Development and Physiological Characteristics. Izdatelstvo Akademii Nauk SSSR, Moscow, Russia.

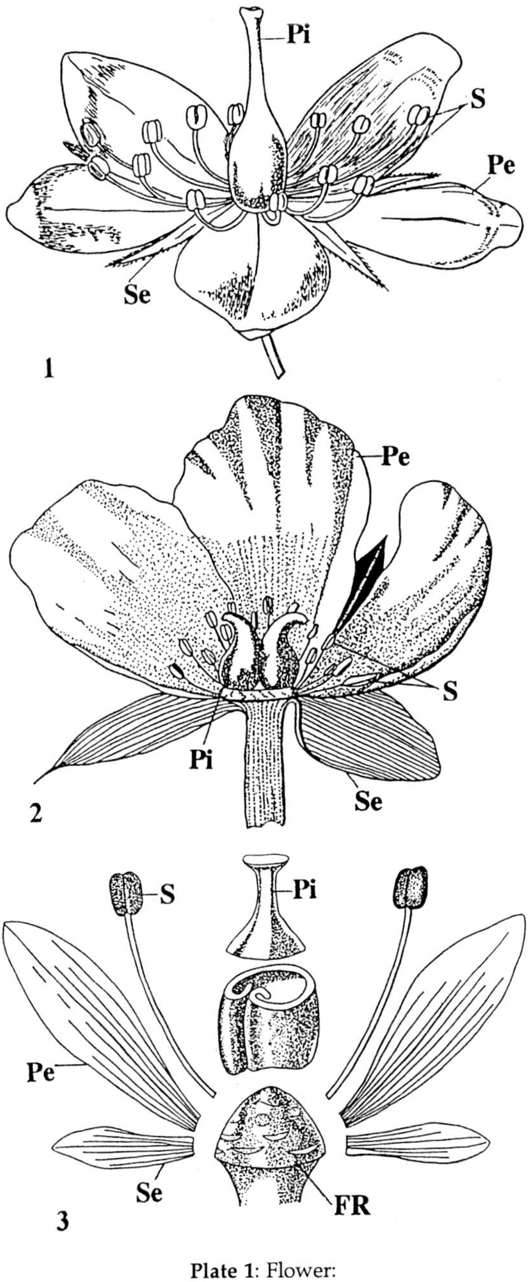

Plate 1: Flower:

1—perfect bisexual flower; 2 – flower of peony (longitudinal section with numerous stamen and two pistils, others removed); 3 – separate parts of a flower (diagrams). FR – flower receptacle, Pe – petal, Pi – pistil, S – stamen, Se – sepal.

1-3 – after Kursanov *et al.*, 1940.

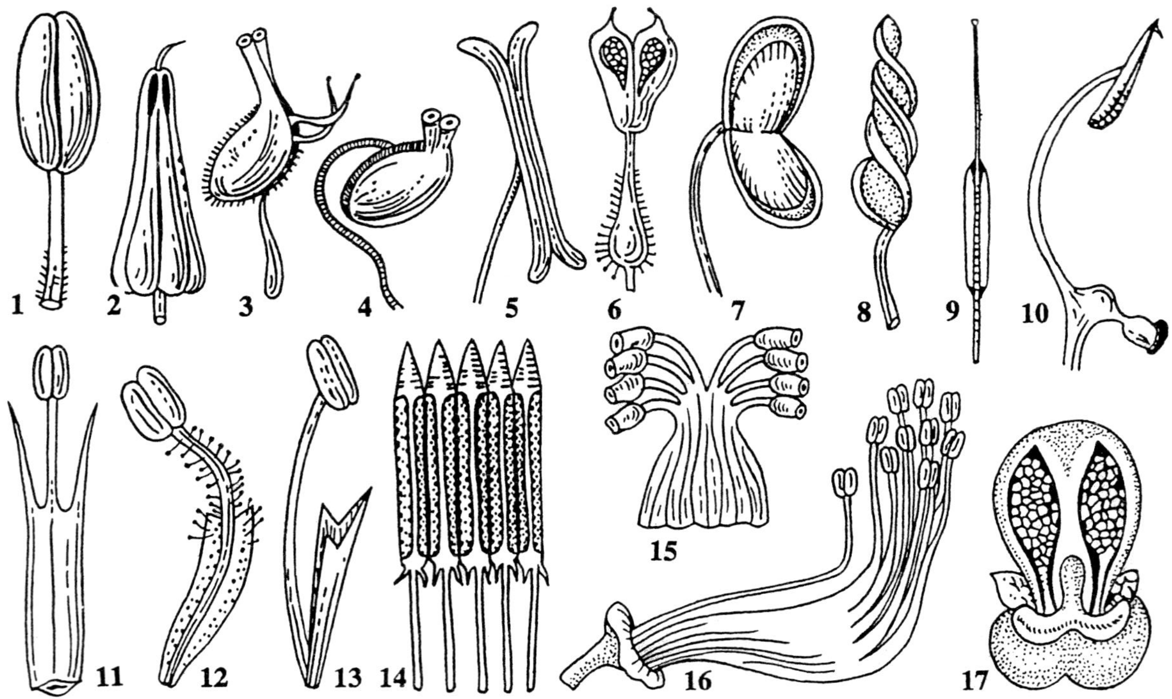

Plate 2: Stamens of different plants.

1 – *Hyoscyamus niger*, 2 – *Galanthus nivalis*, 3 – *Vaccinium uliginosum*, 4 – *Pyrola grandiflora*, 5 - *Anthoxanthum odoratum*, 6 – *Arctous alpina*, 7 – *Galeopsis angustifolia*, 8 – *Erythraea erythraea*, 9 – *Paris quadrifolia*, 10 – *Salvia pratensis*, 11 – *Alluim* sp., 12 – *Aconitum* sp., 13 – *Borago* sp., 14 – *Taraxacum officinale* (5 synantherous stamens), 15 – *Polygala* sp. (8 stamens with united filaments), 16 – *Pisum sativum* (9 stamens with united lower parts of filaments and one free stamen), 17–*Platanthera bifolia* (a stamen and a stigma).

1-17 – modified from Gylarov, 1989.

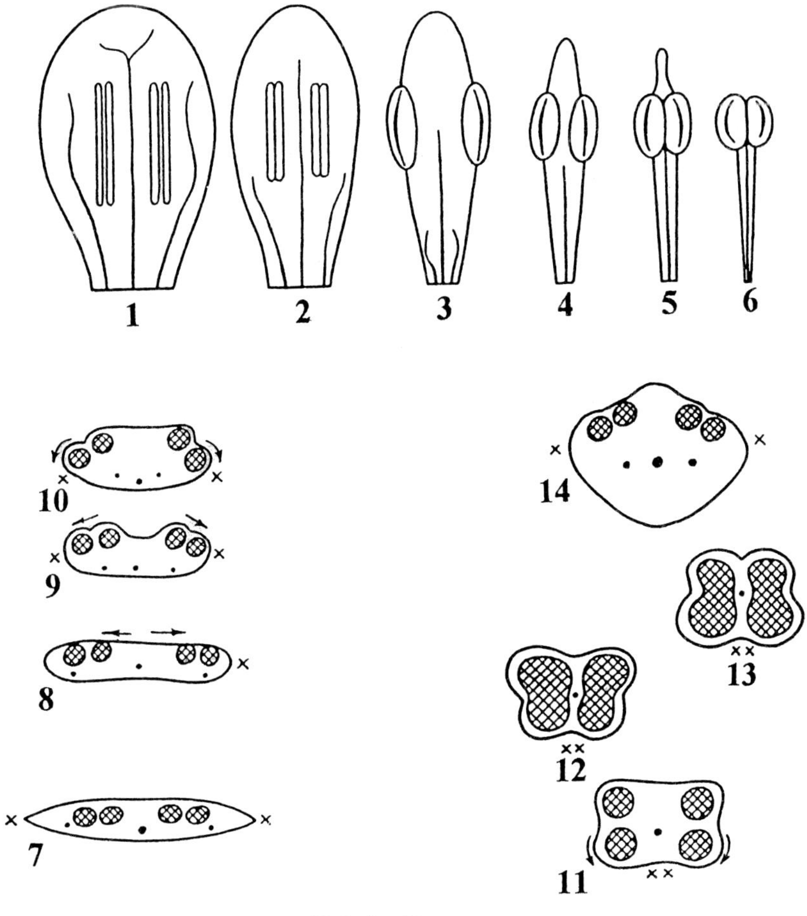

Plate 3: Stamen.

1-6 – Stages of stamen (microsporophyll) evolution from a primitive microsporophyll (*Degeneria vitiensis*) (1) to a highly specialized type (6); 7-19 – transformation of position of stamen margins (xx) and sporangia from submerged (7) to emerging (12, 13, 18, 19)): 7-13 – transformation of type of bursting (7 – introrse; 11 – lartrorse; 12 – extrorse; 13 – introrse); 14-19 – different stages, observed in *Magnolia* and *Liriodendron* (14, 15, 16 – *Magnolia maingayi, M. nitida, M. grandiflora* respectively, introrse bursting, slightly emerging sporangia; 18 – *Liriodendron,* extrorse bursting, emerging sporangia, change of their position accompanied by change in position of minute outer vessels; 19 – *Magnolia acuminata,* lartrorse-introrse bursting, emerging sporangia); 20-31 – diagram of stamen structure: 20-25 – longitudinal sections of stamens (see successive release of anther from blade); 26-31 – transections of anthers showing development of anther accompanied by reduction of lateral tissues and connective tissue; 32-35 – transections of anthers (see transformation of connective shape during flowering): 32, 33 – *Iris pumila,* transformation determined by differential growth (32 – immature anther, 33 – mature one); 34 - *Nymphaea colorata,* transformation determined by a cambium-like growth and a typical comb develops on the connective (mature anther); 35 – staminode of *Hydrostemma longifolia* (Barclayaceae).

1-5 – after Takhtajan, 1980; 6-34 – modified from Eames, 1961; 35 – after Batygina, orig.

16

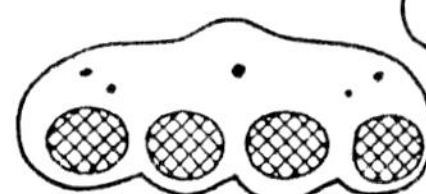
18

19

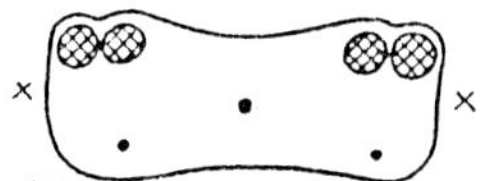
15

17

25

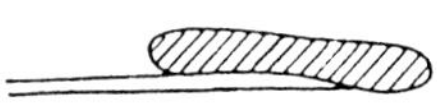
24

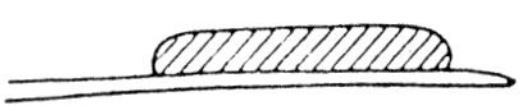
23

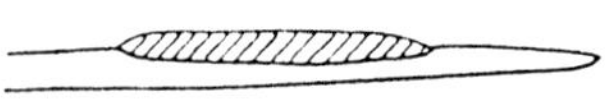
22

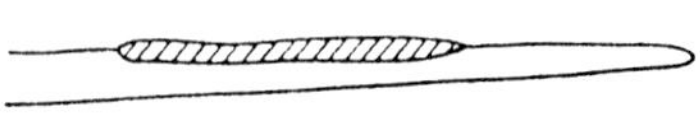
21

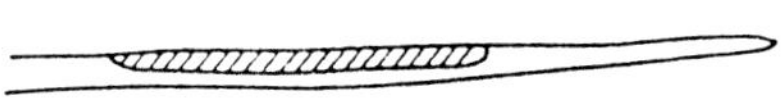
20

31

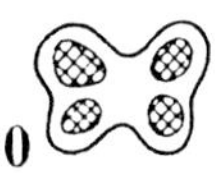
30

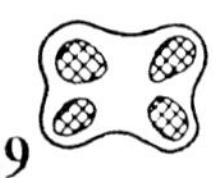
29

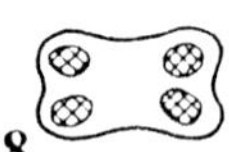
28

27

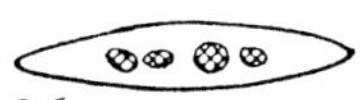
26

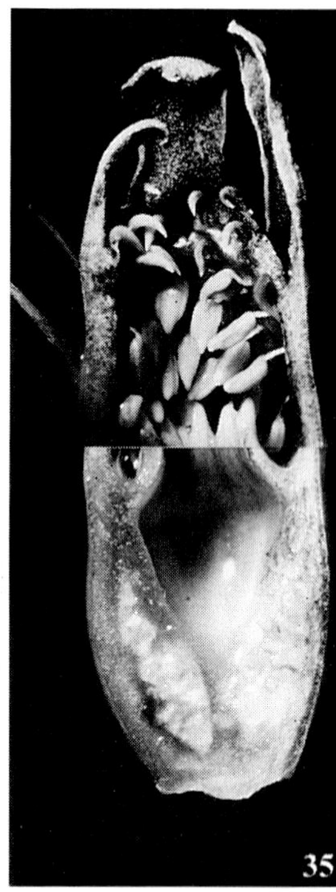
35

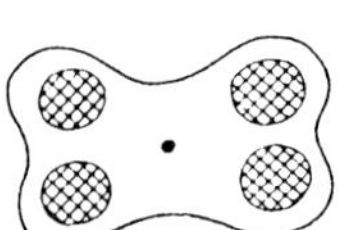
32

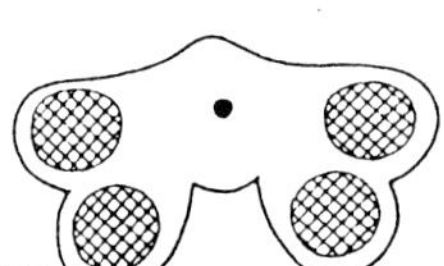
33

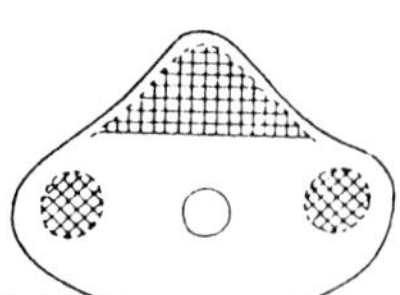
34

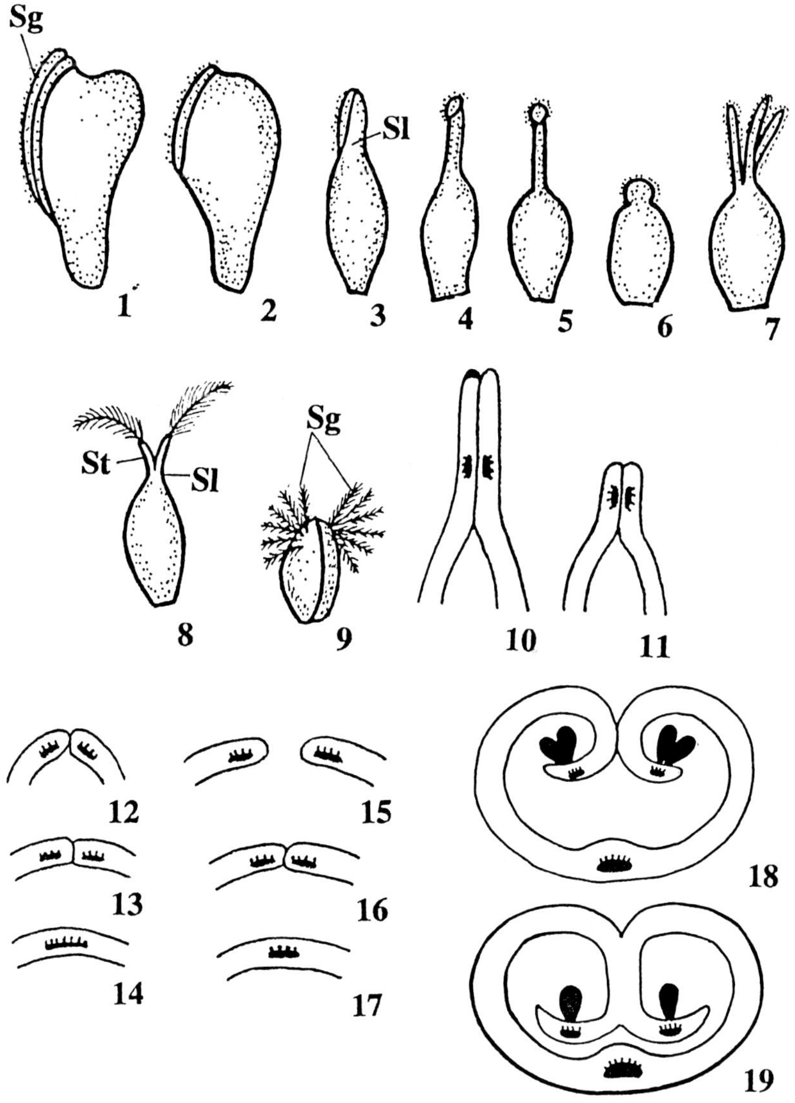

Plate 4: Gynoecium. Carpel. Placenta.

1-9–types of stigma: 1 – sutural stigma of an apocarpous gynoecium ('stigmatic comb'); 2 – sutural stigma with united margins of carpel; 3 – deccurent stigma; 4 – terminal stigma; 5 – capitate stigma; 6 – sessile capitate stigma; 7 – three filiform stigmas; 8 – pinnate stigma; 9 – racemose stigma; 10-18 – two ways of linking of carpel margins (diagram): 10-14 - carpel folds lengthwise, its margins coming into contact by their ventral surfaces and vascular bundles of carpel margins half-inverted (10 – vast marginal area; 11 – reduction of marginal area; 12-14 – stages of concretion of conductive trend); 15-17 – carpel folds up, its margins drawing closer together and conductive bundles slightly inverted: note stages of approaching and adjoining; 17, 18 – folding of carpel, dorsal surface of one-half its blade contacting that of another, bundles inverted, carpel margins remain free; 19-26 – main types of gynoecium and placentation: 19 – ancestral form with opened megasporophyll and submarginal arrangement of ovules;

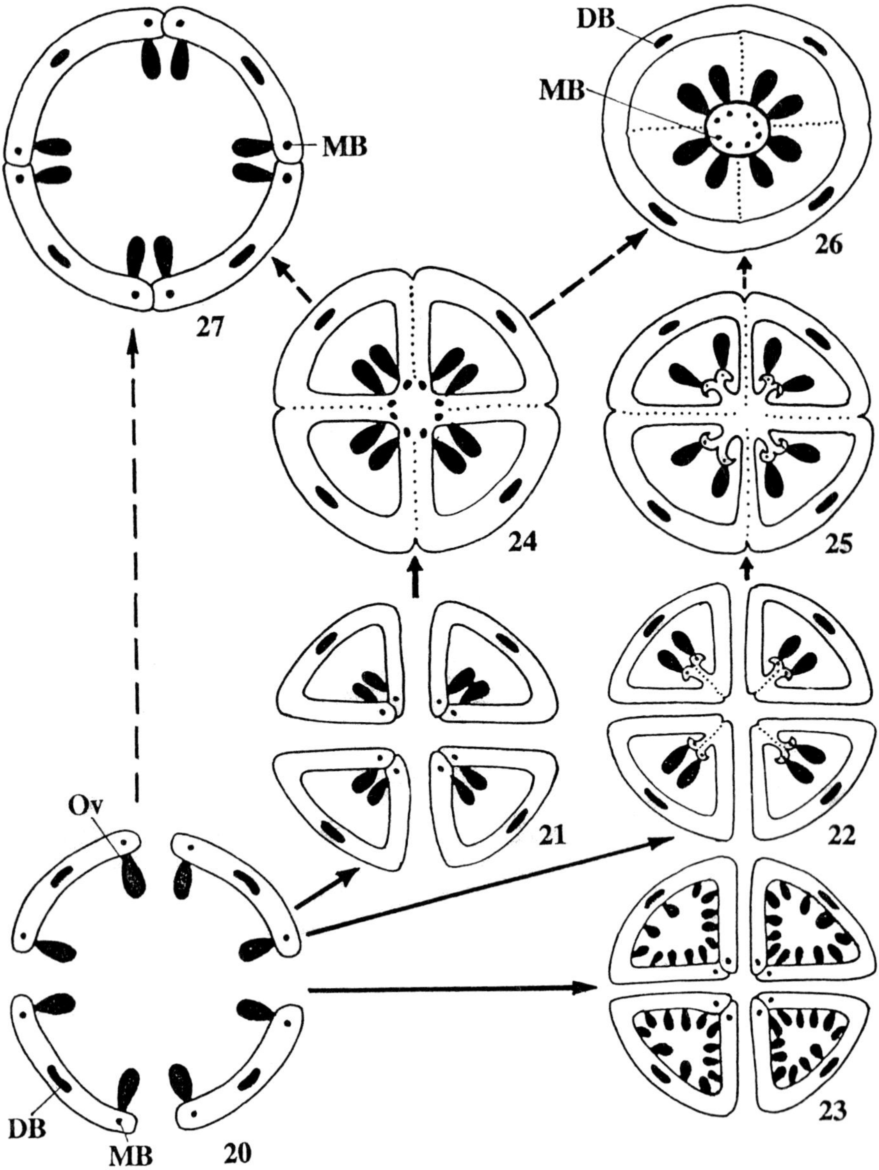

Plate 4 (*contd.*)

20-22 – apocarpous gynoecium (20, 21 – sutural-angular placentation, 22 – laminal placentation); 23-24 – coenocarpous gynoecium with sutural-angular placentation; 25 – lysicarpous gynoecium with columnar central placentation; 26 – paracarpous gynoecium with parietal placentation (20, 22, 23, 26 – margins of folded carpel flat, 21, 24 – margins of folded carpel intorted). DB–dorsal vascular bundle, MB – marginal vascular bundle, Ov – ovule, Sg – stigma, Sl – stylodium, St - style.

1-9, 19-26 – after Korchagina, orig.; 10-18 – after Eames, 1961.

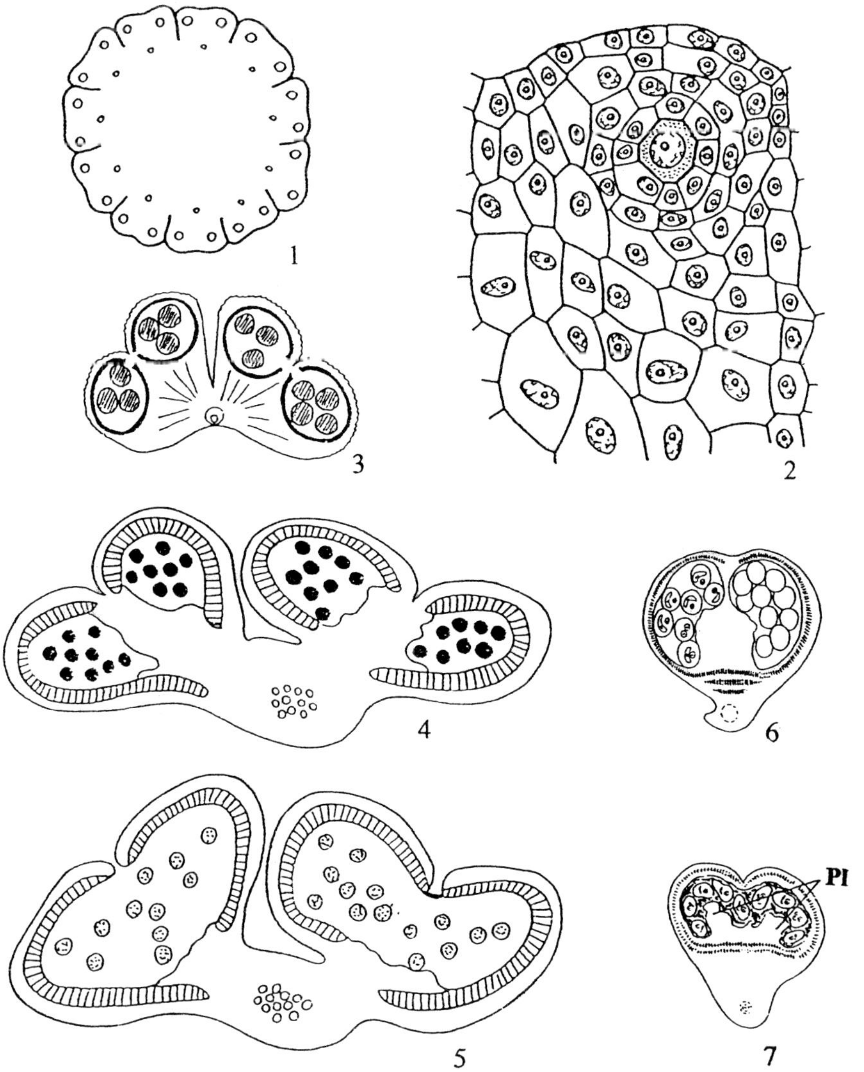

Plate 5: Anther. Microsporangium.

1, 2 – anther of *Myristica macrothyrsa* (Myristicaceae) with numerous microsporangia and separate microsporangium with formed wall; 3 – tetrasporangiate anther of *Scabiosa bipinnata* (Dipsacaceae), which retains individuality of each microsporangium up to its bursting; 4, 5 – tetrasporangiate anther of *Phytolacca americana* (Phytolaccaceae) with thecae moved apart; septa between microsporangia in thecae destroyed (5); 6, 7 – bisporangiate anther of *Calathea louisae* and *Ctenanthe lubbersiana* (Marantaceae), septa between microsporangia destroyed; 1-7 – cross-sections; Pl – placentoid.

1, 2 – after Mametyeva and Kamelina, 1981; 3 – after Kamelina, 1980; 4, 5 – after Kamelina, 1983b; 6, 7 – after Kamelina, 1990.

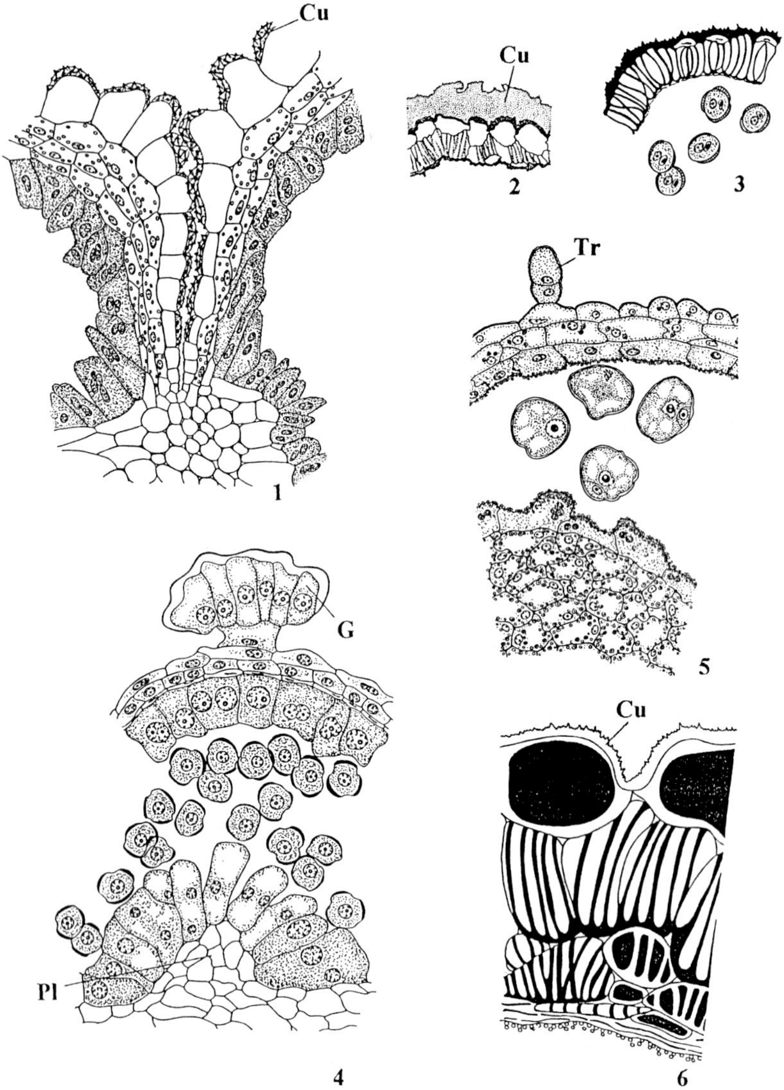

Plate 6: Epidermis.
1 – fragment of anther of *Coriaria myrtifolia* (Coriariaceae), cells of epidermis with stellate cuticle; 2 – epidermis with very thick cuticle in *Peganum harmala* (Peganaceae); 3 – in mature anther of *Nitraria komarovii* (Nitrariaceae) epidermis not persisting; 4 – fragment of anther of *Stachys trinervis* (Lamiaceae), note peltate glandules on epidermis and heteromorphic tapetum; 5 – fragment of anther of *Incarvillea potaninii* (Bignoniaceae), note trichomes on epidermis and heteromorphic tapetum with orbicules, cells of placentoid contains starch; 6 – large cells of mature anther epidermis in *Brexia madagascariensis* (Brexiaceae) filled with tannin, endothecium and persisting middle layers show fibrous thickenings, profile of epidermis cuticle denticulate; Cu – cuticle, G – glandule, Pl – placentoid, Tr – trichome.

1 – after Kamelina, 1985b; 2 – after Kapil and Ahluwalia, 1963; 3 – after Kamelina, 1985c; 4 – after Kamelina and Dzevaltovsky, 1987; 5 – after Kamelina *et al.*, 1990; 6 – after Kamelina,1988a.

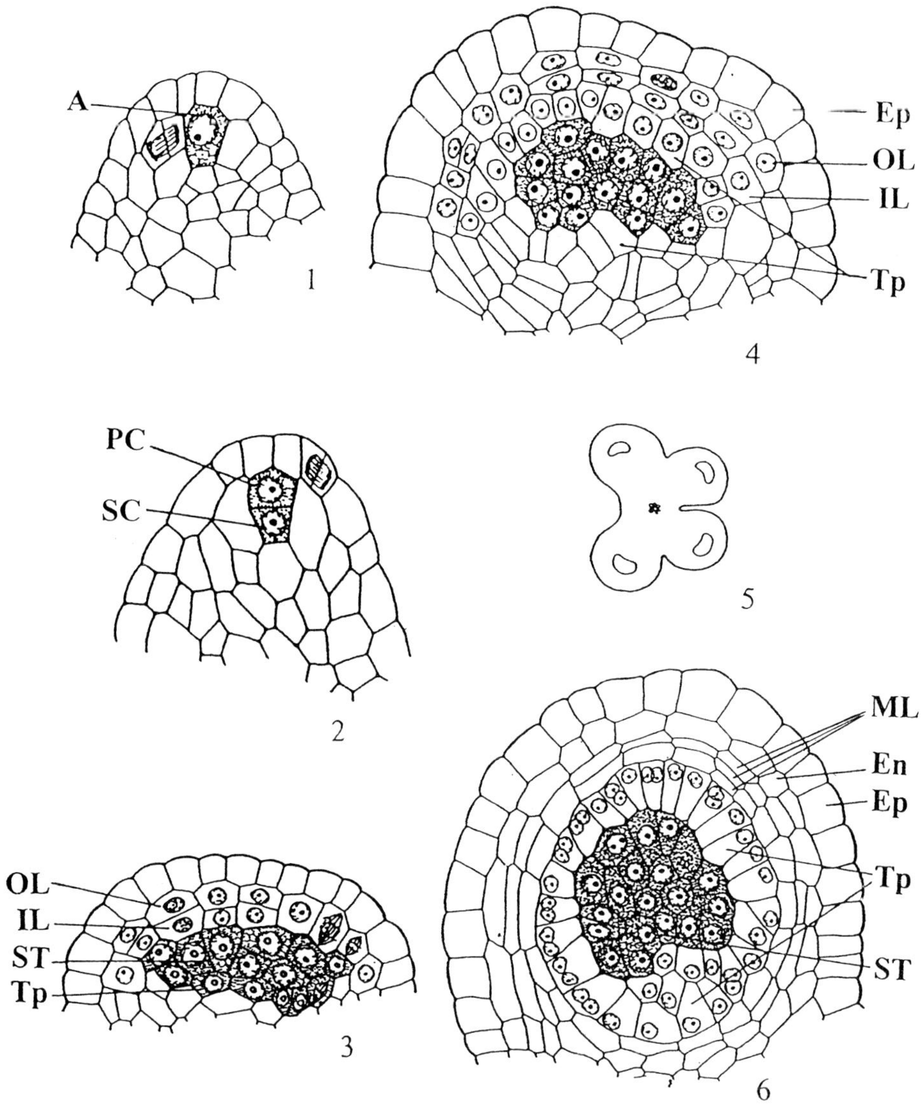

Plate 7: Microsporangium wall.
1-15 – development of microsporangium wall in *Eleutherococcus senticosus* (Araliaceae) (1-6) and *Triticum aestivum* (Poaceae) (7-15). Tapetum differentiation at connective side from microsporangium meristem (1-5) and from parietal layer (7-10), completely formed microsporangium wall (6, 11), mature microsporangium wall (15). A – archesporium, En – endothecium, Ep – epidermis, IL – inner layer, ML – middle layer, OL – outer layer, PC – parietal cell, PL – parietal layer, SC – sporogenous cell, ST – sporogenous tissue, Tp–tapetum. 1-6 – after Anisimova, 1987; 7-15 – after Batygina, 1987.

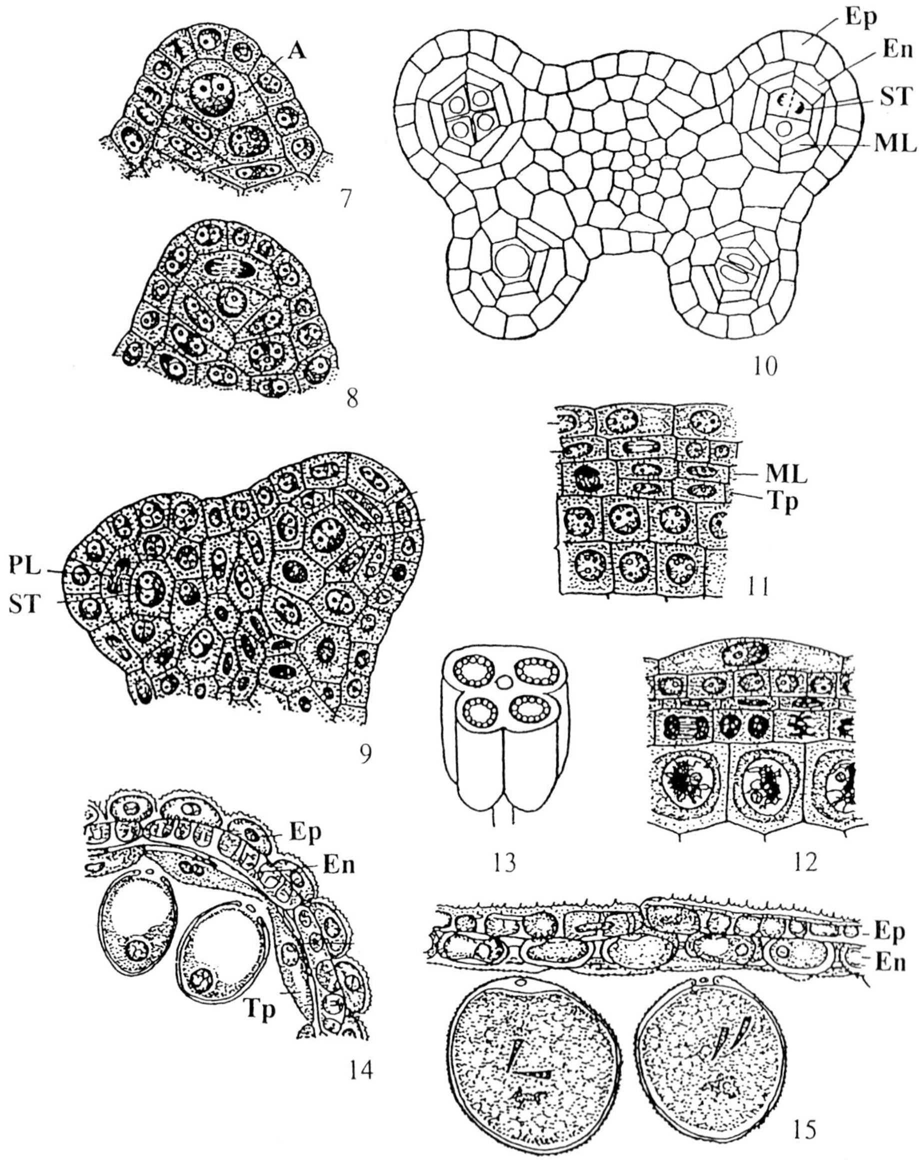

Plate 7 (*Contd.*)

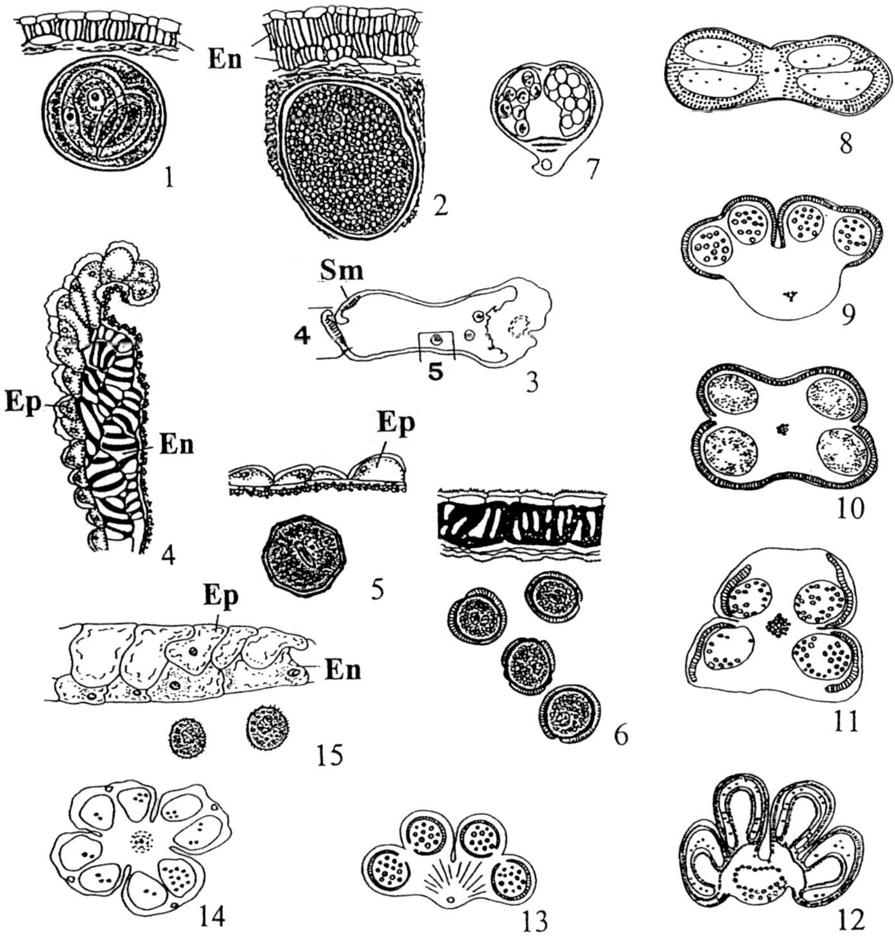

Plate 8: Endothecium. Fibrous thickenings.

1 – wall of mature anther with one-layered endothecium in *Calathea louisae* and two-layered endothecium in *Maranta leuconeura* (Marantaceae) (2); 3-5 – theca of mature anther of *Incarvillea potaninii* (Bignoniaceae), endothecium with fibrous thickenings arises only in stomium area (3, 4) and rest of wall consists of epidermis and tapetal membrane with orbicules under it (5); 6 – endothecium of mature anther wall in *Nandina domestica* (Berberidaceae) with very massive fibrous deposits (1-6 – cross-sections); 7-14 – anthers showing various number and location of microsporangia, different variants of fibrous layers arrangement in anthers: *Calathea louisae* (7), *Bubbia hoveana* (Winteraceae) (8), *Calycanthus fertilis* (Calycanthaceae) (9), *Daphniphyllum macropodum* (Daphniphyllaceae) (10), *Eupomatia laurina* (Eupomatiaceae) (11), *Ostrowskia magnifica* (Campanulaceae) (12), *Scabiosa bipinnata* (Dipsacaceae) (13); 14, 15 – endothecium in wall of mature anther of *Zannichellia pedunculata* (Zannichelliaceae) without fibrous thickenings (1-15 - cross-sections); En – endothecium, Ep – epidermis, Sm – stomium.

1, 2, 7 – after Kamelina, 1990; 3-5 – after Kamelina *et al.*, 1990; 6 – after Kamelina and Tuchina, 1982; 8 – after Freiberg, 1981; 9-11 – after Kamelina, 1981d; 12 – after Kamelina and Zhinkina, 1989; 13 – after Kamelina, 1980; 14, 15 – after Kamelina and Teryokhin, 1990.

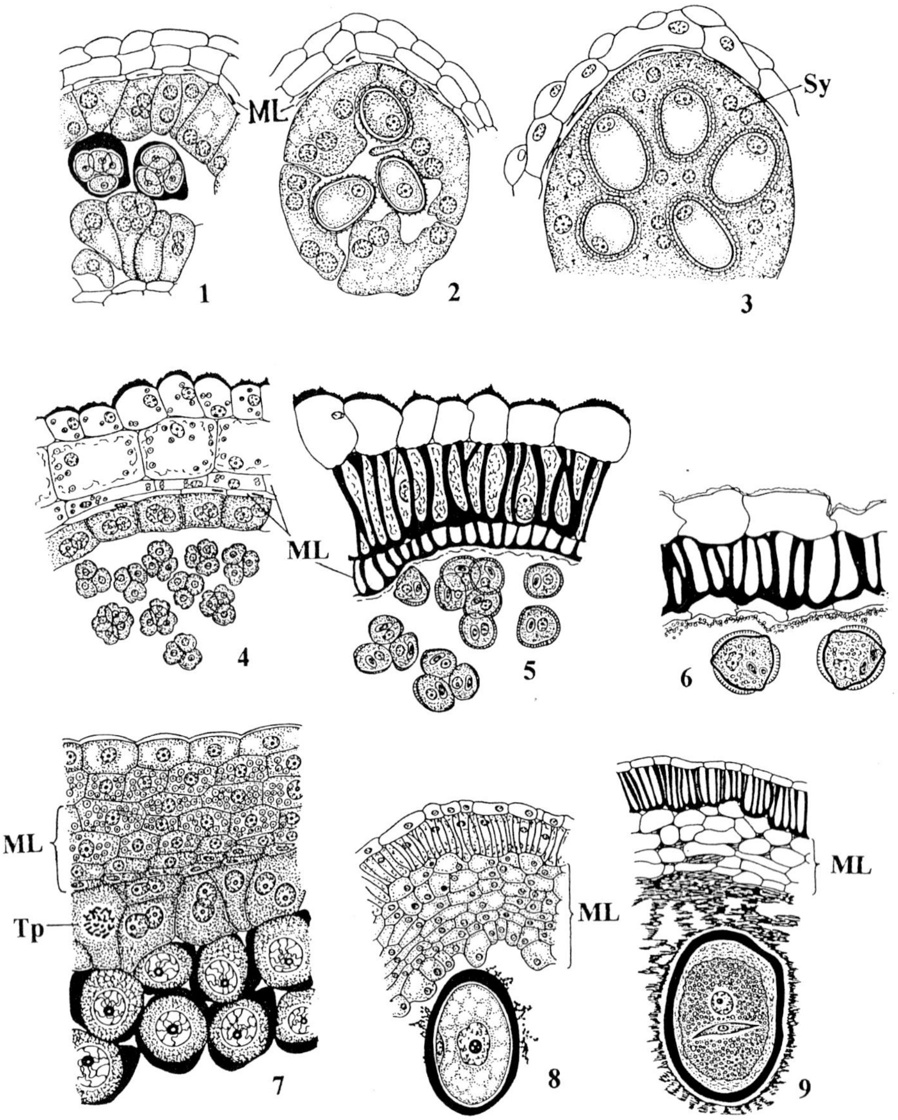

Plate 9: Middle layer.

1-3 – one middle layer in the microsporangium wall in *Leycesteria formosa* (Caprifoliaceae), which degenerates during cell tapetum reorganization into syncytium; 4 – two middle layers in *Datisca cannabina* (Datiscaceae), inner of which collapses while outer persists, its cells with fibrous thickenings (5) differ from those in endothecium; 6 – persistent middle layer in mature anther of *Stegnosperma halimifolium* (Stegnospermataceae) without fibrous thickenings; 7 – numerous middle layers in *Nandina domestica* (Berberidaceae), do not persist in mature anther; 8 – numerous middle layers in *Ctenanthe lubbersiana* (Marantaceae), excreting slime into microsporangium cavity (9); ML – middle layer, Sy – syncytium, Tp – tapetum.

1-3 – after Kamelina *et al.*, 1987; 4, 5 – after Kamelina and Proskurina, 1985; 7 – after Kamelina and Tuchina, 1982; 8, 9 – after Kamelina, 1990.

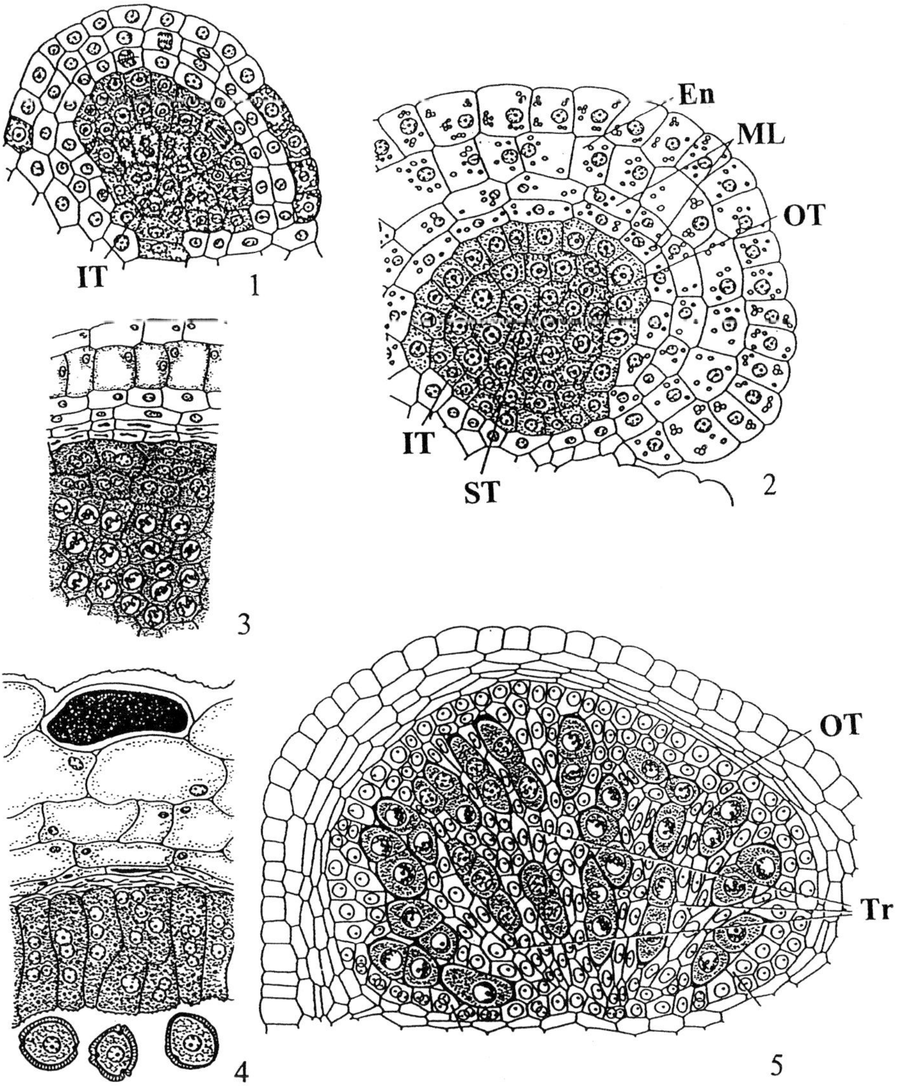

Plate 10: Tapetum.

1 – heteromorphic tapetum in *Escallonia rubra* (Escalloniaceae); 2 – morphologically homogeneous tapetum in *Datisca cannabina* (Datiscaceae); 3 – multilayered tapetum with binuclear cells in *Daphniphyllum macropodum* (Daphniphyllaceae); 4 – tapetum with multinuclear cells and orbicules in *Brexia madagascariensis* (Brexiaceae); 5 – trabeculae of tapetal origin in anther of *Gentiana lutea* (Gentianaceae); En – endothecium, IT – inner part of tapetum, ML – middle layer, OT – outer part of tapetum, ST – sporogenous tissue, Tr – trabeculae.

1 – after Kamelina, 1984; 2 – after Kamelina, 1983a; 3 – after Kamelina, 1981c; 4 – after Kamelina, 1988a; 5 – after Shamrov, 1988.

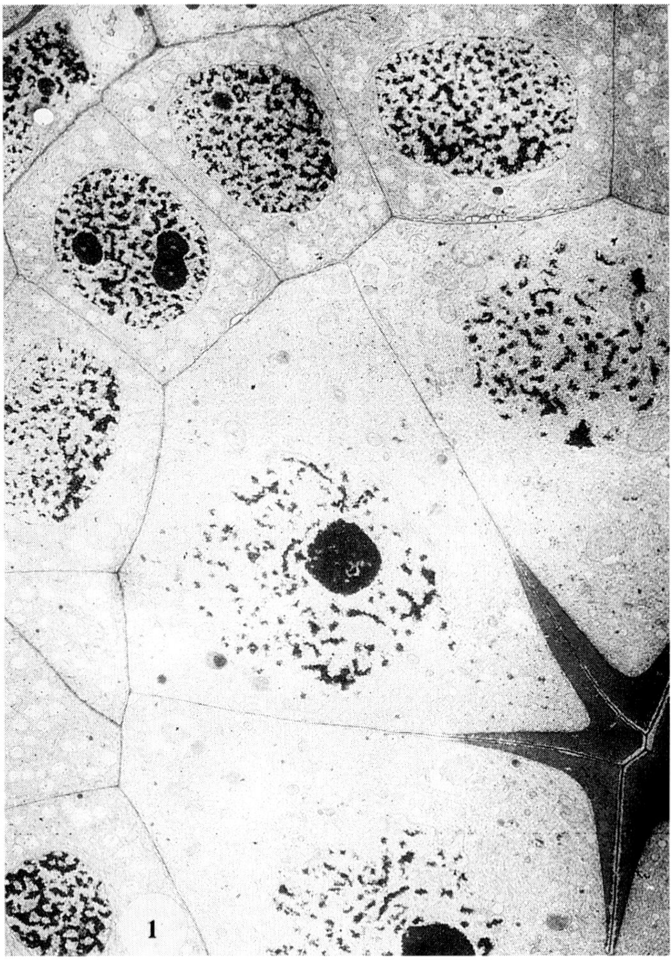

Plate 11: Tapetum and orbicules.

1 – cross-section of anther of *Triticum aestivum* (TEM), see the stellate structure of callose and sporogenous cells surrounded by tapetum (×22,000); 2-5 – orbicules (Ubisch bodies) of various shape (SEM): 2 – from round-triangular to moniliform in *Physopsis lachnostachya* (Chloranthaceae) (X1500), 3 – laminate with rough edge in *Euphorbia falcata* (×4500), 4 – compactly arranged in parallel rows in *Euphorbia bivonae* (×4500), 5 – globular or annular in *Nymphaea* (×230); 6 – complicated congestions of orbicules on pollen grain wall of *Oxalis pes-caprae* (SEM, ×10,000); 7 – part of tapetal cell at the stage of late tetrad (TEM), note small globular bodies (pro-orbicules) in depressions of plasmalemma; globular bodies without membrane (arrow) and with membrane (arrowhead) in cytoplasm of tapetal cells (×25,000); 8 – fragment of tapetal cell (TEM) with core of an orbicule consisting of plural structures connected with reservoirs of granular ER (arrows), note surface of glycocalyx of tapetal cells (×7000); 9 – part of tapetal cell of *Triticum aestivum* (TEM) with orbicular coat (arrow) and orbicules (arrowhead) (×30,000); ER – endoplasmic reticulum; G – glycocalyx, PO – pro-orbicules; U – core of an orbicule (Ubisch body).

1-3, 6 – after El-Ghazaly, orig.; 4 – after El-Ghazaly and Jensen, 1986a; 5 – after Rowley and Walles, 1987.

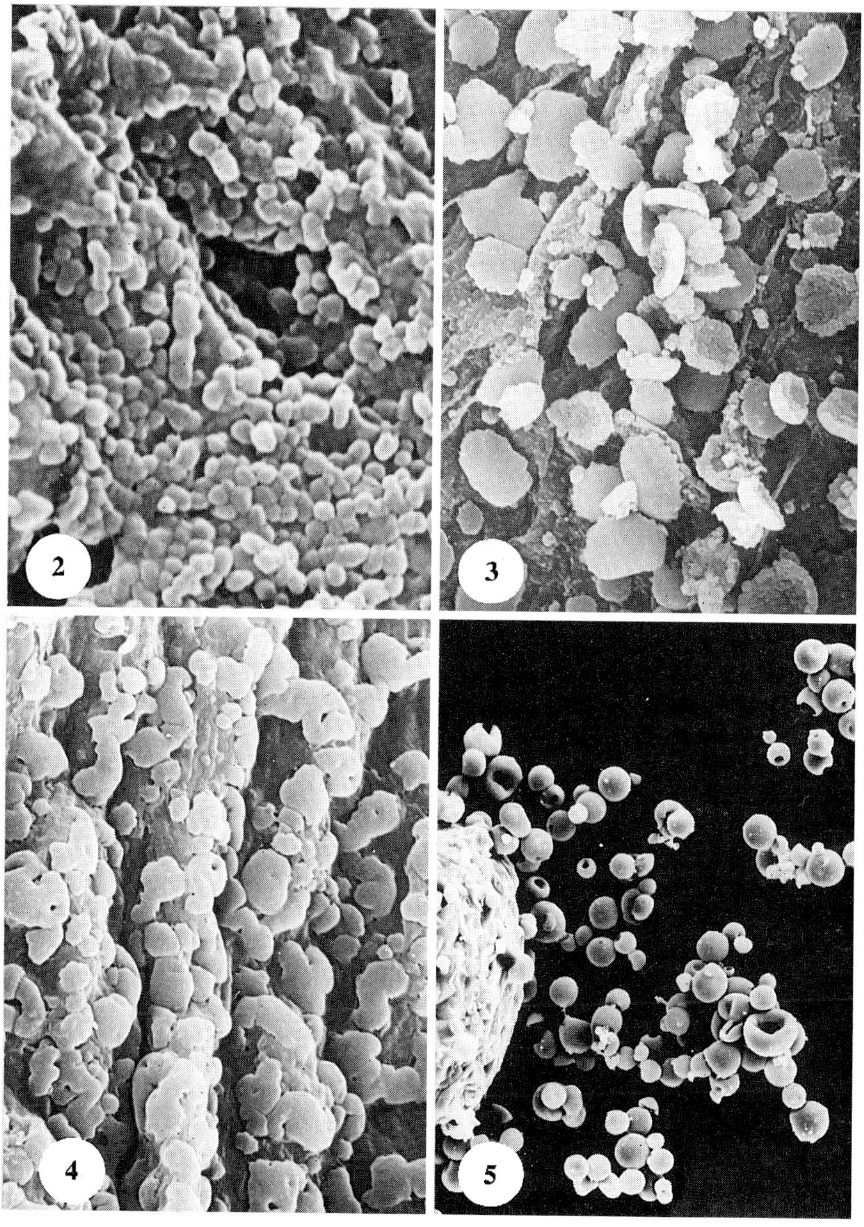

Plate 11 (*Contd.*)

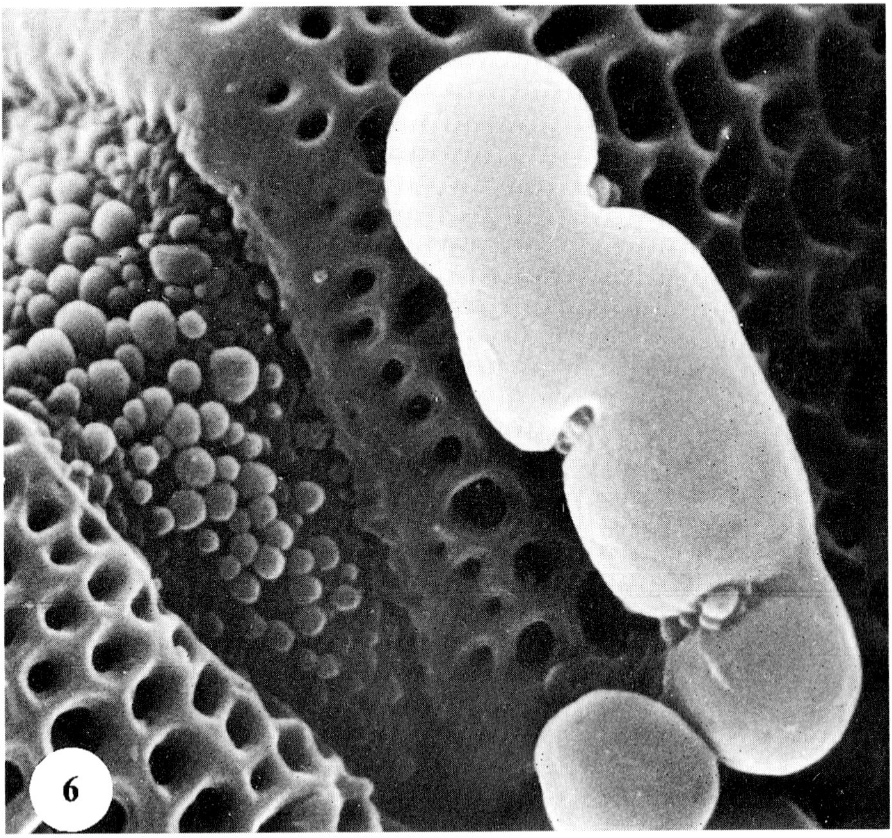

Plate 11 (*Contd.*)

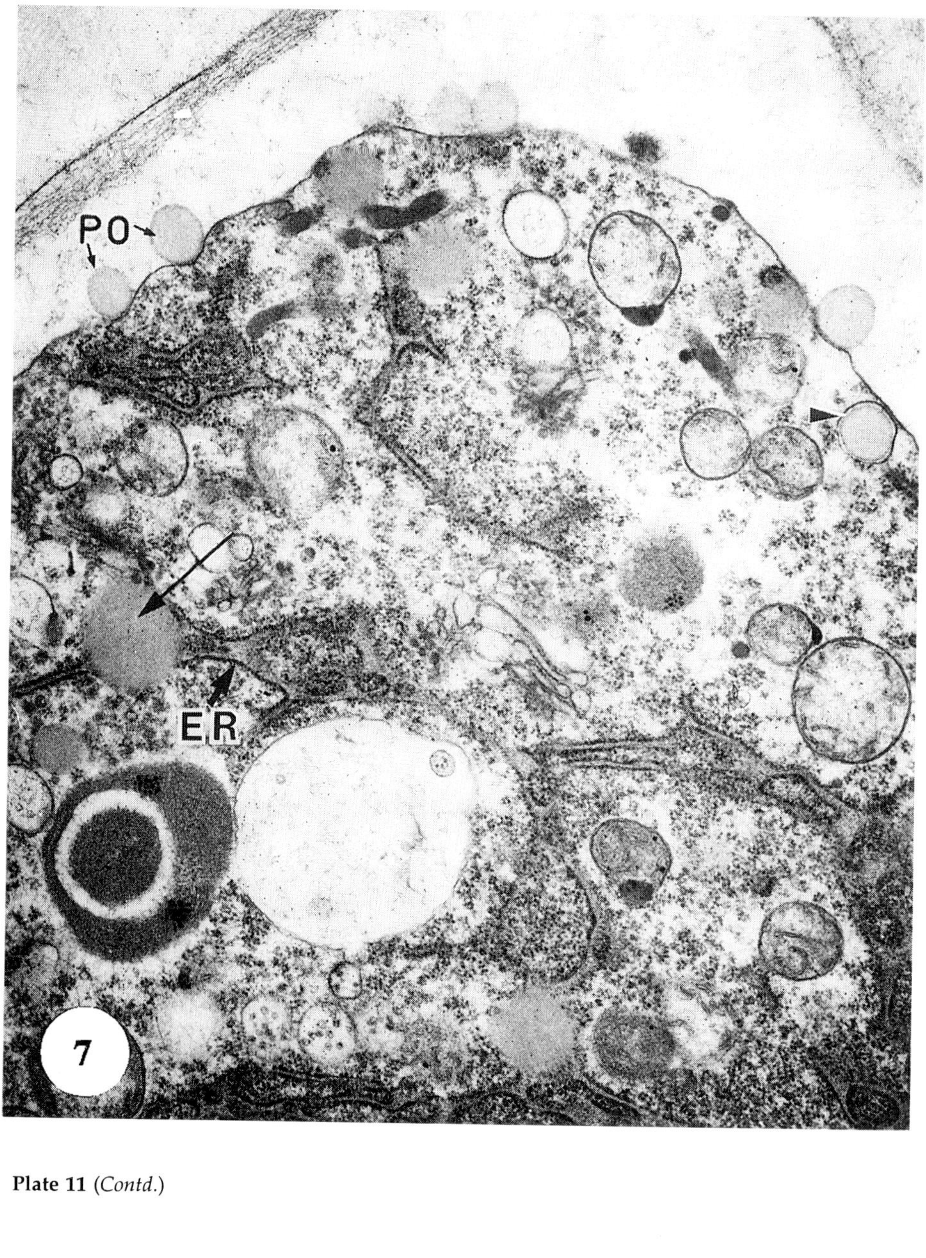

Plate 11 (*Contd.*)

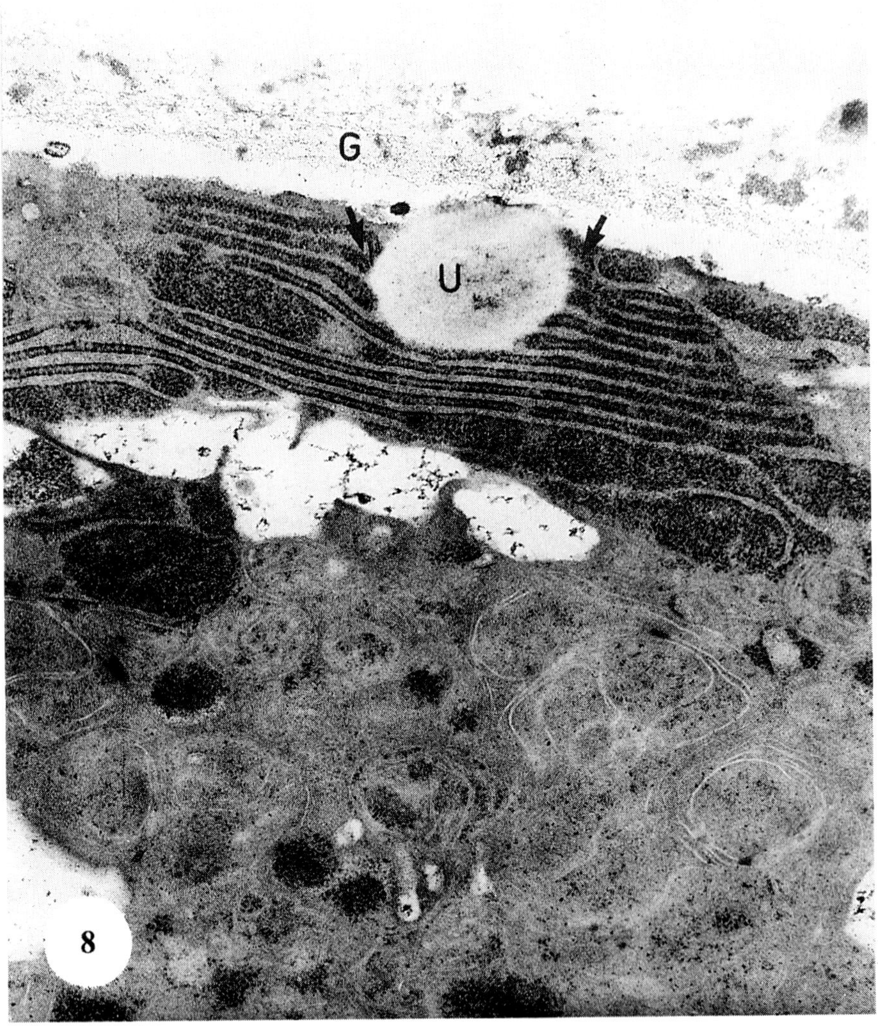

Plate 11 (*Contd.*)

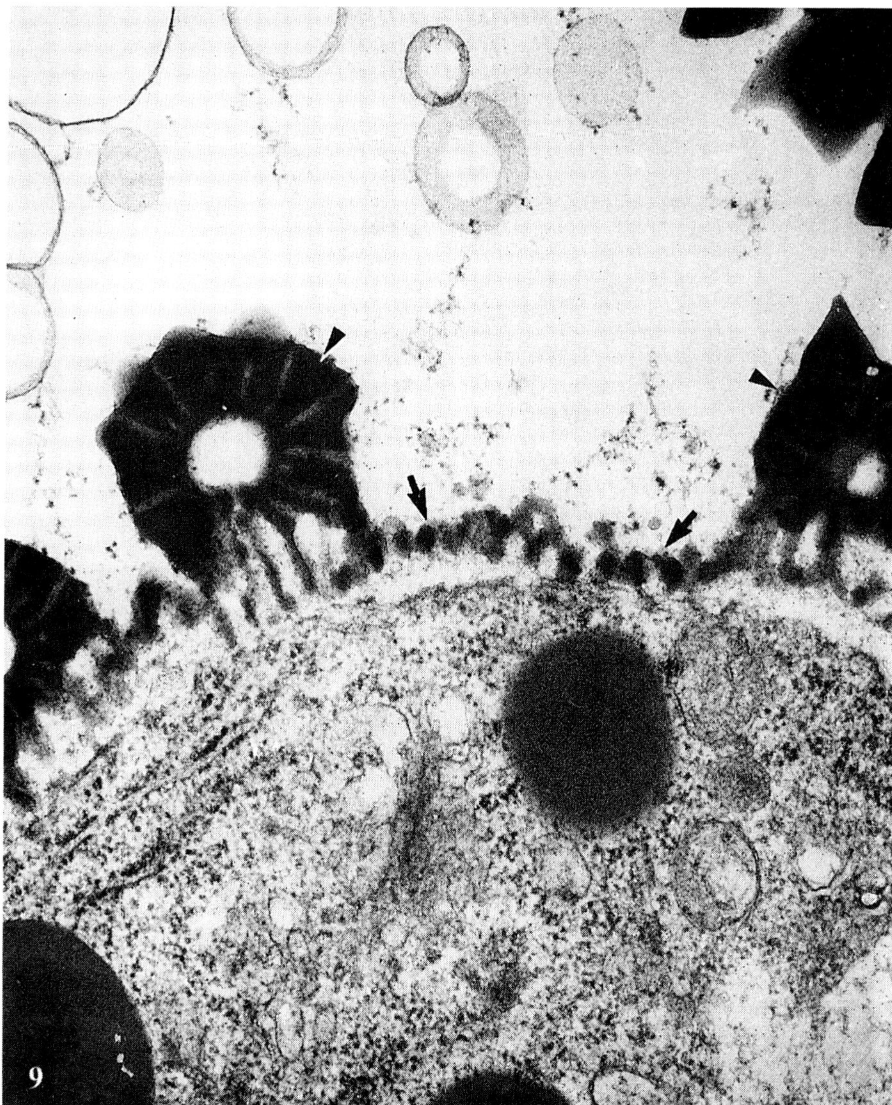

Plate 11 (*Contd.*)

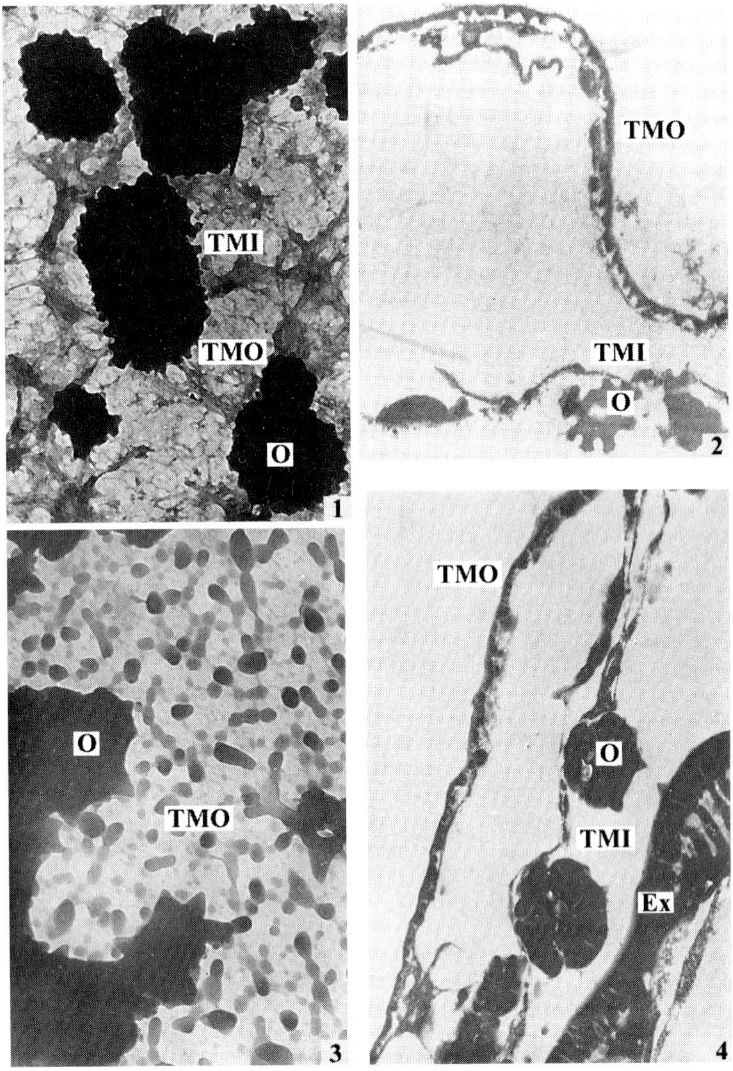

Plate 12: Tapetal membrane.

1-16 – fragments of sporopollenin membrane of tapetal cells in anthers viewed from surface (1, 3, 5, 8, 9, 11, 13, 15) and from cross-sections (2, 4, 6, 7, 10, 12, 14, 16) treated with concentrated sulphuric acid: 1, 2–*Pinus sylvestris*, 3,4–*Beta vulgaris*, 5, 6–*Fragaria vesca*, 7, 8–*Medicago sativa*, 9, 10–*Solanum tuberosum*, 11, 12–*Lycopersicon esculentum*, 13, 14–*Triticum aestivum*, 15, 16– *Helianthus annuus*. (Magnification: 1, 3, 9, 11, 13, 16 – × 15,000; 2 – × 7000; 4 – × 10,000; 5,8 – × 3500; 6 – × 4200; 7 – × 1400; 10 – × 2500; 12 – × 11,200; 14 – × 13,800; 15 – × 7500); Et M – extratapetal membrane, Ex – exine of pollen grain, O – orbicule, TMI – tapetal membrane of inner tangential wall, TMO – tapetal membrane of outer tangential wall, TMR – tapetal membrane of radial wall.

1-4 – after Golubeva, 1983; 5-8–after Ogorodnikova, 1990; 9-12 – after Ogorodnikova, 1976a; 13-16 – after Ogorodnikova, 1976b.

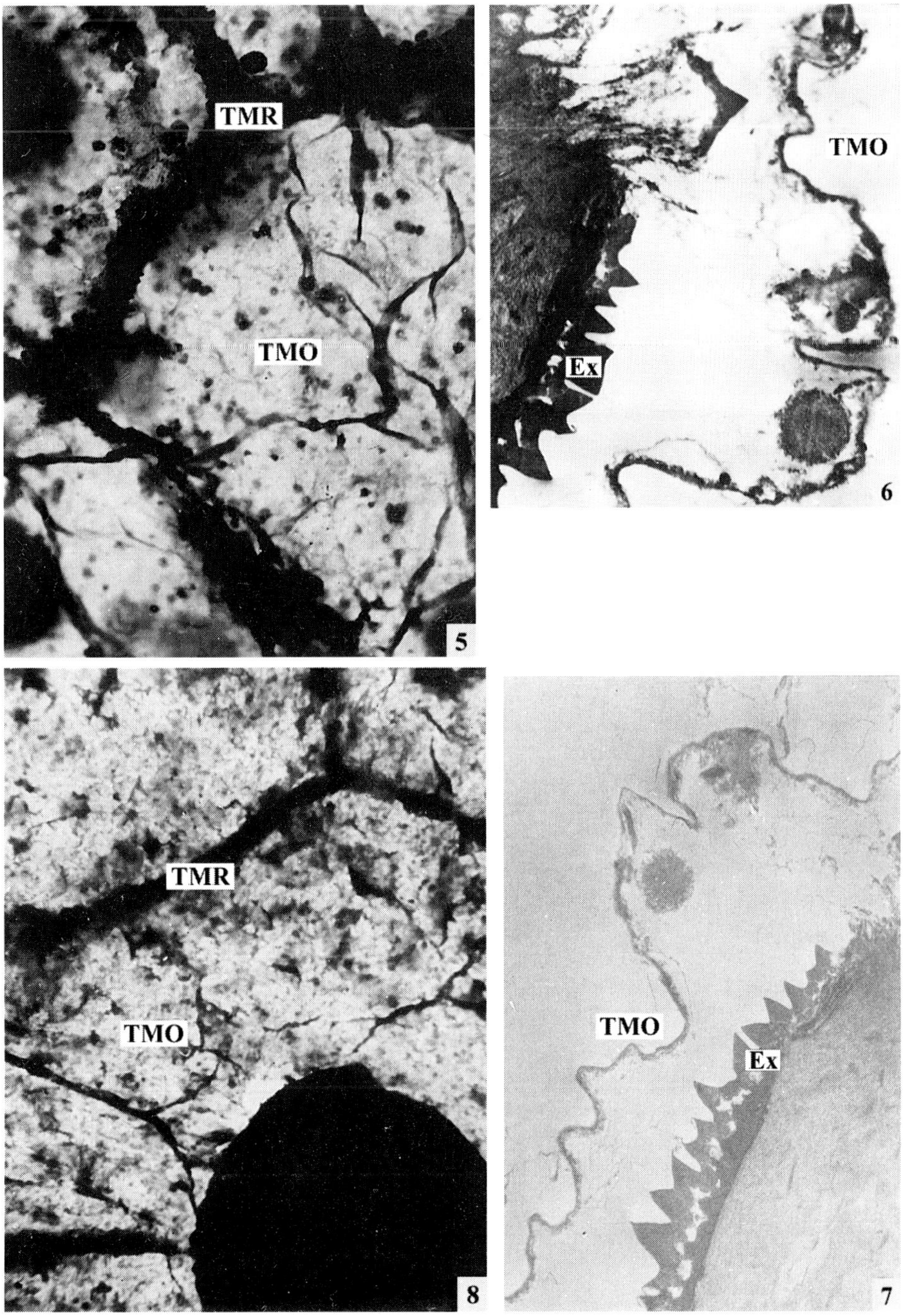

Plate 12 (*Contd.*)

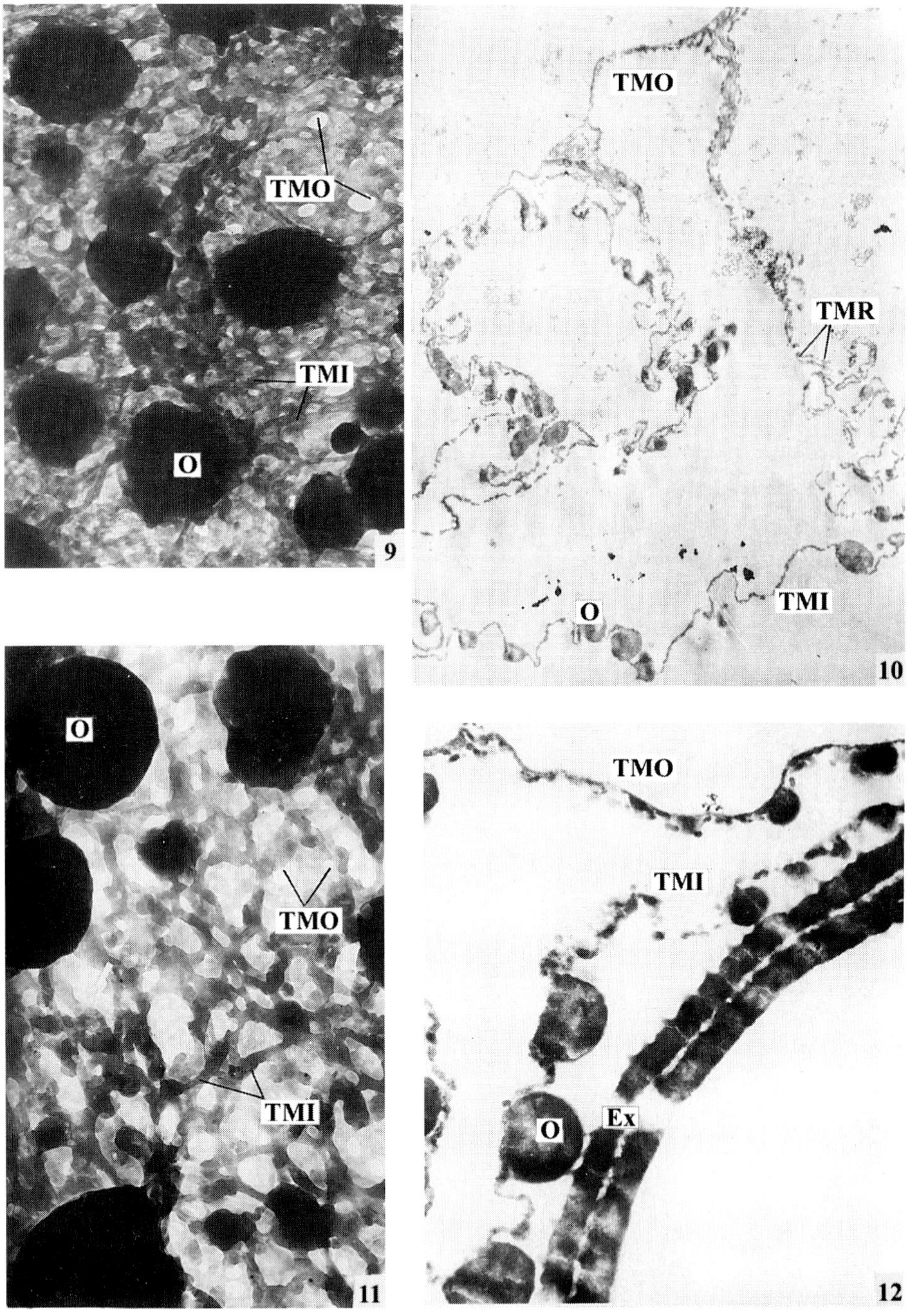

Plate 12 (*Contd.*)

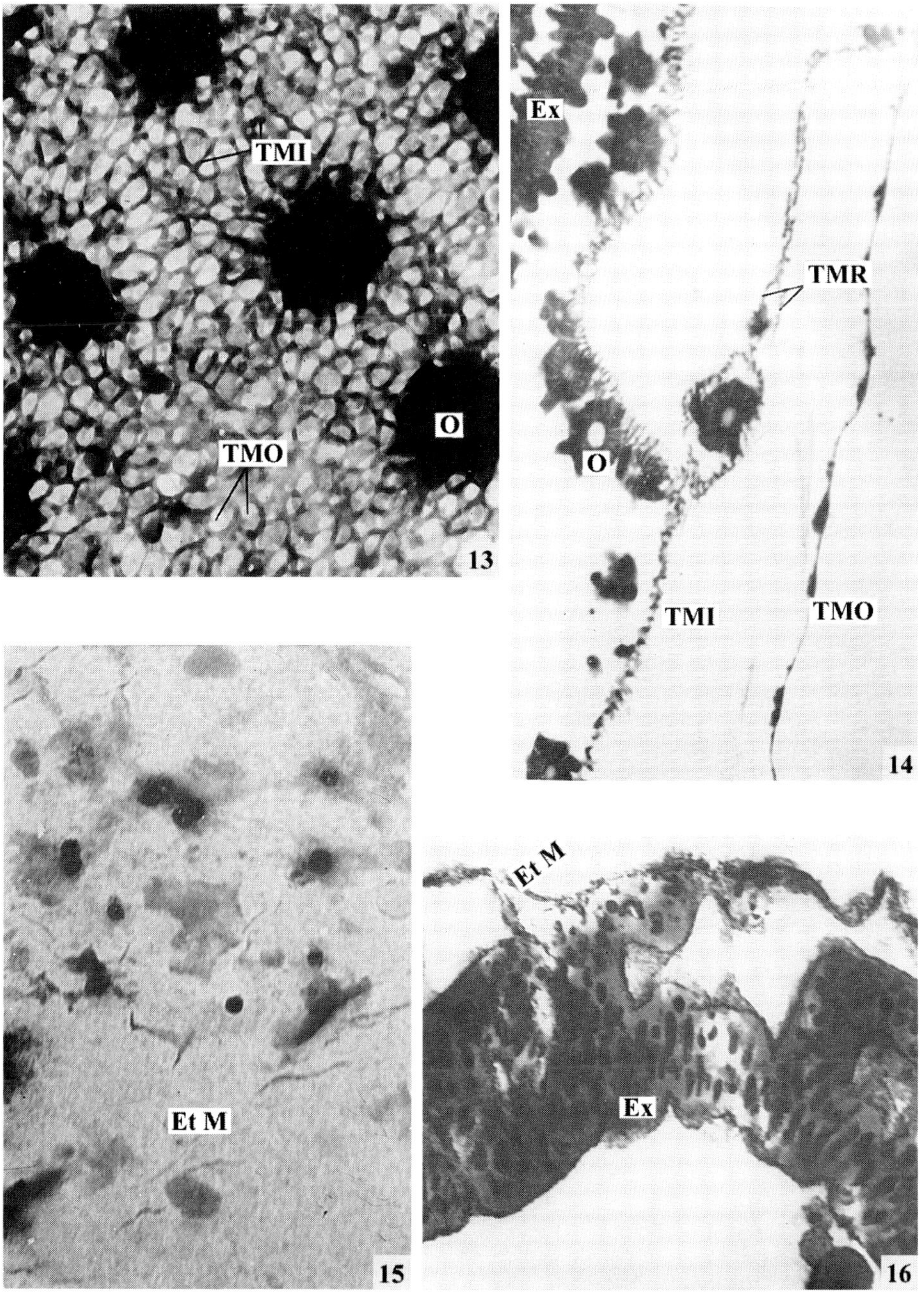

Plate 12 (*Contd.*)

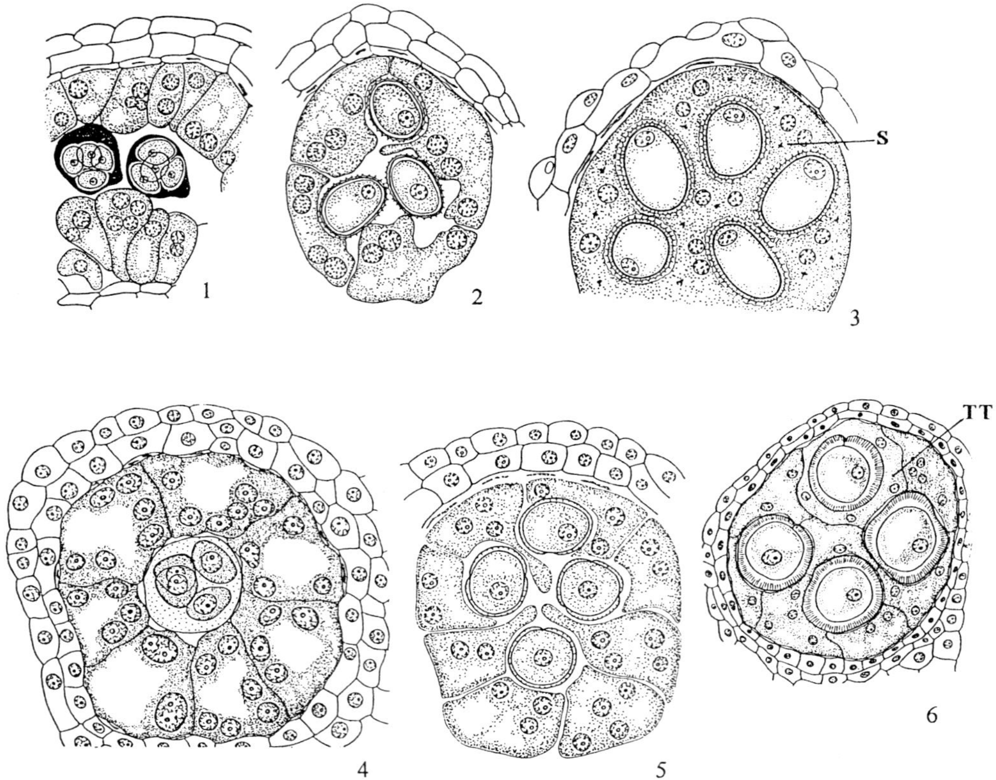

Plate 13: Tapetum reorganization.

1-3 – reorganization of cellular tapetum into syncytium in *Leycesteria formosa* (Caprifoliaceae); 4– 6 – reorganization of cellular multinuclear tapetum into tapetal tissue in species of *Scabiosa* (Dipsacaceae); 2, 5 – amoeboid stage of tapetum development (1-6 – cross-sections); S – syncytium, TT – tapetal tissue.

1-3 – after Kamelina *et al.*, 1987; 4-6 – after Kamelina, 1987a.

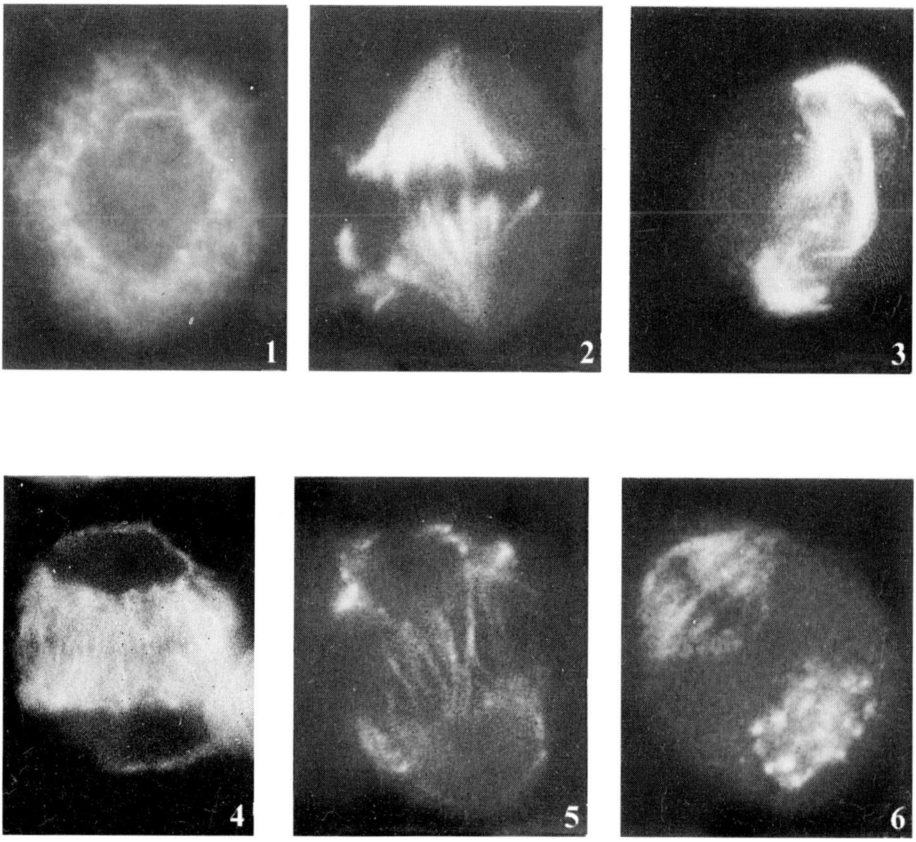

Plate 14: Microsporogenesis and immunofluorescence of cytoskeleton.
1-8 – microsporogenesis in *Nicotiana*, fluorescence of microtubule rows, revealed with indirect immunocytochemical method: 1 – prophase I, diplotene, the net of microtubules concentrating near the nucleus; 2 – metaphase I, a spindle, chromosomes in a non-fluorescing equatorial plane; 3 – late anaphase I, a system of polar microtubules and fluorescing caps of kinetochore bands; 4 – telophase I, microtubules of phragmoplast; 5 – interkinesis, the band of microtubules around nuclei, some microtubules of phragmoplast still persist between nuclei; 6 – metaphase II, spindles of microtubules at longitudinal section and at cross-section, phragmoplast has almost degenerated; 7 – telophase II, tetrapolar secondary phragmoplast; 8, 9 – successive microsporogenesis; 8 – dyad; 9 – isobilateral tetrad of microspores, less usual than tetrahedral (this photo was kindly given by Prof. J. Cebrat); 10 – tetrad, callose fluorescence; 11 – simultaneous microsporogenesis, dyad of *Nymphaea* – equatorial congestion of organelles; 12 – metaphase I, microsporocyte of *Malva*, organelle aggregation around meiotic apparatus; 13 – microsporocyte of *Malva* after meiosis I, aggregation of organelles around interkinetic nuclei.

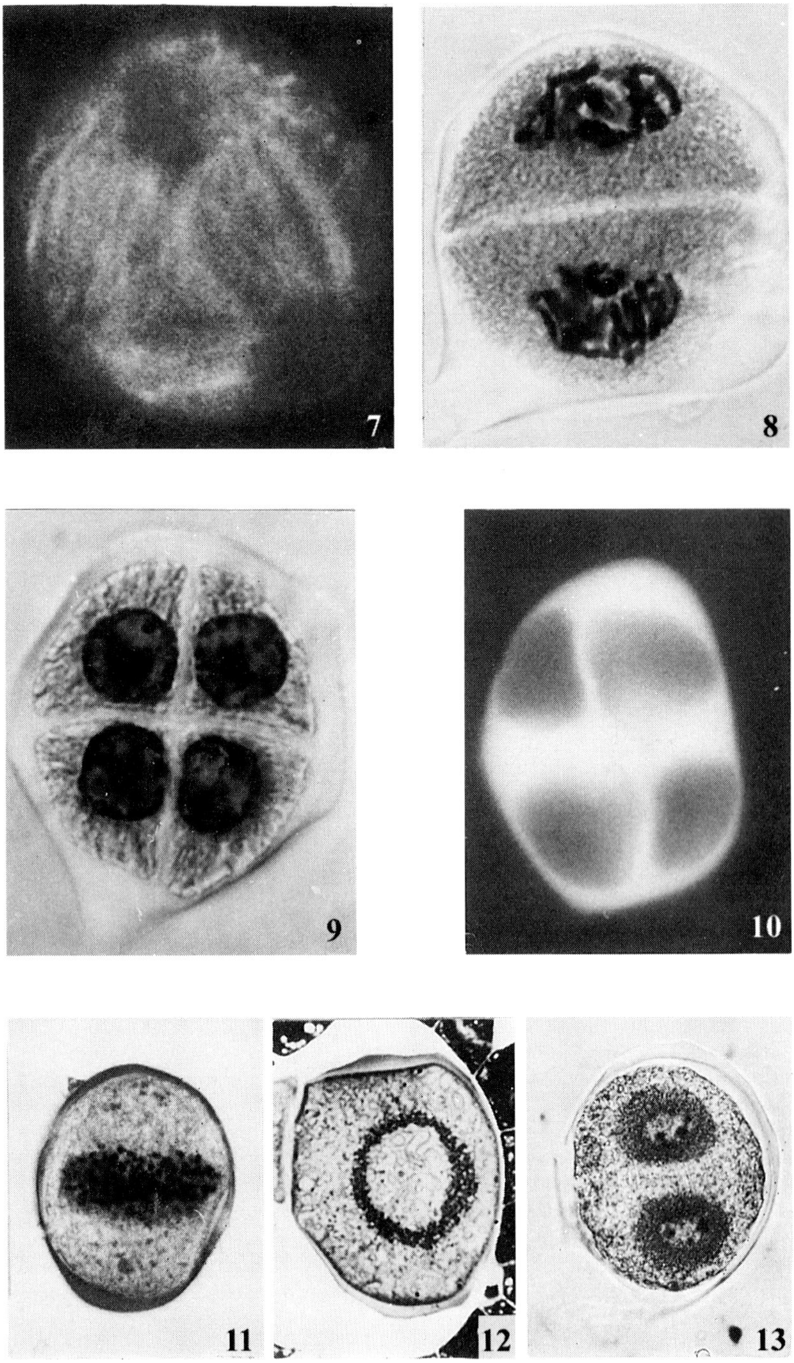

Plate 14 (*Contd.*)

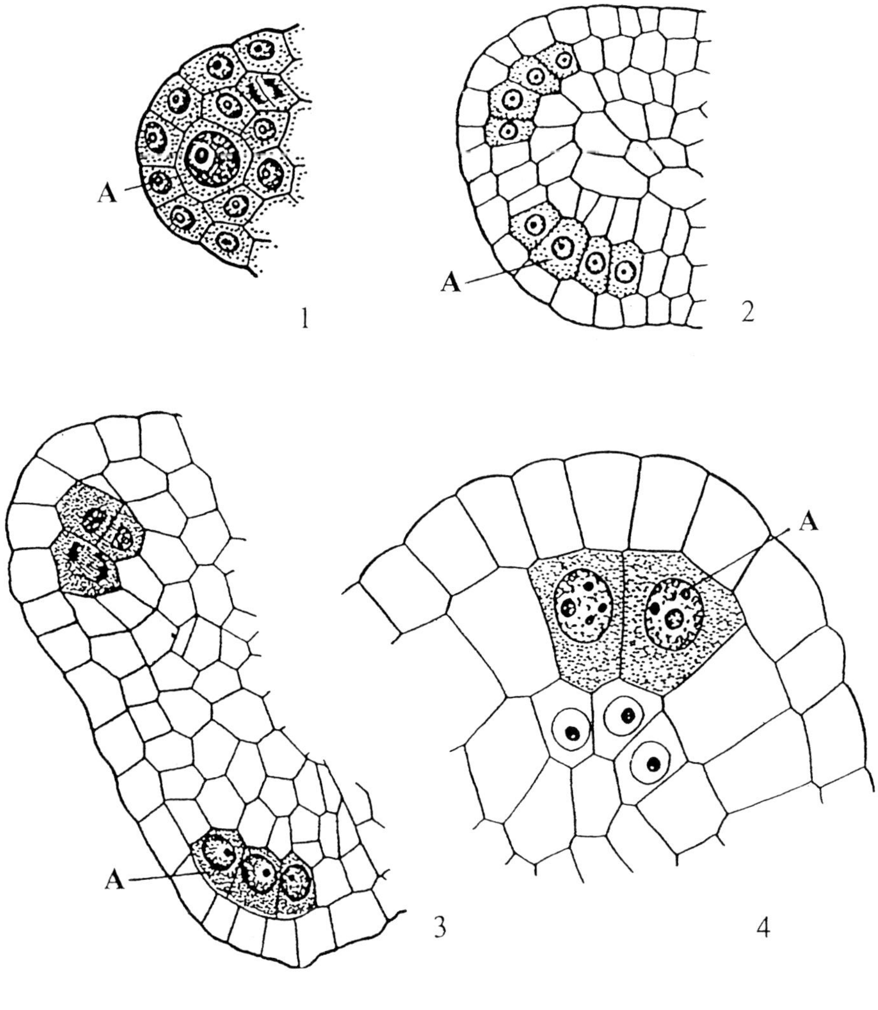

Plate 15: Archesporium.

1-4 – archesporium differentiation in anthers of *Zizyphus jujuba* (1), *Linum perenne* (2), *Allium caspium* (3) and *Helianthus annuus* (4); A – archesporium.

1 – after Shevchenko, orig.; 2 – after Rao, 1966; 3 – after Shamrov, 1990a; 4 – after Joshi and Pantulu, 1941.

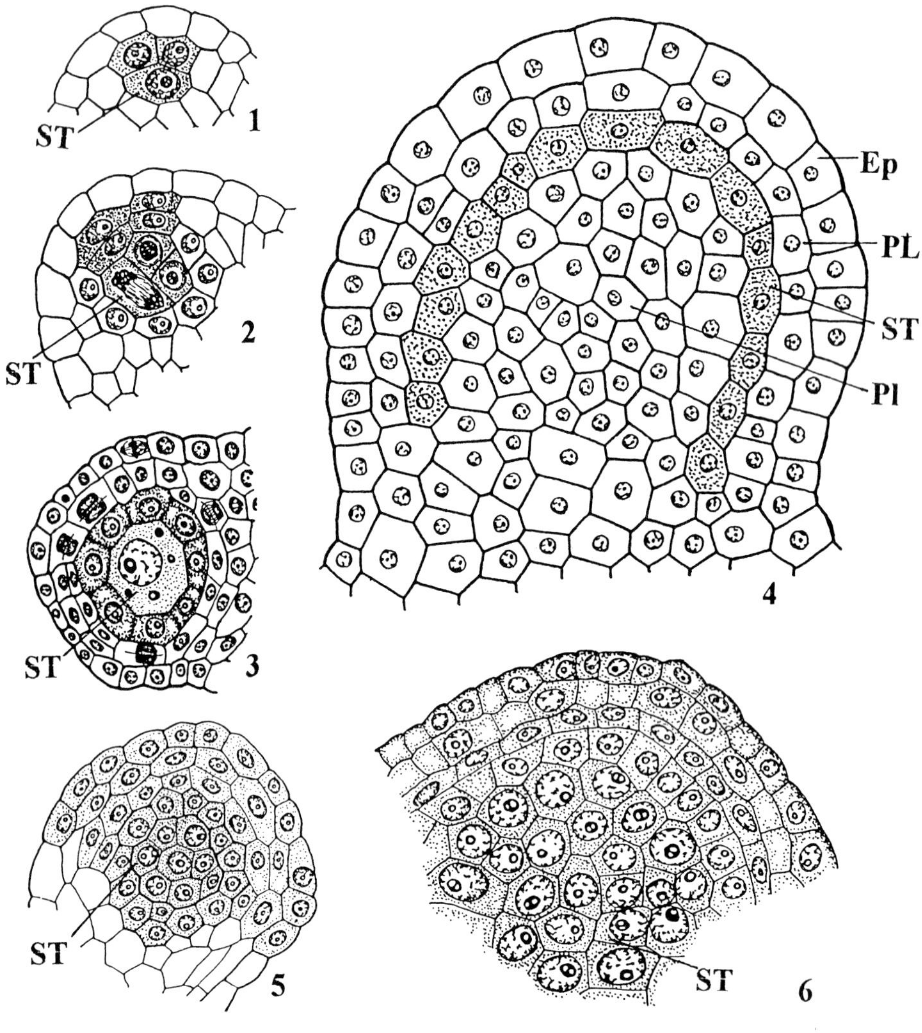

Plate 16: Sporogenous tissue.

1-6 – sporogenous tissue formation in *Allium caspium* (1, 2), *Scabiosa lucida* (3); *Ostrowskia magnifica* (4), *Davidia involucrata* (5) and *Rhoeo discolor* (6); Ep – epidermis, Pl – placentoid, PL – primary parietal layer, ST – sporogenous tissue.

1, 2 – after Shamrov, 1990a; 3 – after Kamelina, 1987a; 4 – after Kamelina and Zhinkina, 1989; 5 – after Shevchenko and Kamelina, 1987; 6 – after Freiberg, 1990.

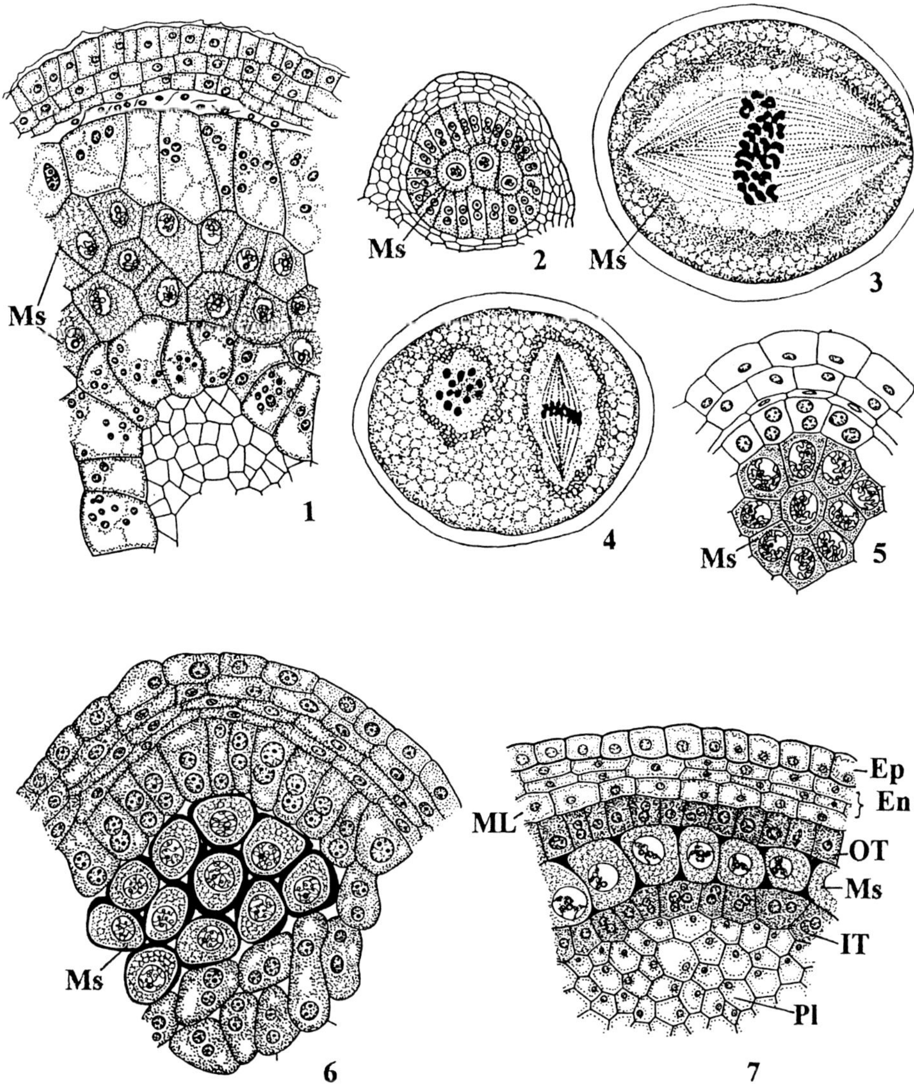

Plate 17: Microsporocyte.
1-7 – microsporocytes at different stages of development in *Campsis radicans* (1), *Linum perenne* (2), *Gossypium hirsutum* (3, 4), *Reseda lutea* (5), *Pittosporum tobira* (6) and *Ostrowskia magnifica* (7); En – endothecium, Ep – epidermis, IT – inner tapetum, ML – middle layer, Ms – microsporocyte, OT – outer tapetum, Pl – placentoid.

1 – after Freiberg, 1987a; 2 – after Rao, 1966; 3-4 – after Kamalova *et al.*, 1983; 5 – after Chaban, 1983; 6 – after Zhinkina and Kamelina, 1985; 7 – after Kamelina and Zhinkina, 1989.

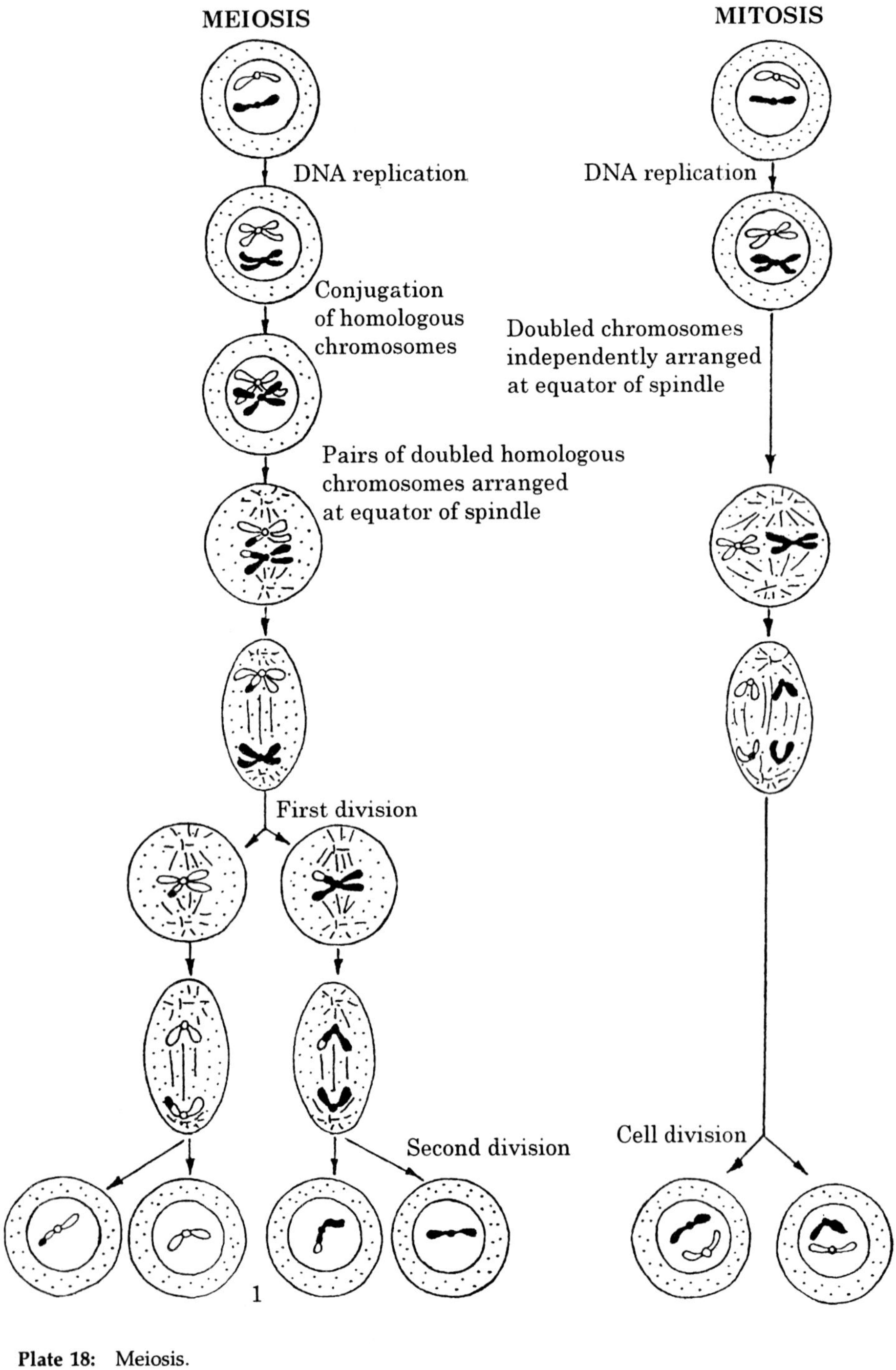

Plate 18: Meiosis.

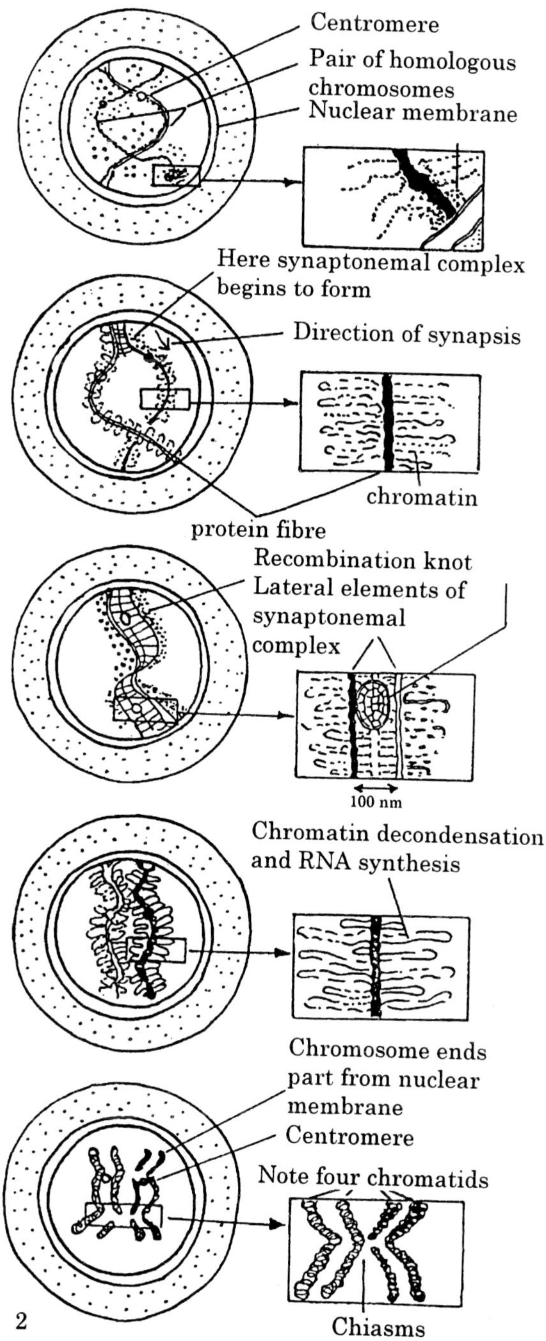

Plate 18 (*Contd.*)

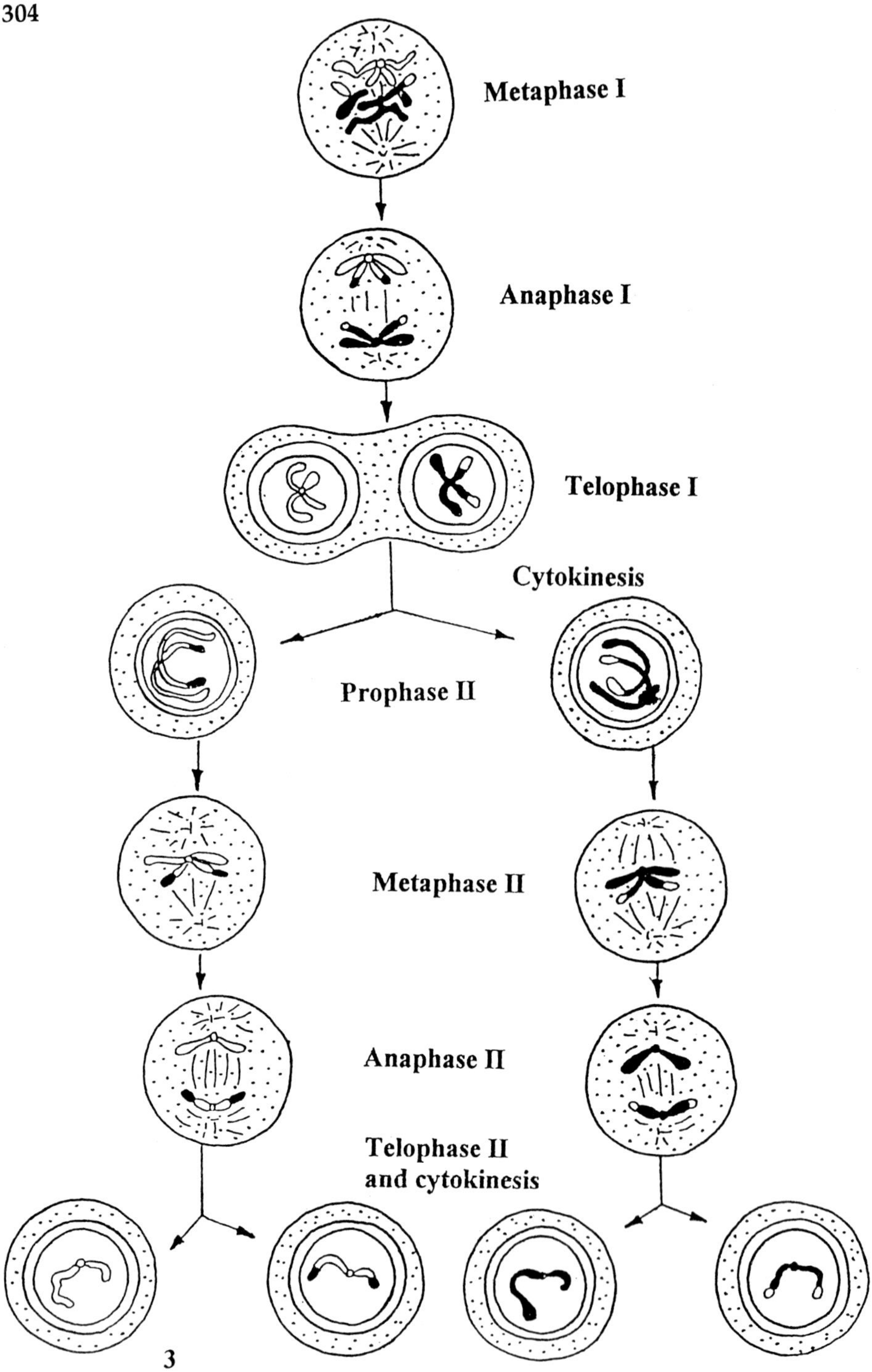

Plate 18 (*Contd.*)

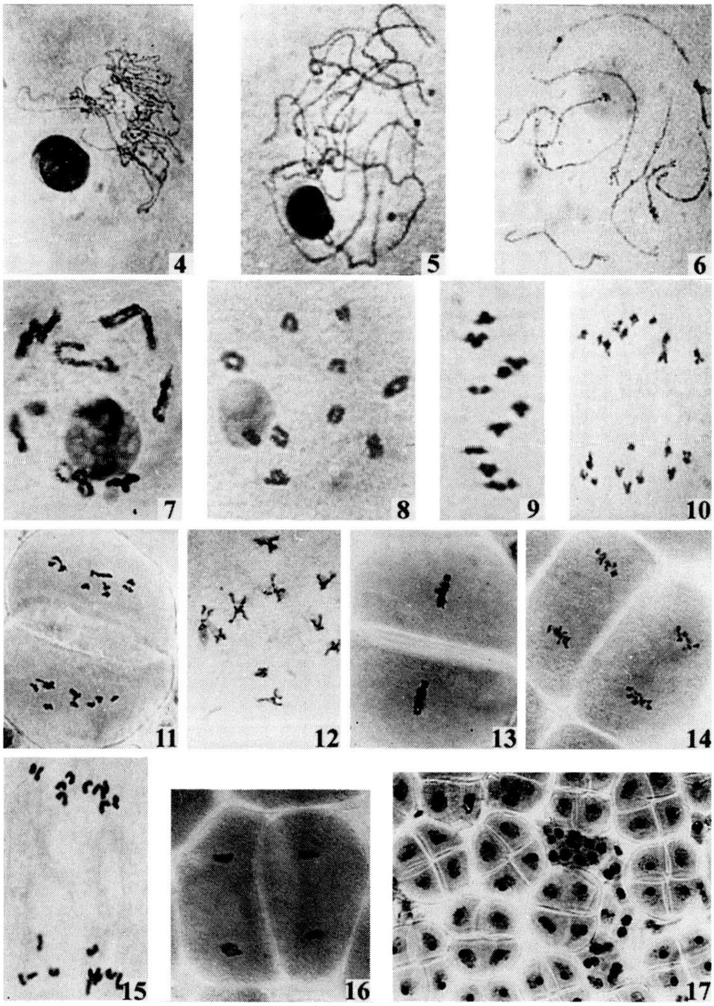

Plate 18: Meiosis. (*Contd.*)

1 – meoisis compared to usual mitosis (diagram). For the purpose of simplicity only one pair of homologous chromosomes is shown. Conjugation of homologous chromosomes takes place only in meiosis; since every chromosome is doubled before conjugation and consists of two sister chromatids, two cell divisions are necessary to form haploid gametes. Thus four haploid gametes arise from every diploid cell which undergoes meiosis. 2,3 – diagrams showing visible changes of two homologous chromosomes during meiosis. Here the events taking place in mammalian cells are presented, though a similar picture can be observed in cells of many other organisms also: 1 – five stages in prophase I of meiosis; 3 – rest of meiosis stages. For explanation see text; 4-17 – picture of typical meiosis in maize (microphotos, magnification 90 × 10, except dyad and tetrad stages (20 × 10): 4-8 – prophase I, 4 – zygotene, 5-6 – pachytene, 7 – diplotene, 10 bivalents with chiasmata distinctly visible, 8 – diakinesis: note 10 closed bivalents and a nucleus; 9 – metaphase I, 10 bivalents have lined up in equatorial plane, at some bivalents kinetochore fibres directed to poles visible; 10 – anaphase I, 10 chromosomes move to each pole, every chromosome consisting of two chromatids connected in the centromere area; 11 – stage of dyads, general view; 12, 13 – prometaphase II, chromosomes in same state as in anaphase I, as is perfectly shown at the specially magnified chromosomes of this stage (13); 14 – anaphase II, centromeres of sister chromatids divide, well seen in photo 15; 16 – stage of tetrads, 17 - general view of final stage of meiosis.

1-3 – after Alberts *et al.*, 1987; 4-17 – after Golubovskaya, 1988.

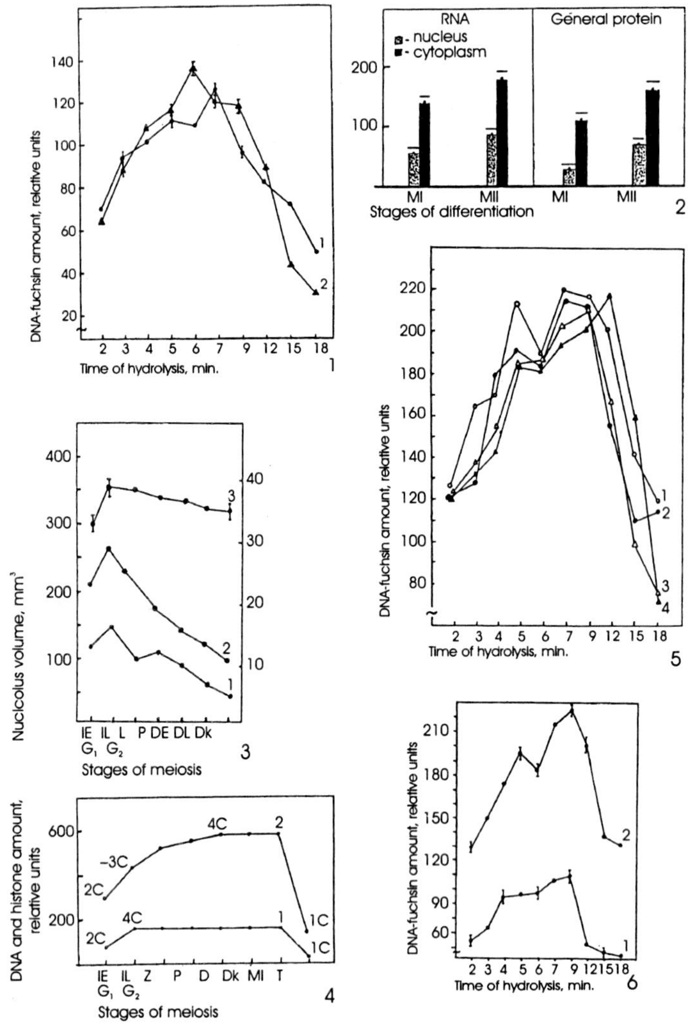

Plate 19: Cytochemistry of meiosis.

1 – hydrolysis curves for nuclei of sporogenous tissue cells of *Lilium candidum* at stages of beginning (1) and end of mitoses (2); 2 – content of RNA and (general?) protein in nucleus and cytoplasm of sporogenous tissue cells at stages of beginning (MI) and end of mitoses (MII); 3 – volume changing curves of the nucleolus (1) and RNA content in the nucleolus (2) and in nucleus (3) during meiosis in *Lilium candidum*; 4 – changing content of DNA (1) and histone (2) in microsporocytes of *Lilum candidum* during course of meiosis; 5 – hydrolysis curves for microsporocytes of *Lilium candidum* at stages of zygotene (1), pachytene (2), diakinesis (3) and metaphase I (4); 6 – hydrolysis curves for microsporocyte nuclei of *Lilium candidum* at stages of early (1) and late premeiotic interphase (2); C – stage; D – diplotene, DE – early diplotene, Dk – diakinesis, DL – late diplotene, IE – early interphase, IL – late interphase, L – leptotene, MI – metaphase I, MII – metaphase II, P– pachytene, T – tetrads, Z – zygotene.

1-5, 6 (2, 3) – after Bugara, orig.; 6(1) – after Reznickova, 1984.

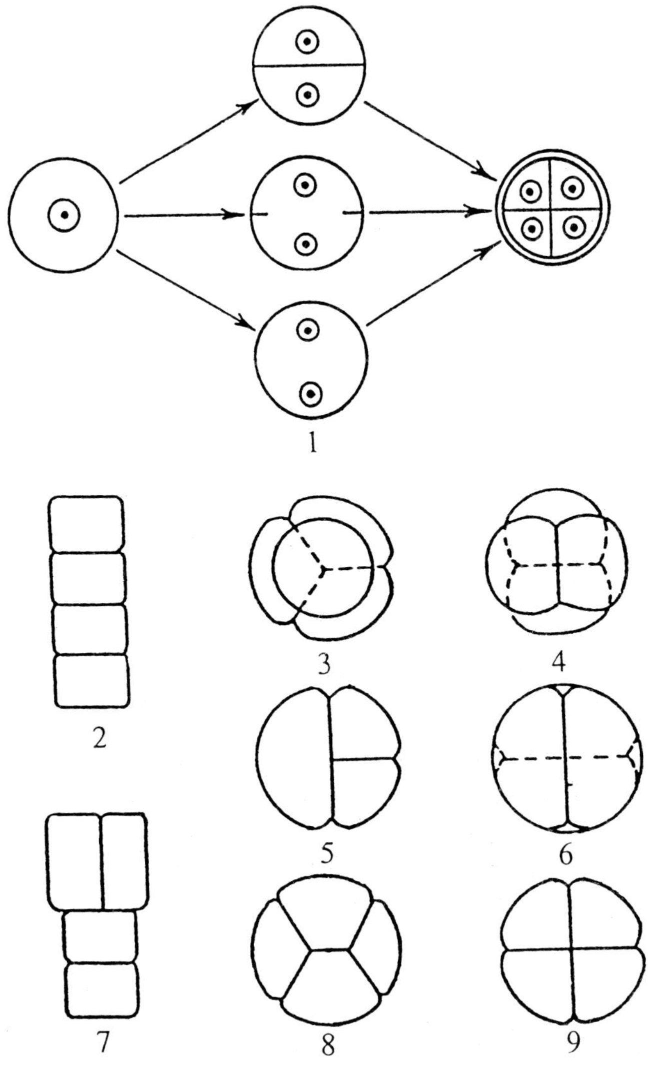

Plate 20: Microspore tetrads.

1 – microspore tetrad formation types: successive, simultaneous and intermediate; 2-9 – types of microspore tetrads: 2 – linear, 3, 4 – tetrahedral, 5, 6 – decussate, 7 – T-shaped, 8 – rhomb-shaped, 9 – isobilateral; 10-24 – types of microspore tetrads in different positions (schematic): 10-15 – tetrahedral, 16-19 – decussate, 20, 21 – isobilateral. Thin lines show outlines of microspore cell, just formed after meiosis; 22-41 – microspore tetrad formation in *Haloxylon aphyllum* (22-28), *Plumbago europaeum* (29-34), *Reseda lutea* (35-38) and *Orobanche pallidiflora* (39-41). A – microspore, B – 2 microspores, one completely covered by another on the slide, C – directions of turn of tetrad around its vertical (a-b) and horizontal (c-d) axes.

1 – after Kamelina, orig.; 2-24 – after Sladkov and Grevtsova, 1991; 22-28 – after Konycheva and Kadyrova, 1983; 29-34 – after Korobova, 1983; 35-38 – Chaban, 1983; 39-41 – after Teryokhin, 1987b.

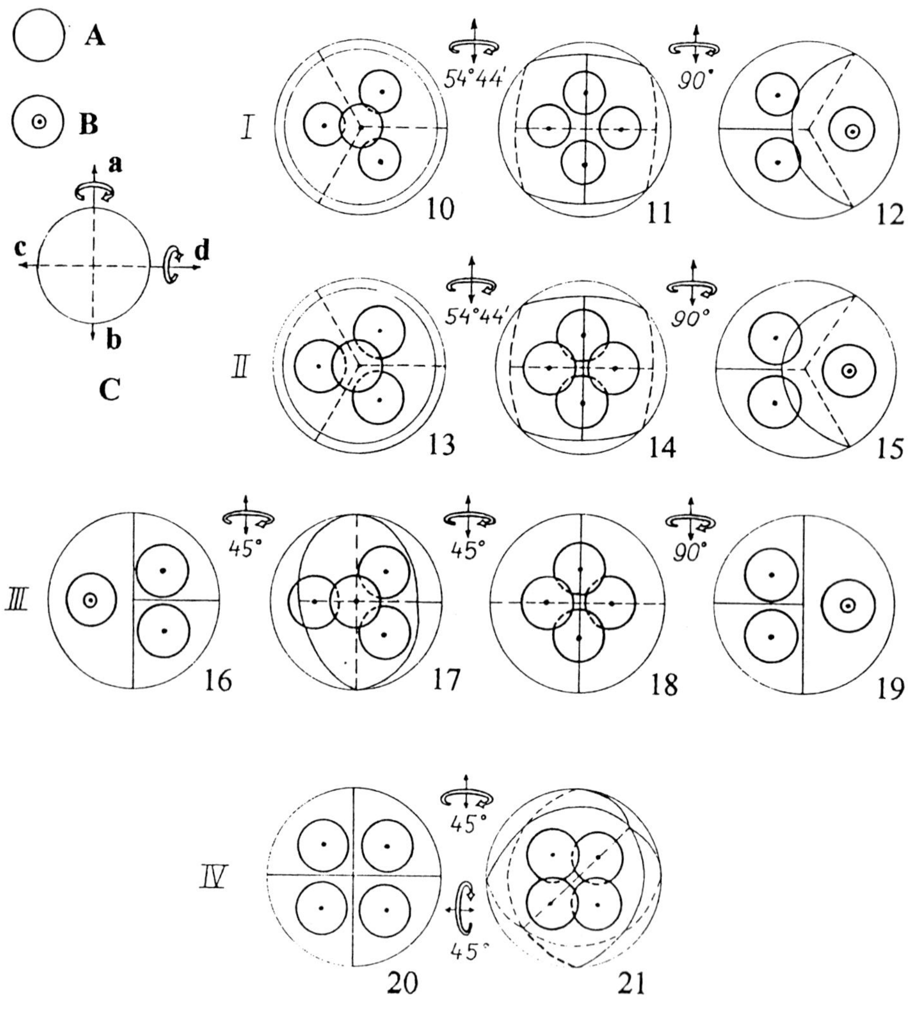

Plate 20 (*Contd.*)

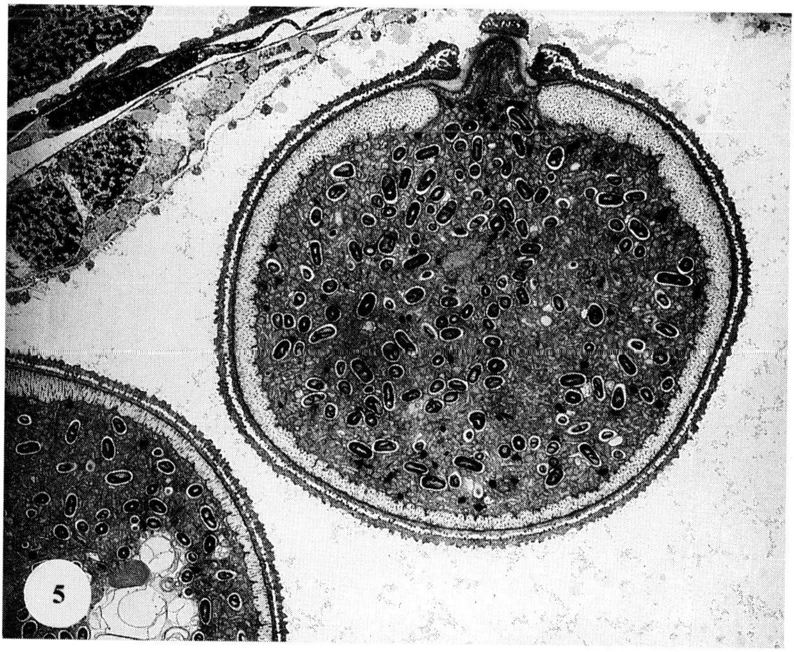

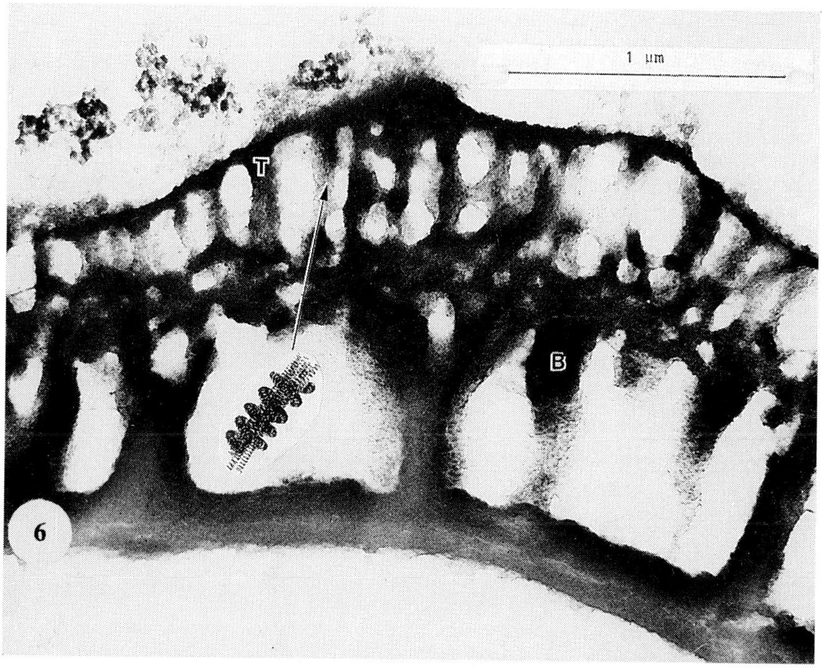

Plate 24 (*Contd.*)

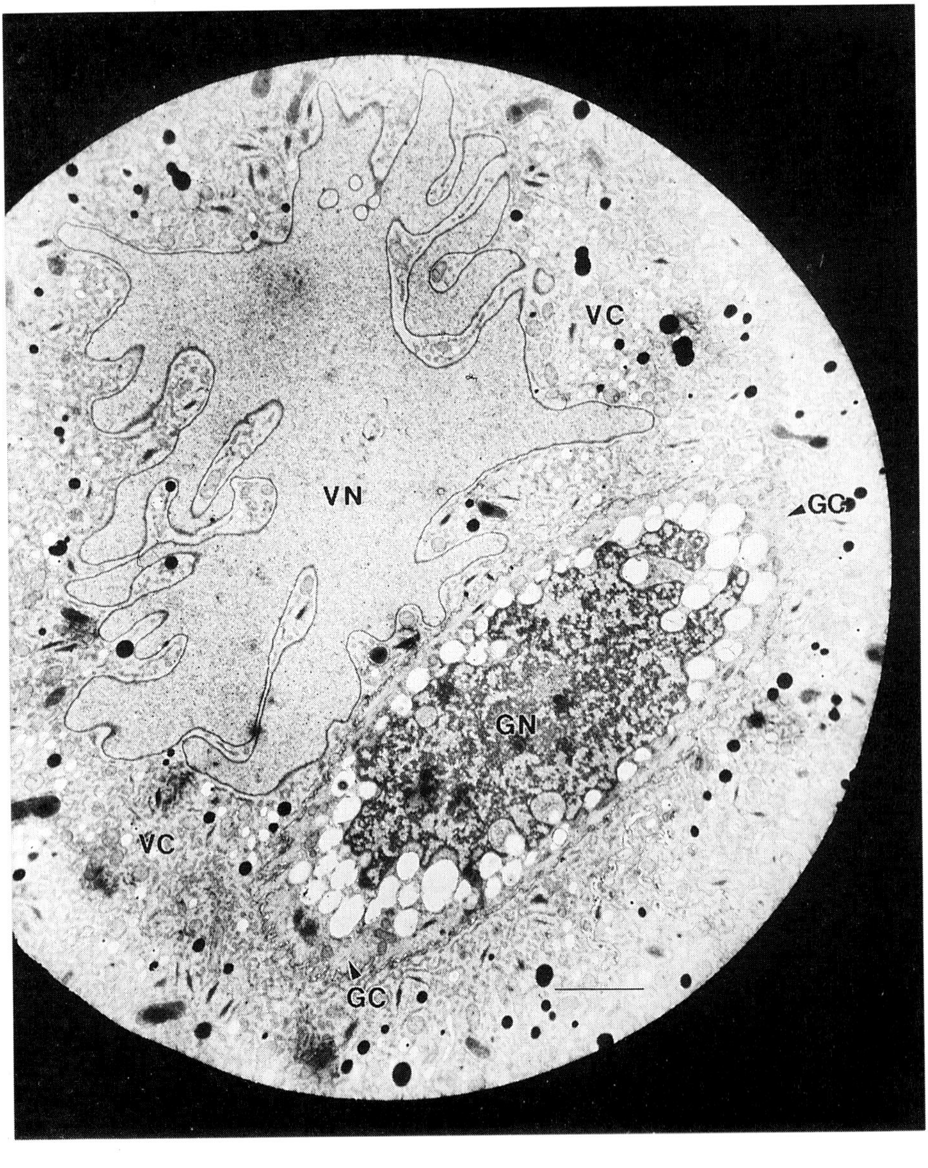

Plate 25: Ultrastructure of generative cell.

Cytoplasm of vegetative cell, nucleus of vegetative cell, generative cell and its nucleus in *Hippeastrum vitatum* (TEM); scale – 3 mcm; GC – generative cell, GN – nucleus of generative cell, VC – vegetative cell, VN – vegetative nucleus.

After Mogensen, 1986.

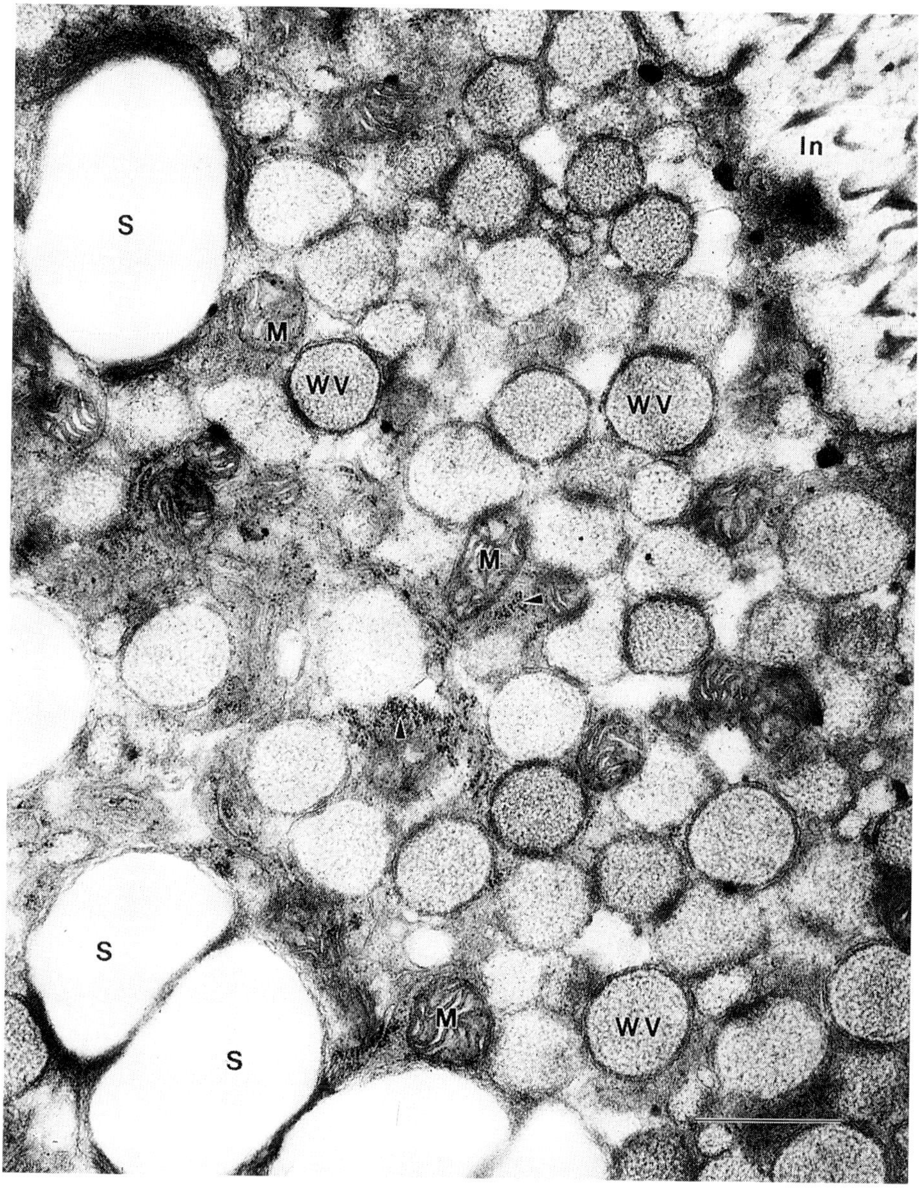

Plate 26: Ultrastructure of vegetative cell.

Part of a vegetative cell of wheat pollen grain with numerous vesicles, containing precursors of cell wall substance, mitochondria, plastids with starch, ribosome clusters (arrow heads without stems) and the intine of pollen grain wall (TEM); scale – 1 mcm; In – intine of pollen wall, M – mitochondria, S – amiloplast, WV – vesicles containing the precursors of cell wall components. After Mogensen and Ladyman, 1989.

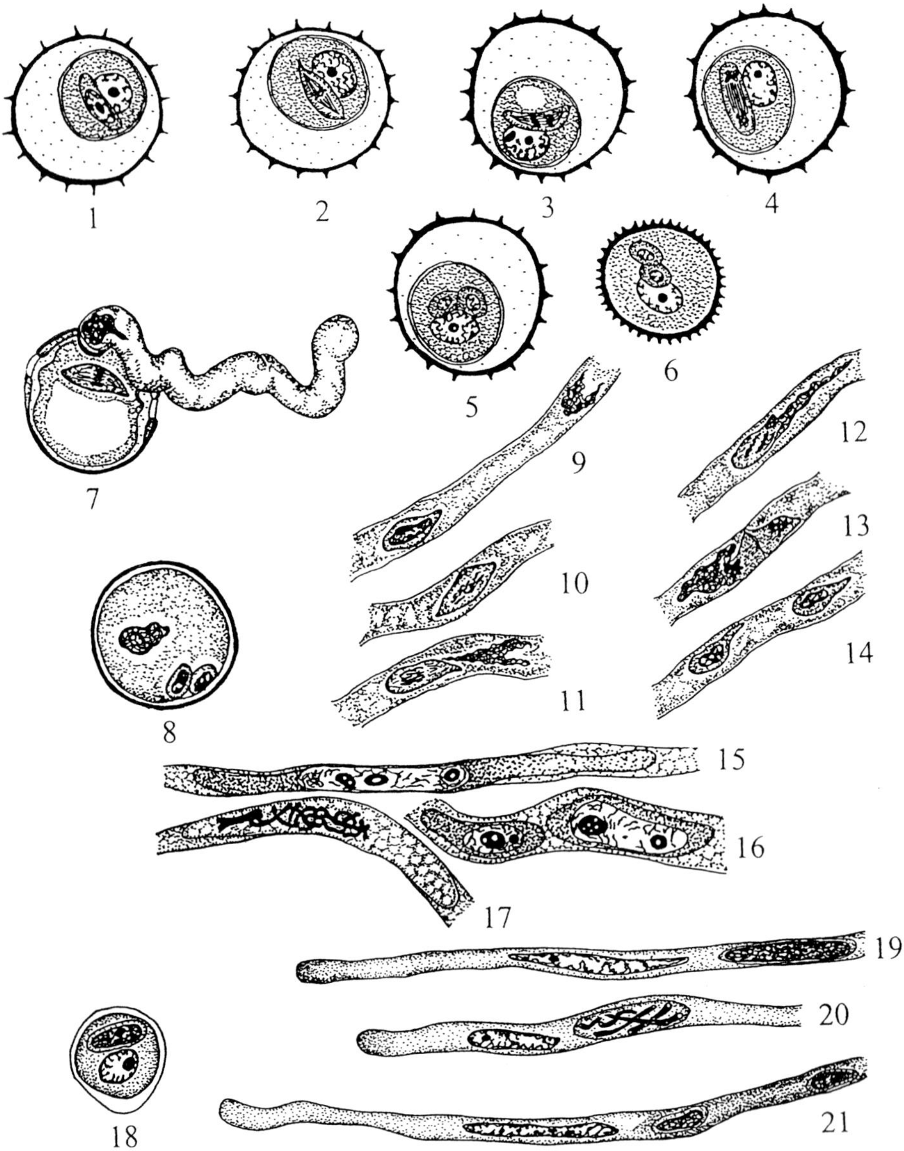

Plate 27: Sperm cell.

1-6 –generative cell division and sperm formation in pollen grain of *Eminium alberti;* 7, 8 – pollen grains of *Campanula rapunculoides* from a stigma (7 – germinating pollen grain with dividing generative cell, 8 – three-celled pollen grain); 9-14 – spermiogenesis in pollen tube of *C. rapunculoides;* 15-17 – spermiogenesis in pollen grains of *Tulipa;* 18-21 – two-celled pollen grain and pollen tubes, illustrating spermiogenesis in *Trillium camschatcense*.

1-6 – after Korobova and Zhinkina, 1990; 7-14 – after Korobova and Zhinkina, 1987; 15-17 – after Belyaeva, 1973; 18-21 – after Naumova, 1990.

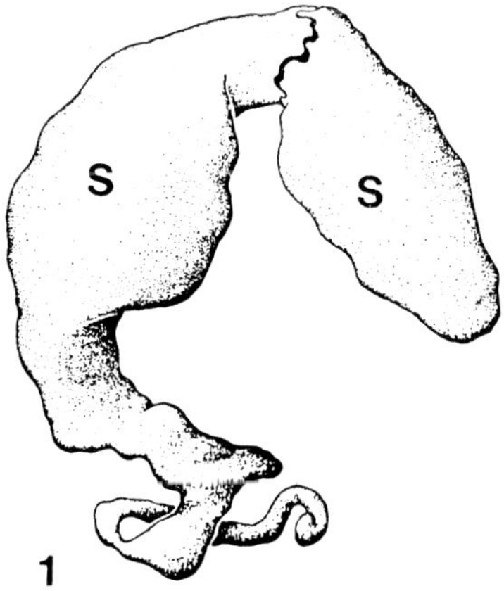

1

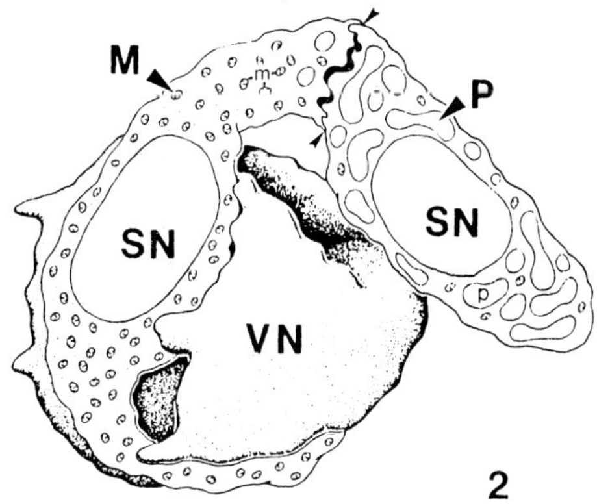

2

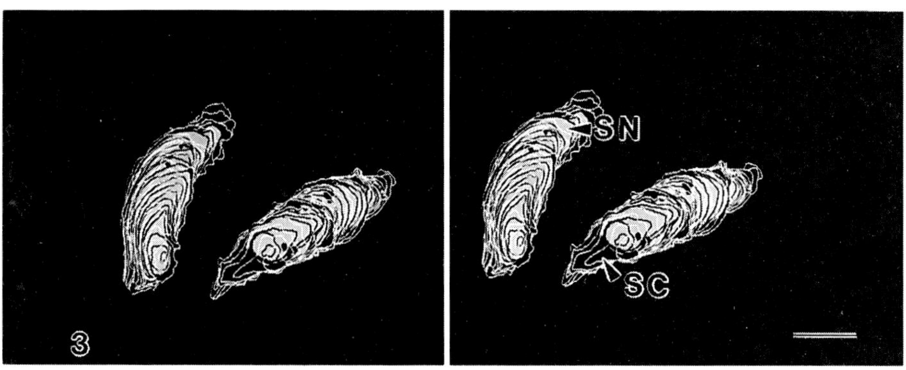

Plate 28: Sperm dimorphism.

1 – computer reconstruction of sperms of *Plumbago;* note differences between sperms in shape and size, and also their connection by common wall; 2 – inner appearance of two sperms (Fig. 1) together with vegetative nucleus; note differences between sperms in nuclei size, plastid and mitochondria amount and close connection between larger sperm and vegetative nucleus; arrowheads without stems point to common cell wall of sperms; 3 – stereoscopic pair of computer reconstruction of both sperms on the base of series of ultrathin sections of a mature pollen grain of barley: note similarity of sperms in shape and size, and also that they are not connected to each other and none is connected with vegetative nucleus; scale – about 3 mcm; M – mitochondria, P – plastid, S – sperm, SC – sperm cytoplasm, SN – sperm nucleus, VN – vegetative cell nucleus.

1, 2 – after Russell, 1984; 3 – after Mogensen and Rusche, 1985.

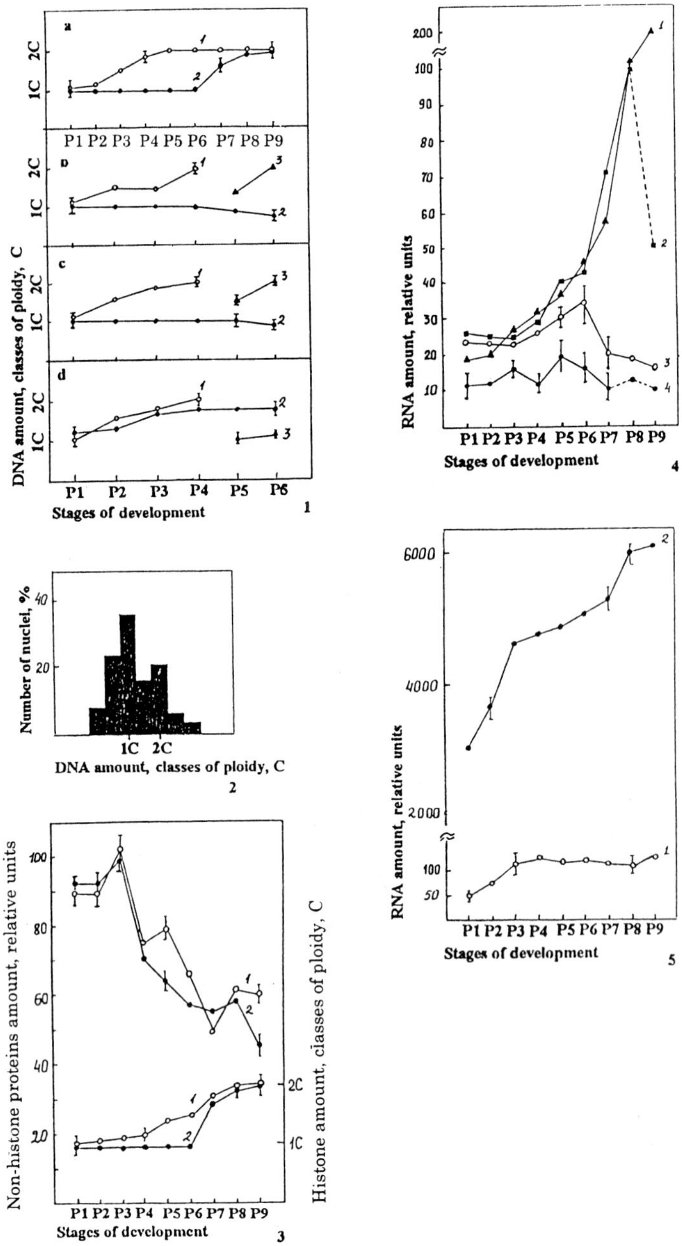

Plate 29: Pollen grain cytochemistry.

1 – DNA transformation curve in the process of pollen grain development in *Lilium candidum* (a), *Salvia sclarea* (b), *Lavandula angustifolia* (c) and *Coriandrum sativum* (d) (1 – generative cell, 2 – vegetative cell, 3 – sperms); 2 – histogram of *Coriandrum sativum* pollen grains distribution by DNA content in vegetative cell nucleus in the day of anther opening; 3 – histon and non-histon proteins content changing curves for the nuclei of generative (1) and vegetative (2) cells of developing pollen grain of *Lilium candidum*; 4 – RNA content changing curves for chromatin of nuclei of vegetative (1) and generative (2) cells and in a nucleolus (3, 4) in the process of pollen grain development in *Lilium candidum*; 5 – the changes of RNA content in the cytoplasm of generative (1) and vegetative (2) cells of developing pollen grain of *Lilium candidum*.

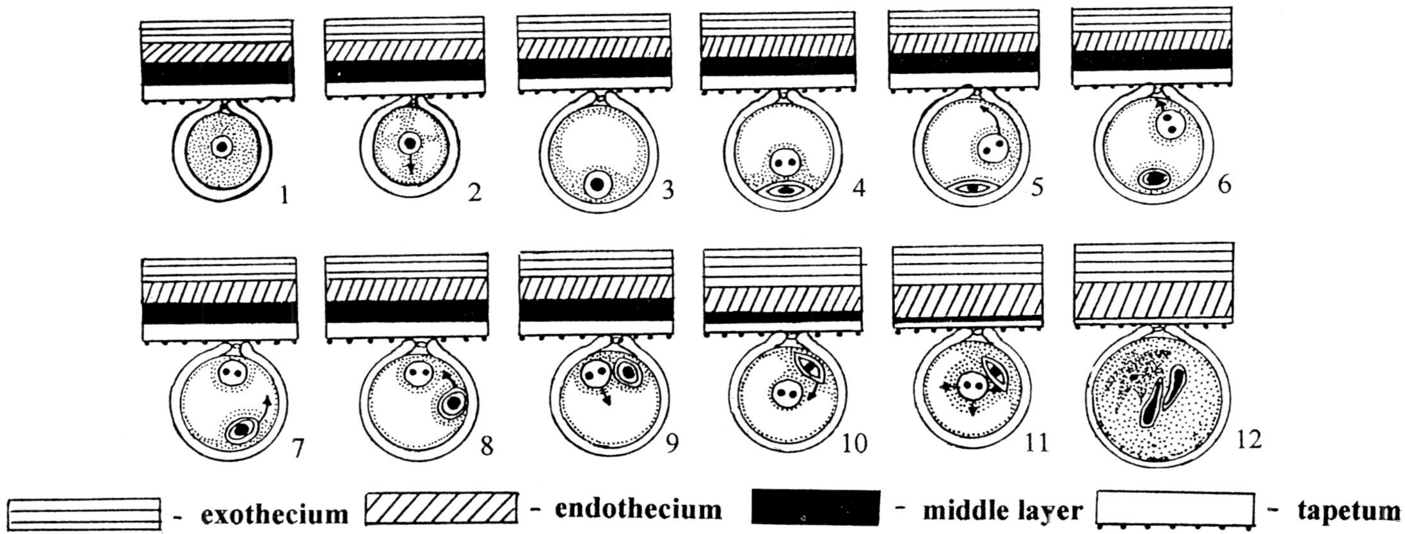

Plate 30: Nucleus and cell movement in developing pollen grains.

1-59 – main stages of pollen grain development and movement of nuclei of generative and vegetative cells in it in grasses: 1-12 – schematized (arrows show direction of nuclei and cell movement); 13-59 – *Triticum aestivum*, fixed (13-49) and living (50-59) material: 13-21 – microsporogenesis, 22-49 – formation of three-celled pollen grain, 50 – cytoplasm of microspore not vacuolized, 51 – beginning of vacuolization: small vacuoles arise, 52, 53 – end of vacuolization, nucleus locates by inner end of microspore, vacuole at outer end; microspore at end of growth period; nucleus situated beside inner end of the pollen grain, opposite pore; 54-55 – pollen grain just after first mitosis; vegetative cell nucleus in its initial position near generative cell; 56 – callose in wall of generative cell (under ultra-violet lighting); 57 – end of movement of vegetative cell nucleus, situated near pore; 58, 59 – end of generative cell movement; see starch grains in cytoplasm of vegetative cell. Pore with operculum well visible in all photos; Cl – callose, D – dyad of microspores, GC – generative cell, NG – nucleus of generative cell, NM – nucleus of microspore, NS – sperm nucleus, NuV – nucleolus of vegetative cell, NV – nucleus of vegetative cell, Op – operculum, P – pore, SC – sperm cytoplasm, Sp – sperms, St – starch, TM – tetrad of microspores, V – vacuole, VC – vegetative cell, Ex – exine, In - intine.

1-49 – after Batygina, 1987a; 50-59 – after Romanov, 1970c; Oryol, unpubl. data.

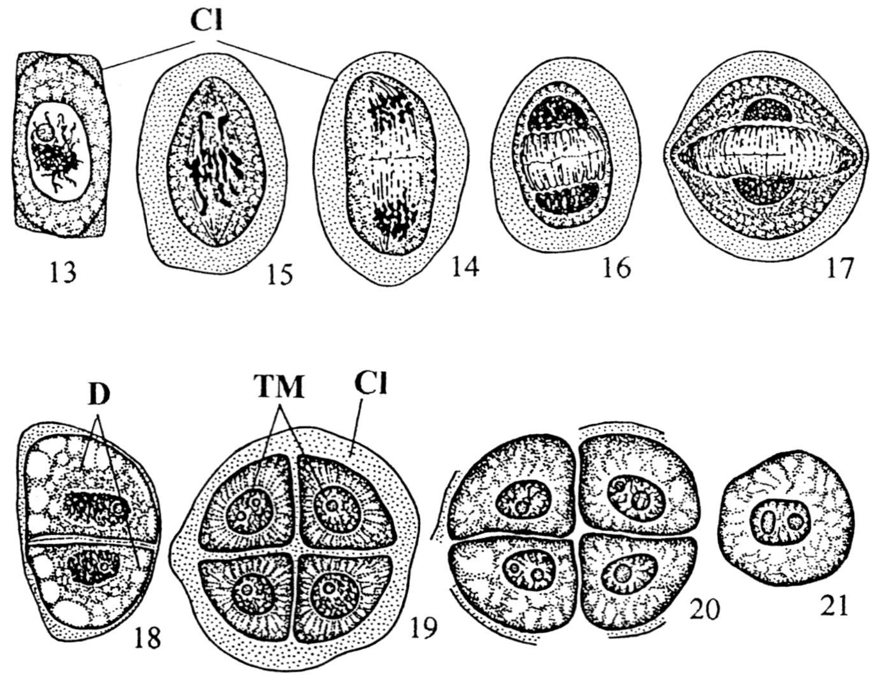

Plate 30 (*Contd.*)

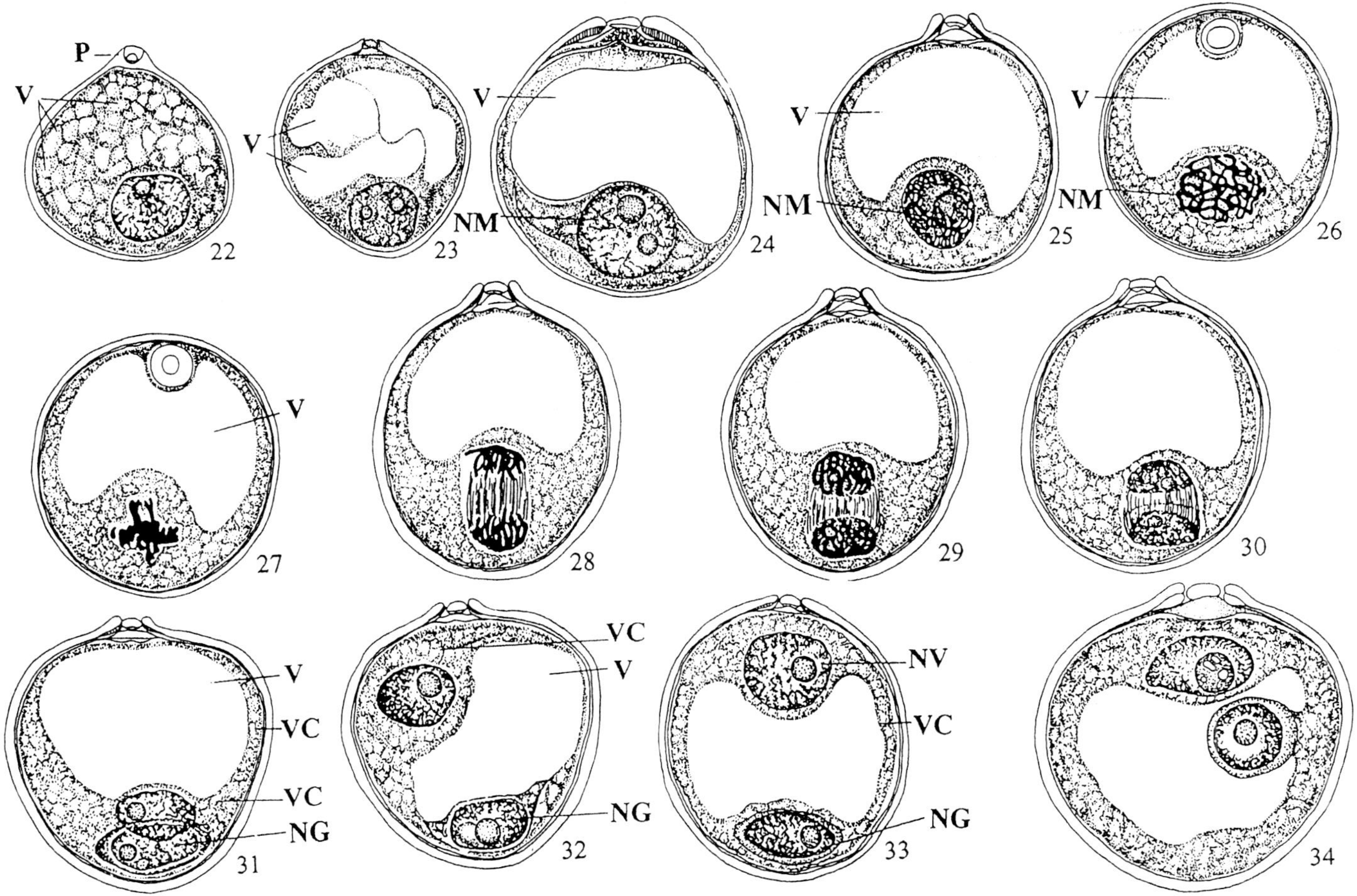

Plate 30 (*Contd.*)

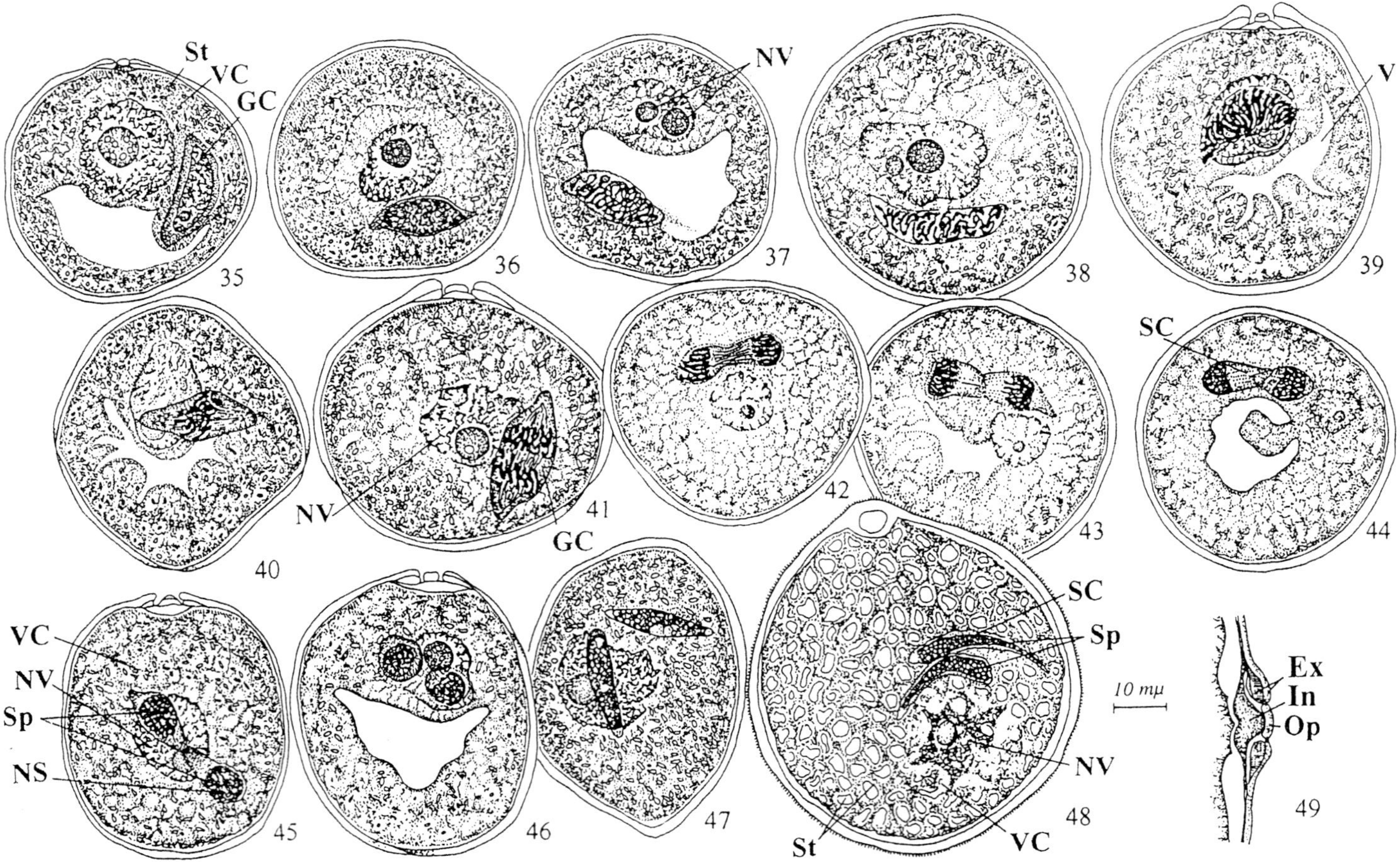

Plate 30 (*Contd.*)

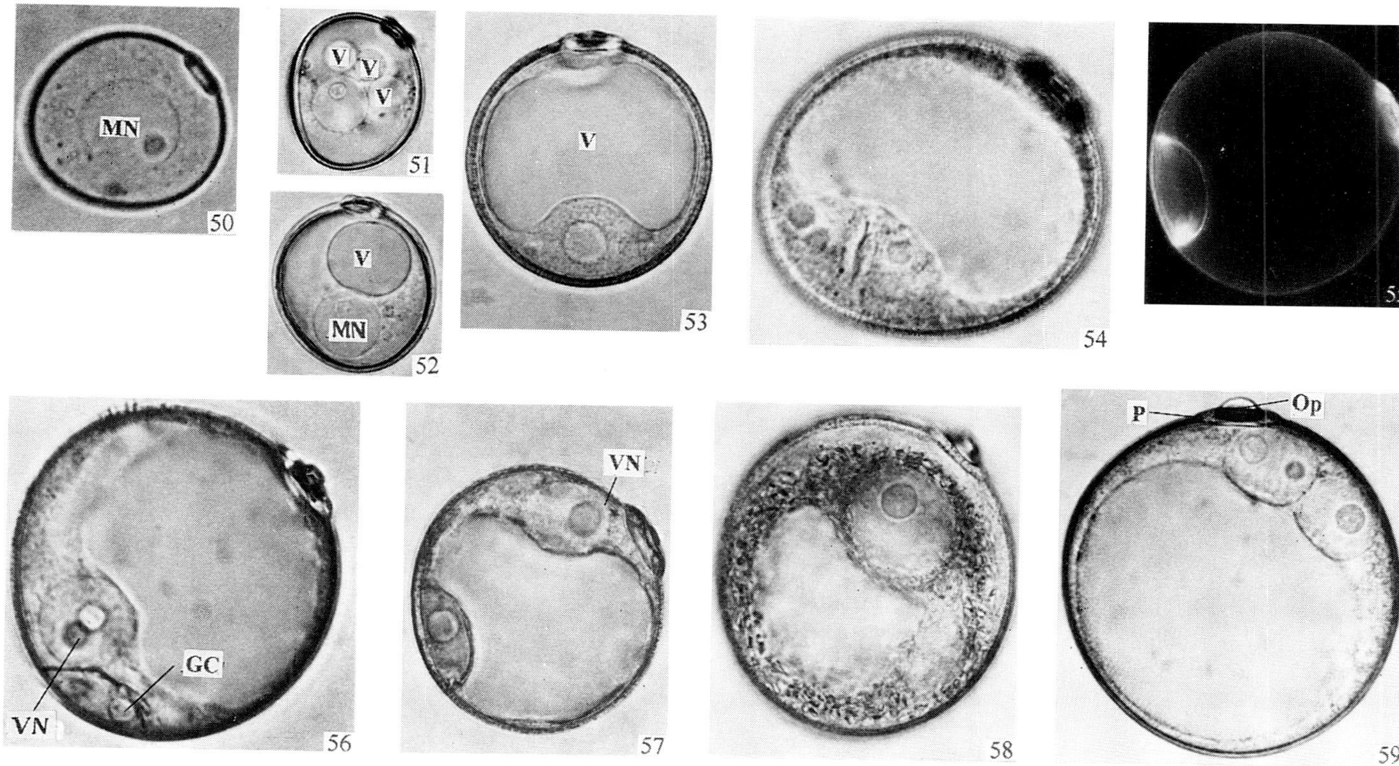

Plate 30 (*Contd.*)

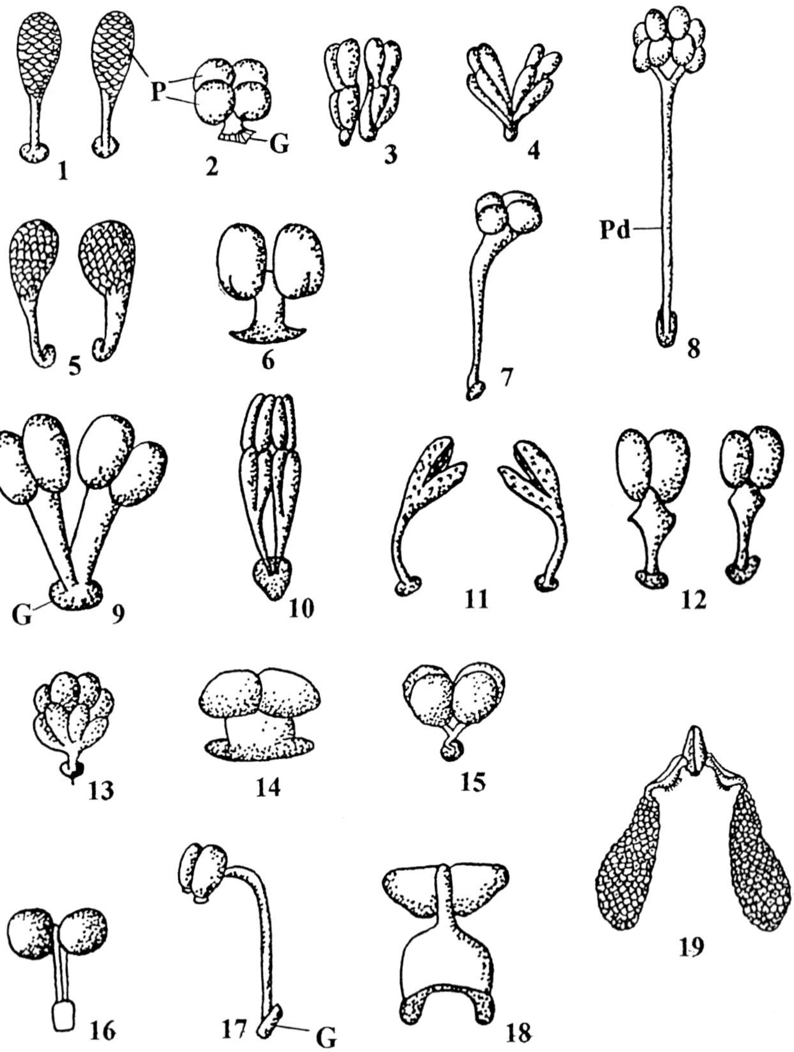

Plate 31: Pollinium.

1-18 – pollinia in orchids: 1-4 – number of pollinia in an anther: 1 – two, 2 – four, 3, 4 – eight; 5 - 8 – number of pollinia on one pedicle: 5 – one, 6 – two, 7 – four, 8 – eight; 9-12 – pollinia, attached to an adhesive glandule: 9, 10 – joined, 11, 12 – separate. Pollinium pedicle: 6 – quadrangular, 8 – long and thin, 9 – inversely conical, 11 – solid pedicle, 12 – hastate, 13 – short, 14 – thick, 15 – filiform, 16 – straight, 17 – bent. Adhesive glandule of a pollinium (retinaculum): 2 – flat, 4 – elliptic, 9 – disc-shaped, 10 – triangular, 13 – obovate, 15 – ovate, 17 – narrow and oblong-quadrangular. Pollinium shape: 2 – rounded, 4 – terete, 18 – broad-triangular. Pollinium texture: 1, 11, 15 – granular; 19 – pollinia in Asclepiadaceae; 20, 21 - pollen tetrads in *Victoria cruciana, Drimis winteri*; 22-34 – pollen and complex pollen formations of orchids: 22, 23 – single pollen grain of *Cypripedium insigne* before and after germination, 24-32 – non-germinated and germinated pollen grains in tetrads of *Calanthe veithii*, 33, 34 – non-germinated and germinated pollen grains in massula in *Platantera bifolia*, 35 - arrangement of three types of exine at surface of pollen grains connected in massula in *Peristylus spiranthes* (schematic); G – glandule of a pollinium, M – massula, P – pollinium, Pd – pollinium pedicle, T – tetrad, Tp – tapetum, I – type of I exine, II – type of II exine, III – type of III exine.

1-18 – after Fedorov and Artjushenko, 1975; 19 – after Tzvelev, 1981; 20 – after Khanna. 1967; 21 – after Bhandari and Venkataraman, 1968; 22-34 – after Poddubnaya-Arnoldi, 1976; 35 – after Zee and Siu, 1990.

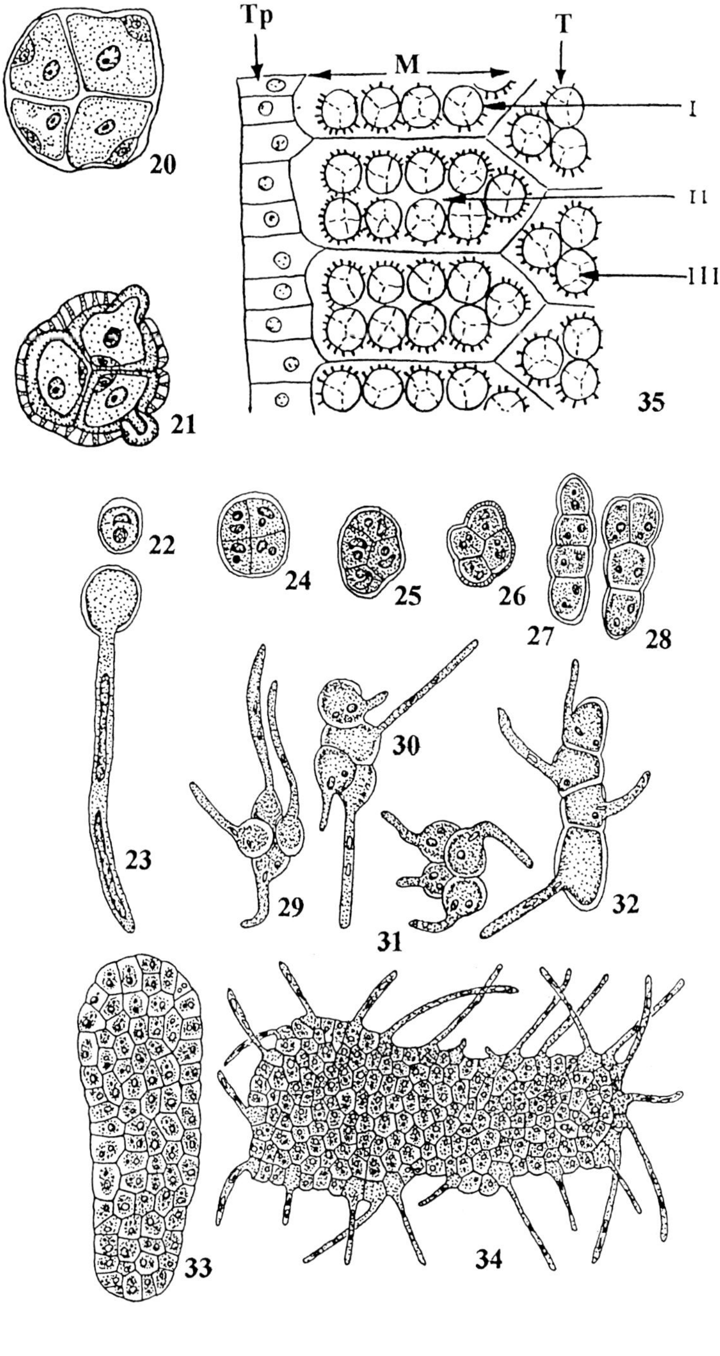

Plate 31 (*Contd.*)

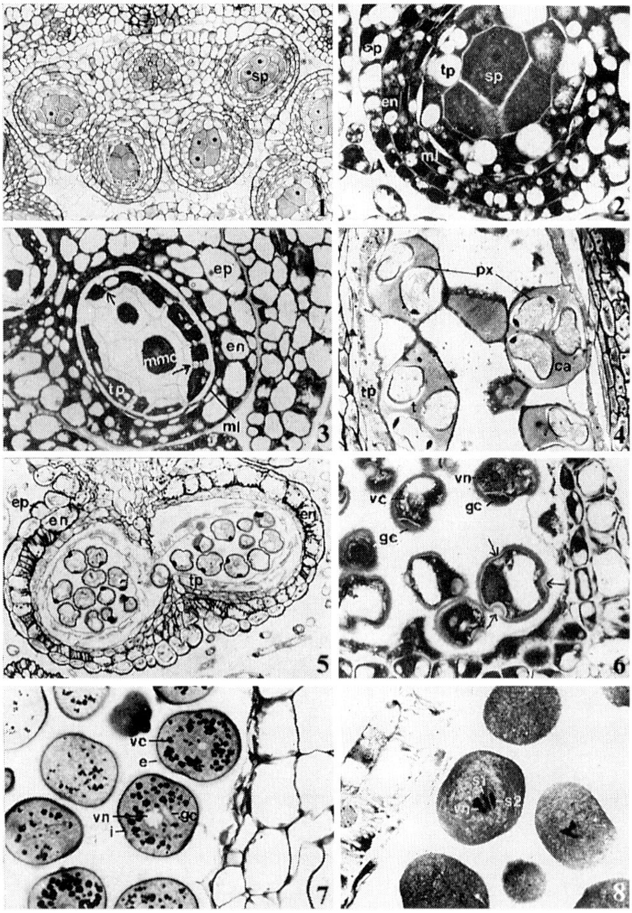

Plate 32: Anther: Histochemical and ultrastructural profile.

1-8 – *Procopis juliflora*. Light micrographs of cross-sections of anthers. 1 – stained with pyronin Y; 2, 3, 6, 8 – stained with Coomassie brilliant blue; 4, 5, 7 – stained with Periodic acid-Schiff's reagent (PAS). 1 – a tetrasporangiate anther showing sporogenous cells with RNA-rich cytoplasm and prominent pyroninophilic nucleoli. (× 250); 2 – a locule to show anther wall differentiation into epidermis, endothecium, two distinct middle layers and innermost vacuolate tapetum. Sporogenous cells are characterized by large nuclei and cytoplasm rich in nuclear proteins (× 750); 3 – anther locule at microspore mother cell stage. Intercellular connections between adjacent tapetal cells (arrows) are prominent (× 600); 4 – portion of anther locule to show microspore tetrads. Tapetal cell walls reveal dissolution. Each cell of the tetrad reveals blueprint of the primexine that stains with PAS reagent. Around the tetrads, cellulosic mother cell walls persist and individual microspores show development of incipient apertures (× 600); 5 – anther locules showing endothecium with well-developed, U-shaped, PAS positive, fibrous thickenings (× 300); 6 – portion of anther locule at early maturation pollen grain stage. Mitosis results in formation of small generative cell near wall of microspore. Three distinct germinal apertures with poorly developed exine evident (× 750); 7 – mature pollen grains at the time of anthesis. Exine reveals faint hue, whereas intine stains intensely. Generative cell spindle-shaped, possesses thin, PAS positive wall and is embedded in vegetative cell cytoplasm (× 750); 8 – generative cell divides precociously into two elongate sperm cells which stain intensely for total proteins. Vegetative nucleus and the pair of sperm cells in close association (× 600); ca – callose, e – exine, en – endothecium, ep – epidermis, gc – generative cell, i – intine, ml – middle layers, mmc – microspore mother cell, px – primexine, s1, s2 – sperm cells, sp – sporogenous cell, t – terad, tp – tapetum, vc – vegetative cell, vn – vegetative nucleus.

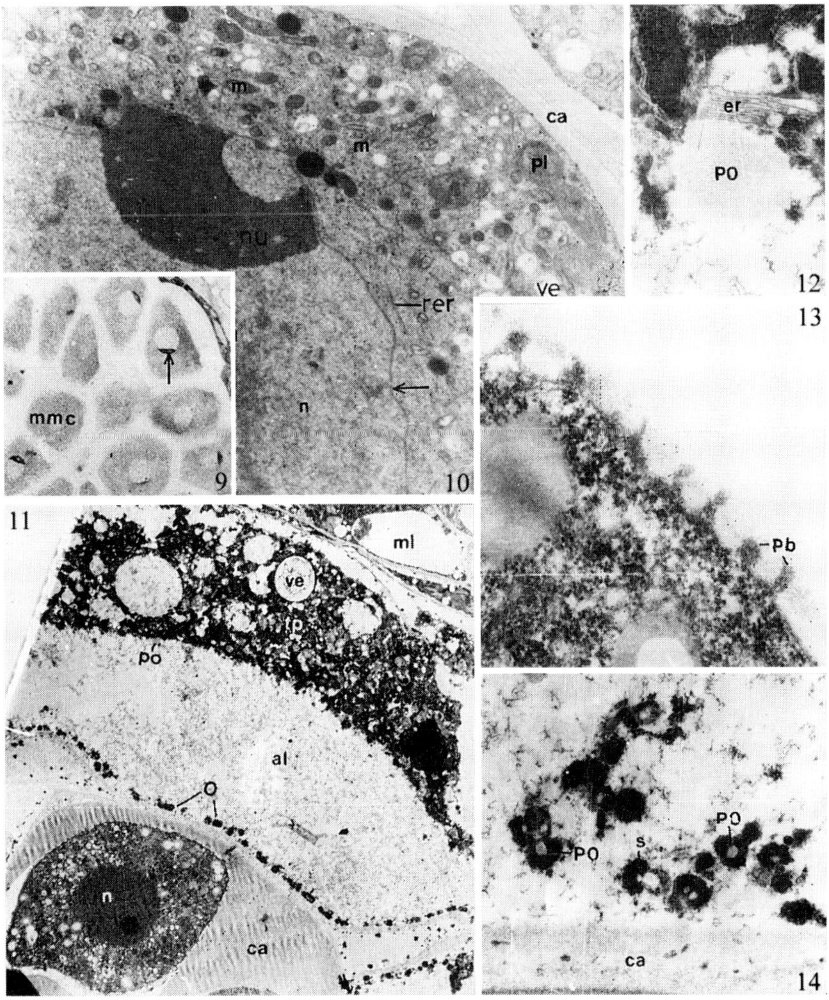

Plate 32 (*Contd.*)

9, 10 – *Sesbania speciosa;* 11-14 – *Prosopis juliflora.* 9 – light micrograph of portion of transection of anther stained with pyronin Y to show feebly stained cytoplasm and pyroninophilic nucleoli of microspore mother cells. Crescent-shaped, darkly stained polar nucleolar-cap noteworthy (× 1000); 10 – electron micrograph of portion of microspore mother cell during early prophase (pachytene stage). Crescent-shaped nucleolar-cap adpressed to nuclear membrane (arrow). Cell cytoplasm shows high profiles of rough endoplasmic reticulum, ribosomes, pleomorphic mitochondria, plastids and vesicles (× 19,025); 11 – electron micrograph of anther locule showing portion of middle layer, tapetum and microspore tetrads. Tapetal cytoplasm dense and extremely vesicular. Pro-orbicules continuosly extruded through tapetal cell membrane into anther locule where they are coated with sporopollenin and transformed into orbicules (× 4350); 12 – portion of tapetal cell from 11 magnified to show pro-orbicule closely associated with stacked cisternae of endoplasmic reticulum (× 39,000); 13 – peripheral portion of microspore tetrad. Probacules appear as small, square to wedge-shaped blocks and radiate from microspore plasmalemma (× 38,750); 14 – portion from 11 magnified to show accretion of dense plaques of sporopollenin around the pro-orbicules (× 18,900); a – anther locule, ca – callose, m – mitochondrion, ml – middle layer, mmc – microspore mother cell, n – nucleus, nu – nucleolus, o – orbicule, pb – probacule, pl – plastid, po – pro-orbicule, rer – rough endoplasmic reticulum, s – sporopollenin, tp – tapetum, ve – vesicle.

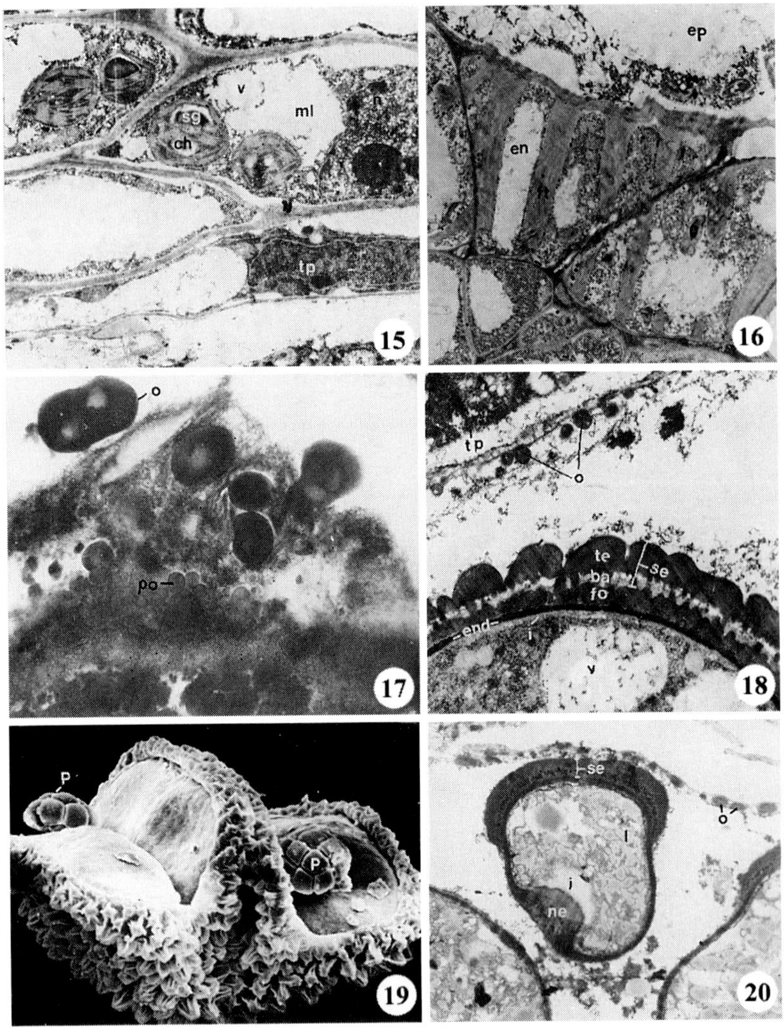

Plate 32 (*Contd.*)

15-18 – *Prosopis juliflora;* 19, 20 – *Albizzia lebbek;* 15 – electron micrograph of portion of middle layer cells at microspore tetrad stage. Many chloroplasts with well developed thylakoids and starch grains are evident (× 13,250); 16 – endothecium at vacuolate microspore stage show well-developed U-shaped thickenings with distinct fibrillar texture (× 4875); 17 – periphery of degenerating tapetum showing complex orbicules lining anther locules at uninucleate microspore stage (× 22,275); 18 – portion of vacuolate microspore adjacent to degenerative tapetum. Microspore cytoplasm becomes progressively less dense and more vacuolate, with concomitant increase in thickness and complexity of exine. Pollen wall well differentiated into sexine (tectum and bacula) and nexine 1 (foot-layer). Nexine 2 (endexine) is more electron-dense than nexine 1. Intine development still in progress (× 18,900); 19 – scanning electron micrograph of anther aplit open to show dehiscing polyads (× 550); 20 – transmission electron micrograph of portion of polyad to show lipid-rich pollen grains. Exine bilayered with sexine confined to only outer surface of polyad, whereas nexine attached to intine layer and surrounds each pollen grain. Exine with outer circular surface due to contact pressure with orbicules (× 3800); ba – bacula, ch – chloroplast, en – endothecium, end – endexine, fo – foot-layer, ft – fibrous thickening, i – intine, ml – middle layer, n – nucleus, ne – nexine, o – orbicule, p – polyad, po – pro-orbicule, se – sexine, sg – starch grain, te – tectum, tp – tapetum.

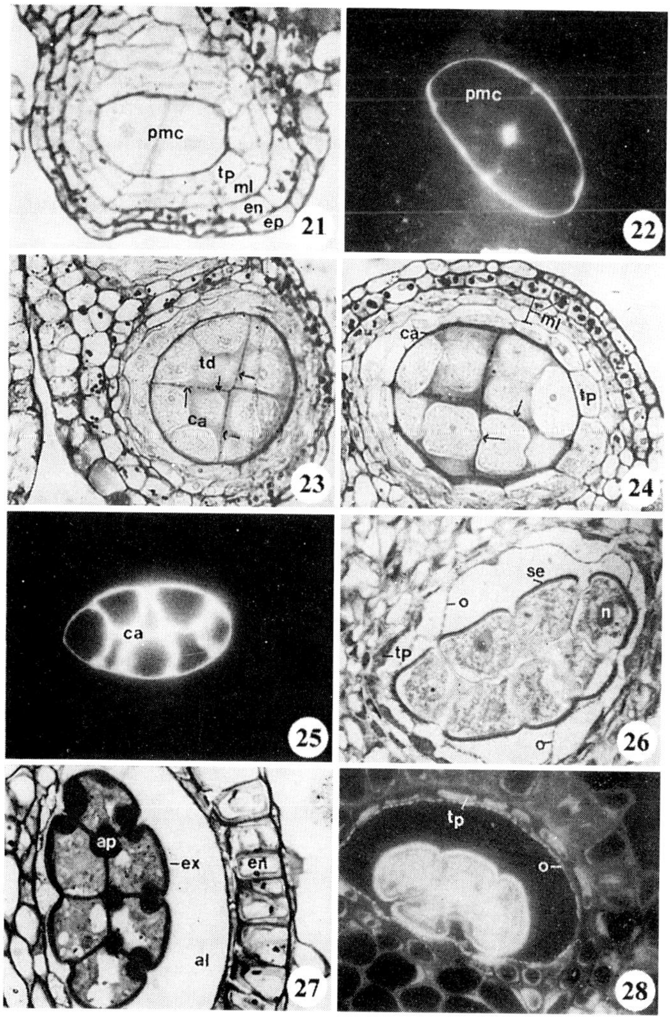

Plate 32 (*Contd.*)

21-28 – *Albizzia lebbek*. Transections of anthers. 21, 23, 24, 27 – light micrographs stained with PAS reagent; 26 – stained with Schiff's reagent; 22, 25, 28 – fluorescence micrographs; 21 – anther locule to show two pollen mother cells. Anther wall layers reveal prominent epidermis, endothecium, middle layer and tapetum. (× 250); 22 – anther locule at 4-celled pollen mother cell stage. Continuous callosic wall envelops pollen mother cells. Fluorescing point at the centre reveals junction where pollen mother cell walls meet. (× 225); 23, 24 – anther locules to show tetrads. Tetrads enveloped by callosic wall which shows cytomictic channels (arrows) on inner walls. 23 (× 500), 24 (× 600); 25–the individual grains of polyad separated from one another by callose deposition which is thin on outer face and thick between the pollen grains. (× 600); 26 – polyad at uninucleate stage to show distinctly stained nuclei. Tapetum shows lysis and tapetal derived orbicules closely invest the developing polyads. (× 600); 27 – anther locule to show polyad at binucleate stage. Within polyads, a few pores are located on outer face whereas others occur at corner. Endothecium characterized by PAS positive fibrous-thickenings. (× 600); 28 – anther locule to show pollen exine that fluoresces intensely for sporopollenin. Inner tangential walls of tapetum lined with fluorescing orbicules that form a sac around the polyad. (× 600); a – anther locule, ap – aperture, ca – callose, en – endothecium, ep – epidermis, ex – exine, ml – middle layers, n – nucleus, o – orbicules, pmc – pollen mother cells, se – sexine, td – tetrad, tp – tapetum.

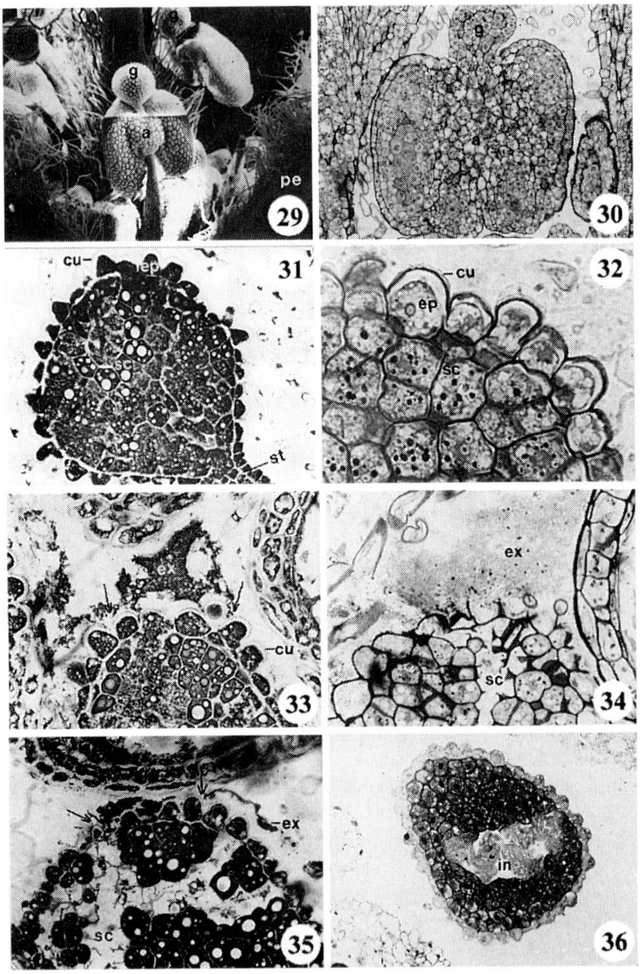

Plate 32 (*Contd.*)

29-36 – *Prosopis juliflora*. 29 – scanning electron micrograph; 30, 32, 34 – light micrographs stained with PAS reagent; 31, 33, 35, 36 – light micrographs stained with Coomassie brilliant blue.; 29 – mature floral bud split open to show anthers surmounted by glands. (× 50); 30 – longitudinal section of young anther with apical gland. Gland is an extension of anther connective, lies above anther locules and forms an appendage. Gland cells bound by well-defined rectangular epidermal cells. (× 250); 31 – mature anther gland shows well-stained secretory cells of globose head rich in proteinaceous inclusions. Epidermal cells densely cytoplasmic and overarched by cuticular layer. (× 250); 32 – portion of anther gland prior to extracellular secretion. Cells of glandular head rich in intracellular polysaccharidic secretions. Prominent cuticular layer veneers epidermal cells and secretory cells gorged with polysaccharide grains. (× 600); 33, 35 – portions of anther glands during secretory phase, to show degenerative protoplasts of secretory cells, whose contents and exudate are protein-rich. When the secretory pressure becomes maximum, the cuticule is distended from the epidermal cell wall and the exudate is forced out through the small cuticular pores (arrows). (× 480); 34 – same, showing secretory cells lysis and exudation of lysed contents along with granules (arrows) that stain positively for polysaccharides. (× 480); 36 – same, to show an insect, perhaps during foraging trapped within anther gland. (× 200); a – anther, cu – cuticle, ep – epidermis, ex – exudate, g – gland, in – insect, pe – petal, sc – secretory cells, st – stalk.

M.R. Vijayaraghavan, Bharti Chaudhry and Rajesh Shah, orig.

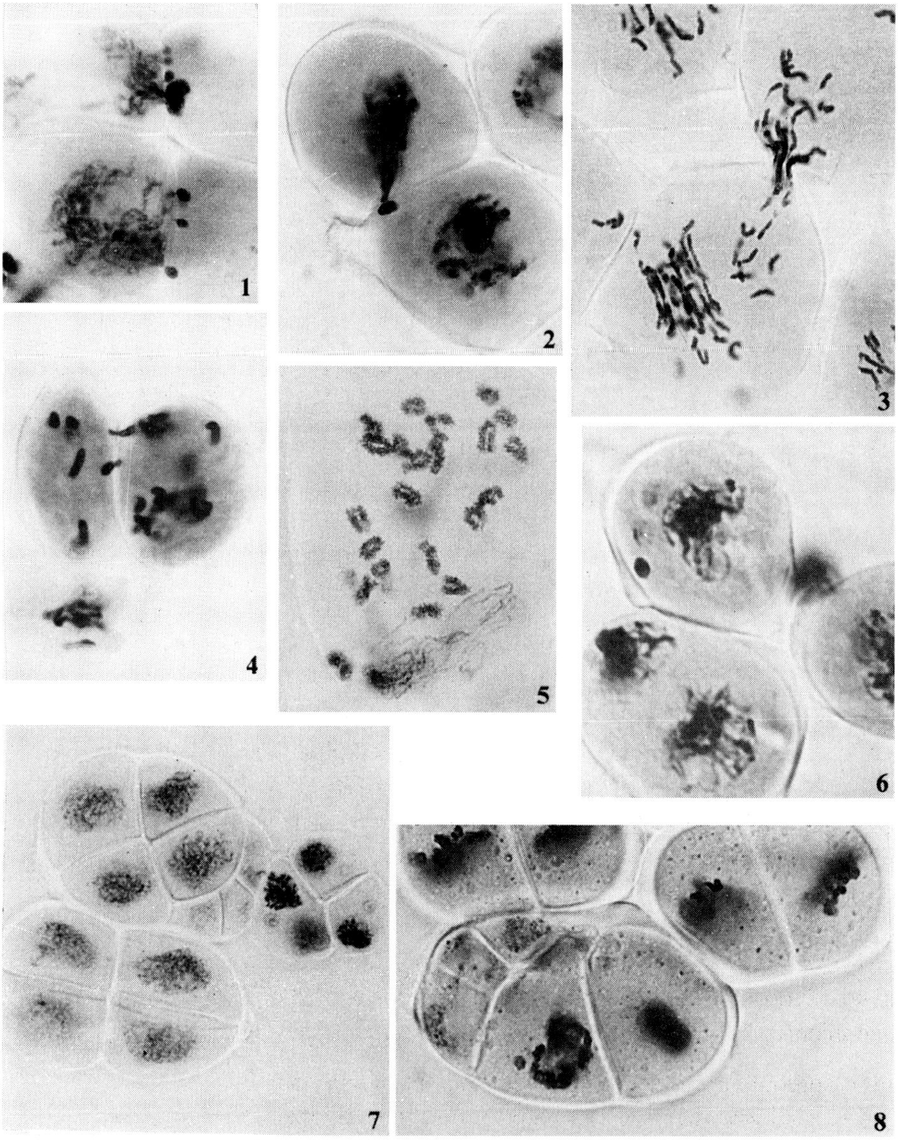

Plate 33: Cytomixis.

1-8 – process of migration of protoplasts in sporogenous cells of wheat-rye amphydiploid hybrids and results of cytomixis: leptonema (1), pachynema (2), metaphase (3, 4), prophase chromosomes and drops of chromatin in diakinesis (5), same in pachynema (6), microcells, separated around chromatin which has migrated and microspore tetrad after losing chromatin (7), same at stage of microspore dyads (8).

1-3 – after Orlova, 1971; 5, 7 – after Orlova, 1976 (see Poddubnaya-Arnoldi, 1976, pp. 126, 127); 4, 6, 8 – after Orlova, orig.

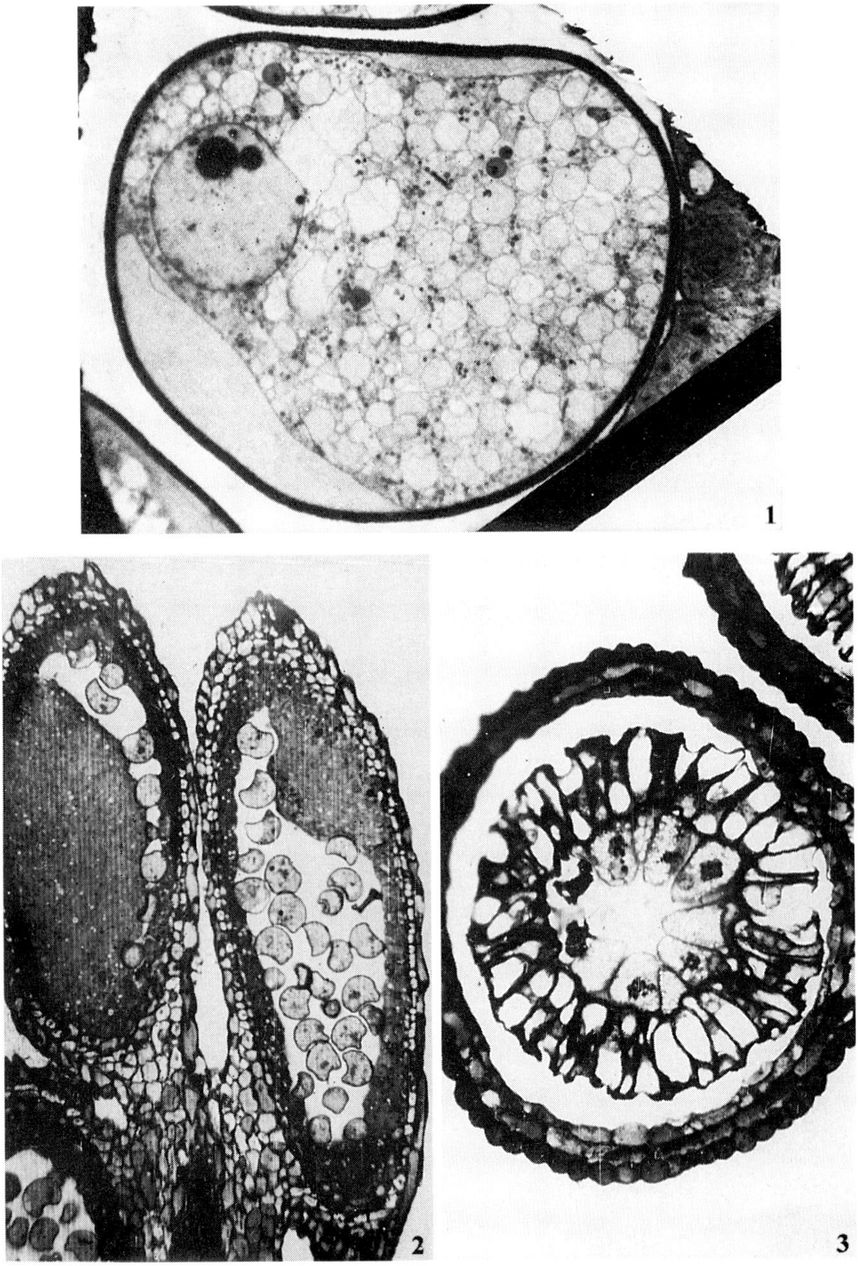

Plate 34: Cytoplasmic male sterility (CMS).

1 – microspore of maize with CMS of Texas type, vacuolization of protoplast; 2 – hypertrophy of tapetum in microsporangia of maize with CMS of Texas type; 3 – microsporangium of maize with 'C' - type CMS, strongly vacuolized tapetal cells with thickened walls squeeze degenerating microspores.

1-3 – after Simonenko, orig.

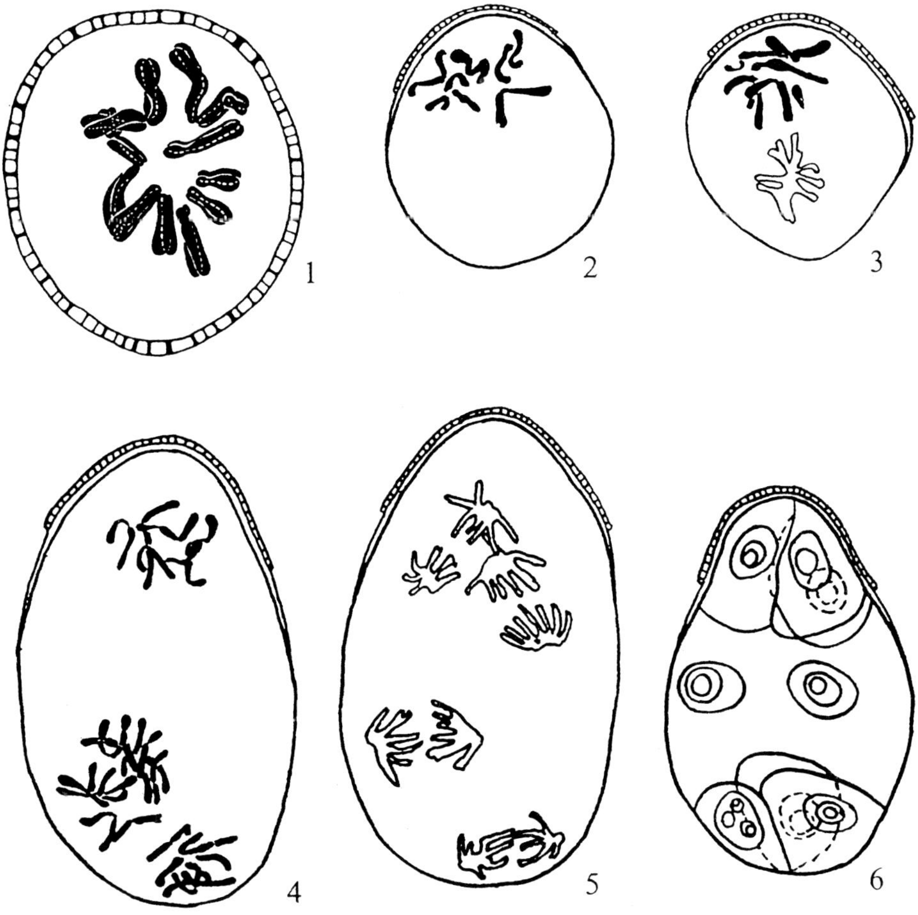

Plate 35: Embryo sacs in anthers.

1-23 – development of embryo sacs from pollen grains in *Hyacinthus orientalis* (1-22) and *Hordeum vulgare* (23): 1, 11, 15 – normal microspores and pollen grains; 2-6, 16-21, 23 – subsequent stages of embryo sac formation from a microspore; 7-10 - formation of embryo sacs of deviating types with different character of nuclei (and cells) distribution to poles (7, 8 – four-, 9 – eight-, 10 – fifteen nuclear embryo sacs); 12-14, 22 – subsequent stages of fertilization of embryo sacs arisen from pollen grains, 24 – microsporogenesis and variations of microspore behaviour in *Heuchera micrantha* (schematic).

1-10, 15-21 – modified from Stow, 1930; 11-14 – modified from Stow, 1934; 23 – after Batygina, orig.; 24 – after Vijayaraghavan and Ratnaparkhi, 1977.

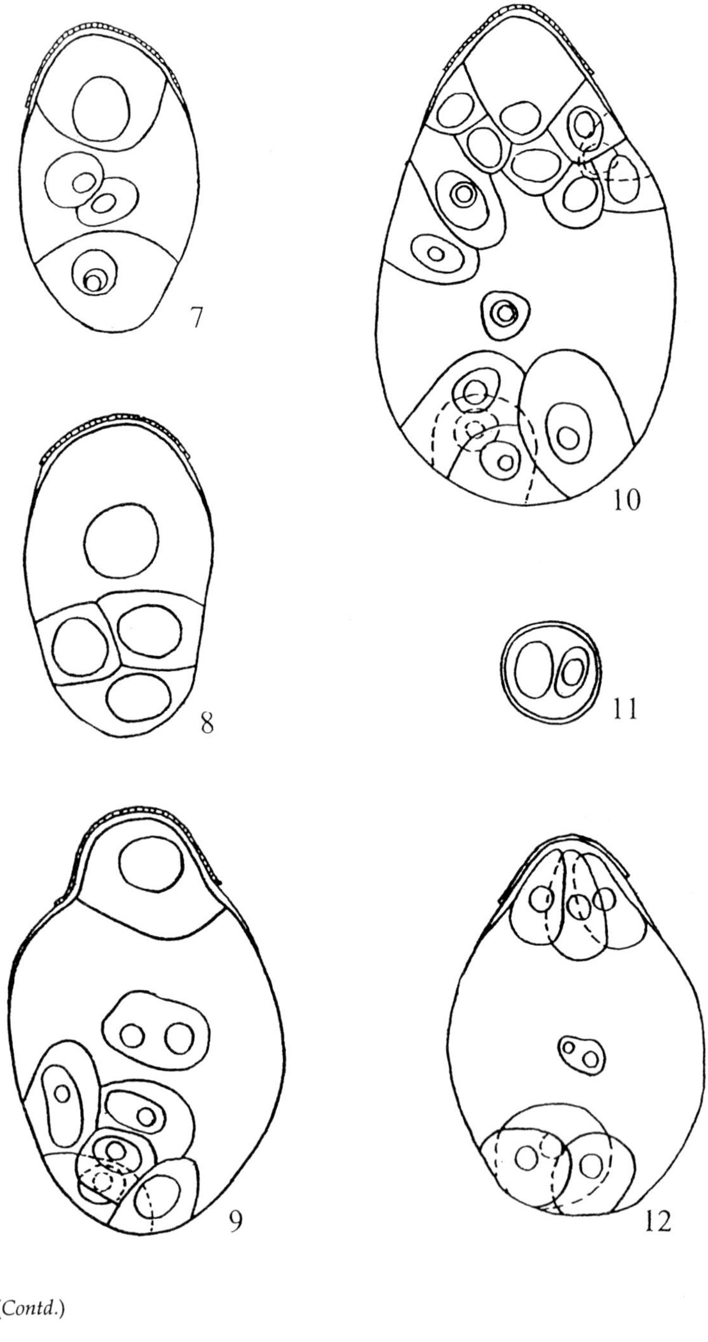

Plate 35 (*Contd.*)

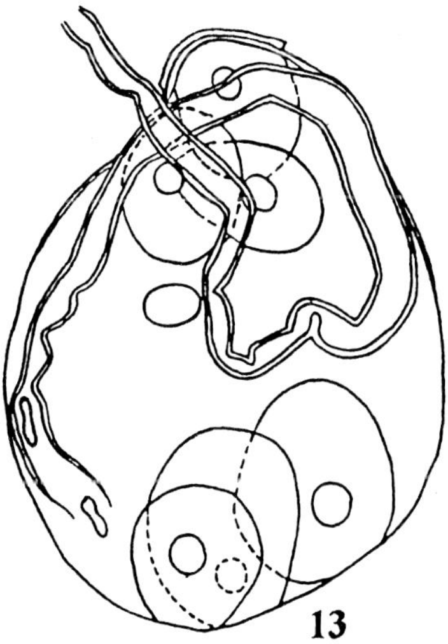
13

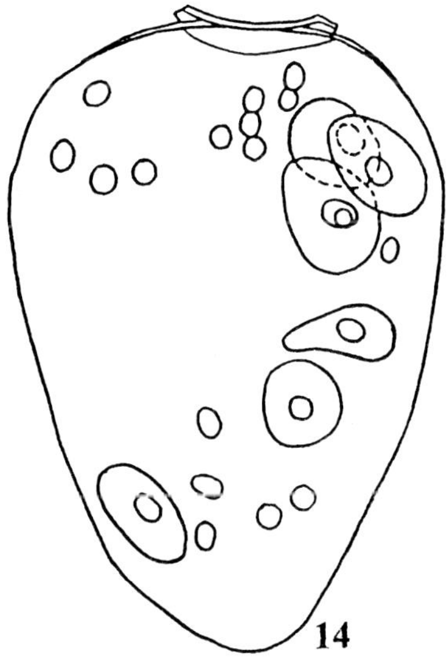
14

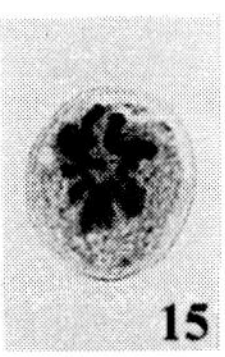
15

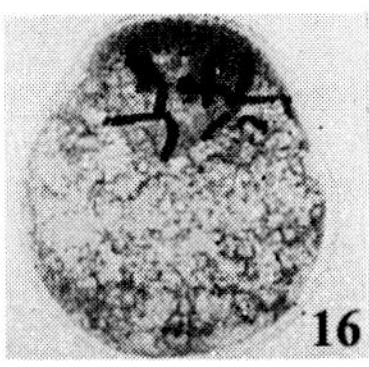
16

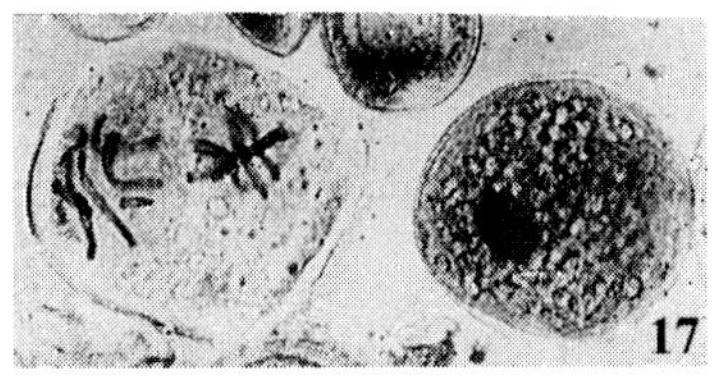
17

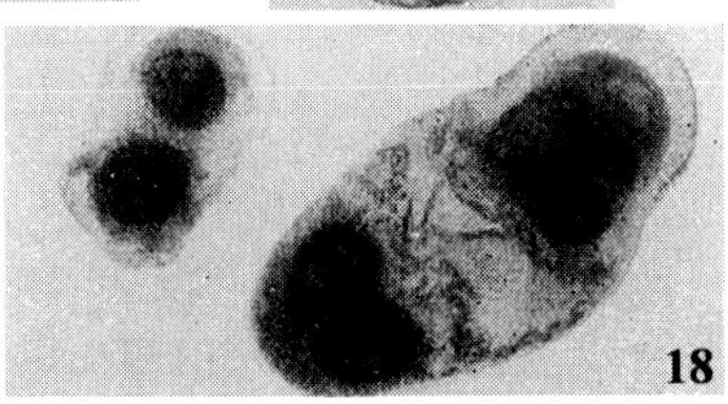
18

19

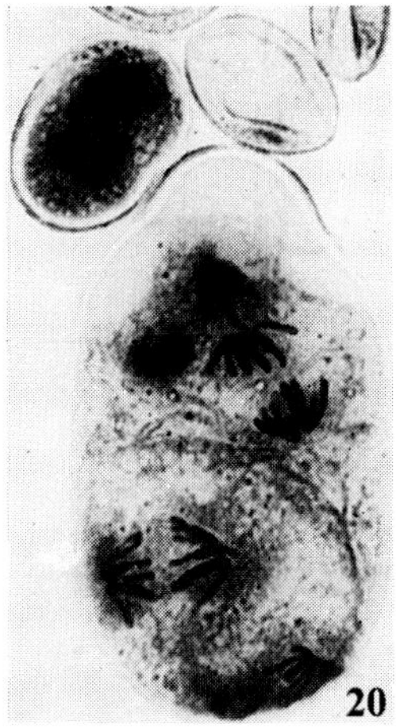
20

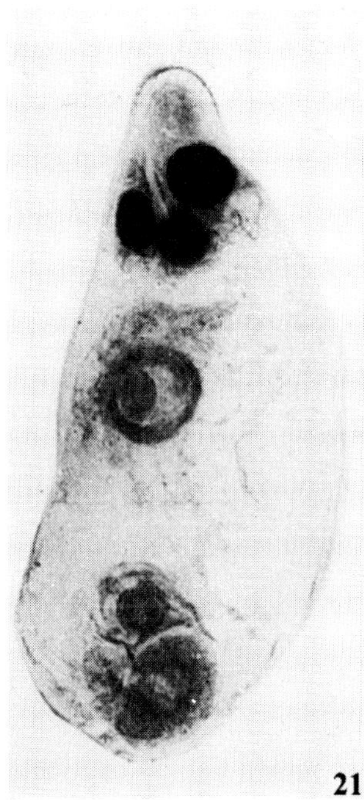
21

Plate 35 (*Contd.*)

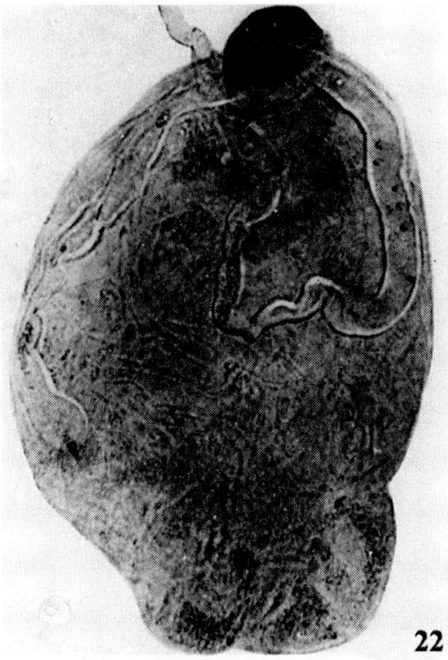

22

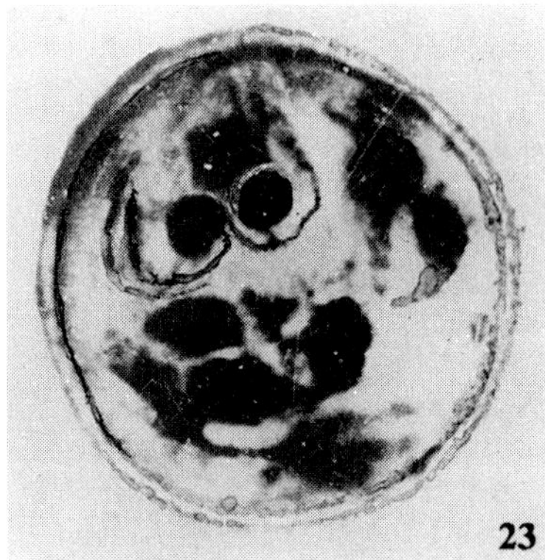

23

Pollen-Embryo Sacs in Heuchera micrantha

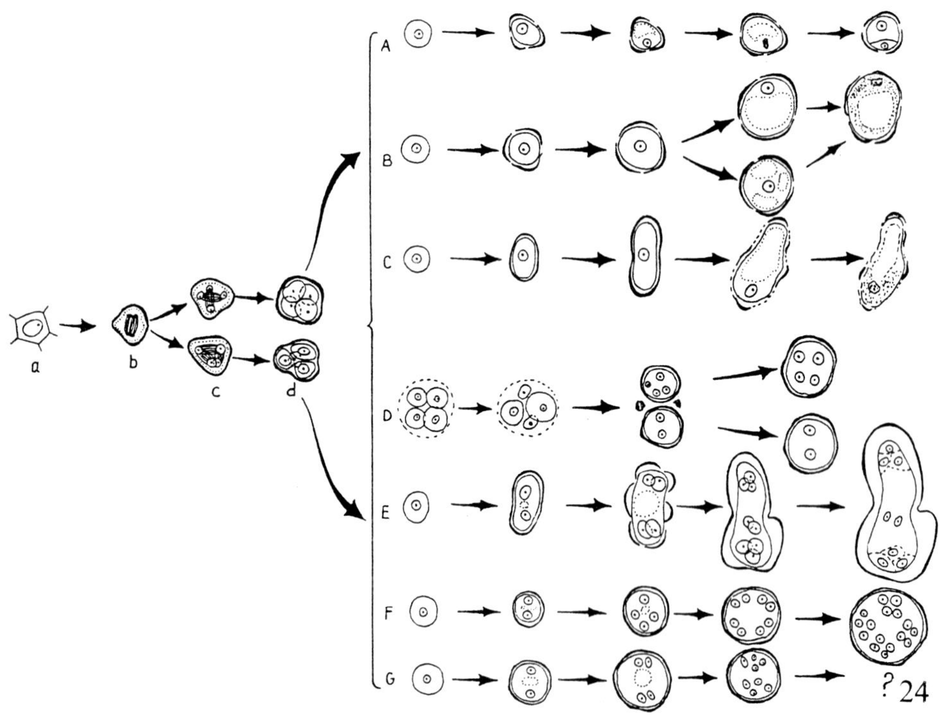

24

Plate 35 (*Contd.*)

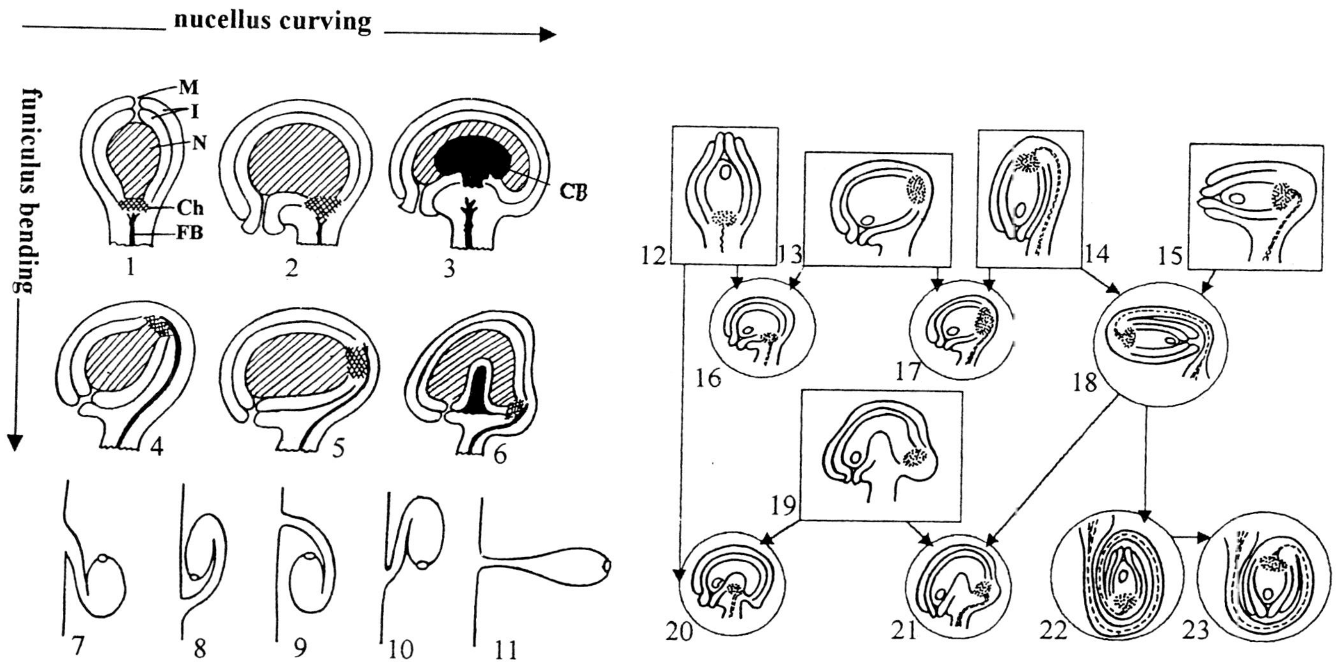

Plate 36: Ovule types.

1-6 – types of ovules: 1-3 – orthotropous series: 1 – orthotropous; 2 – ortho-campylotropous; 3 – ortho-amphitropous; 4-6 – anatropous series: 4 – anatropous; 5 – ana-campylotropous; 6 – ana-amphitropous; 7-11 – ovule types according to their position in an ovary: 7, 8 – epitropous; 9, 10 – apotropous; 11 – pleurotropous; 12-23 – scheme of interconnections of different morphological types of ovules: 12 – orthotropous, 13 – campylotropous, 14 – anatropous, 15 – hemitropous, 16 – ortho-campylotropous, 17 – ana-campylotropous, 18 – hypertropous, 19 – amphitropous, 20 – ortho-amphitropous, 21 – ana-amphitropous, 22 – ortho-circinotropous, 23 – ana-circinotropous; CB – central body, Ch – chalaza, FB – funicular conductive bundle, I – integument, M – micropyle, N – nucellus.

1-6 – after Bocquet, 1959d; 7-11 – after Takhtajan, 1964; 12-23 – modified from Savchenko, 1973.

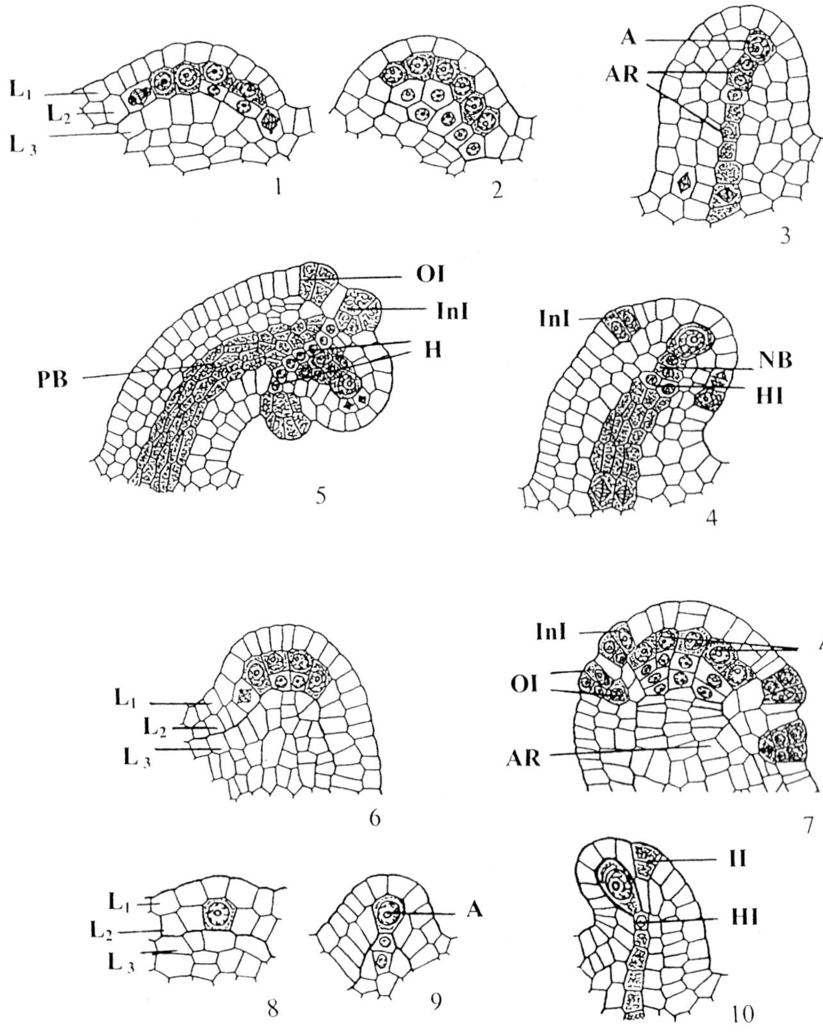

Plate 37: Ovular primordium.

Ovular primordium organization and beginning of ovule structure differentiation: 1-5 – bitegmic crassinucellate ovule with one-celled archesporium in *Nymphaea gigantea*, 6, 7 – bitegmic crassinucellate ovule with multicellular archesporium in *Paeonia lactiflora*, 8-10 – unitegmic tenuinucellate ovule with unicelled archesporium in *Gentiana cruciata*; A – archesporium, AR – axial cell row, H – hypostase, HI – hypostase initials, II – integument initials, InI – inner integument initials, L_1, L_2, L_3 – first, second and the third placental layers respectively, NB – nucellus basal region, OI – outer integument initials, PB – procambial bundle .

1-5 – after Shamrov and Winter, 1991; 6, 7 – after Shamrov, in press; 8-10 – after Shamrov, 1990d.

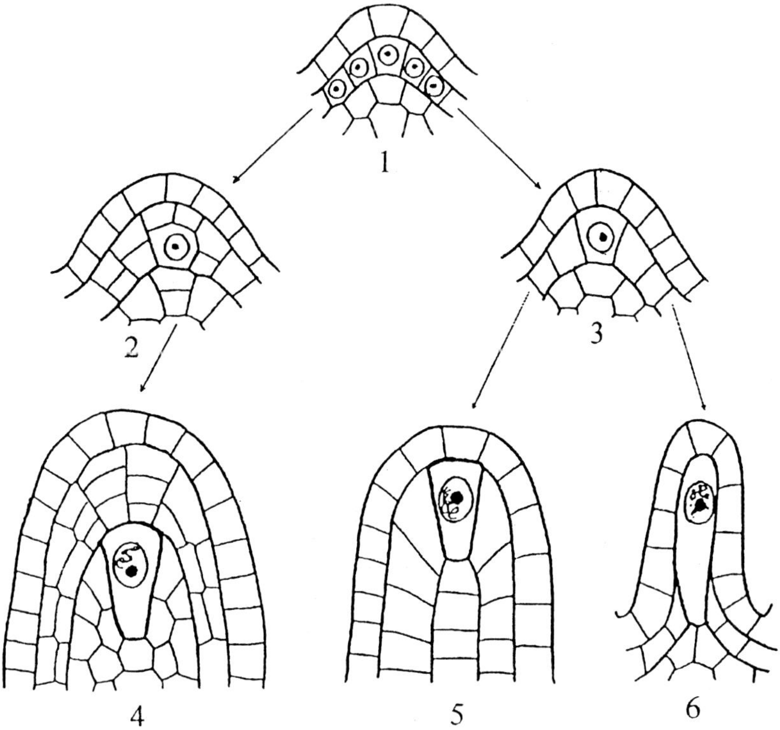

Plate 38: Nucellus.

1-6 – first stages of nucellus formation: thick-walled (1, 2, 4) and thin-walled (1, 3, 5, 6); periclinal cell divisions in subepidermal layer lead to formation of many-layered wall around megasporocyte (2, 4); if they are lacking (3) and growth intensity of megasporocyte differs from that of underlying cells of the third layer, two-layered wall (5), or unilayered one (6) is formed; 7-13 – formation of many-layered nucellus in *Triglochin palustre*; 14-16 – formation of parietal layers and their precocious degeneration in *Smilacina stellata*; 17-19 – formation of unilayered nucellus wall and its prolonged persistence in *Sambucus racemosa* (17) and *S. coreana* (18, 19); 20-24 – formation and degeneration of unilayered nucellus in *Lathraea squamaria*; 25-28 – different degree of development and differentiation of micropylar, lateral and chalazal nucellar zones in *Myriophyllum spicatum* (25), *Houttuynia cordata* (26), *Polygala major* (27), *Sonneratia caseolaris* (28); 29-36 – cell transformation of nucellar epidermis: 29 – elongation of apical cells in *Portulaca grandiflora*, 30, 31 – radial elongation of epidermal nucellar cells in *Drosera rotundiflora*, 32 – periclinal cell divisions in lateral nucellar zone in *Calathea louisae*; 33 – winding outer walls of nucellar epidermal cells and inner walls of integumental cells, filled by tannins in *Peperomia pellucida*, 34-36 – slime deposit in epidermal cells of *Reseda lutea*; A – archesporium, I – integument, II – inner integument, N – nucellus, NC – nucellar cap, NCh – nucellar chalazal zone, NE – nucellar epidermis, NL – nucellar lateral zone, NM – nucellar micropylar zone, OI – outer integument, PC – parietal cells, PT – parietal tissue.

1-11, 20-24 – after Nikiticheva, orig.; 12, 13 – after Nikiticheva and Proskurina, 1990; 14-16 – after Satarova, 1990; 17-19 – after Kamelina *et al.*, 1987 (drawings by N.G. Tuchina); 25, 27 – after Nikiticheva, 1985b, d; 26, 33 – after Nikiticheva, 1981b, c; 28 – after Kamelina, 1985d; 29 – after Batygina, 1983b; 30, 31 – after Teryokhin, 1985; 32 – after Kamelina, 1990; 34-36 – after Chaban and Yakovlev, 1974.

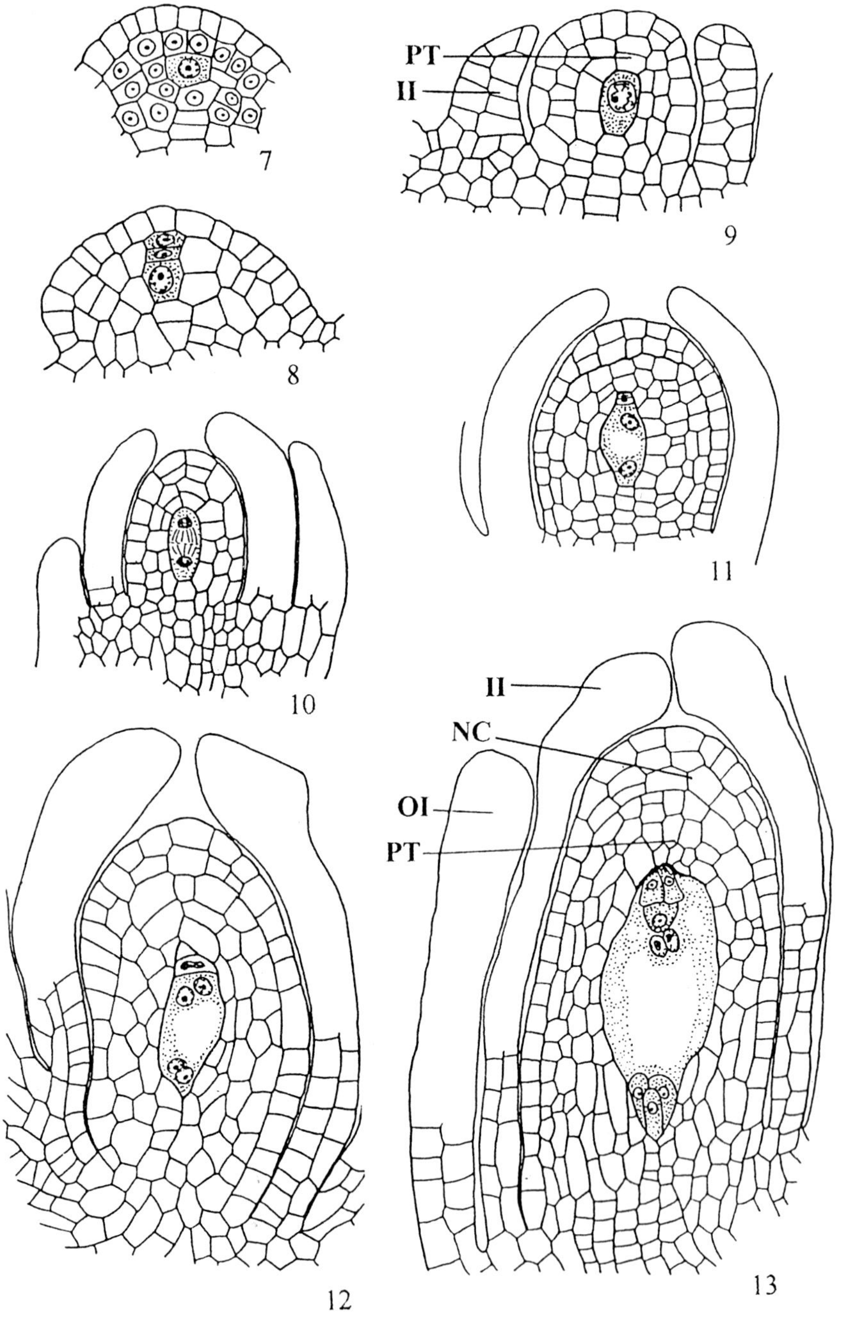

Plate 38 (*Contd.*)

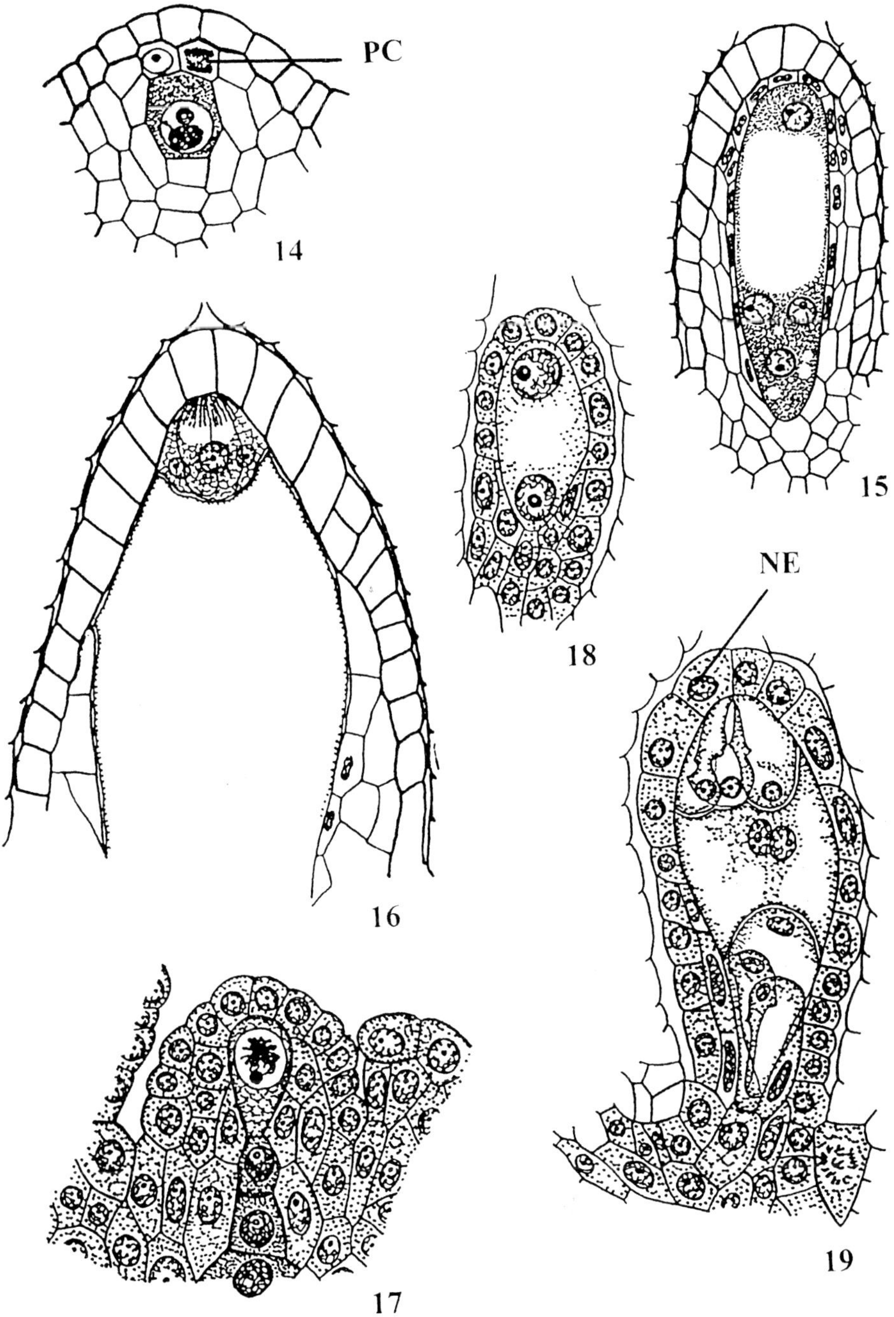

Plate 38 (*Contd.*)

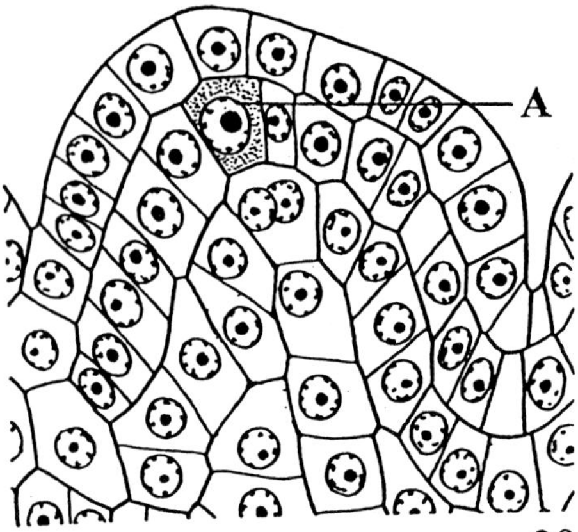

20

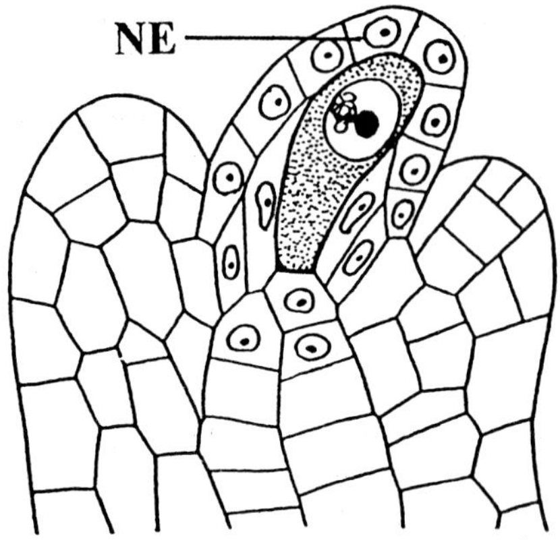

21

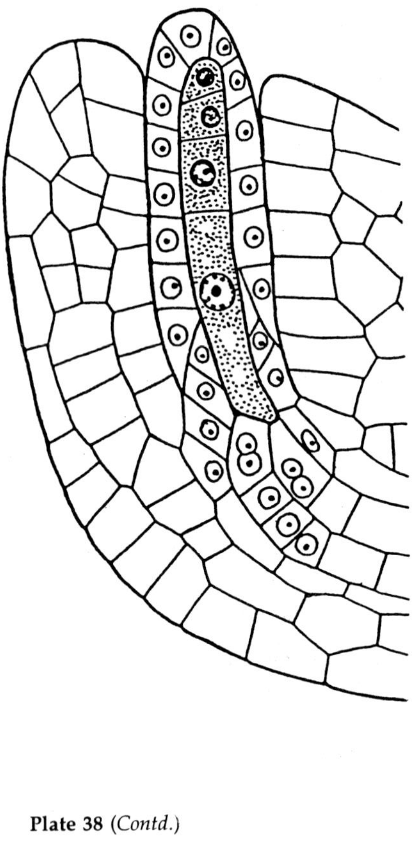

22

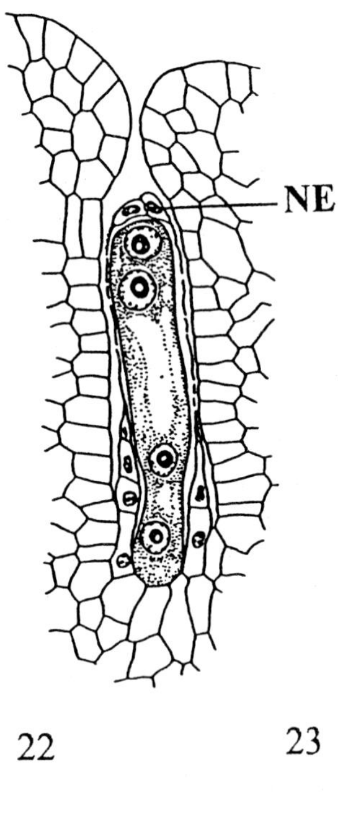

23

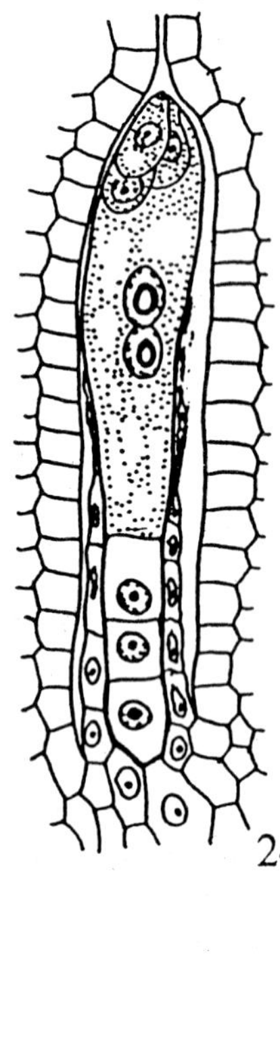

24

Plate 38 (*Contd.*)

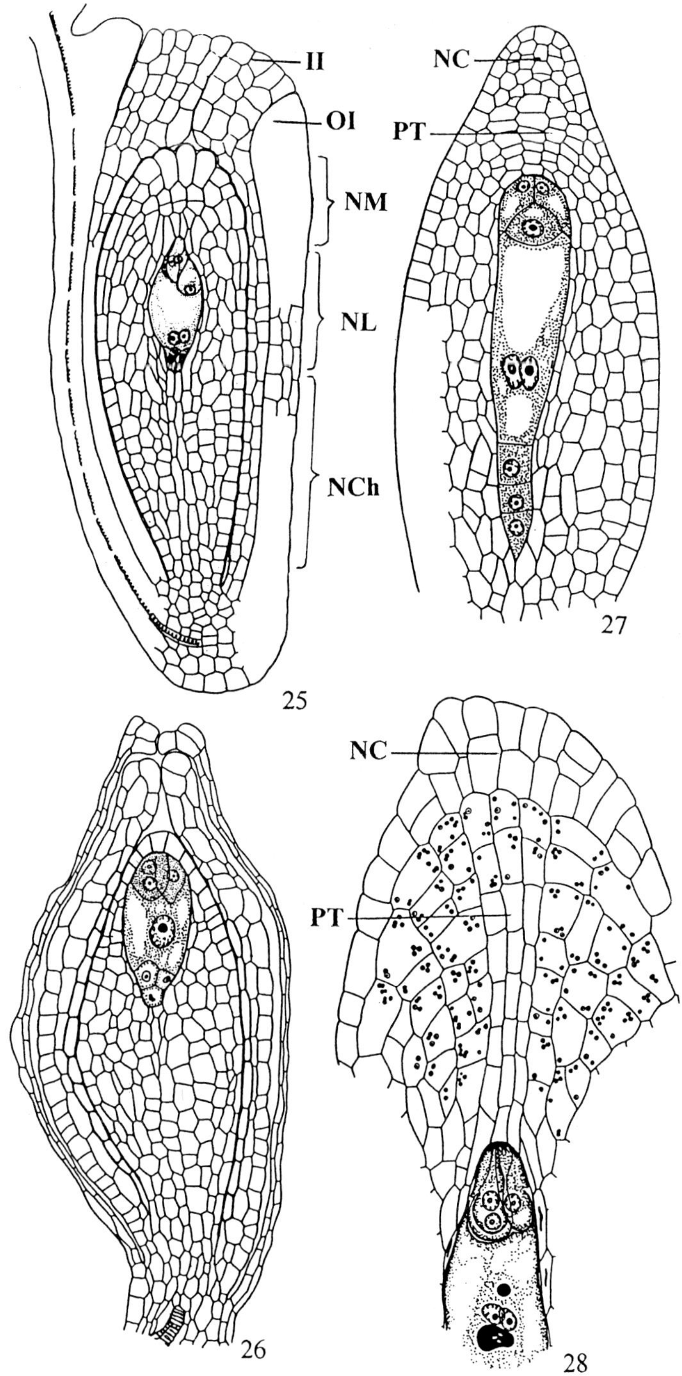

Plate 38 (*Contd.*)

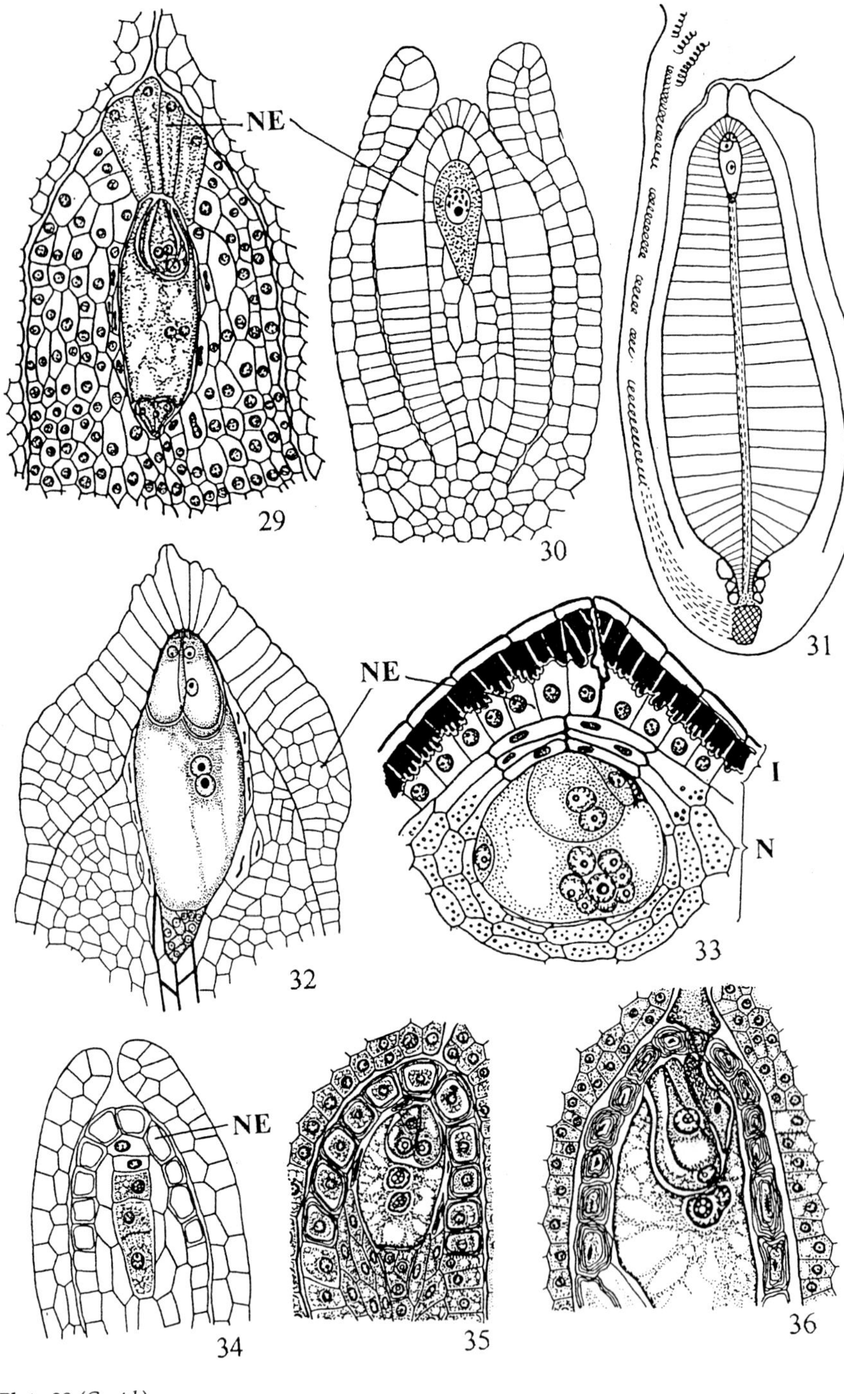

Plate 38 (*Contd.*)

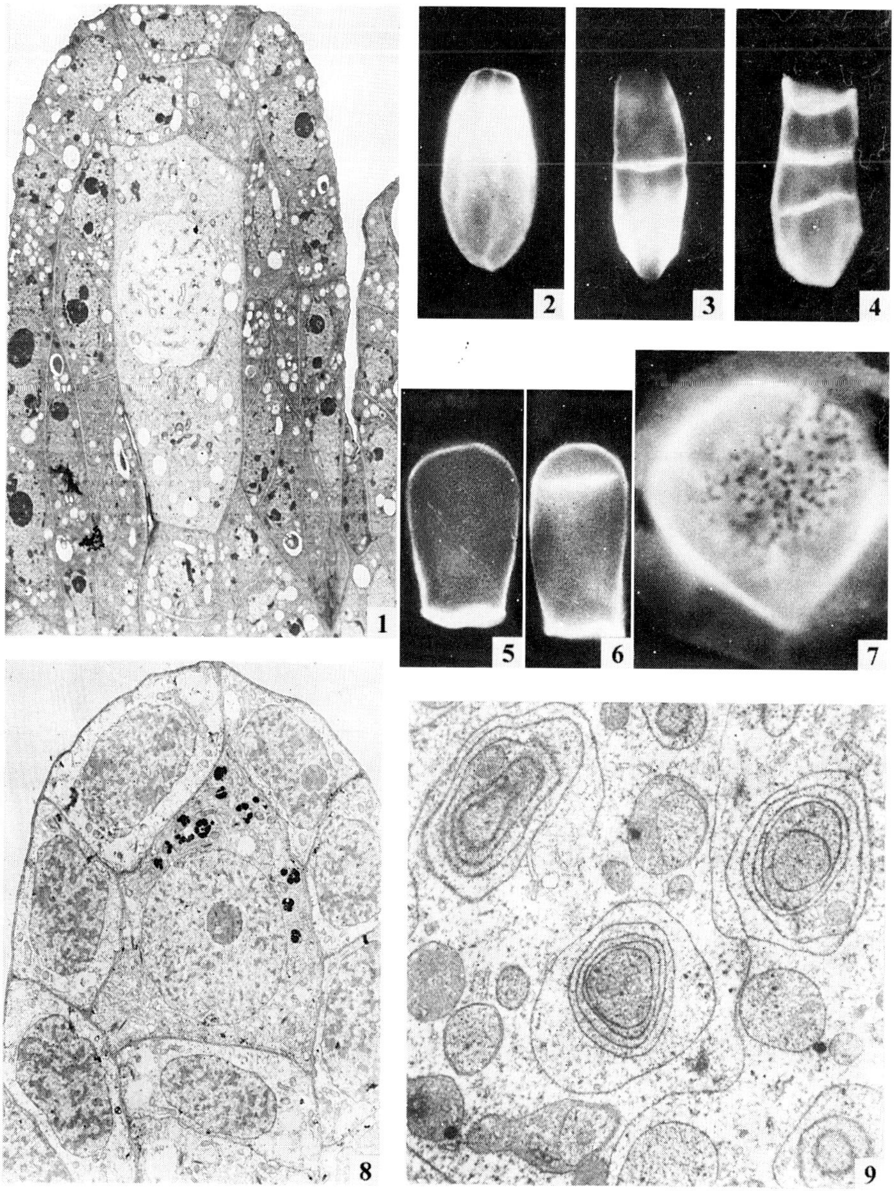

Plate 39: Megasporocyte and megasporogenesis.
1 – nucellus with megasporocyte at stage of prophase I, parietal cell undergoes anticlinal division; 2-4 — megasporogenesis in *Oenothera*, notice fluorescence of callose walls after staining with aniline blue: megasporocyte (2); dyad, callose lacking in micropylar zone (3); megaspore tetrad, fluorescence lacking in micropylar and lateral walls of functional megaspore (4); 5-7 – megasporocytes in *Epipactis* (Orchidaceae), note fluorescence of callose walls: megasporocyte (5), dyad (6), megasporocyte chalazal wall, callose perforated like a sieve (7); 8 – nucellus of *Epipactis*, plastids containing starch grains (dark) concentrate in micropylar part of megasporocyte; 9 – cytoplasmic area of megasporocyte in *Lilium* during prophase I, note multilayered cytoplasmic bodies.

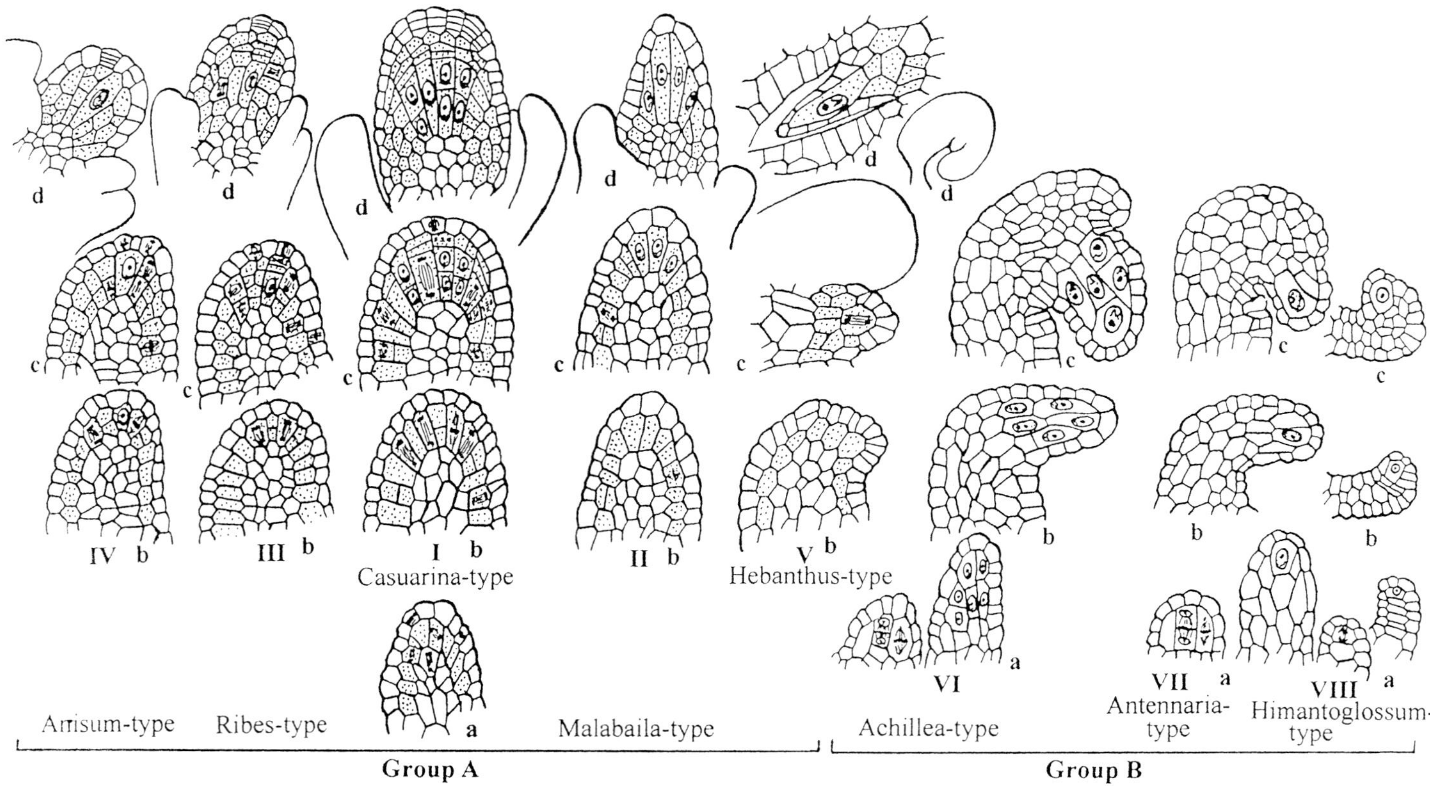

Plate 40: Classification of megasporangium types.
Main types of angiosperm megasporangia: a-d – subsequent stages of their development.

After Kordyum, 1978.

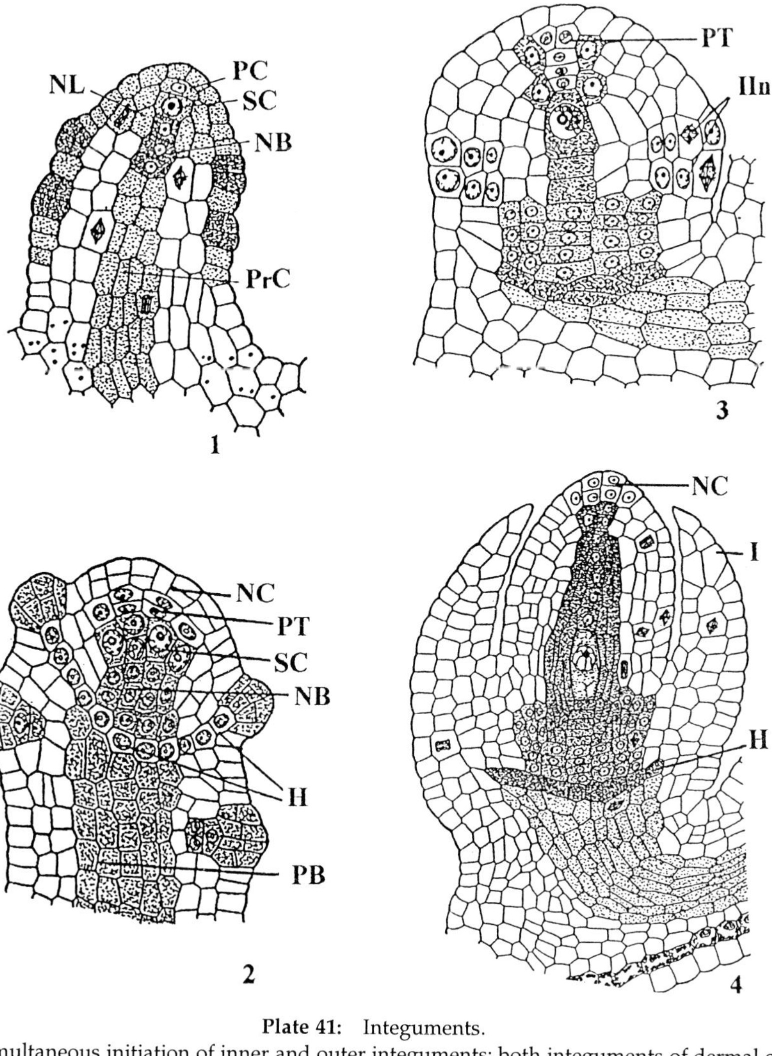

Plate 41: Integuments.

1,2 – simultaneous initiation of inner and outer integuments: both integuments of dermal origin in *Luzula pedemontana* (1), in *Paeonia lactiflora* inner integument of dermal origin, and outer of dermal-subdermal origin (2); 3, 4 – formation of single integument from dermal and subdermal initials in *Ceratophyllum demersum*, 5, 6 – unitegmic ovule in *Gentiana cruciata*, 7 – micropylar collar in *Triantia fugas*, 8 – ovule with three (outer, middle and inner) integuments in *Cananga odorata*, 9 – bitegmic ovule with reduced outer integument in *Cytinus ruber*, 10-13 – possible pathways of transition from bi- to unitegmy: reduction of outer (10), or inner (11) integument, fusion of two integuments (12), elimination of one integument (13); H – hypostase, I – integument, II– inner integument, IIn – integument initials, NB – nucellus basal region, NC – nucellar cap, NL – nucellus lateral region, OI – outer integument, PB – procambial bundle, PC – parietal cells, PrC – procambial cells, PT – parietal tissue, SC – sporogenous cells.

1 – after Shamrov and Anisimova, 1993b; 2-4 – after Shamrov, in press; 5,6 – after Shamrov, 1990d; 7 – after Chikkannaiah, 1962; 8 – after Corner, 1949a; 9 – after Teryokhin and Anisimova, 1981; 10-13 – after Bouman, 1984.

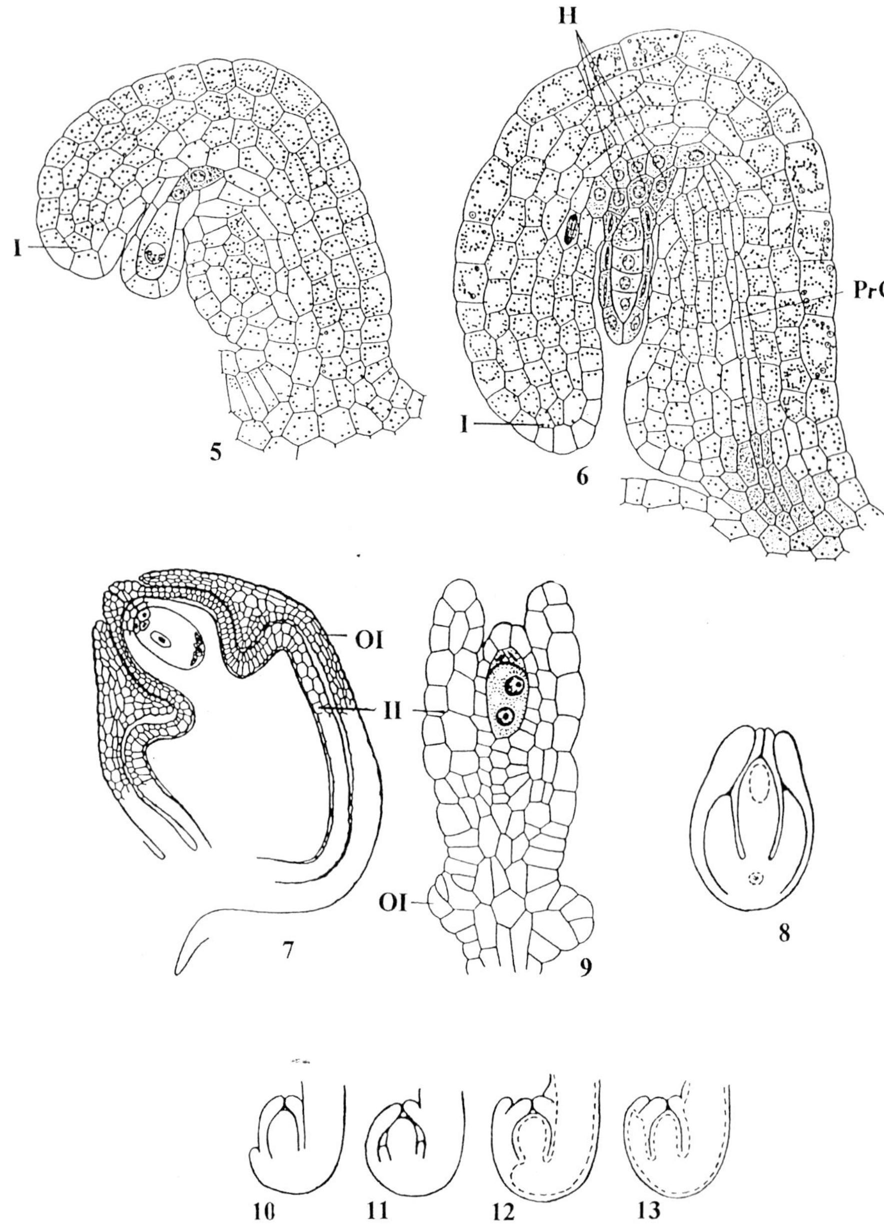

Plate 41 (*Contd.*)

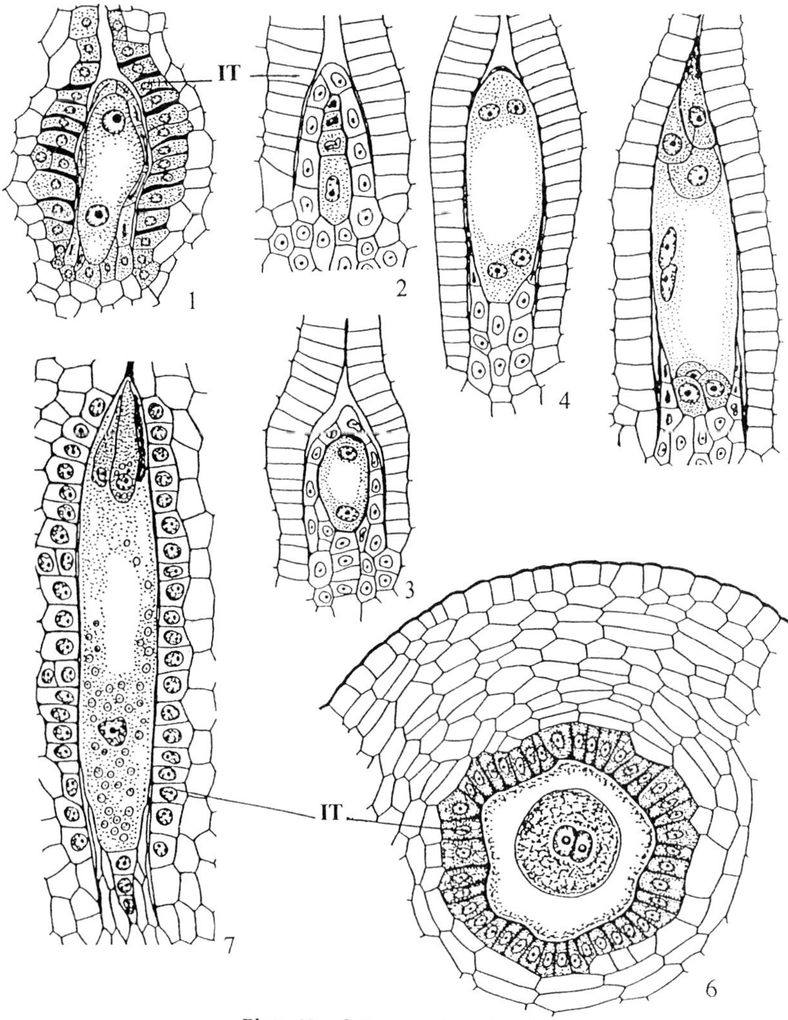

Plate 42: Integumentary tapetum.

1-12 – development and structure of integumentary tapetum: 1 – *Nymphoides humboldiana* (Menyanthaceae), 2-6 – *Halesia carolina*, 7 – *Alniphyllum fortunei* (Styracaceae), 8-12 – *Calendula officinalis* (Asteraceae): diagrams of ultrastructural organization of integumentary tapetum cells at the four-nucleate embryo sac stage (10), mature embryo sac stage (11) and during early embryogenesis (12); 13-18 – integumentary tapetum in ovules before fertilization in Scrophulariaceae (13-18) and Orobanchaceae (19) members: in *Verbascum thapsus* (13), *Melampyrum nemorosum* (14, 15), *Tozzia alpina* (16), *Lathraea squamaria* (17), *Bartsia alpina* (18), *Cistanche flava* (19); 20, 22 – cell fragments of integumentary tapetum at the four-nucleate embryo sac stage (20, see diagram 21) and torpedo embryo stage (22, see diagram 23) in *Valeriana officinalis*: in cells of integumentary tapetum no starch is accumulated compared with integumental parenchyma (20), cell walls of integumentary tapetum, adjacent embryo sac, show cuticle (arrow); Ap – amyloplast, C – cuticle, En – endosperm, H – hypostase, IP – integumental parenchyma, IT – integumentary tapetum, L – lipids, Lp – leucoplast, Ls – lomasoma, M – mitochondria, N – nucellus, Pd – plasmodesma, Pl – plasma membrane, RER – rough endoplasmic reticulum, SG – starch grain, V – vacuole, W – cell wall.

1 – after Batygina and Kolesova, 1987; 2-5, 7 – after Nikiticheva, 1983b; 6 – after Nikiticheva and Shamrov, in press; 8-12 – after Vassilyev and Plisco, 1971; 13-19 – after Nikiticheva, 1979; 20-23 – after Kamelina and Zhukova, 1987.

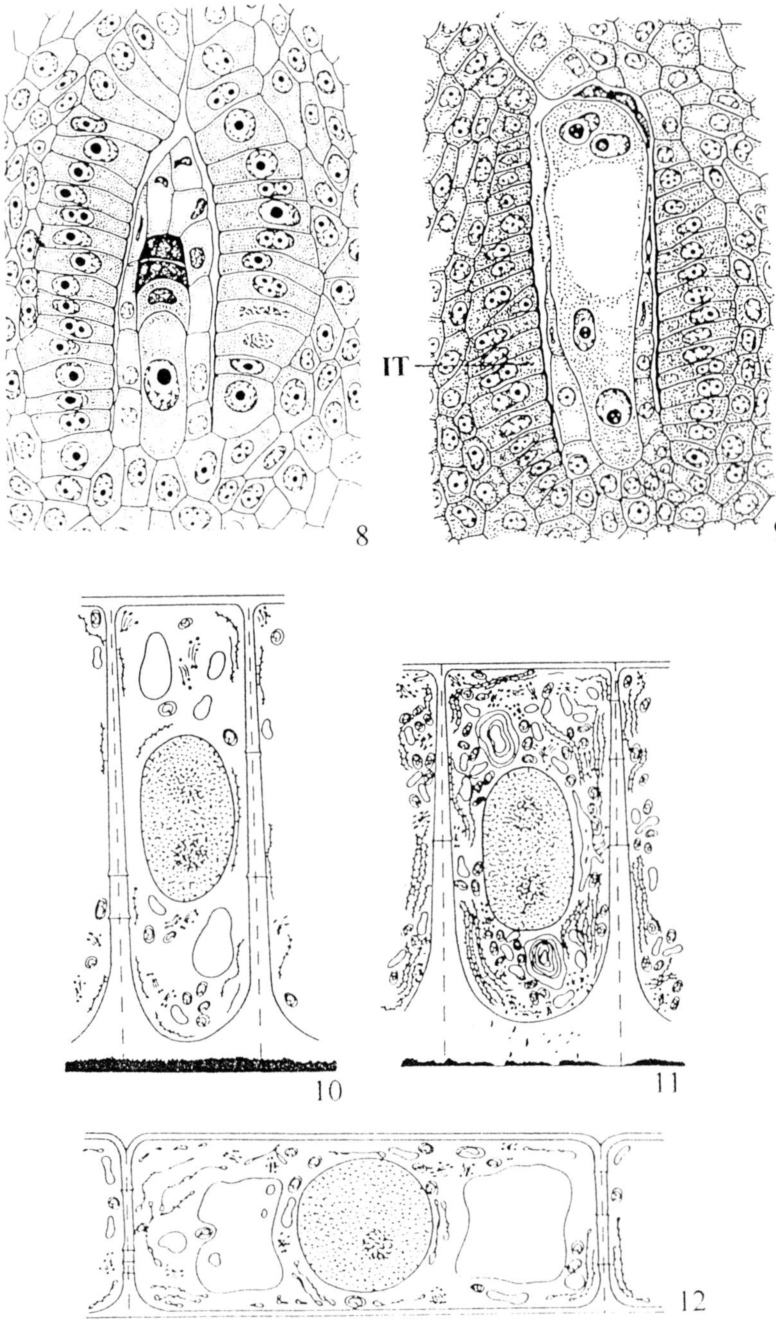

Plate 42 (*Contd.*)

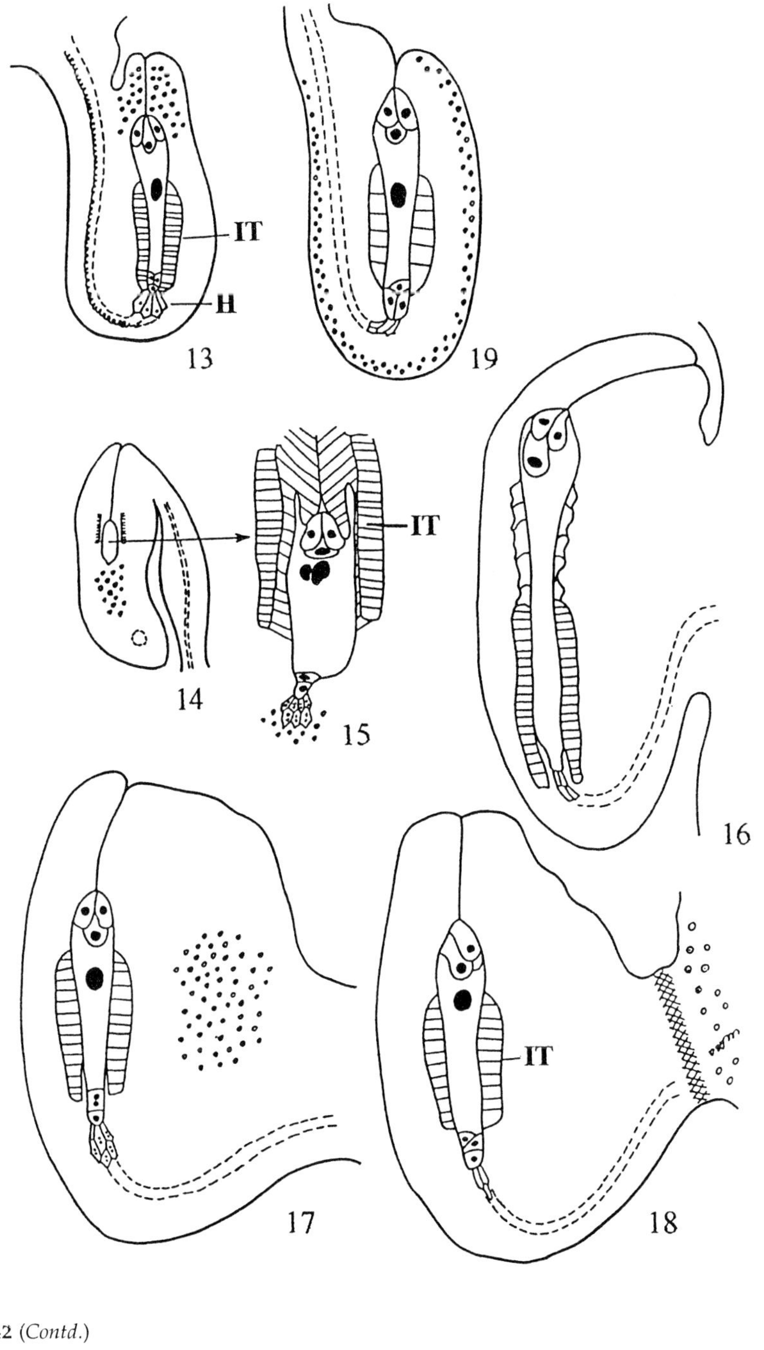

Plate 42 (*Contd.*)

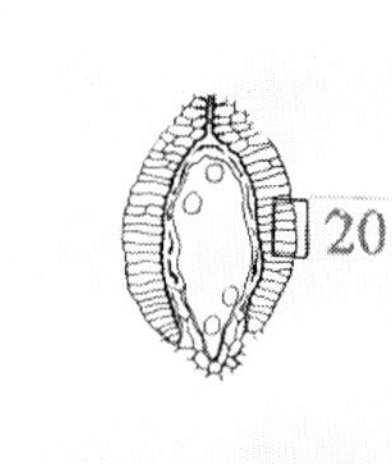

21

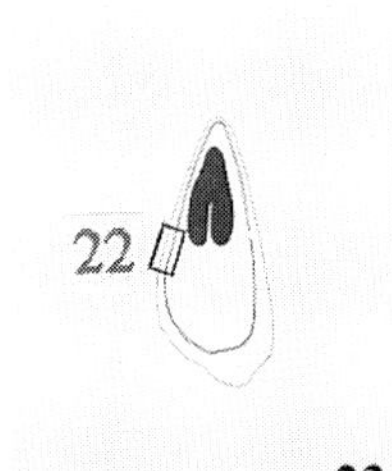

23

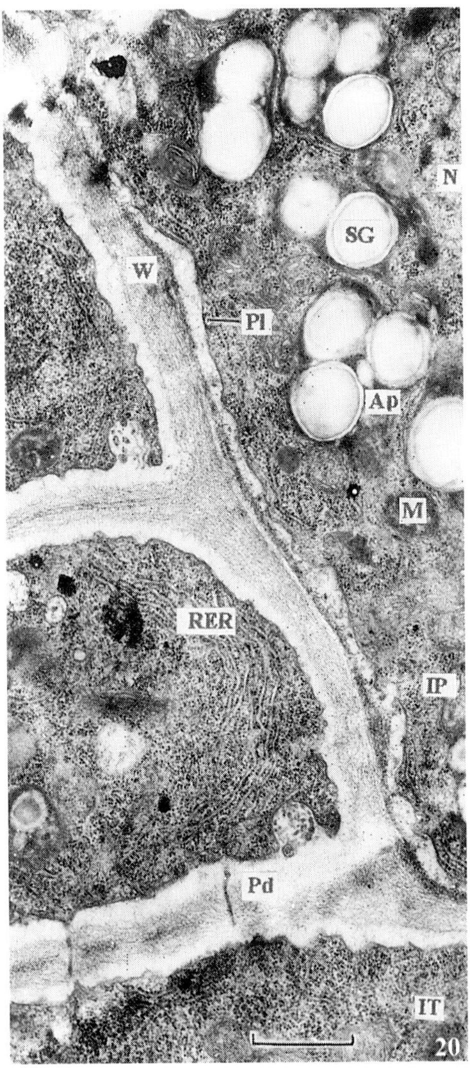

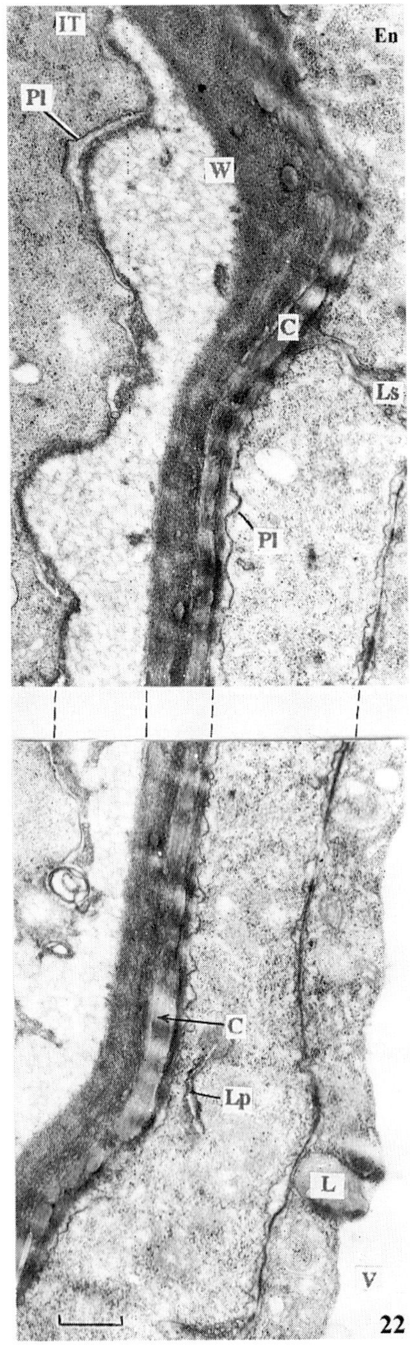

Plate 42 (*Contd.*)

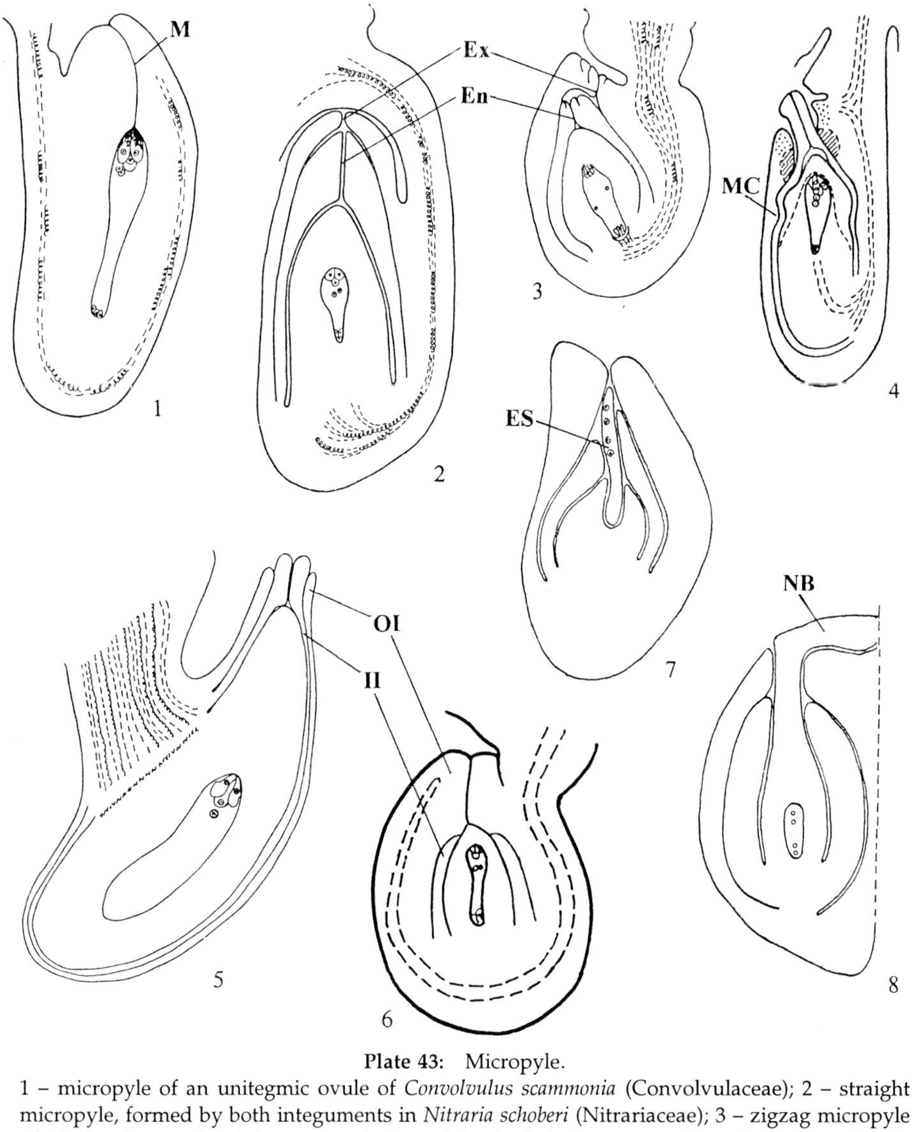

Plate 43: Micropyle.

1 – micropyle of an unitegmic ovule of *Convolvulus scammonia* (Convolvulaceae); 2 – straight micropyle, formed by both integuments in *Nitraria schoberi* (Nitrariaceae); 3 – zigzag micropyle of *Shepherdia argentea* (Elaeagnaceae), formed by both integuments; 4 – micropylar collar in mature ovule of *Calathea louisae* (Marantaceae); 5 – micropyle, formed by inner integument in *Dianthus caryophyllus* (Caryophyllaceae); 6 – micropyle, formed by outer integument in *Paliurus spina christi* (Rhamnaceae); 7 – protrusion of an embryo sac into micropyle of *Arechavaletaia uruguayensis* (Flacourtiaceae); 8 – protrusion of nucellar beak into micropyle of *Chrozophora obliqua* (Euphorbiaceae); En – endostome, ES – embryo sac, Ex – exostome, II – inner integument, M – micropyle, MC – micropylar collar, NB – nucellar beak, OI – outer integument.

1 – after Nikiticheva and Proskurina, 1987; 2 – modified from Kamelina, 1985c; 3 – after Proskurina, 1990; 4 – modified from Kamelina, 1990a; 5 – modified from Kamelina and Yakovleva, 1982; 6 – modified from Batygina *et al*., 1987; 7 – after Ventura, 1938; 8 – after Bor and Kapil, 1976.

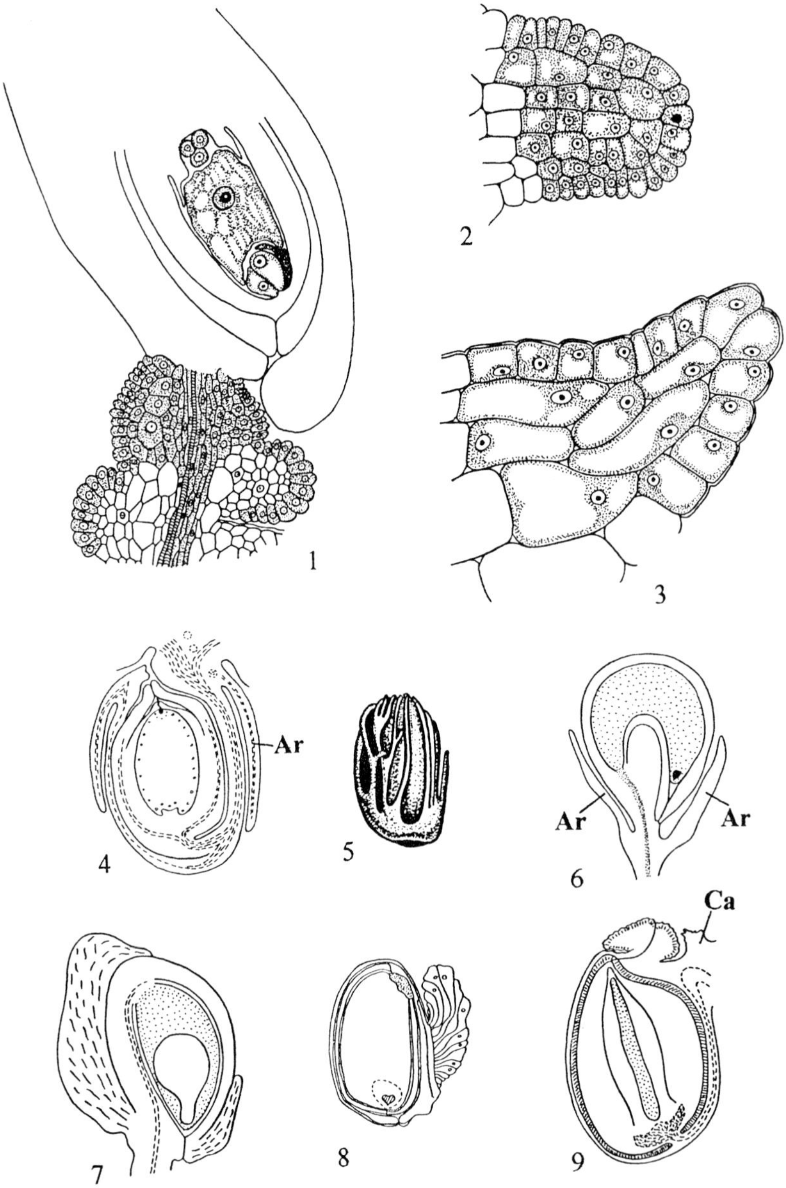

Plate 44: Aril.

1-3 – structure of aril in *Ribes nigrum* at different stages of ovule and seed development; 4, 5 – *Myristica fragrans*: longitudinal section of a seed with developing aril at stage of multicelled proembryo (4) and general appearance of seed with aril (5); 6 – developing seed of *Acrotrema arnottianum* with an aril; 7, 8 – developing seed of *Stachyurus praecox* (7) and *Chelidonium majus* (8) with arilloid; 9 – almost mature seed of *Euphorbia pilosa* with exostomal arilloid (caruncle); Ar – aril, Ca – caruncle.

1-3 – after Komar, 1965a; 4 – after Mametyeva and Kamelina, 1981; 5 – after Velenovsky, 1913; 6 – Swamy and Periasamy, 1955; 7, 9 – after Corner, 1976; 8 – after Crété, 1937.

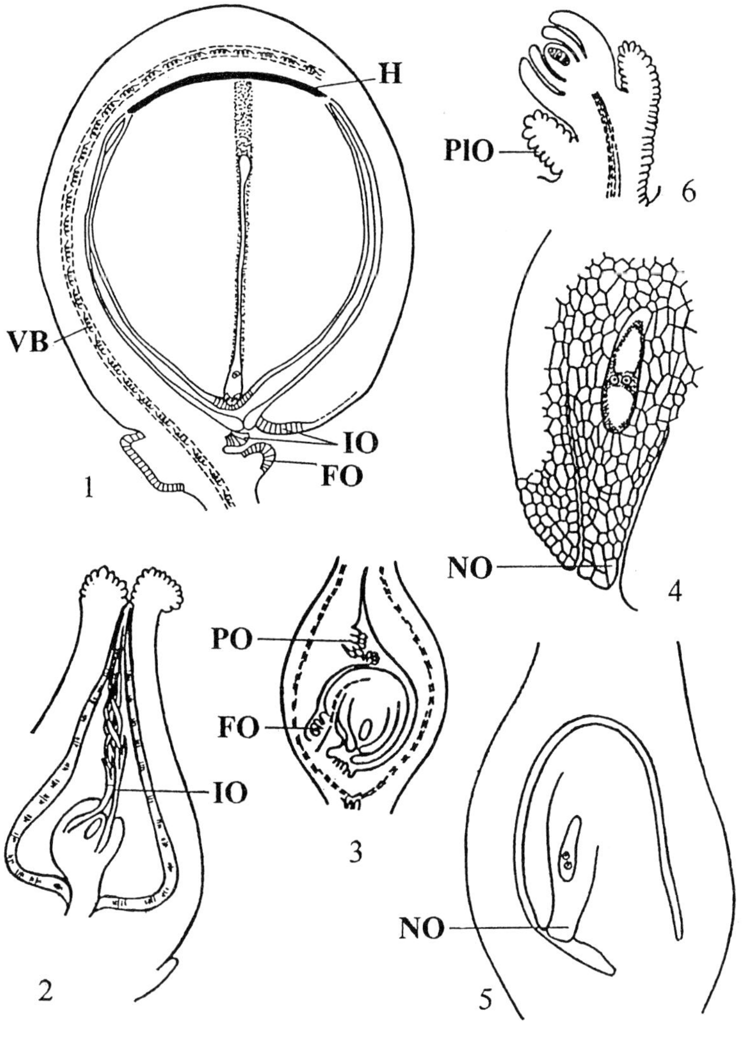

Plate 45: Obturator.

1 – funicular and integumentary (originating from outer integument) obturators in *Victoria amazonica*, 2 – integumentary (originating from inner integument) obturator in *Leukosyke capitellata*, 3 – funicular and parietal obturators in *Ruscus hypophyllum*, 4,5 - nucellar obturator in *Boerhavia repanda*, 6 – placental obturator in *Allium caspium*; FO – funicular obturator, H – hypostase, IO – integumentary obturator, NO – nucellar obturator, PlO – placental obturator, PO – parietal obturator, VB – vascular bundle.

1 – after Shamrov and Winter, 1991; 2 – after Fagerlind, 1941a; 3 – after Savchenko and Komar, 1965; 4, 5 – after Bhargava, 1932; 6 – after Shamrov, 1990a.

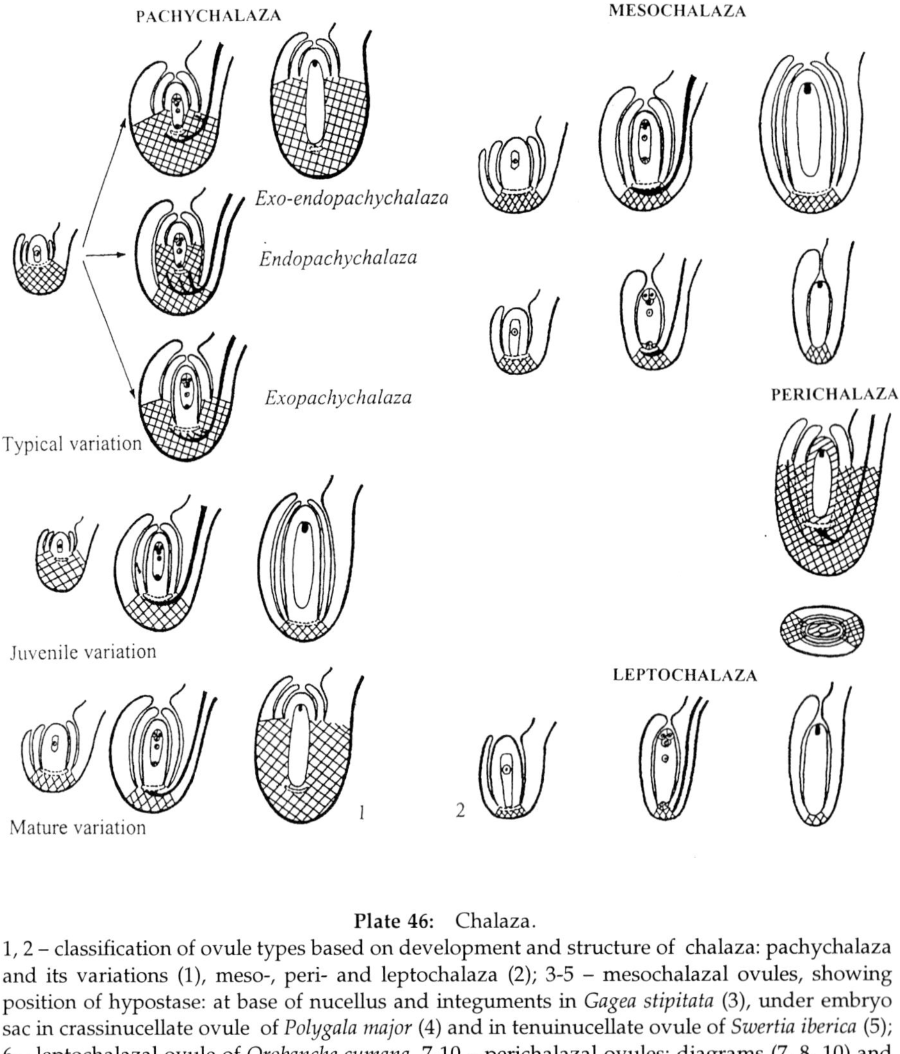

Plate 46: Chalaza.

1, 2 – classification of ovule types based on development and structure of chalaza: pachychalaza and its variations (1), meso-, peri- and leptochalaza (2); 3-5 – mesochalazal ovules, showing position of hypostase: at base of nucellus and integuments in *Gagea stipitata* (3), under embryo sac in crassinucellate ovule of *Polygala major* (4) and in tenuinucellate ovule of *Swertia iberica* (5); 6 – leptochalazal ovule of *Orobanche cumana*, 7-10 – perichalazal ovules: diagrams (7, 8, 10) and in *Cissus witiginea* (9); 11-14 – pachychalazal ovules: diagrams (12-14) and in *Myristica montana* (11); 15 – exopachychalazal ovule of *Nelumbo nucifera*; 16-18 – endopachychalazal ovules: diagrams (17, 18) and in *Bruguiera sexangula* (16); 19, 20 – pachychalazal ovule (juvenile variation) of *Scheuchzeria palustris*; Ch – chalaza, F – funiculus, H – hypostase, II – inner integument, OI – outer integument.

1, 2, 3, 5 – after Shamrov, in press; 4 – after Nikiticheva, 1985d; 6 – after Anisimova, 1983a; 7, 8 – after Periasamy, 1962b; 9 – after Periasamy, 1962a; 10, 12-14, 17, 18 – after Boesewinkel and Bouman, 1984; 11 – after Mametyeva and Kamelina, 1981; 15 – after Batygina *et al.*, 1983; 16 – after Nikiticheva and Yakovlev, 1985; 19, 20 – after Nikiticheva and Proskurina, 1992.

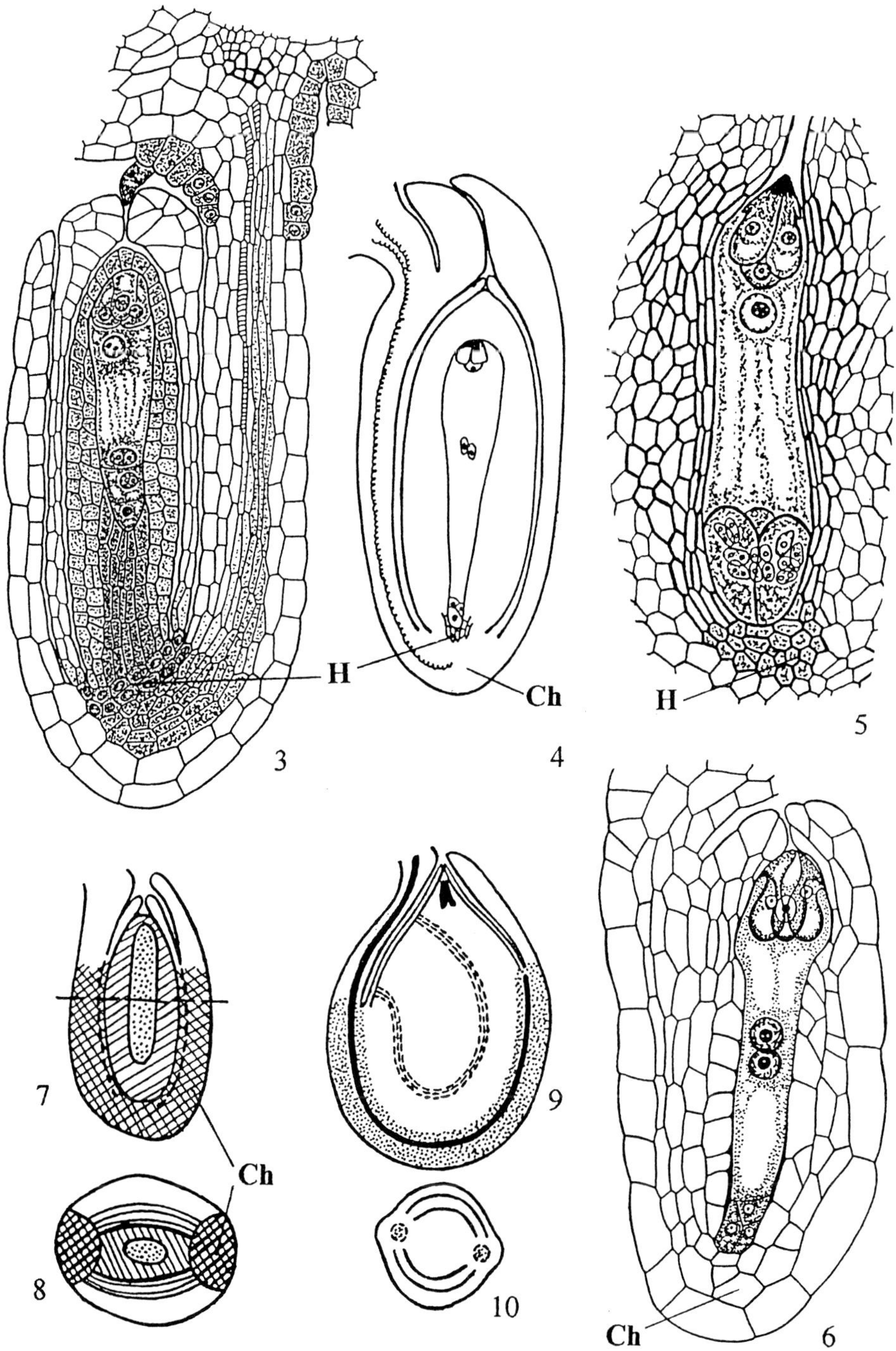

Plate 46 (*Contd.*)

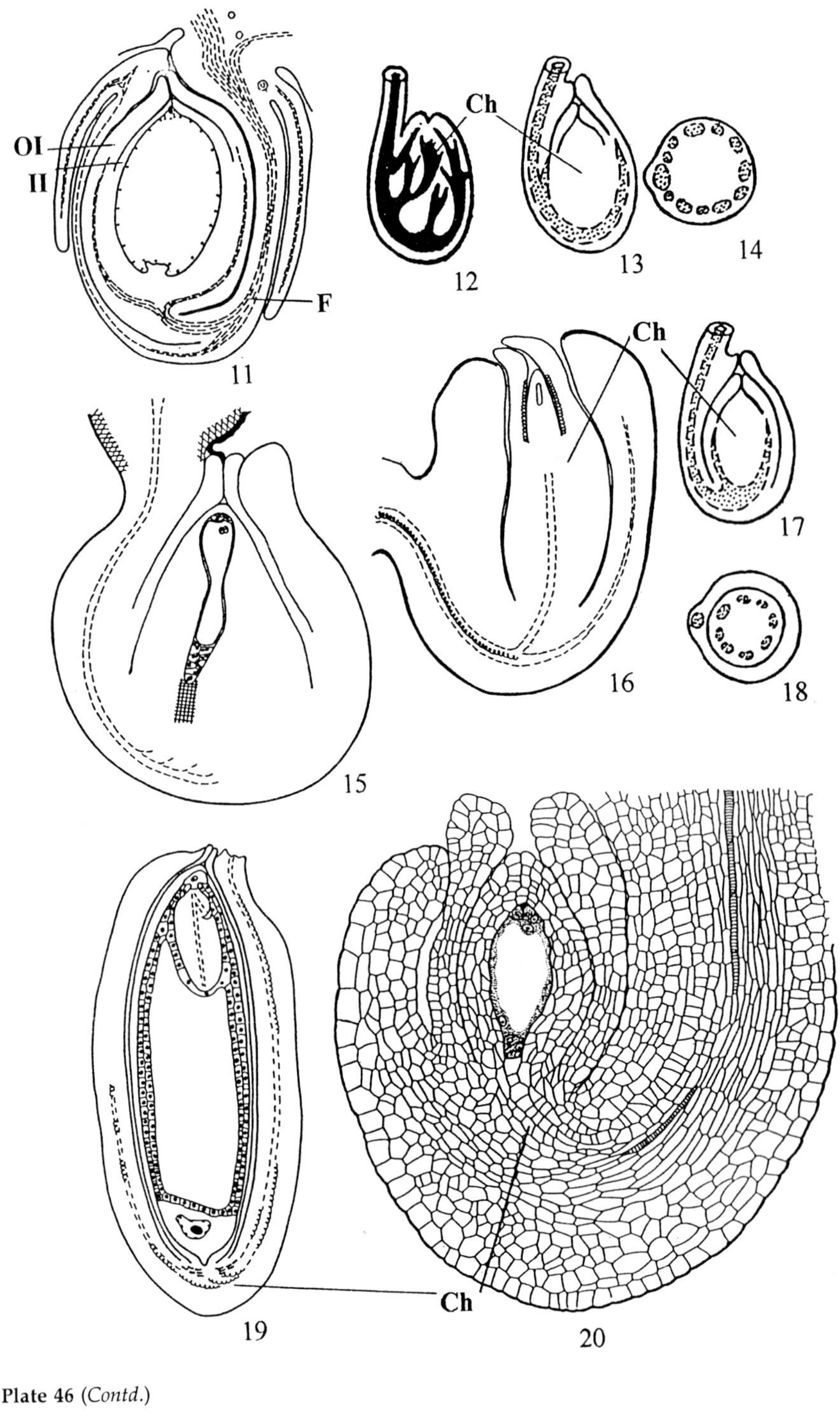

Plate 46 (*Contd.*)

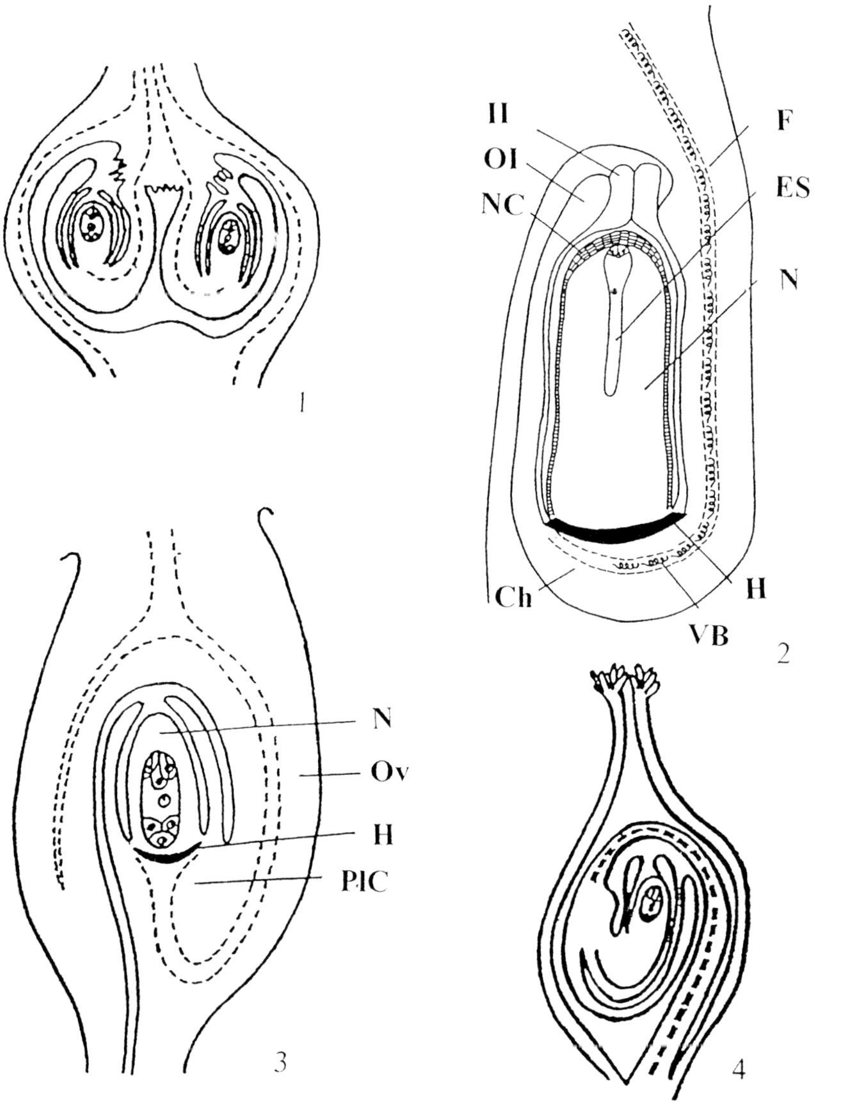

Plate 47: Funiculus.

1 – sessile ovule in *Luzula pedemontana*, 2 – funicular ovule in *Nuphar lutea*, 3 – afunicular ovule in *Ceratophyllum demersum*, 4 – circinate ovule in *Psilostachys sericea*; Ch – chalaza, ES – embryo sac, F – funiculus, H – hypostase, II – inner integument, N – nucellus, NC – nucellar cap, OI – outer integument, Ov – ovary, PlC – placentochalaza, VB – vascular bundle.

1 – after Shamrov and Anisimova, 1993b; 2 – after Winter and Shamrov, 1991; 3 – after Shamrov and Batygina, 1988; 4 – after Kapil and Vasil, 1963.

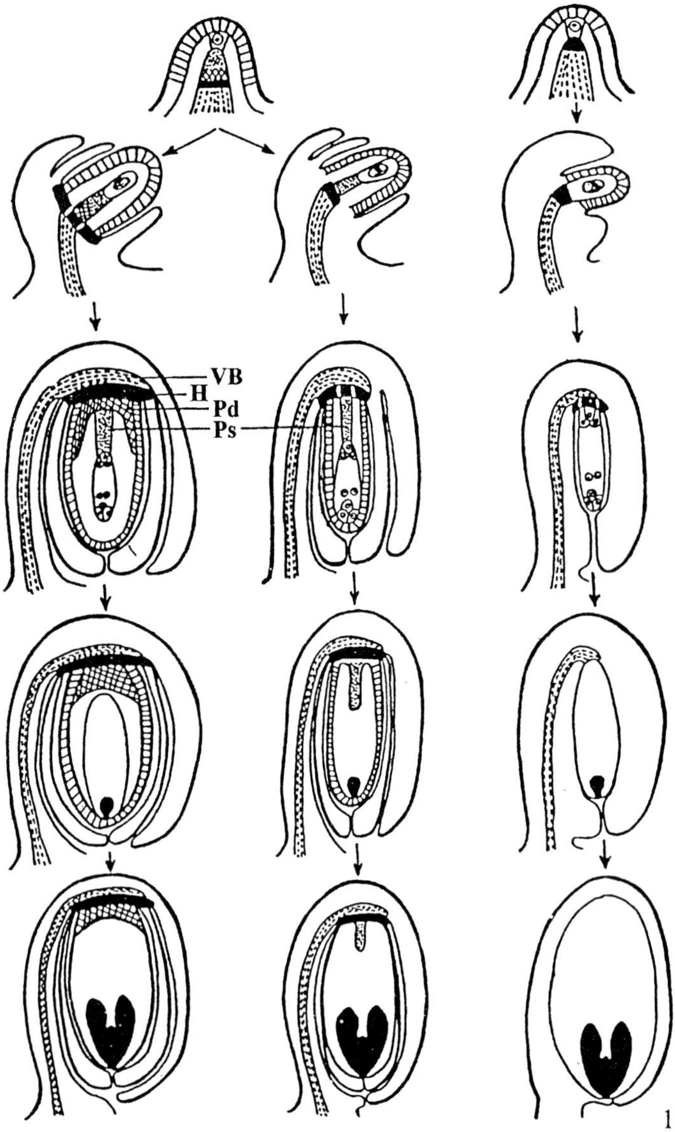

Plate 48: Ovule basal structures.

1 – differentiation of postament, podium, hypostase and vascular bundle from a meristem of basal zone of ovular primordium and their subsequent development in ovules of crassinucellate (left row), tenuinucellate (right row) and intermediate (middle row) types; formation of ovular structures: 2-5 – 'hypostase' (=postament, in our view) in *Mayaca fluviatilis*, 6 – 'hypostase' (=postament and hypostase in our view) in *Brucea amarissima*; 7 – 'hypostase' (=postament and possibly hypostase in our view) in *Datisca cannabina*; 8, 11 – hypostase in *Passerina filiformis* and *Antigonon leptopus*; 9, 10 – postament in *Geissoloma marginatum*; ES – embryo sac; H – hypostase, Pd – podium, Ps – postament, VB – vascular bundle.

1 – after Batygina and Shamrov, in press; 2-5 – after Venturelli and Bouman, 1986; 6 – after Nair and Sukumaran, 1960; 7 – after Boesewinkel, 1984; 8 – after Guèrin, 1916; 9,10 – after Stephens, 1909; 11 – after Solntseva, 1983.

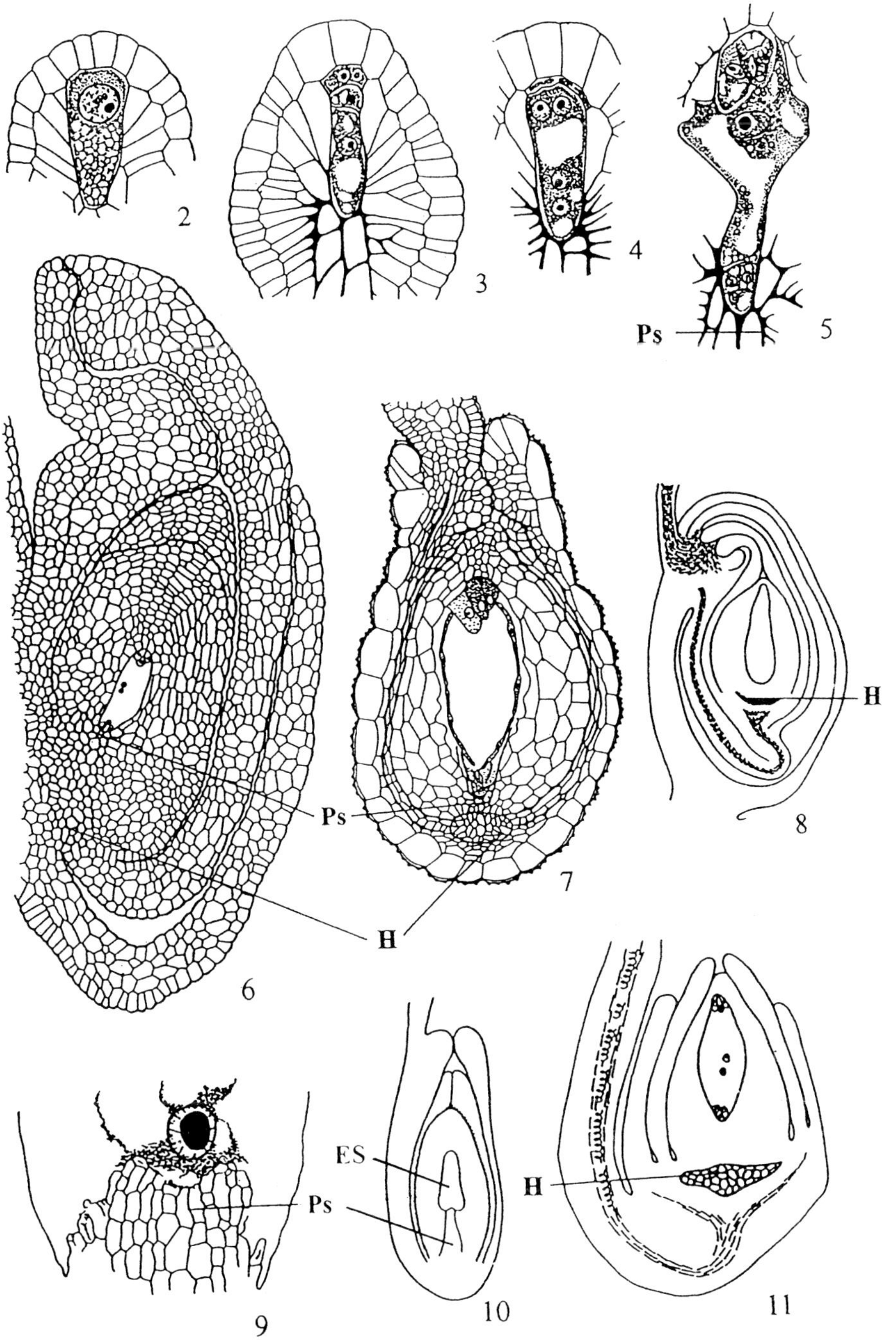

Plate 48 (*Contd.*)

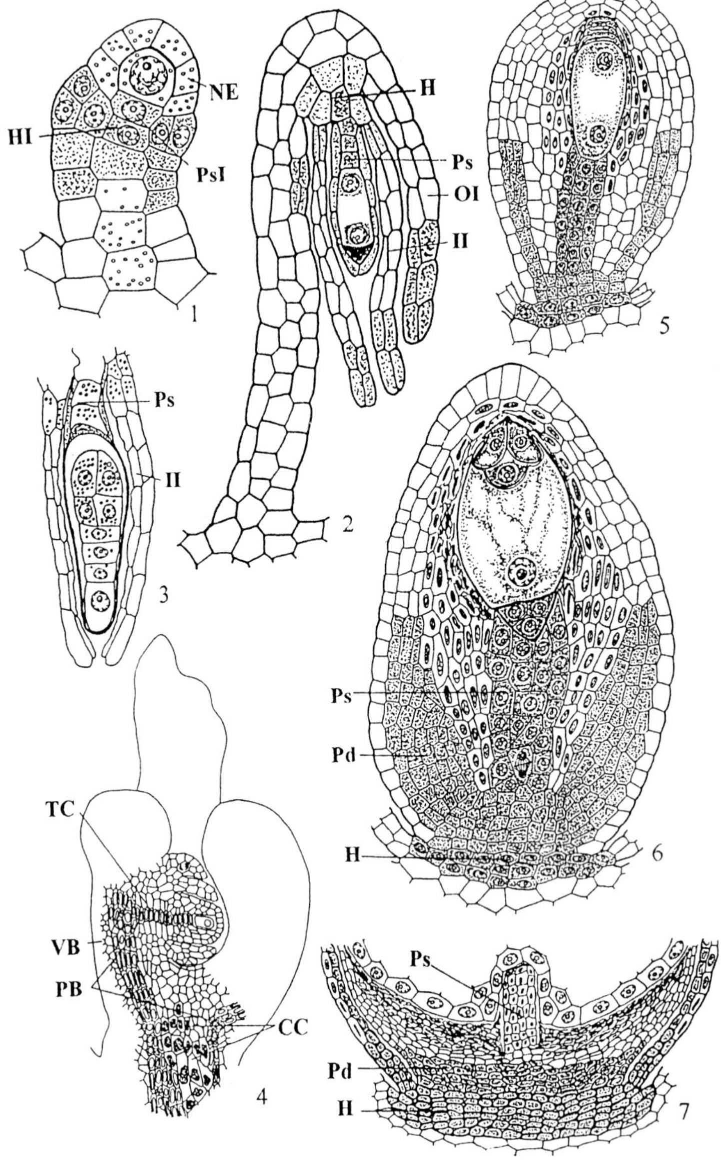

Plate 49: Postament.

1-3 – formation of postament and hypostase in tenuinucellate ovule of *Gymnadenia conopsea*; 4 – postament (tabular cells) in ovule of intermediate type in *Triticum aestivum*; 5-7 – formation of postament, podium and hypostase in crassinucellate ovule of *Ribes aureum*; CC – coating cells, H – hypostase, HI – hypostase initials, II – inner integument, NE – nucellar epidermis, OI – outer integument, PB – procambial bundle, Pd – podium, Ps – postament, Ps I – postament initials, TC – tabular cells, VB – ventral vascular bundle.

1-3 – after Shamrov and Nikiticheva, 1992; 4 – after Batygina, 1974b; 5-7 – after Shamrov, in press.

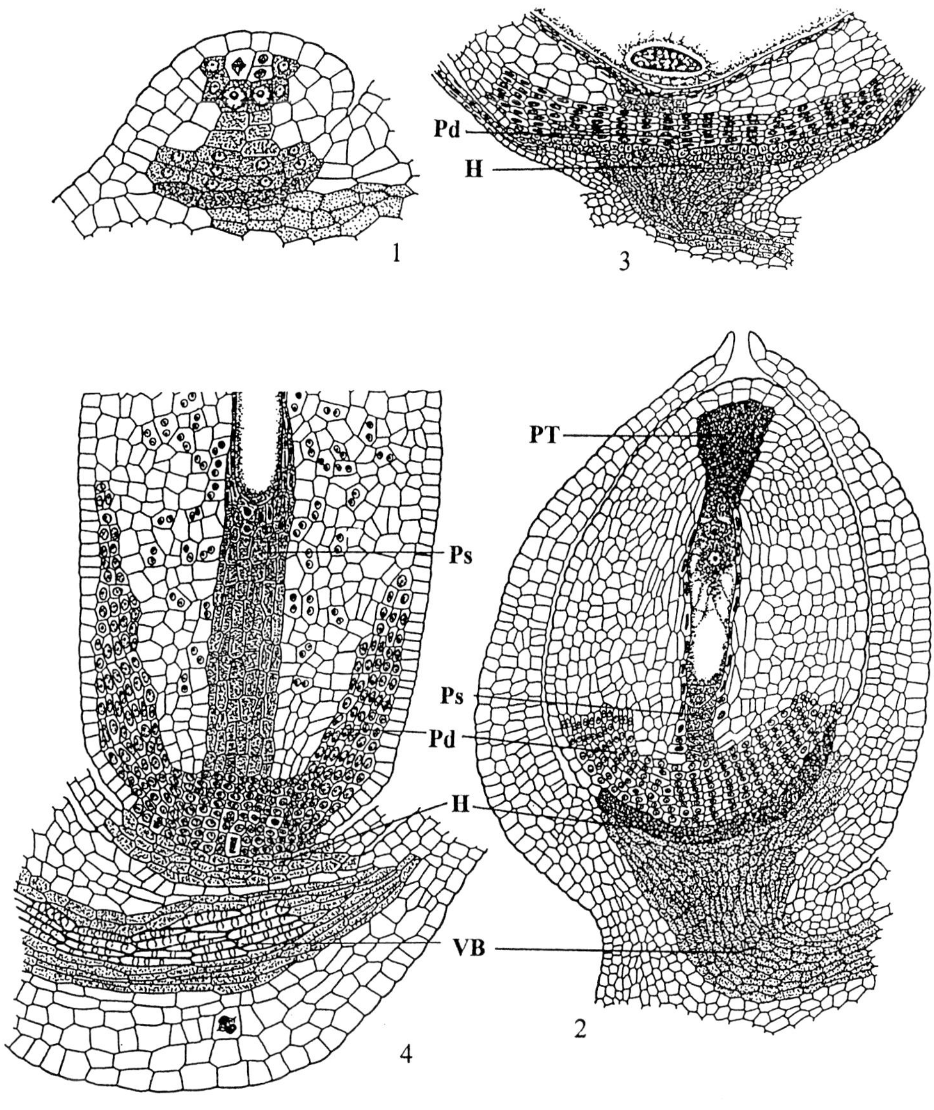

Plate 50: Podium.

1-6 – podium formation in *Ceratophyllum demersum* (1-3), *Nuphar lutea* (4) and *Luzula pedemontana* (5,6); 7-9 – transformation of basal zone of nucellus into postamento-podium in *Gagea stipitata;* Ch – chalaza, ChC – chalazal cell of endosperm, FO – funicular obturator, H – hypostase, II – inner integument, NL – nucellar lateral region, OI – outer integument, Op – operculum, PB – procambial bundle, Pd – podium, PlO – placental obturator, PPd – postamento-podium, Ps – postament, PsI – postament initials, PT – parietal tissue, R – raphe, VB – vascular bundle.

1-4, 7-9 – after Shamrov, in press; 5, 6 – after Shamrov and Anisimova, 1993b.

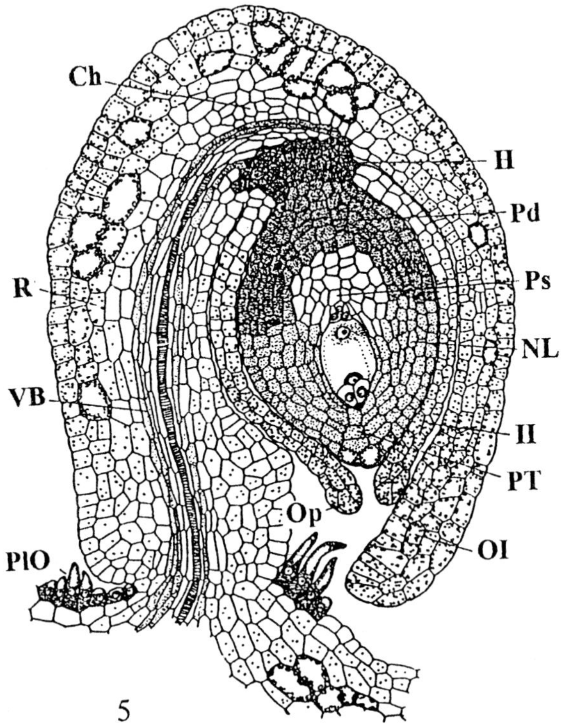

5

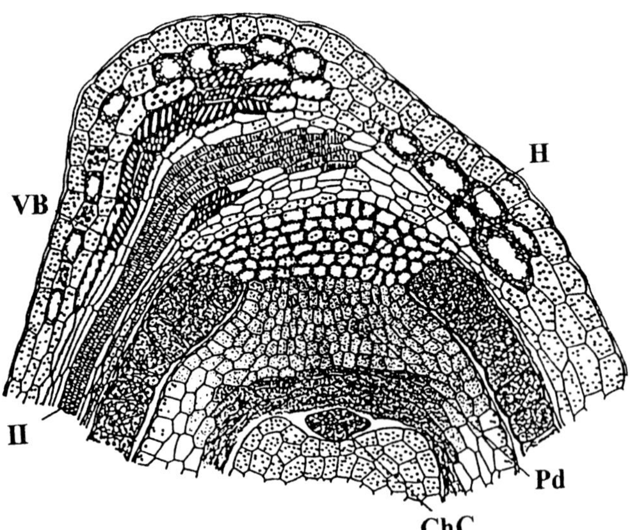

6

Plate 50 (*Contd.*)

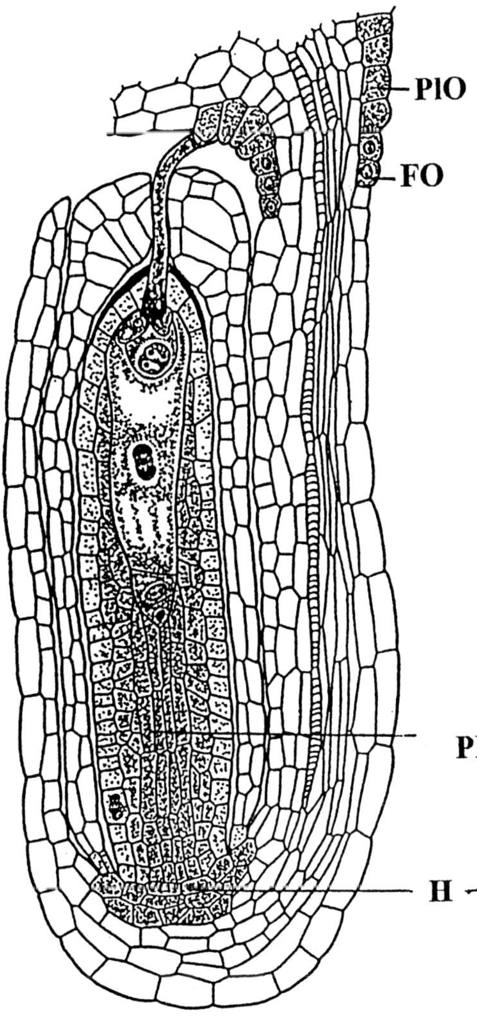

7

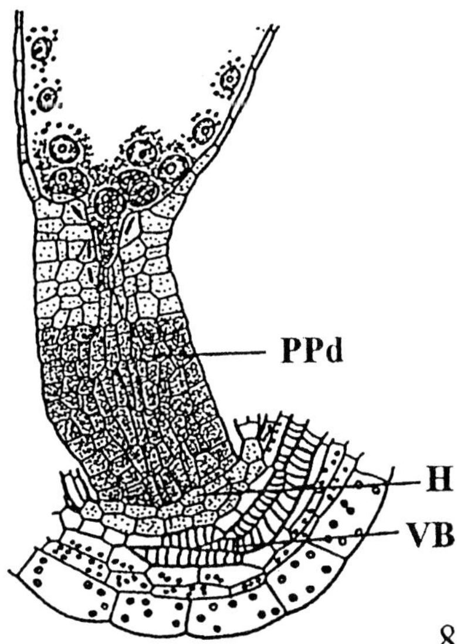

8

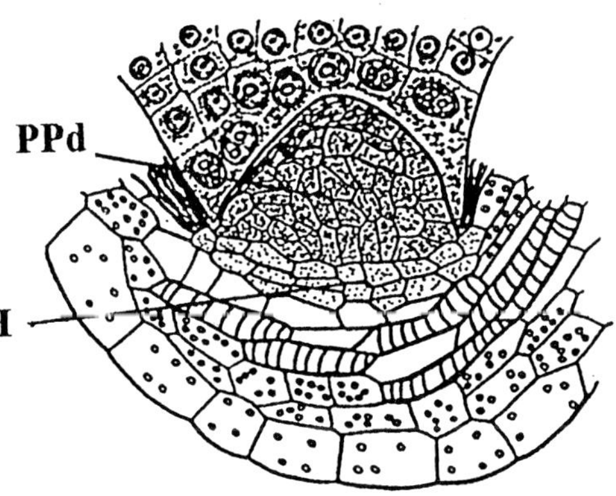

9

Plate 50 (*Contd.*)

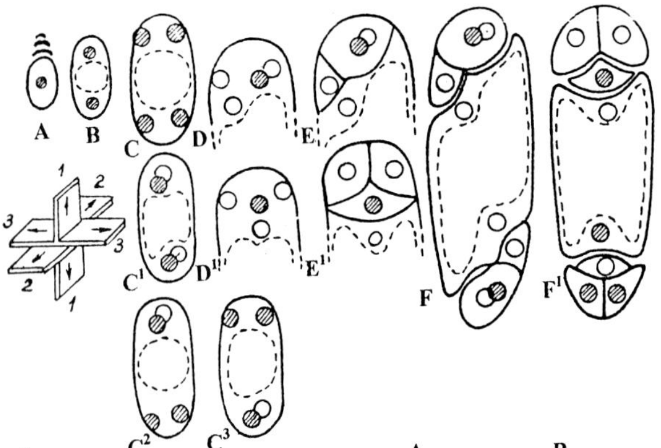

1

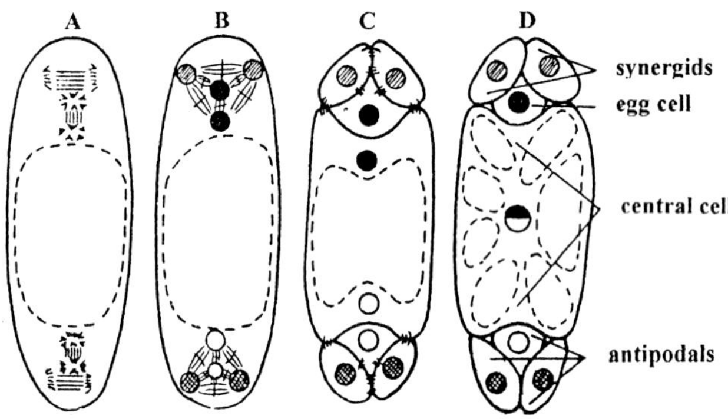

2

Plate 51: Embryo sac (female gametophyte): origin and development.

1 – normal type of embryo sac: scheme of development and beginning of formation: shaded nuclei are nearest to the observer. Three perpendicular planes of division (movement) of nuclei during female gametophyte development are shown on the left. Megaspore (A) gives rise to binuclear coenocyte (young female gametophyte) (B) with a vacuole in its central part. Four-nuclear coenocytic gametophyte (C) arises as a result of second division which plane is perpendicular to that of the first division; nuclei shown at their most distant position (C^1 – the same gametophyte turned 90°), in 'square position' (C^2); in 'tetrahedric position' (C^3). D – upper (micropylar) part of coenocyte after third division of nuclei in perpendicular planes (D^1 – same picture turned 90°). Nuclei disposed at tops of tetrahedron, which corresponds to their maximal moving away from each other. E – upper part of the coenocyte under cellularization (E^1 – same picture turned 90°); walls have arisen between all nuclei to form an egg apparatus and upper polar nucleus; two upper cells with sister nuclei appear to be synergids (at E they lie one under the other, at E^1 they are seen at one plane); third cell the egg, which is smallest at this phase; nuclei move away from the wall to the centre in all three cells. F – general picture of a newly formed embryo sac (F^1 – same picture turned 90°); shown above is the micropylar end, three cells of egg apparatus visible (at F synergids and antipodals lie one under the other), below is the chalazal end, three antipodals visible; both polar nuclei stay in the common cytoplasm of the central cell; later they will draw to its dynamic centre closer to each other.

2 – scheme of last phases of embryo sac development, showing the origin of nuclei as a result of simultaneous mitoses and manner of cell wall setting between them. Everything is shown projected to the plane of the drawing. Sister nuclei pairs are indicated with the same colouring: A – telophase of last (third) mitosis in female gametophyte (spindles of division perpendicular to each other in each pair, the centre of the coenocyte occupied by a large vacuole); B – later phase (cell plates are settled between just-formed nuclei disposed in tops of tetrahedron; they arise in primary spindles connecting sister nuclei and in secondary spindles which form between non-sister nuclei as well); C – just-formed embryo sac (cells have separated from each other - egg apparatus above, antipodals below and central cell with both free polar nuclei and large vacuole appears in centre); D – last phase of embryo sac development (polar nuclei have fused in centre of central cell to form secondary nucleus).

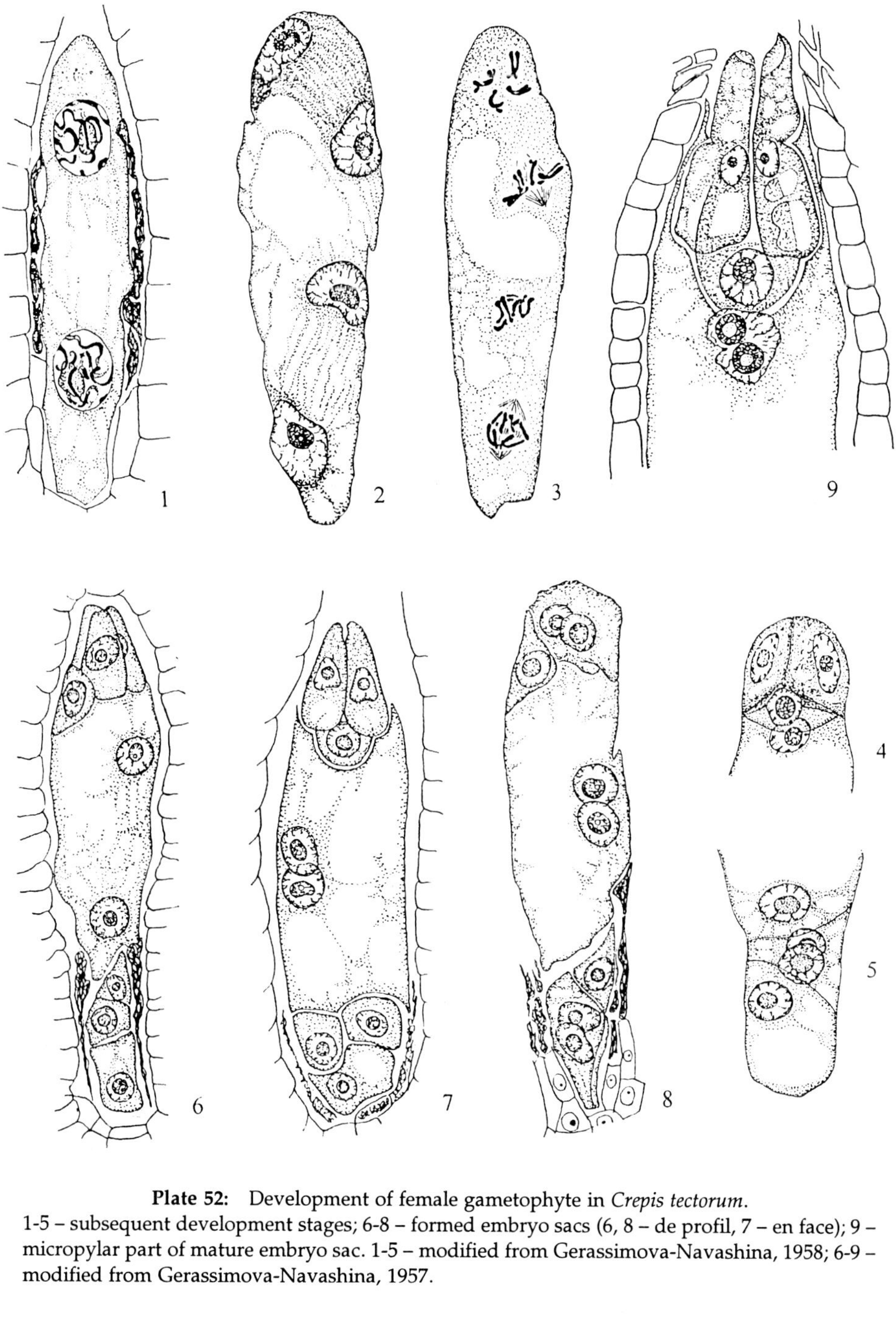

Plate 52: Development of female gametophyte in *Crepis tectorum*.
1-5 – subsequent development stages; 6-8 – formed embryo sacs (6, 8 – de profil, 7 – en face); 9 – micropylar part of mature embryo sac. 1-5 – modified from Gerassimova-Navashina, 1958; 6-9 – modified from Gerassimova-Navashina, 1957.

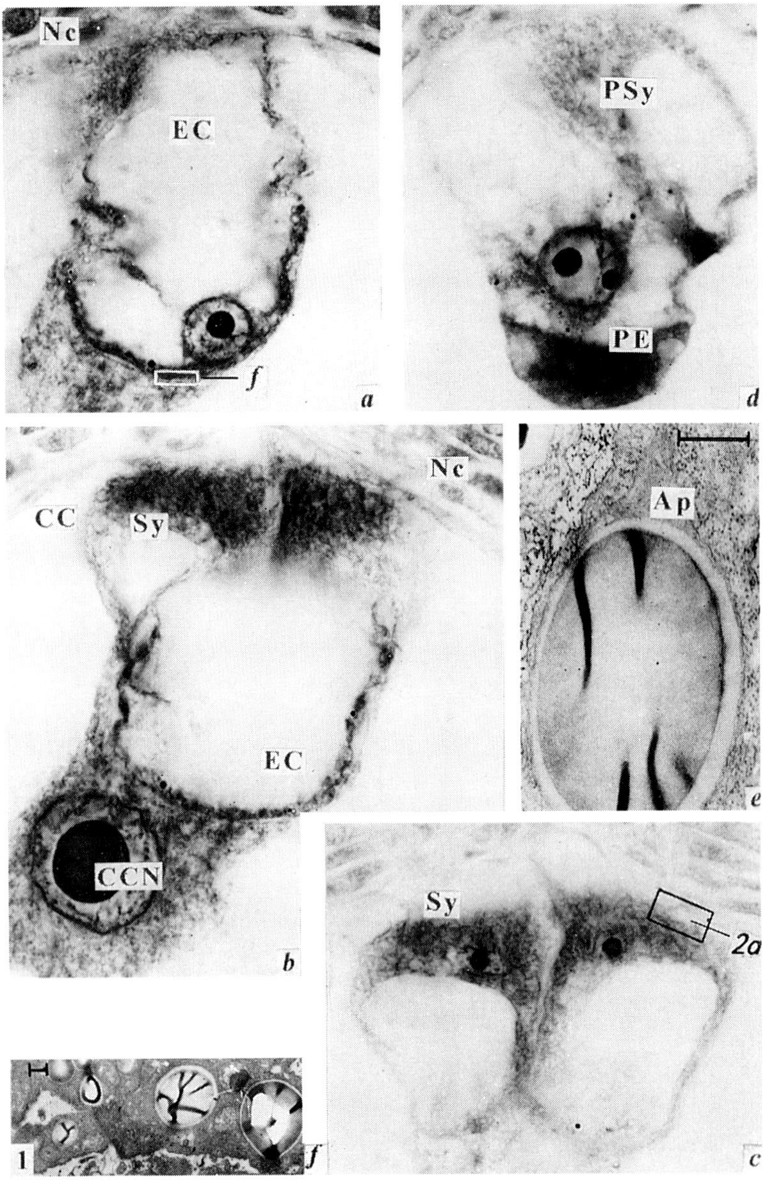

Plate 53: Egg cell.

1 – micropylar part of embryo sac of *Alcea rosea* before fertilization (a-c, e, f) and after fertilization (d): a-c – three serial sections of egg apparatus; d – two-celled proembryo and persistent synergid; e – amyloplast of egg cell; f – fragment of apical part of egg cell at boundary with central cell; cell wall lacking between them, plasma membranes of both cells sinuous; 2a, b – fragments of synergid basal part in *A. rosea*; scale – 1 μm; Ap – amyloplast, CC – central cell, CCN – central cell nucleus, EC – egg cell, FA – filiform apparatus, Nc – nucellus, PE – proembryo, PSy – persistent synergid, Sy – synergid.

1a – after Kamalova *et al.*, 1983; 1b-f, 2 – after Zhukova, unpublished data.

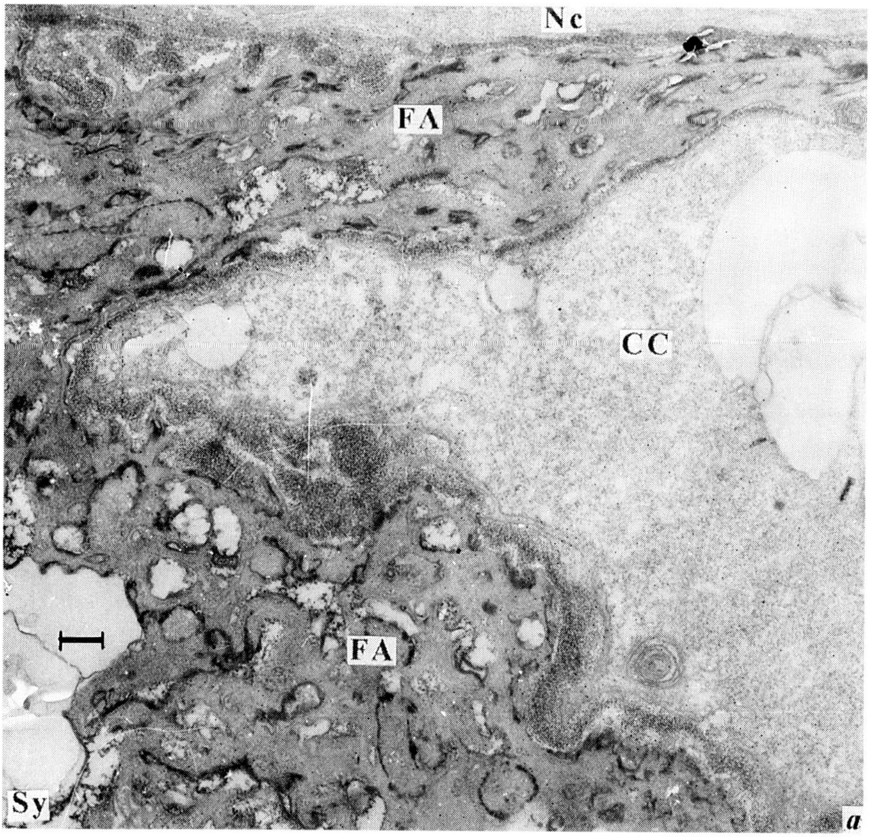

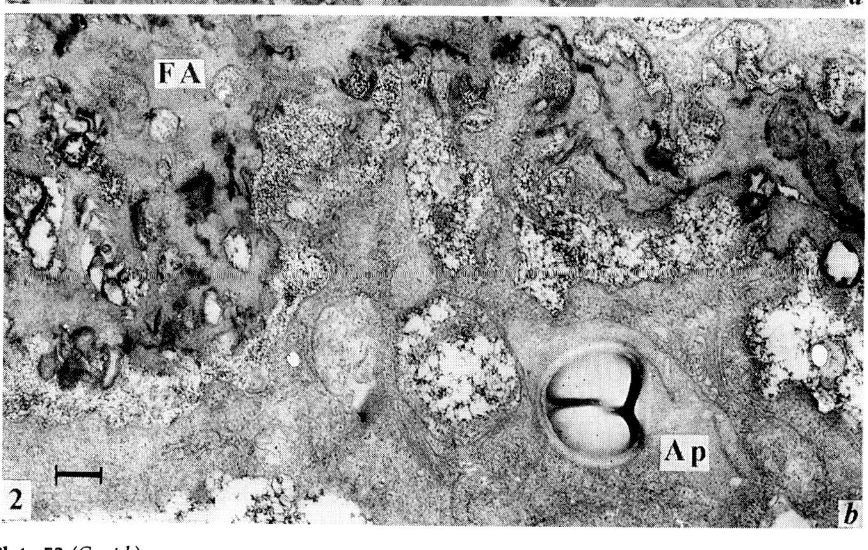

Plate 53 (*Contd.*)

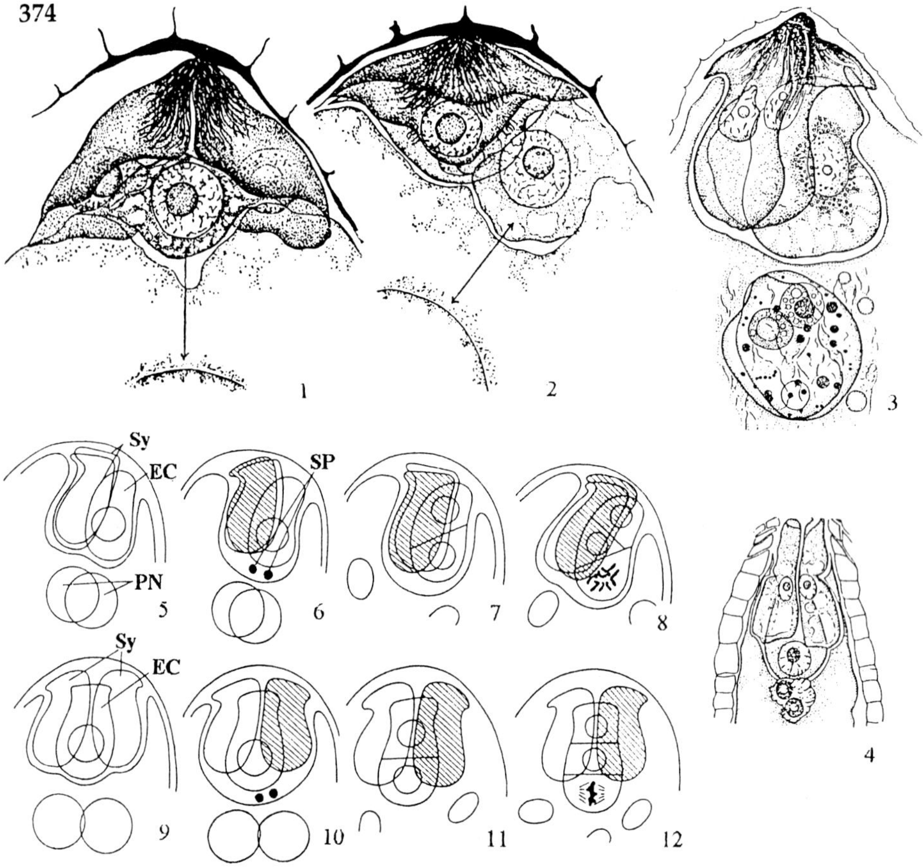

Plate 54: Synergids.

1, 2 – micropylar part of embryo sac of *Scilla sibirica* (1 – en face, 2 – de profil), arrows show direction of further sperm movement; 3 – synergids of *Triticum aestivum* with dental outgrowths; 4 - funnel-shaped synergids of *Crepis tectorum*; 5-12 – schemes of micropylar part of embryo sac of *Triticum* before fertilization (5, 9), during fertilization (6, 10) and at early embryogenesis (7, 8, 11, 12) (5-8 - dorsiventral sections and 9-12 – bilateral sections); 13 – synergids with haustorial outgrowths in *Ursinia anthemoides*; 14 – persistent synergid and four-celled embryo of *Argyropsis candida*; 15 – synergids of *Eucalyptus cinerea* similar to egg cells; 16 – diagram of micropylar part of embryo sac of *Linum usitatissimum*, illustrating arrangement of synergids and egg cell; 17-22 – some subsequent stages of formation and destruction of giant persistent synergid of *Allium nutans*; 23 – fragment of *Valeriana officinalis* embryo sac at stage of zygote and two-celled endosperm; basal part of persistent synergid with stored starch shown; cell wall of micropylar cell of two-celled endosperm at boundary with base of persistent synergid possesses wall projections; 24 – fragment of apical part of persistent synergid in embryo sac of *Epipactis atrorubens* at stage of multicelled proembryo and coenocytic endosperm; tip of synergid shows lobe-like outgrowths, its cell wall interrupted; Ap – amyloplast, EC – egg cell, En – endosperm, FA – filiform apparatus, IT – integumentary tapetum, N – nucleus, Nl – nucleolus, PE – proembryo, PN – polar nuclei, PSy – persistent synergid, SP – sperm cell, Sy – synergid.

1, 2 – after Gerassimova-Navashina, 1958; 3, 5-12 – after Batygina, 1974b; 4 – after Gerassimova-Navashina, 1957; 13 – after Vidajati, 1973; 14 – after Vorsobina and Solntseva, 1990; 15 – after Polunina, 1959; 16 – after Vazart, 1965; 17-22 – after Stozharova and Poddubnaya-Arnoldi, 1977; 23 – after Kamelina, Zhukova, 1987; 24 – after Zhukova, unpublished data.

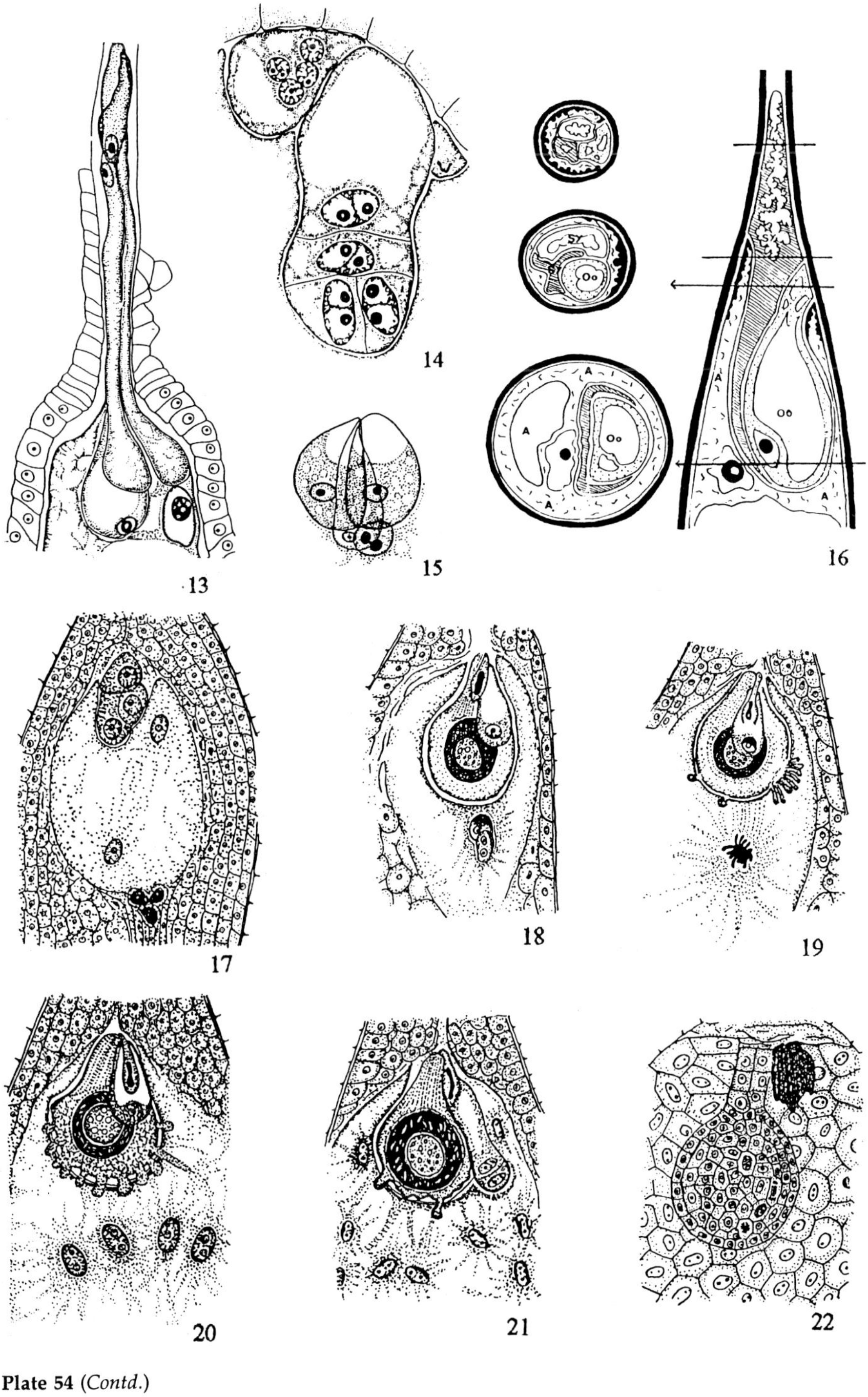

Plate 54 (*Contd.*)

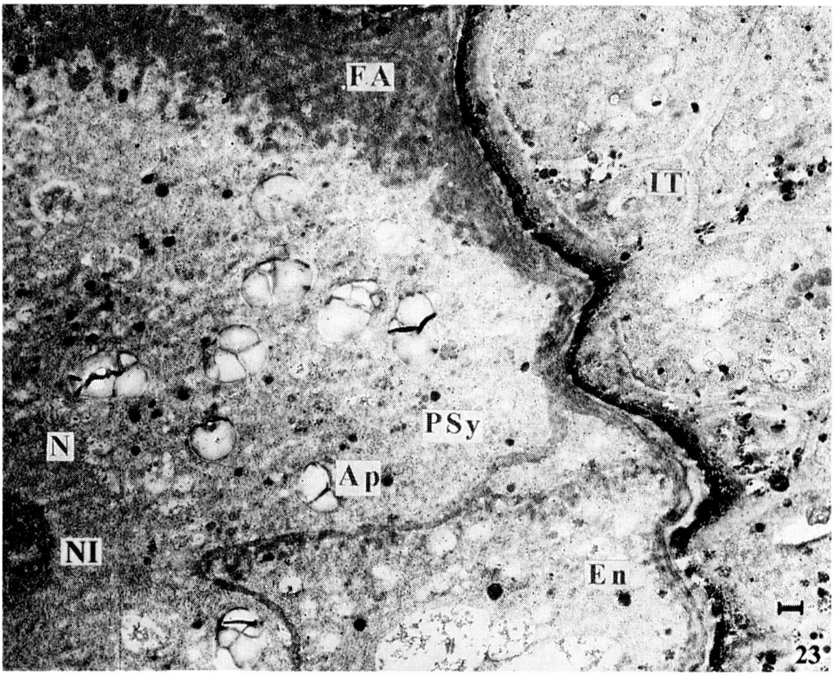

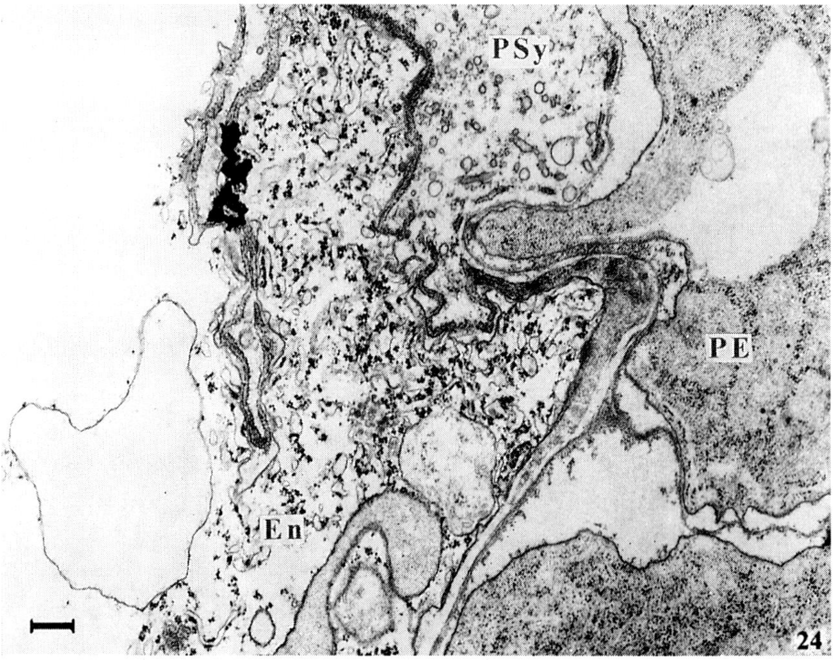

Plate 54 (*Contd.*)

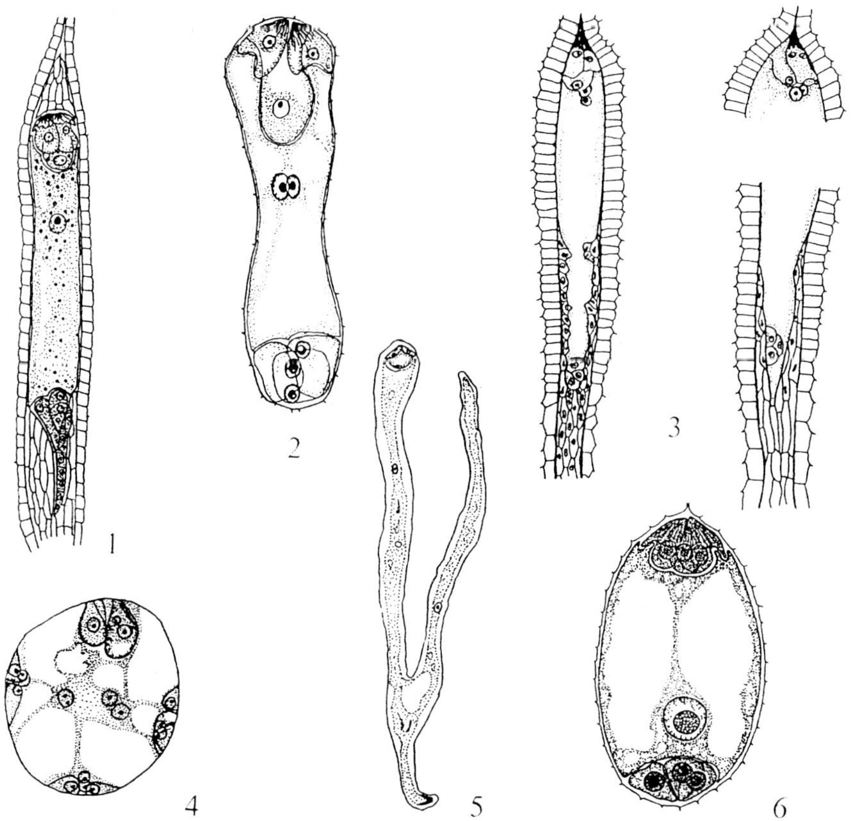

Plate 55: Central cell.

1-6 – diversity in shape of the central cell of mature embryo sac of flowering plants and in size ratio of this cell and other elements: *Davidia involucrata* (1), *Paliurus spina christi* (2), *Linum liburnicum* (3), *Malpighia glabra* (4), *Olax stricta* (5), *Galanthus nivalis* (6). Examples of variable position of central cell secondary nucleus shown (1, 6); location of nucleus at its chalazal part observed in species with Helobial type of endosperm development (6); localization of polar nuclei, which do not complete fusion before fertilization, also varies (2-4). Figure 3 illustrates one of the variants of central cell size increase by local growth at its chalazal pole during embryo sac maturation, which leads to displacement of antipodals to a lateral position. Embryo sac with central cell forming lateral outgrowth presented in Fig. 5. In majority of species studied starch is the main storage substance of the central cell; 7-10 – fragments of longitudinal sections of *Alcea rosea* ovule at stage of mature embryo sac (7-9) and diagram of ovule (10), showing central cell of very elongated shape and illustrating correlation of its size with that of other cells of the embryo sac; 11, 12 – fragments of the central cell protoplast of *A. rosea* near egg apparatus; mitochondria ameboid in shape, Golgi apparatus in state of hypersecretion, plastids lack starch; arrow points at protein crystal of a plastid (12); tubes of plastid and concentrically arranged thylacoids also shown; scale – 1 μm; CC – central cell, CCN – central cell nucleus, D – dictyosome, EA – egg apparatus, Lp – leucoplast, M – mitochondria.

1 – after Shevchenko and Kamelina, 1987; 2 – after Batygina *et al.*, 1987; 3 – after Nikiticheva, 1985c; 4 – after Singh, 1961; 5 – after Agarwal, 1963; 6 – after Ostapenko and Khvedynich, 1987; 7-12 – after Zhukova, unpublished data.

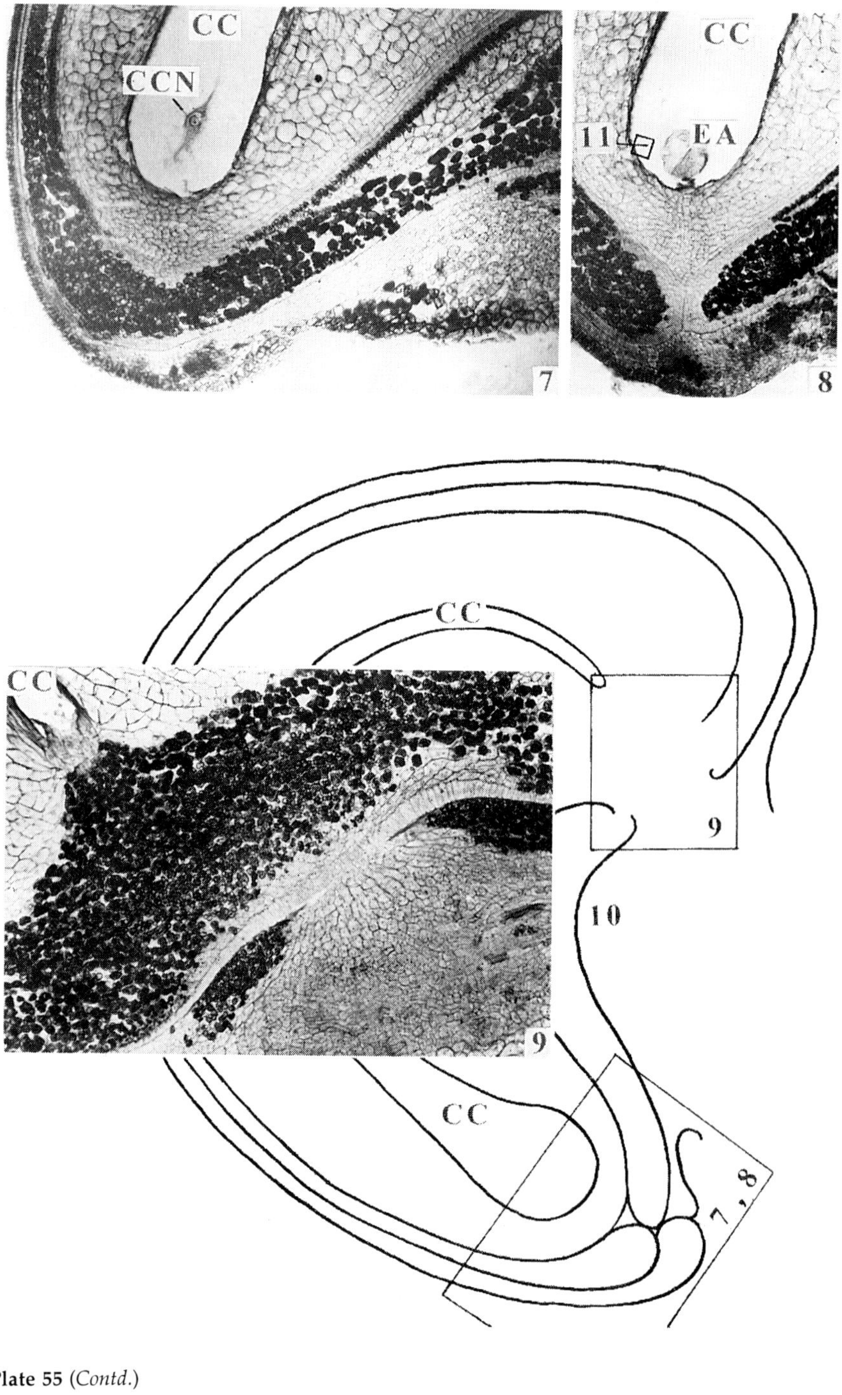

Plate 55 (*Contd.*)

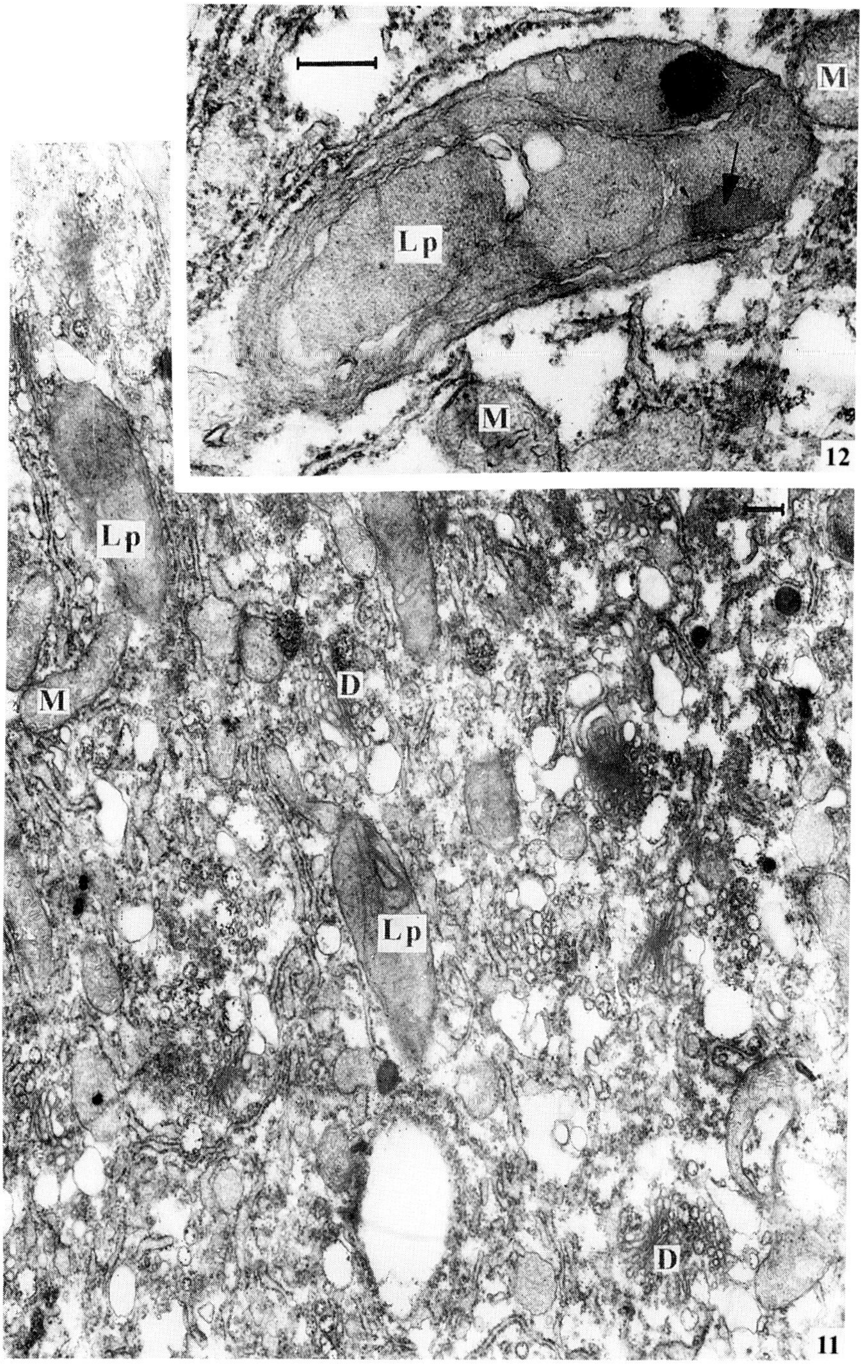

Plate 55 (*Contd.*)

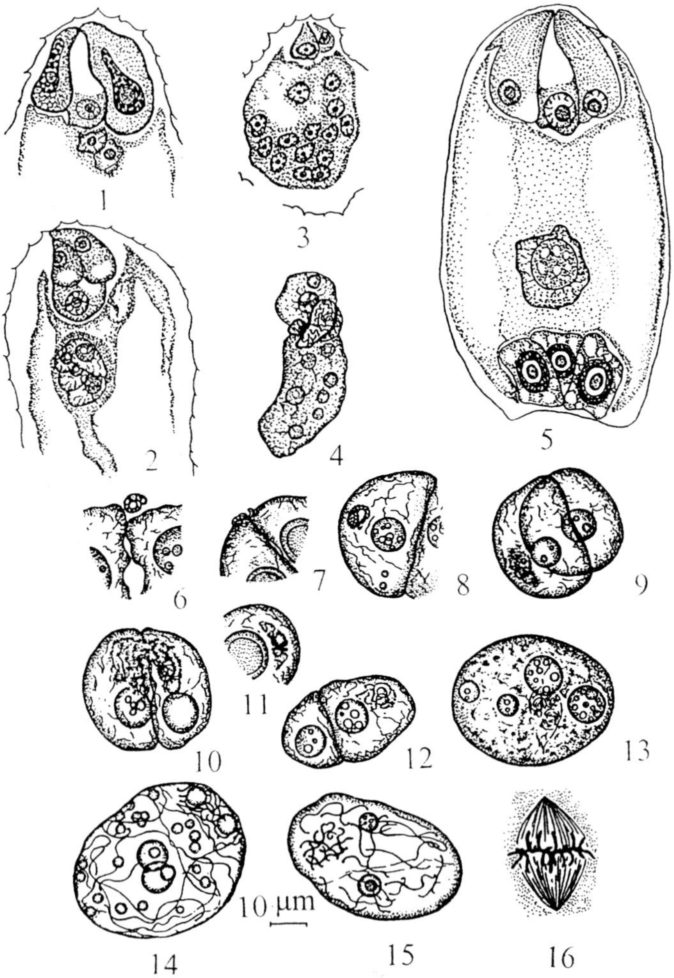

Plate 56: Polar nuclei.

1 – micropylar region of *Allium schoenoprasum* embryo sac, chalazal and micropylar polar nuclei of same structure; 2 – four polar nuclei in central cell of *Plumbago capensis* embryo sac, developing according to Plumbago-type; 3 – embryo sac of *Peperomia rezediflora* with disturbed process of cellularization, there are 14 nuclei at the central cell; 4 – haploid micropylar and triploid chalazal polar nuclei in contact with sperm cell nucleus in *Lilium caucasicum*; 5 – embryo sac of *Galanthus nivalis*, central cell shows secondary nucleus (= fusion nucleus); 6-16 – subsequent stages of central cell polar nuclei fusion in embryo sac of wheat during their union with sperm cell nucleus; appearance of great amount of additional nucleoli at nucleus of primary endosperm cell shown, these nucleoli differing very much from nucleoli discharged by sperm cell nucleus in size and colour: 6, 7 – sperm cell at area of contact of two polar nuclei, 8, 9, 11, 12 – sperm cell nucleus fuses with one polar nucleus, 10 – sperm cell nucleus fuses simultaneously with both polar nuclei, 13-16 – nucleus of primary endosperm cell at interphase (13), prophase (14, 15), metaphase (16), (6-10, 13-15 – *T. aestivum*, 11, 12, 16 – *T. monococcum)*.

1-5 – after Gvaladze, 1976; 6-16 – after Batygina, 1974b.

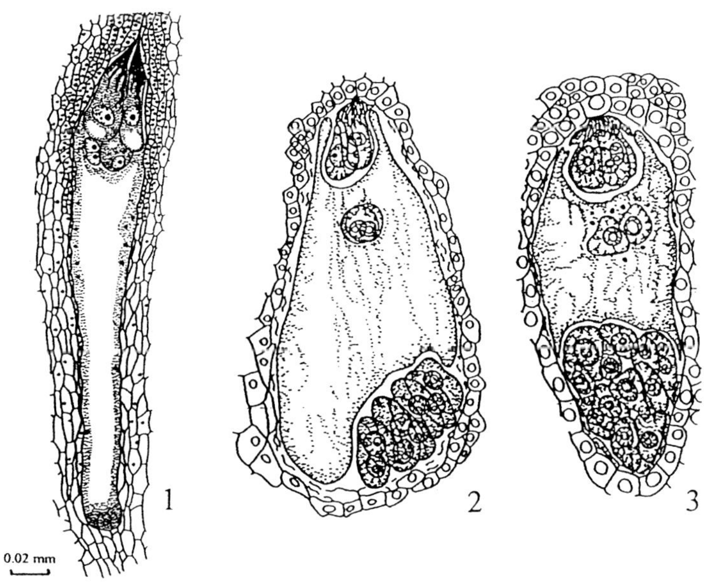

Plate 57: Antipodal cells.

1 – mature embryo sac of *Convolvulus scammonia*; 2, 3 – embryo sac of *Triticum* (2 – de profil, 3 – en face); 4,5 - most widespread variants of cell position at three-celled antipodal complex of mature embryo sac (4 – *Eupomatia laurina*, 5 – *Royena lucida*); 6 - antipodals imitating egg apparatus (embryo sac of *Scilla sibirica*), structure of antipodals persists after fertilization also; 7 – increase in number of antipodals (up to 8) and nuclei in each of them (up to 5) during embryo sac maturation in *Nelumbo nucifera* (a, b) and in process of its fertilization (c); 8 – Oenothera-type of embryo sac development without antipodals in *Ludwigia palustris*; 9 – persistent antipodals at one of final stages of seed development in *Cynomorium songaricum*; 10 – embryo sac of *Gentianella lingullata*; one of rare variants of antipodal arrangement in embryo sac of flowering plants; 11 – some moments of embryo sac formation in *Anemone nemorosa* (longitudinal sections): a – uninuclear antipodals, b – antipodals with four nuclei, c – secondary uninuclear antipodals at embryo sac under fertilization, 12 – diagram of embryo sac chalazal part before fertilization in *Aconitum napellus* (third antipodal not shown); 13-15 – fragments of cytoplasm in bladder part of antipodals of *A. napellus*, invaginations of leucoplasts (arrow) and concentric membrane formations produced by their envelope (double arrow) shown; 16 – fragments of antipodal protoplasts of mature embryo sac of *Epipactis atrorubens*, leucoplasts (starch not found); 17, 18 – persistent antipodals of *Valeriana officinalis* after division of primary endosperm nucleus during the onset of cellularization (17, also see diagram at Fig. 18) and at stage of two-celled endosperm (18); small starch grains visible in plastids; scale – 1 µm; AC – antipodal cell, Ap – amyloplast, CC - central cell, D – dictyosome, ECS – extracellular space, En – endosperm, Lp – leucoplast, N – nucleus, Nc – nucellus, PB – paramural body, PCW – projections of cell wall, SG – starch grain, V – vacuole, W – cell wall.

1 – after Nikiticheva and Proskurina, 1987; 2, 3 – after Batygina, 1974b; 4 – after Kamelina, 1981d; 5 – after Nikiticheva, 1983a; 6 – after Gerassimova-Navashina, Batygina, 1958; 7 – after Batygina *et al.*, 1982; after Titova, unpublished data, 8 – after Batygina, 1985a; 9 – after Teryokhin *et al.*, 1975; 10 – after Shamrov, 1988; 11 – modified from Yoffe, 1968; 12-14 – after Zhukova and Sokolovskaya, 1977; 15, 16, 18 – after Zhukova, unpublished data; 17 – after Kamelina and Zhukova, 1987.

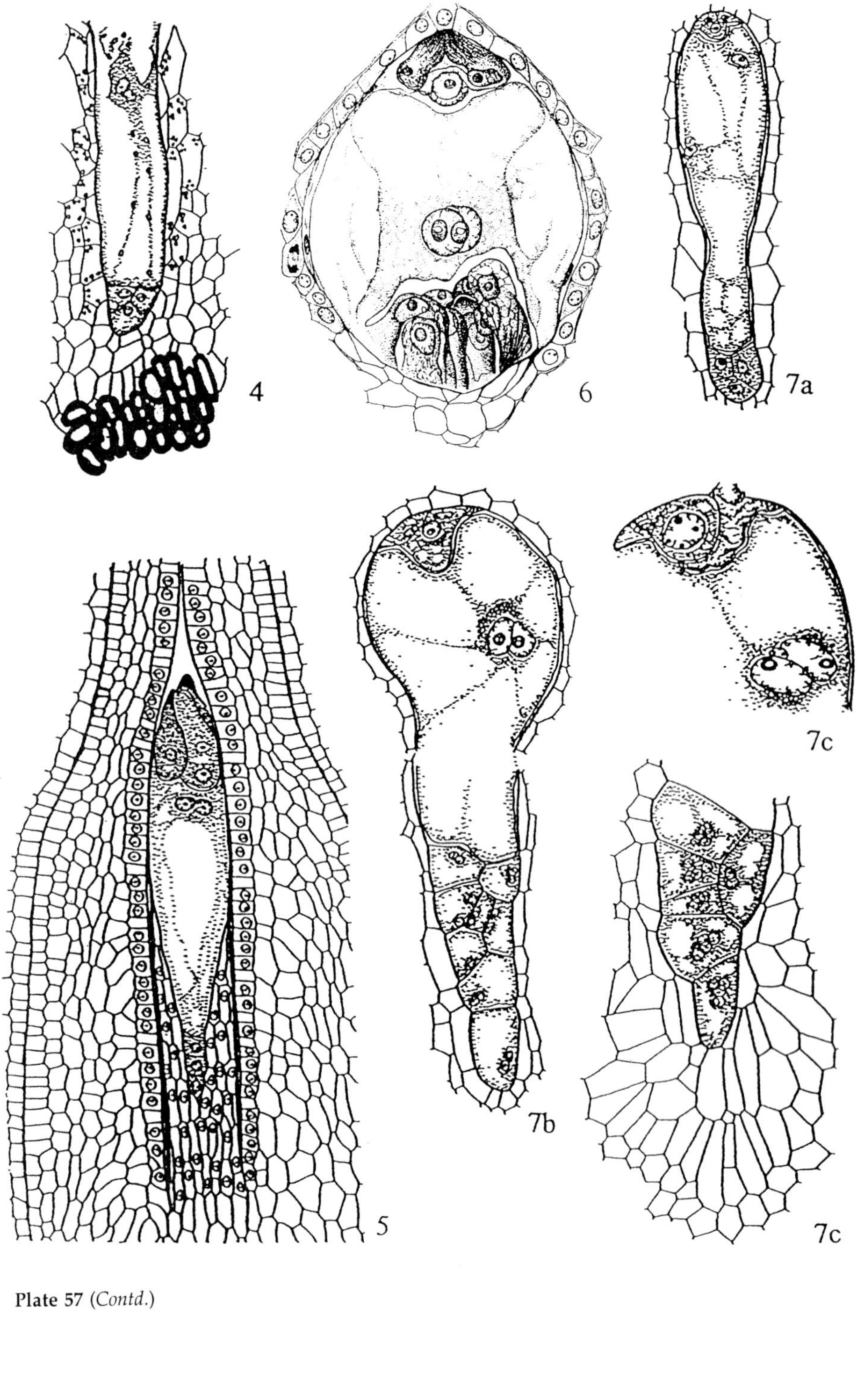

Plate 57 (*Contd.*)

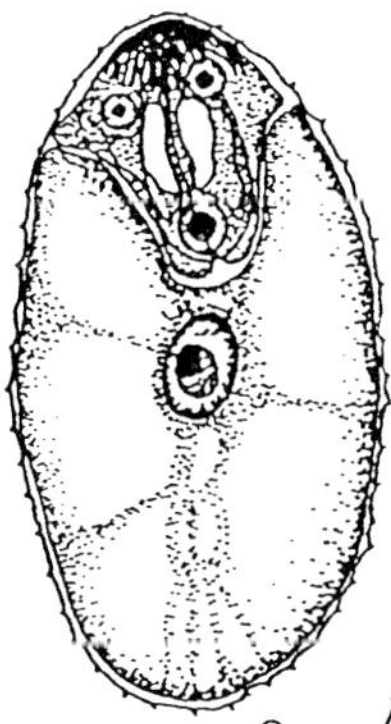

8

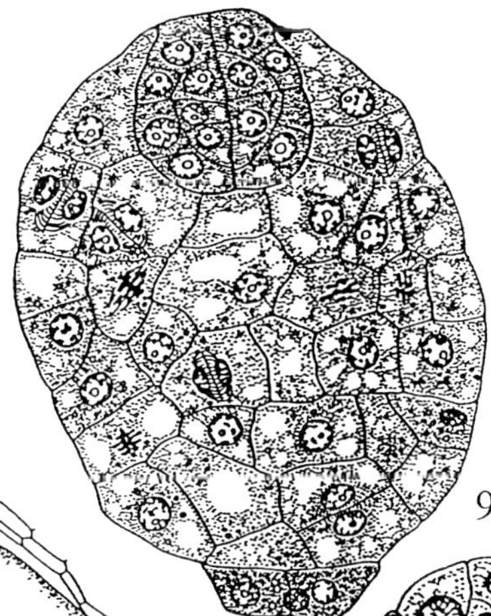

9

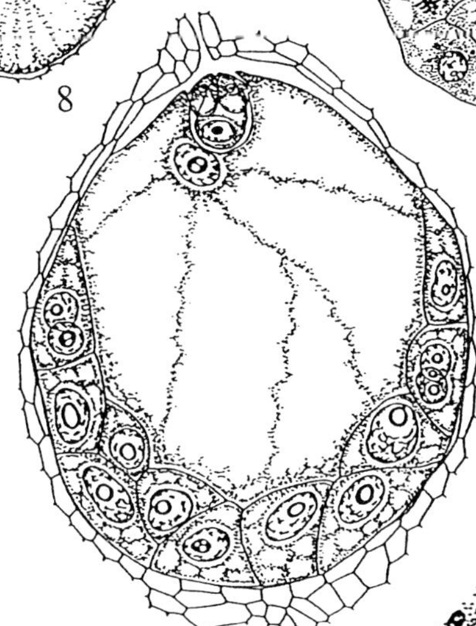

10

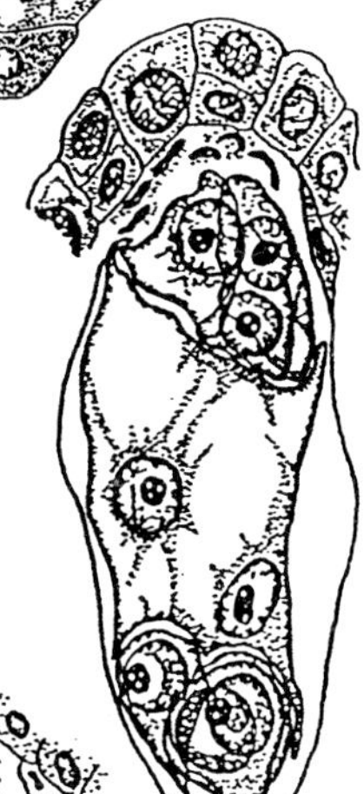

11a

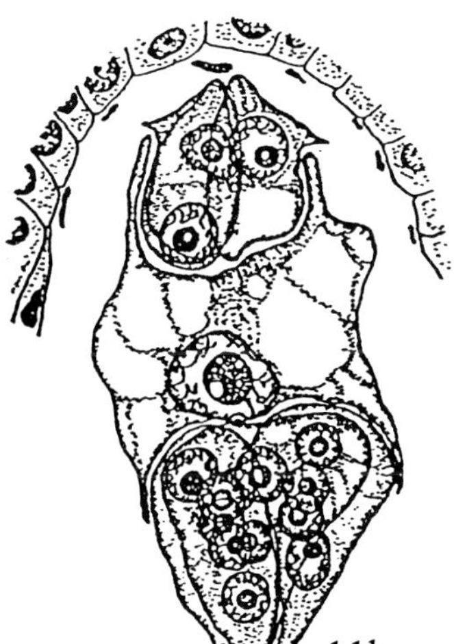

11b

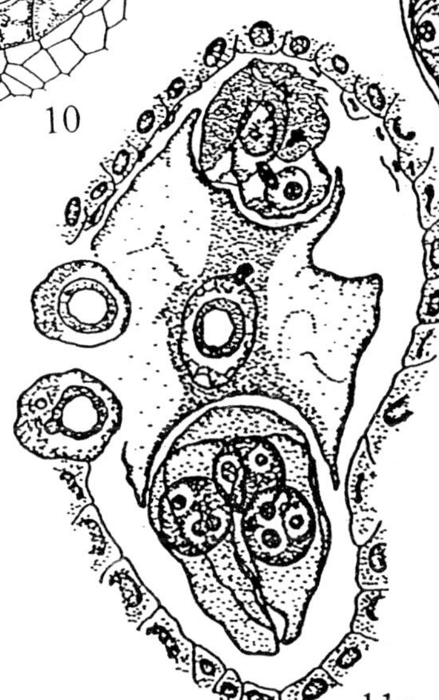

11c

Plate 57 (*Contd.*)

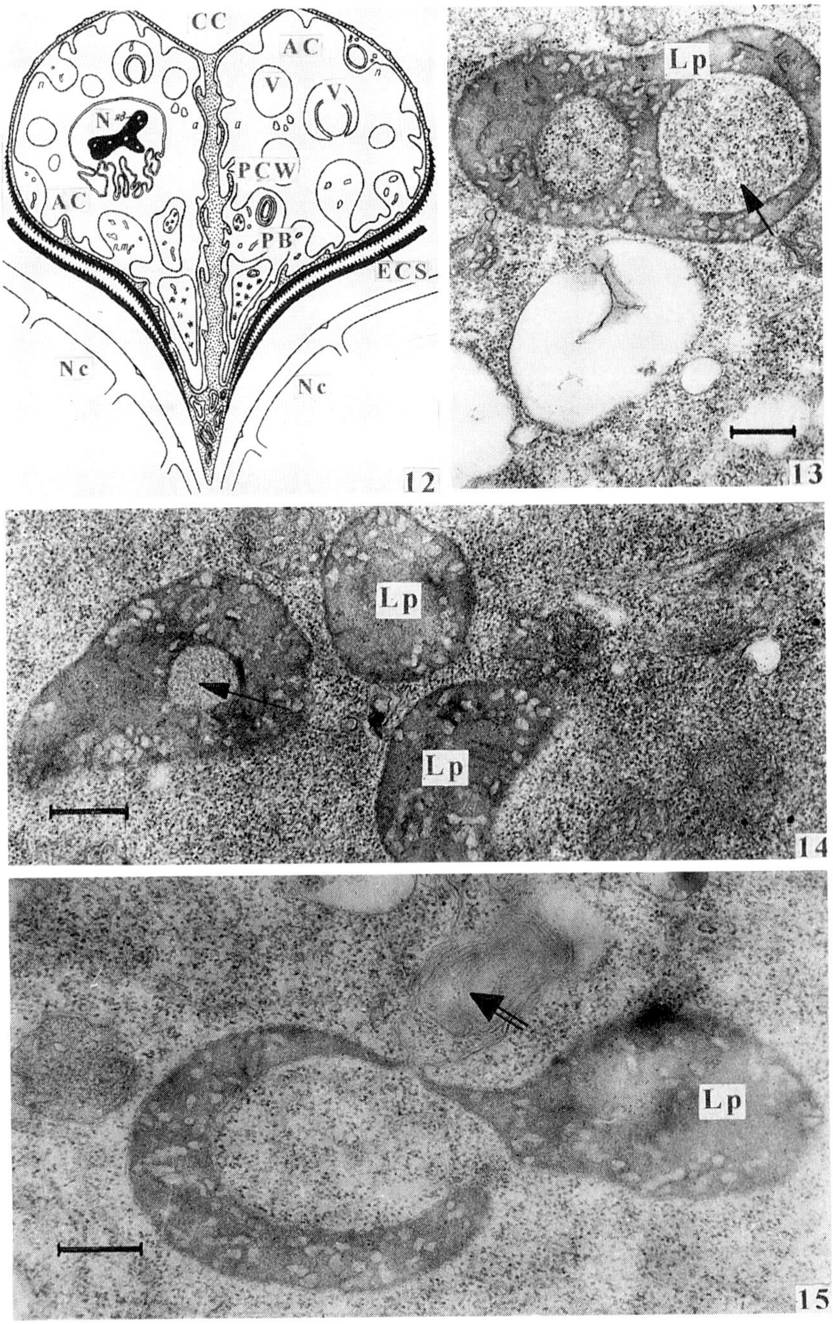

Plate 57 (*Contd.*)

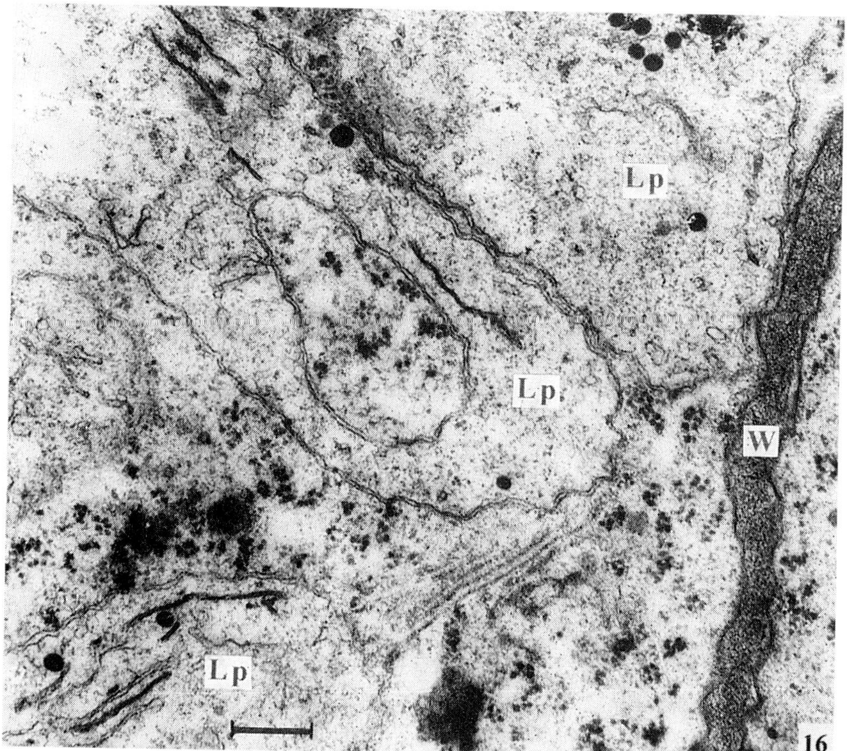

16

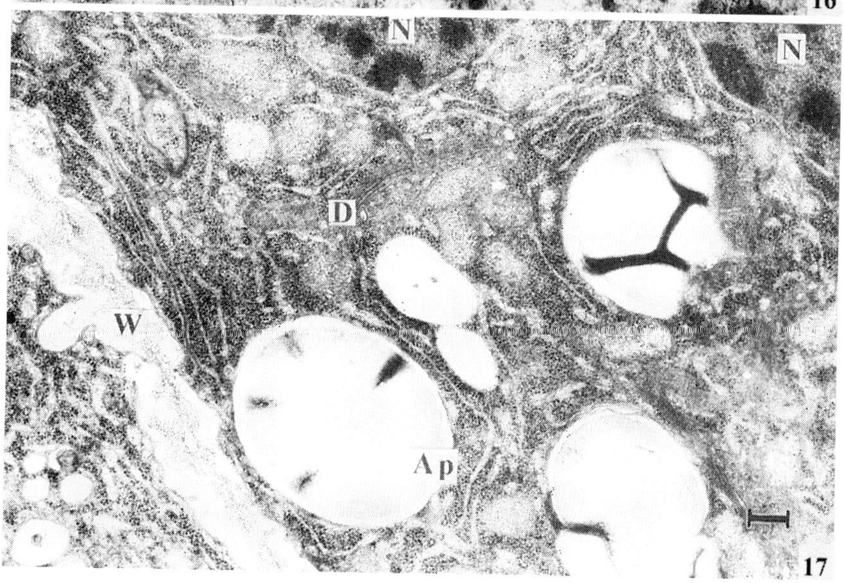

17

Plate 57 (*Contd.*)

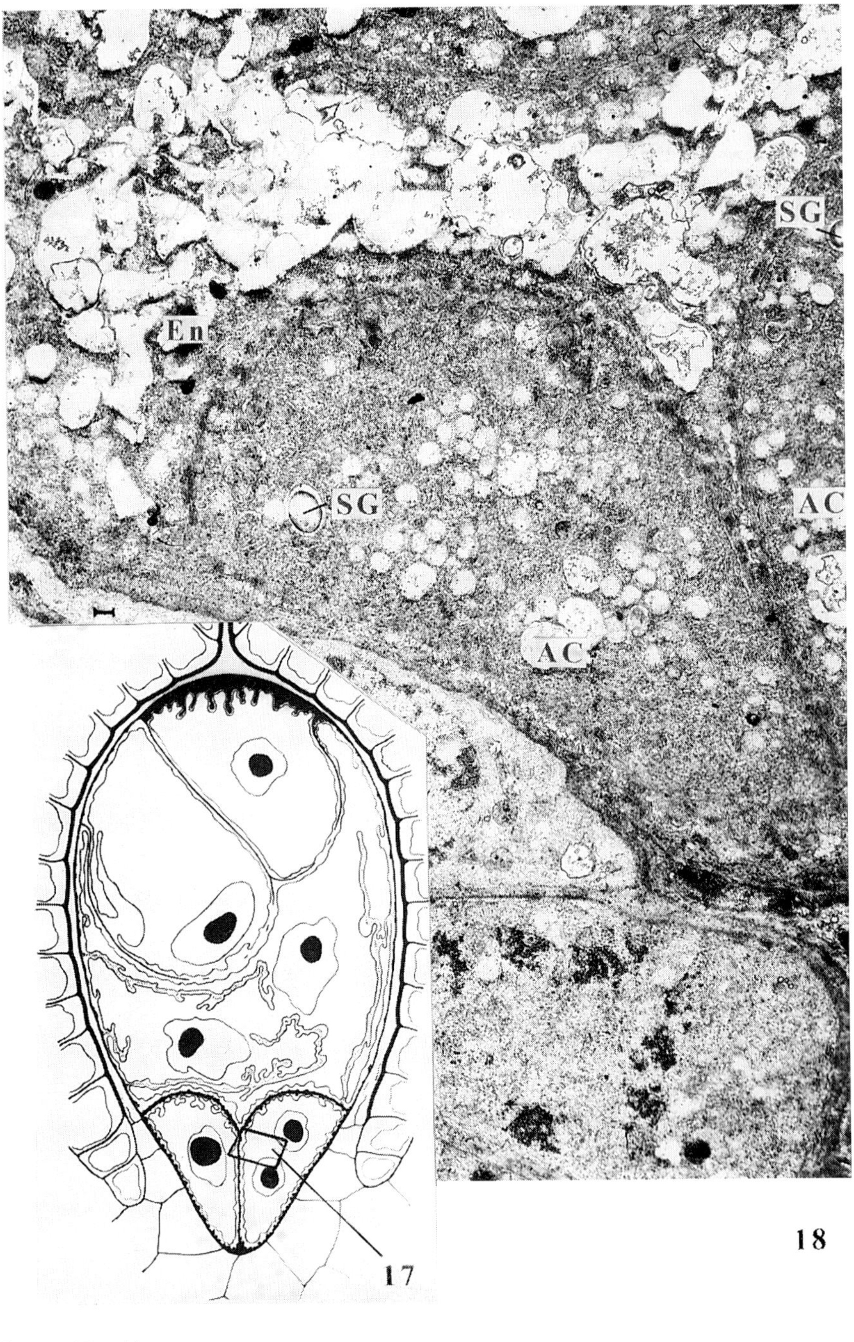

Plate 57 (*Contd.*)

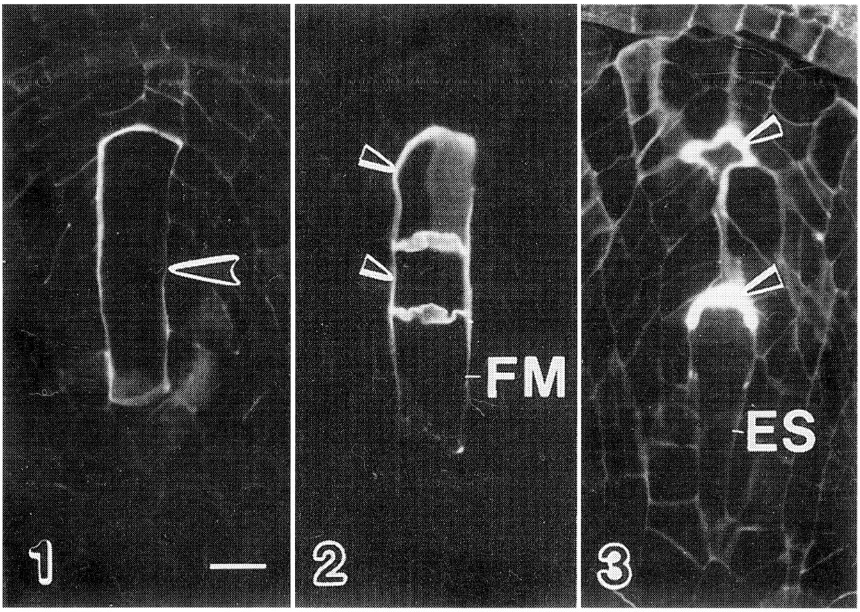

Plate 58: Fine structure of female gametophyte development.

1-3 – localization of callosic cell walls during megaspore and dyad cell abortion in *Zea mays*, stained using decolorized aniline blue and observed with epifluorescence microscopy. scale – (Micropylar pole upright in all Figures) scale – 10 mm. 1 – megasporocyte, callosic wall indicated at arrowhead; 2 – triad of meiotic products, showing callosic wall in degenerating dyad and megaspore cells (unlabelled arrowheads) and functional megaspore; 3 – two-nucleate megagametophyte showing remaining callosic walls of the abortive meiotic cells (unlabelled arrowheads); 4-7 – transmission electron micrographs of megagametophyte development in *Zea mays*. scale – 5 µm: 4 – megasporocyte. Nucleus displaced to micropylar end in mid-prophase I; most heritable organelles located chalazal to the nucleus; 5 – functional megaspore. Vacuolization begins with small widely distributed vacuoles. Nucleus central to chalazal and heritable organelles perinuclear; 6 – two-nucleate megagametophyte, with two degenerating meiotic cells (arrowheads), which remain callose positive (Fig. 3). Central vacuole forms by coalescence of smaller vacuoles. Cytoplasmic density greater than surrounding cells; 7 – four-nucleate megagametophyte and degenerating meiotic cells (arrowhead). Vacuoles merge and form compartments apparently involved in lytic activity; separate chalazal and micropylar vacuoles also visible; ES – megagametophyte (= embryo sac), FM – functional megaspore.

1-7 – after Russell, 1979.

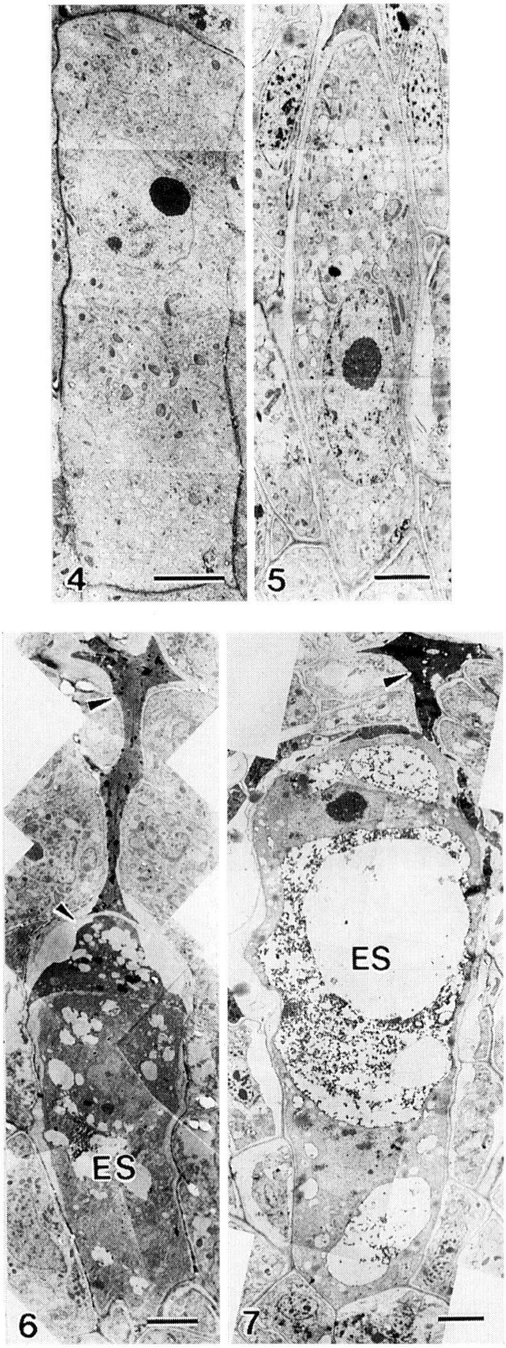

Plate 58 (*Contd.*)

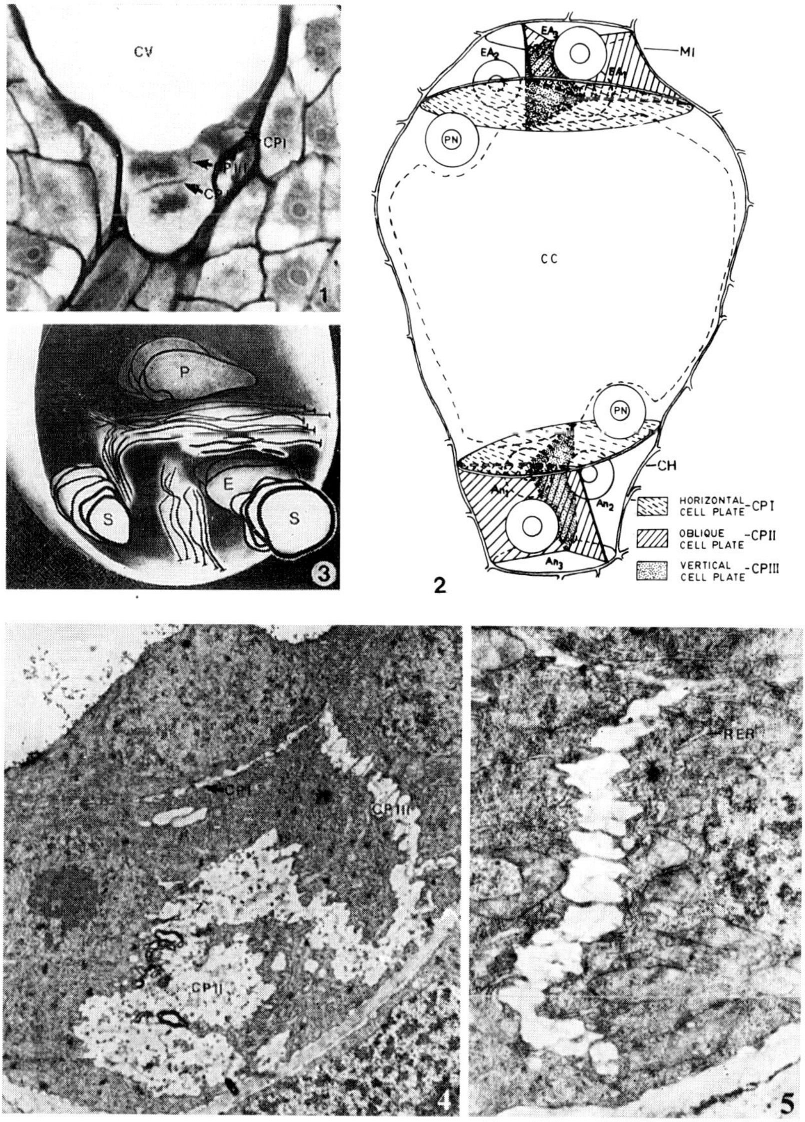

Plate 59: Cytokinesis in embryo sac.

1 – chalazal end of embryo sac at stage of early telophase of third mitotic division; note three very PAS-positive cell plates. Two cell plates (Cpi and CPII) visible between chromatin masses of two dividing nuclei, while third (CP III) perpendicular to first two (CM), thinner, shorter and localizes between them; 2 – 3D-reconstruction of embryo sac after end of cytokinesis; 3 – reconstruction of barley embryo sac at early stage of cell wall formation in egg apparatus; segments of cell walls marked with short strokes connect with embryo sac wall, segments without strokes end in cytoplasm; 4, 5 – chalazal area of *Ranunculus sceleratus* embryo sac (longitudinal section, TEM): 4 – three cell plates (CPI, CPII and CPIII) visible, 5 – near cell plates numerous cisternae of rough ER seen (4 – × 8800; 5 – × 17500); An$_1$, An$_2$, An$_3$ – antipodals, CC – central cell, CH – chalazal area, CP – cell plate, CV – central vacuole, E – egg cell or egg nucleus, EA$_1$, EA$_2$, EA$_3$ – cells of egg apparatus, MI – micropylar area, P, PN – polar nuclei, RER – ribosomal endoplasmic reticulum, S – synergid or synergid nucleus.

1, 2 – after Bhandari and Chitralekha, 1989; 3 – after Cass *et al.*, 1986; 4,5 – after Bhandari, orig.

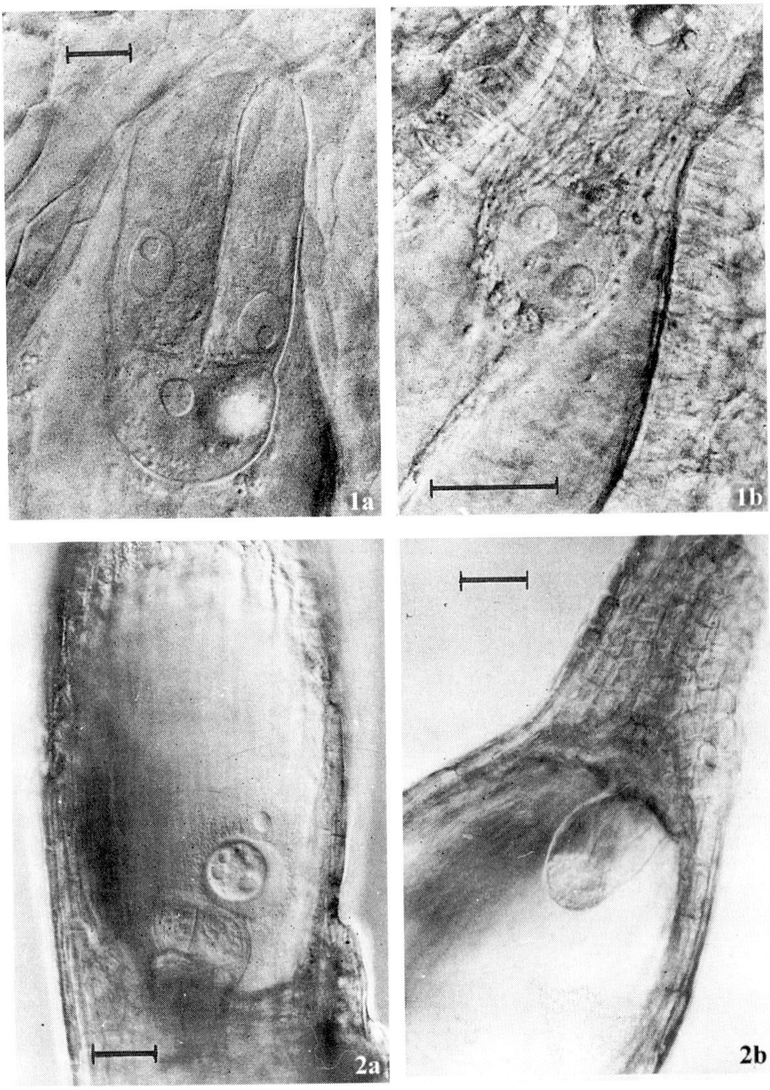

Plate 60: Living embryo sac.

1a – egg apparatus of embryo sac in *Jasione montana* (unfixed material); 1b – part of embryo sac of *Jasione montana* before fertilization with polar nuclei in central part and egg cell in micropylar part; 2a – partially isolated embryo sac of *Galanthus nivalis* during fertilization; secondary nucleus of central cell and sperm nucleus visible in chalazal part besides antipodals; 2b – zygote in partially isolated embryo sac of *Galanthus nivalis*. 3a – population of isolated embryo sacs of *Nicotiana tabacum*, excised by ferment treatment (× 140); 3b – isolated embryo sac with egg, two synergids, central cell and two antipodals (× 410). Scales: 1a – 10 µm, 1b – 50 µm, 2a, 2b – 100 µm.

1a, 1b – after Erdelská; frames from motion picture C 1075/1973, IWF, Göttingen; 2a, 2b – after Erdelská, frames from motion picture C 1463/1982, IWF, Göttingen; 3a, 3b – kindly given by Sun Meng-xiang, Yang Hong-yuan and Zhou Chang.

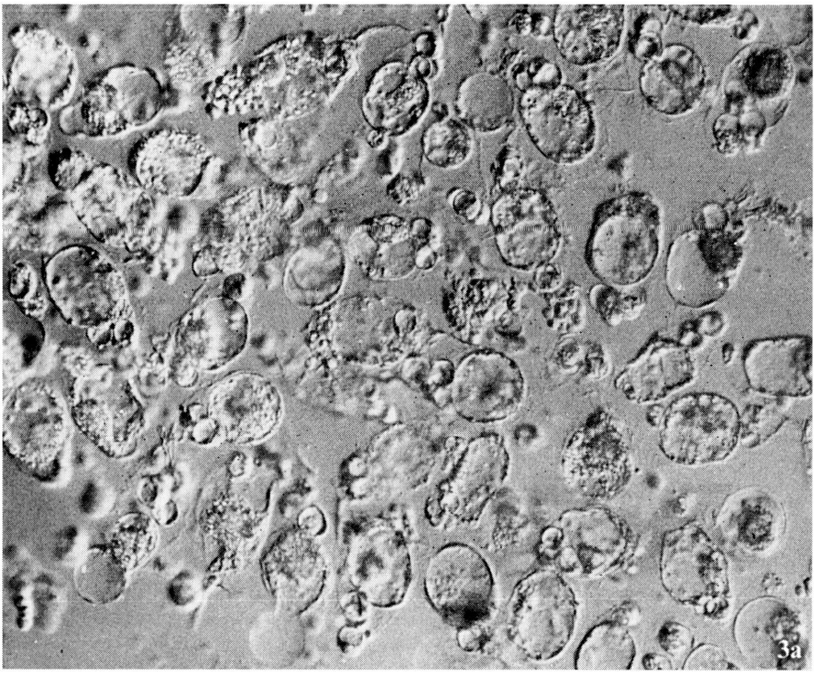

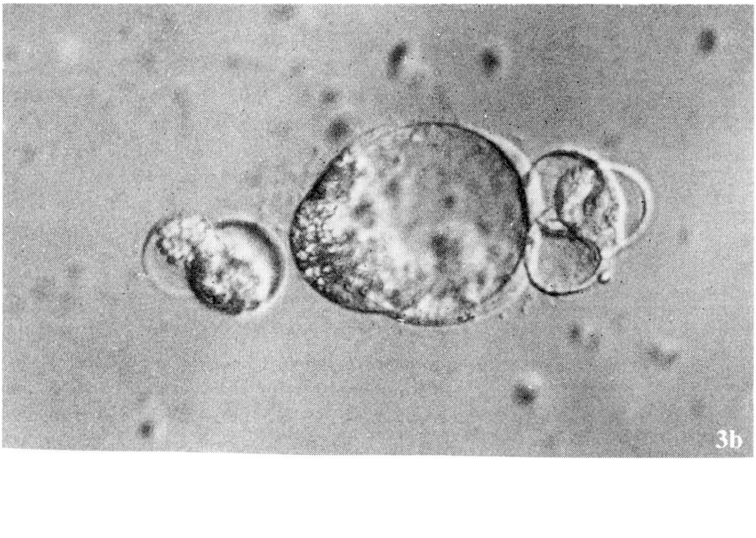

Plate 60 (*Contd.*)

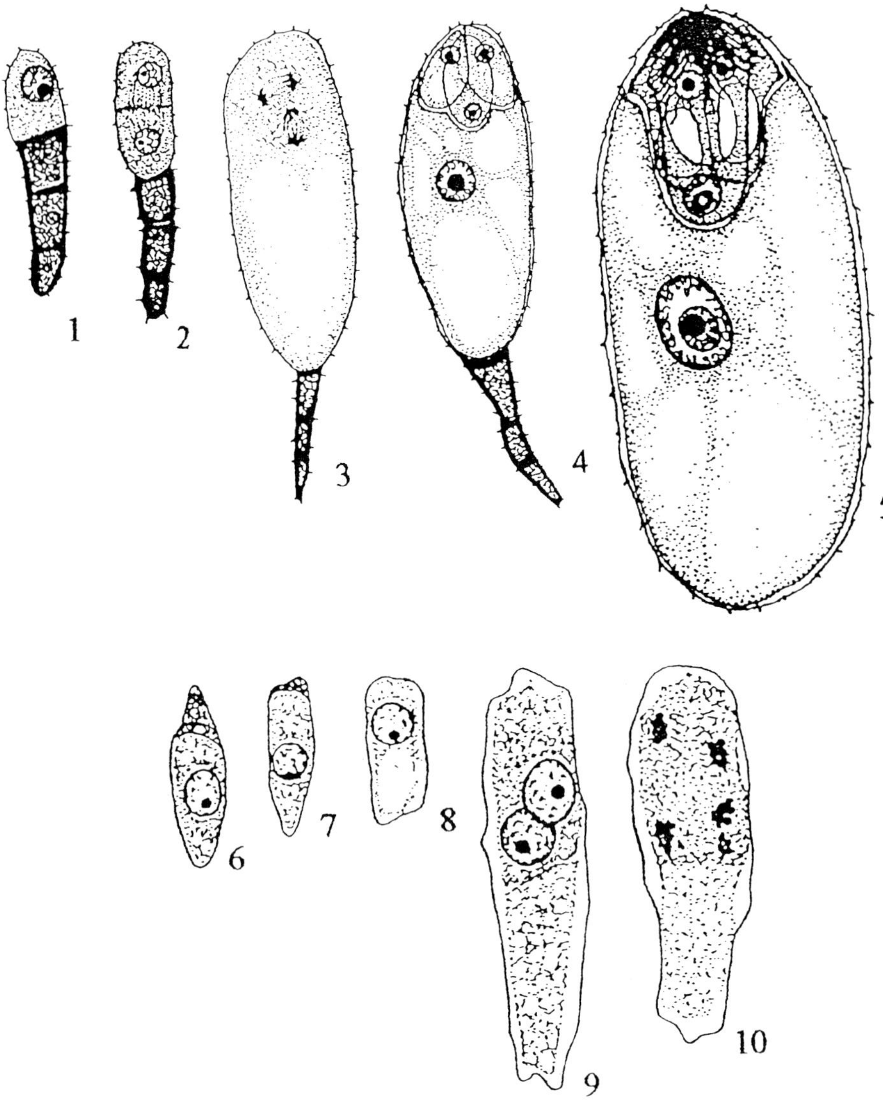

Plate 61: Oenothera-type embryo sac development.

1-58 – development of Oenothera-type embryo sac: according to classic mode—in *Chamaenerion dodonaei* (Onagraceae) (1-5) and Schisandra-variation – in *Schisandra chinensis* (Schisandraceae) (6-22), *Cabomba caroliniana* (Cabombaceae) (23-33), *Nuphar lutea* (Nymphaeaceae) (34-46), *Hydrostemma longifolia* (Barclayaceae) (47-58).

1-5 – after Batygina, 1985a; 6-10 – after Yoschida, 1962; 11-22 – after Swamy, 1964; 23-33 – after Batygina *et al.*, 1982; Titova, 1988; 34-58 – after Winter, 1987.

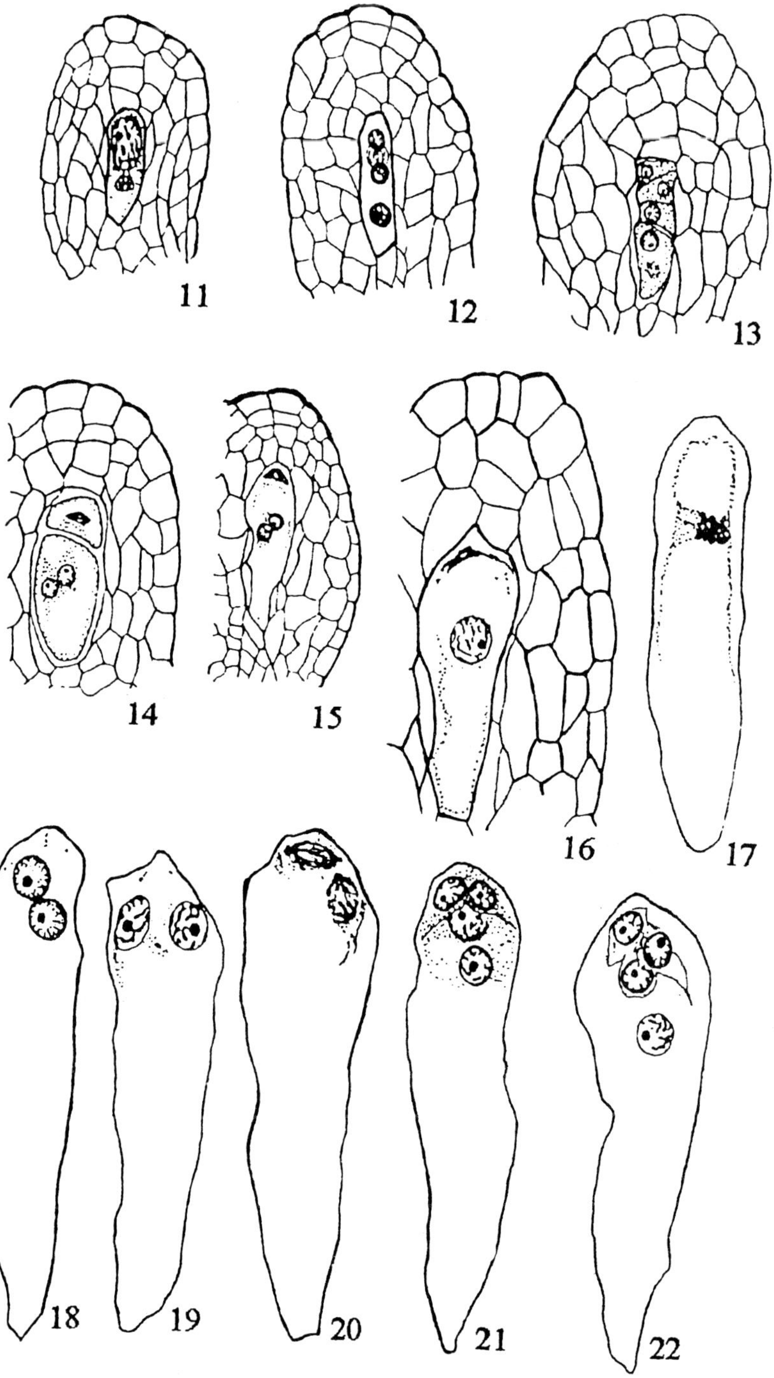

Plate 61 (*Contd.*)

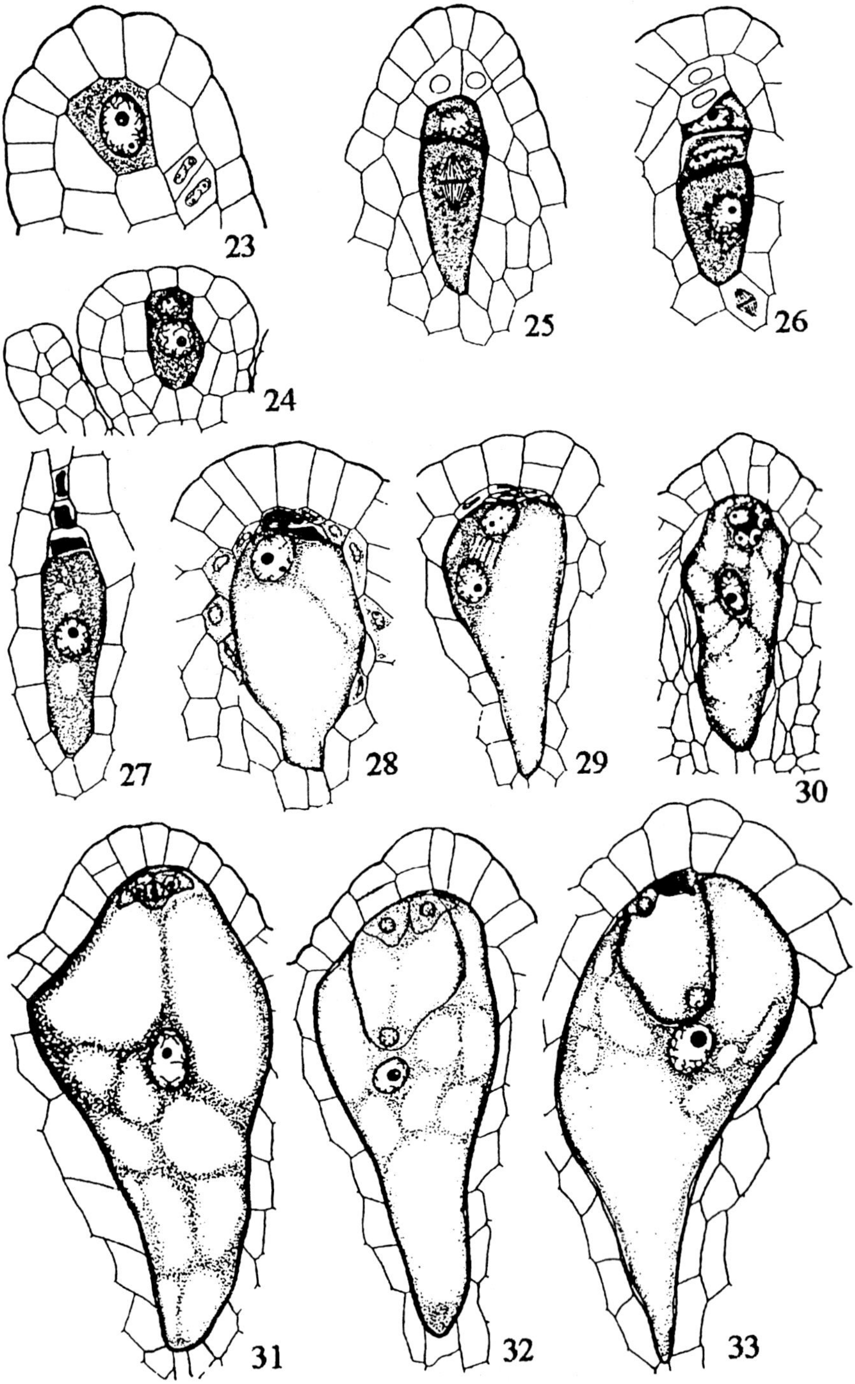

Plate 61 (*Contd.*)

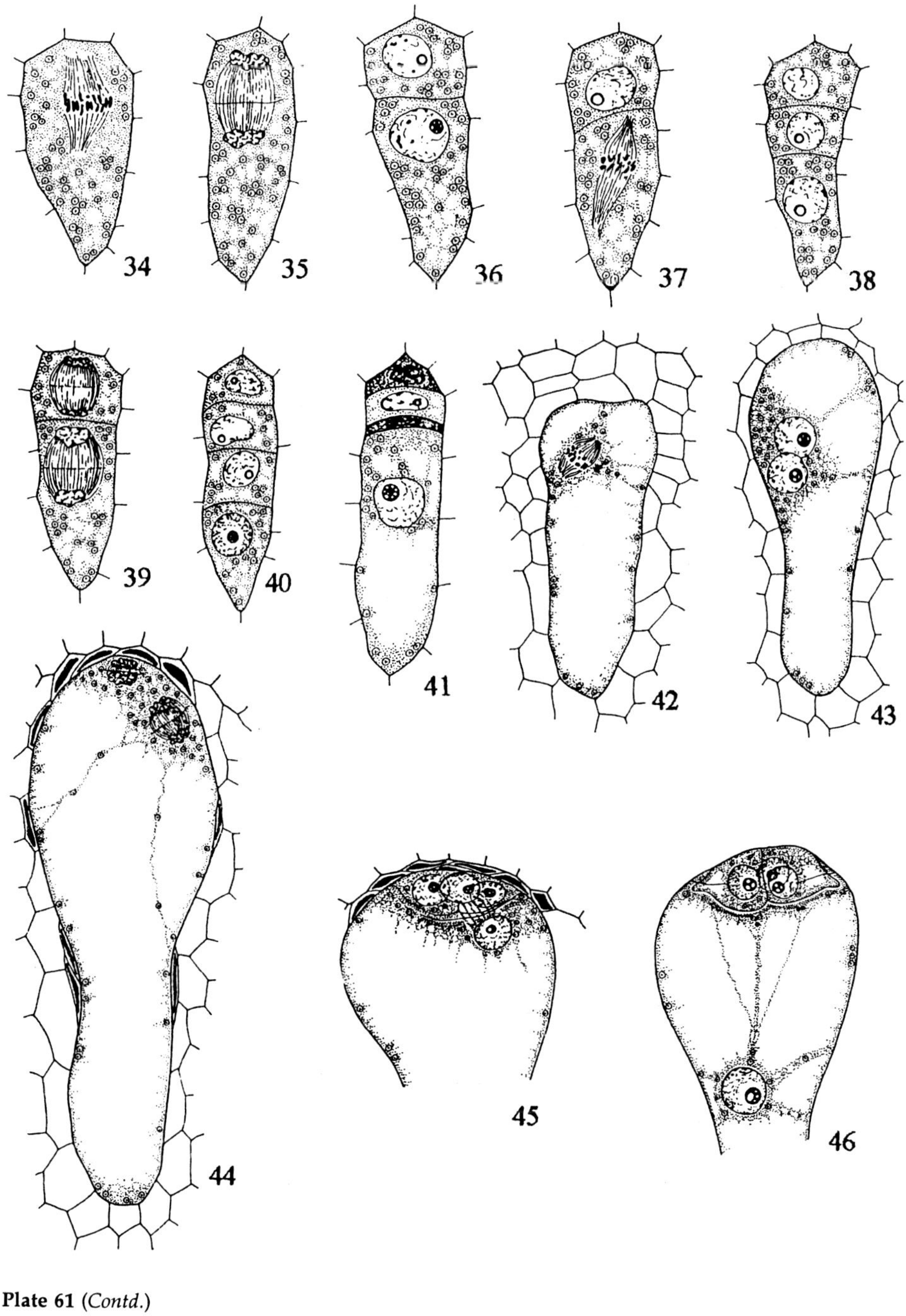

Plate 61 (*Contd.*)

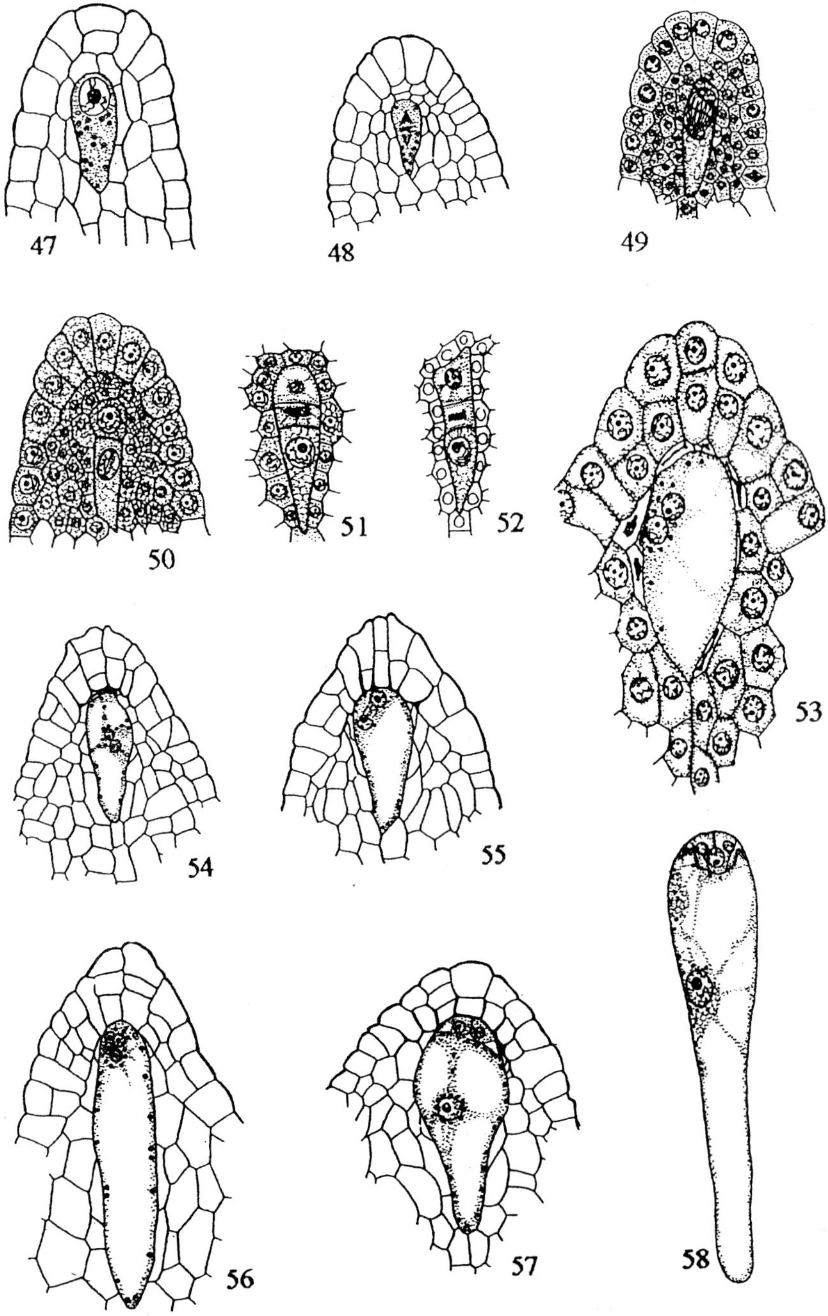

Plate 61 (*Contd.*)

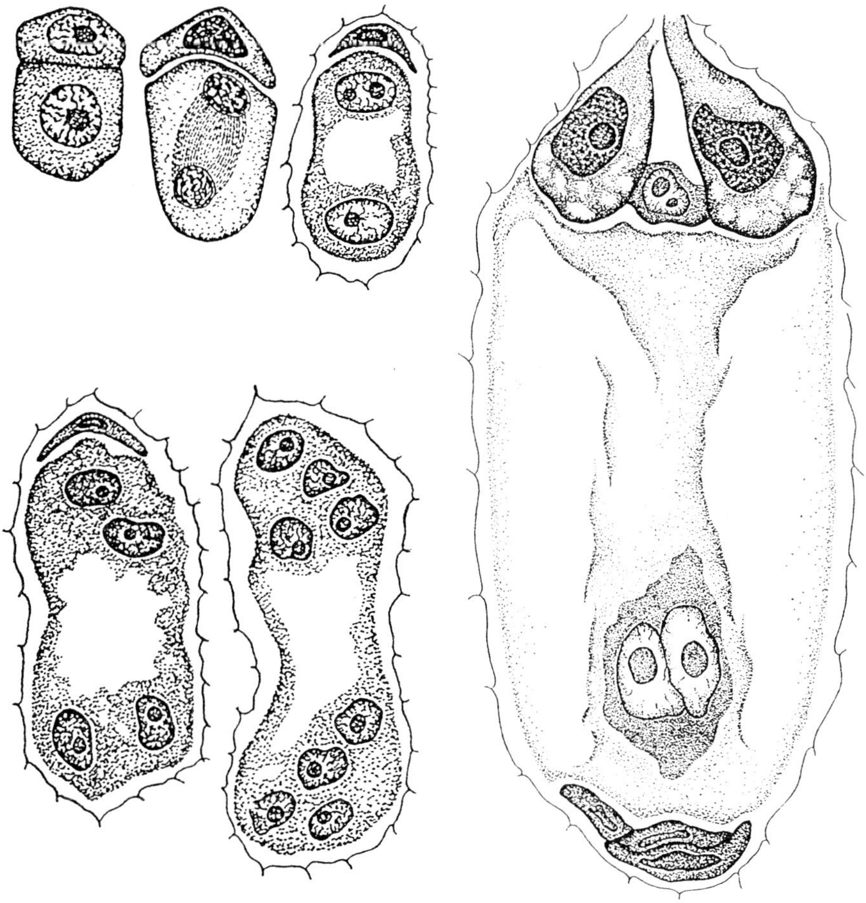

Plate 62: Allium-type embryo sac development.

1-6 – subsequent stages of embryo sac development in *Allium schoenoprasum*: dyad (1); division of dyad lower cell (2); two-, four- and eight-nucleate stages of embryo sac development (3-5); mature embryo sac (6).

1-6 – after Gvaladze, 1976.

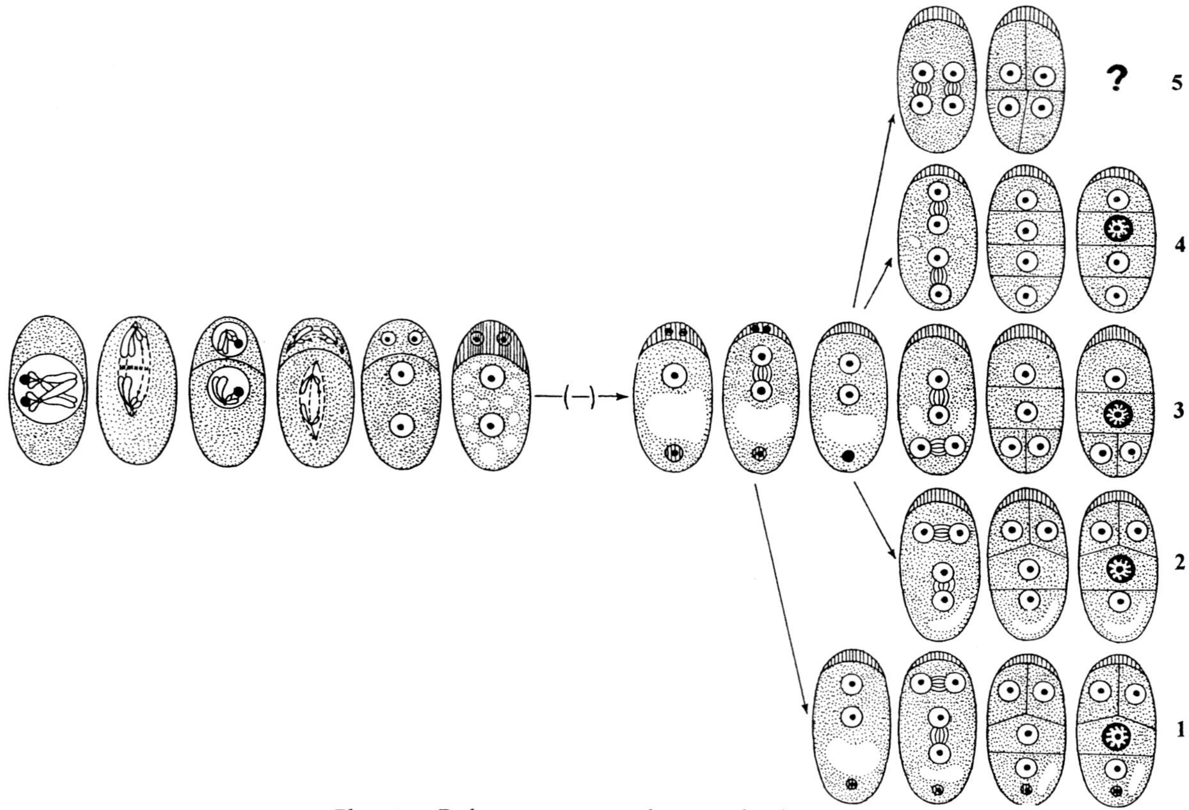

Plate 63: Podostemum-type embryo sac development.

Scheme of formation of different variations of Apinagia – (= Podostemum –) type of embryo sac: 1 – variation with degenerating dyad cell basal nucleus persisting in mature embryo sac; 2-5 – variations of embryo sac in which basal nucleus completely degenerates to end of sac formation (larger shaded nuclei delineate an egg cell in the diagrams).

Modified from Battaglia, 1987b.

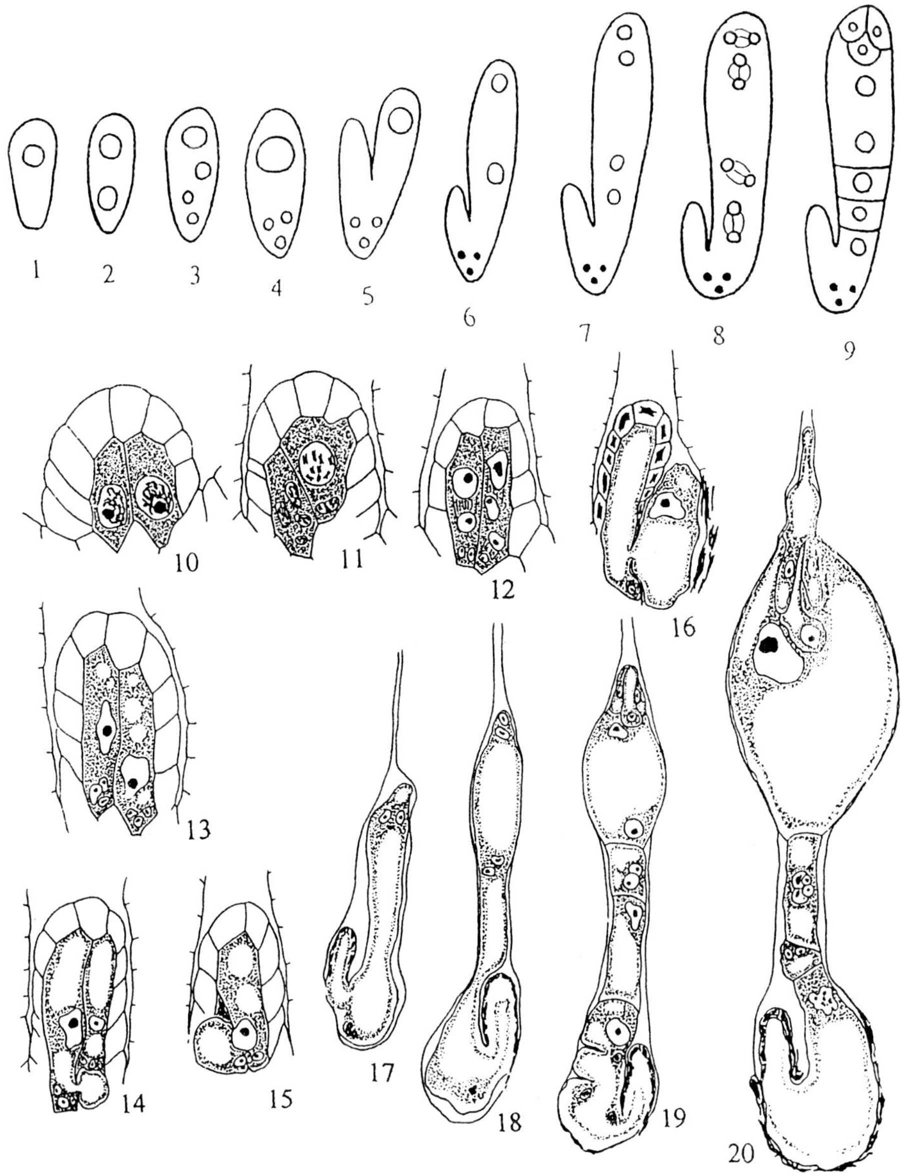

Plate 64: Balsamita-type embryo sac development.

1-9 – sequence of embryo sac development according to Balsamita-type (schematic); 10-20 – megasporogenesis and embryo sac development in *Balsamita vulgaris*: megasporogenesis (10–12), coenocytic tetranuclear megasporocyte with three degenerating chalazal nuclei (13), embryo sac development and formation of chalazal outgrowth (14-18), organized and mature embryo sacs (19, 20) with three-celled egg apparatus, central cell and three antipodals: two multinuclear, the chalazal one with outgrowth contains nucleus and three pycnotic spore nuclei.

1-9 – after Romanov, 1971b; 10-20 – after Fagerlind, 1939c.

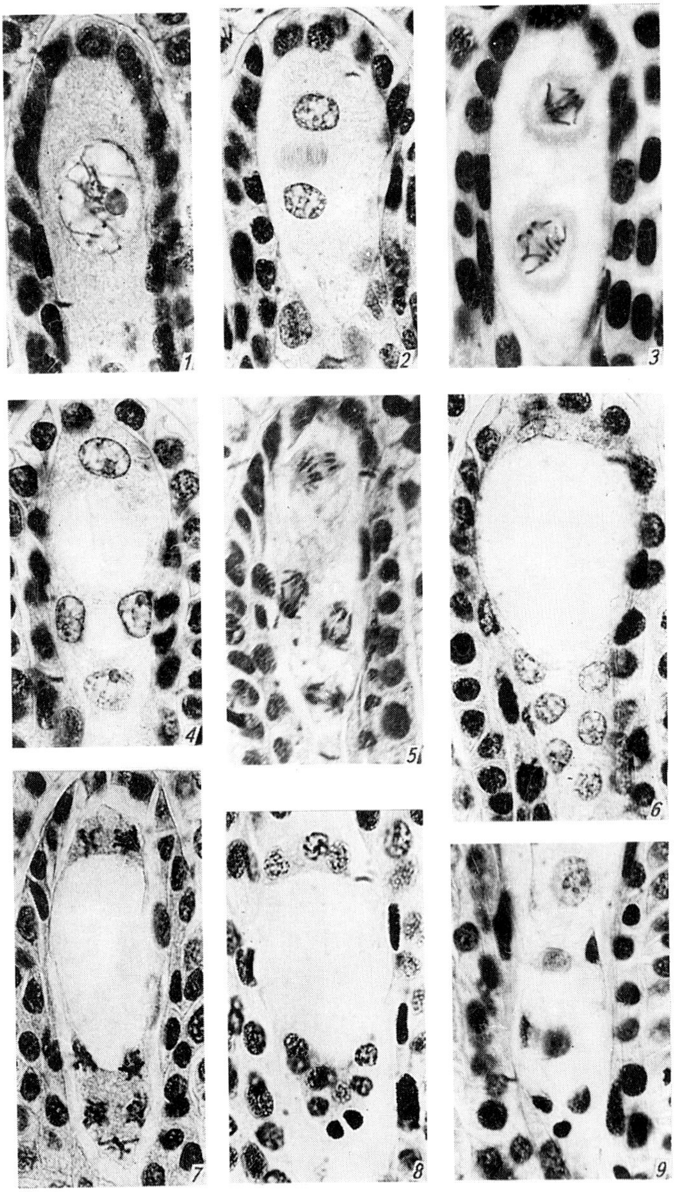

Plate 65: Drusa-type embryo sac development.

1 – megasporocyte; 2 – binucleate coenocyte; 3 – meiosis anaphase II; 4 – megaspore nuclei distribution 1+3; 5 – anaphase of III division; 6 – eight-nucleate embryo sac with 2+6 nuclei distribution; 7 – metaphase of IV division; 8 - late telophase of IV division, increase in depression towards chalazal end; 9 – lower polar nucleus and antipodals in organized embryo sac (1, 7, 8 – *Tulipa korolkovii*; 2, 3, 5, 6, 9 – *T. nitida*; 4 – *T. wilsoniana*).

1-9 – Pechenitsyn, orig.

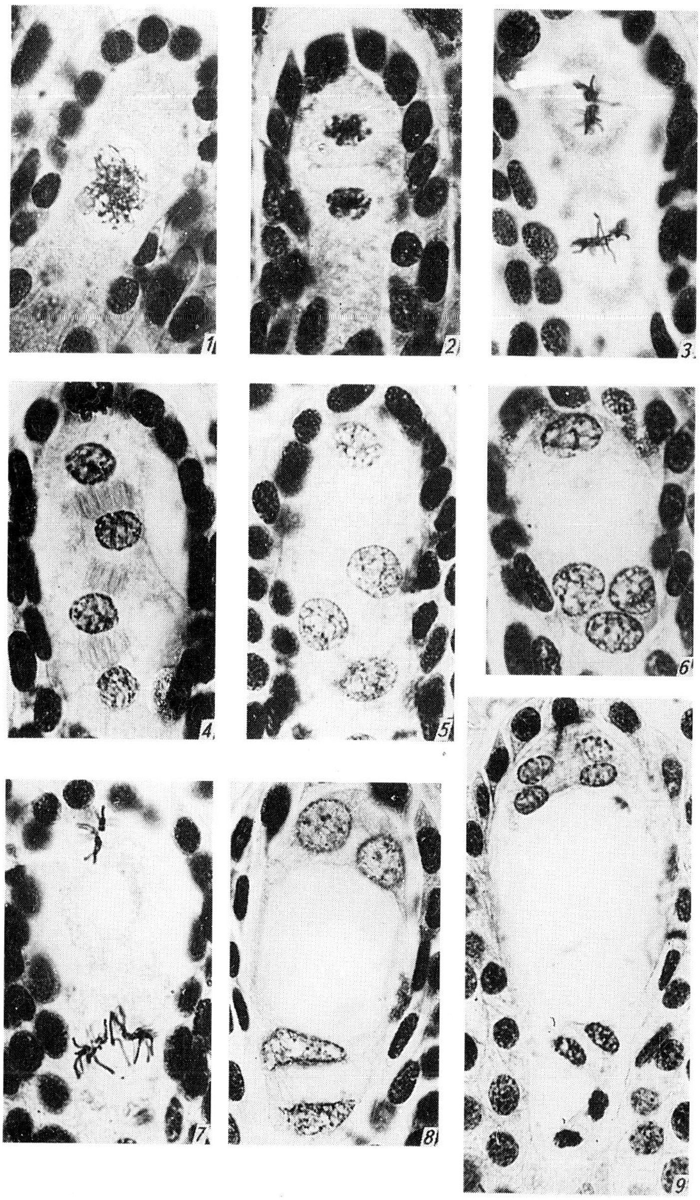

Plate 66: Fritillaria-type embryo sac development.

1 – meiosis prophase I; 2 – meiosis telophase I, cell plate formation; 3 – meiosis metaphase II; 4 – primary four-nucleate embryo sac; 5,6 – megaspore nuclei distribution 1+3; 7 – metaphase of division III; 8 – secondary four-nucleate embryo sac; 9 – late telophase of division IV (1, 2, 6 – *Tulipa anadroma*; 3, 5, 7 – *T. affinis*; 4, 8 – *T. tschimganica*; 9 – *T. julia*).

1-9 – Pechenitsyn, orig.

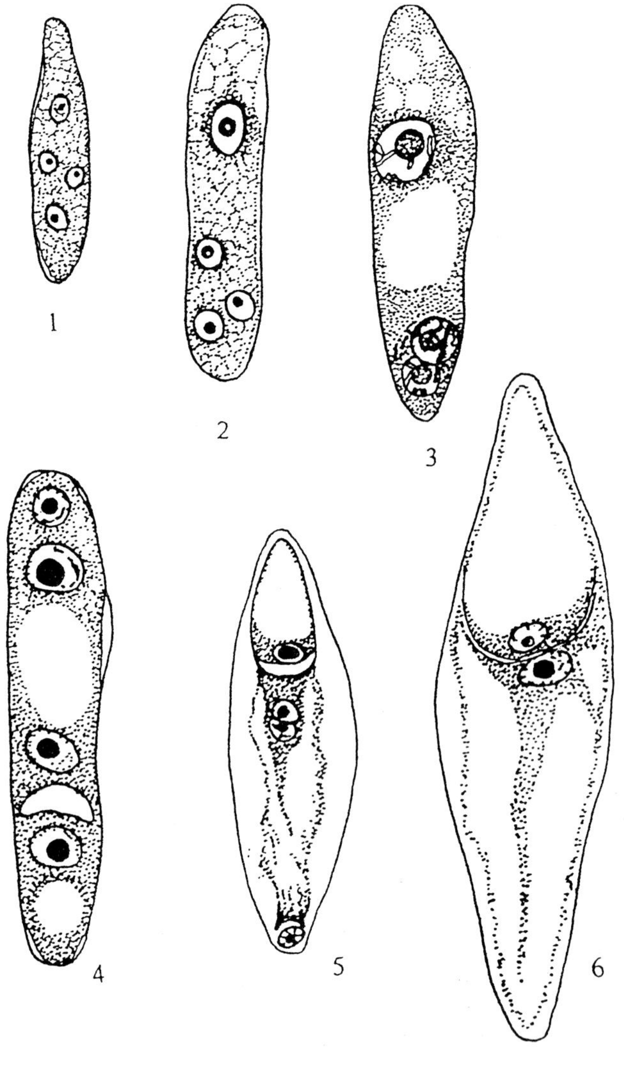

Plate 67: Plumbagella-type embryo sac development.

1-6 – embryo sac development in *Plumbagella micrantha*: 1 – megasporocyte with four megaspore nuclei; 2-4 – formation of two haploid nuclei in micropylar part of embryo sac, fusion of three nuclei and subsequent division with formation of two triploid nuclei in chalazal part; 5 – organized embryo sac with egg cell, central cell and antipodal; 6 – mature embryo sac with egg cell and central cell.

1-4 – after Boyes, 1939; 5, 6 – after Dahlgren, 1916.

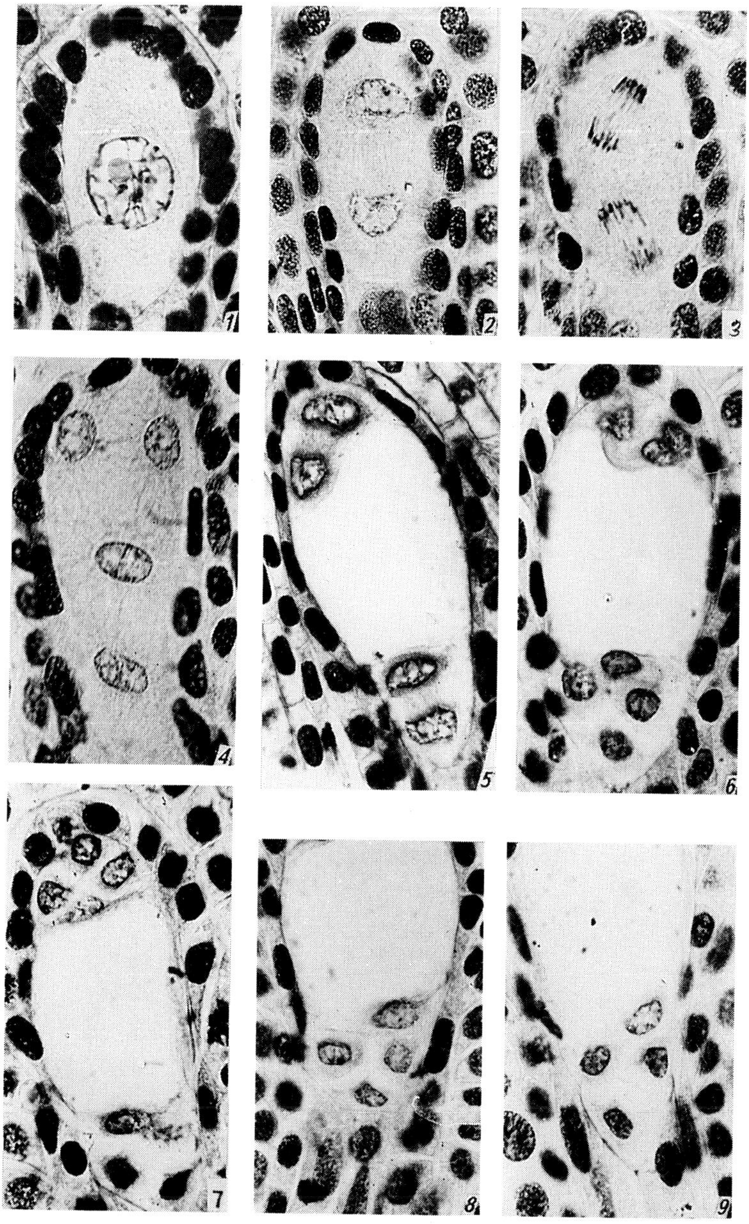

Plate 68: Adoxa-type of embryo sac development.
1 – megasporocyte; 2 – binuclear coenocyte; 3 – meiosis anaphase II; 4 – beginning of megaspore nuclei distribution (2+2); 5 – prophase of division III; 6 – late telophase of division III; 7 – eight-nucleate embryo sac; 8, 9 – chalazal end of organized embryo sac (1, 4 – *Tulipa kolpakovskiana;* 2, 3, 5-9 – *Tulipa tetraphylla*).

1-9 – Pechenitsyn, orig.

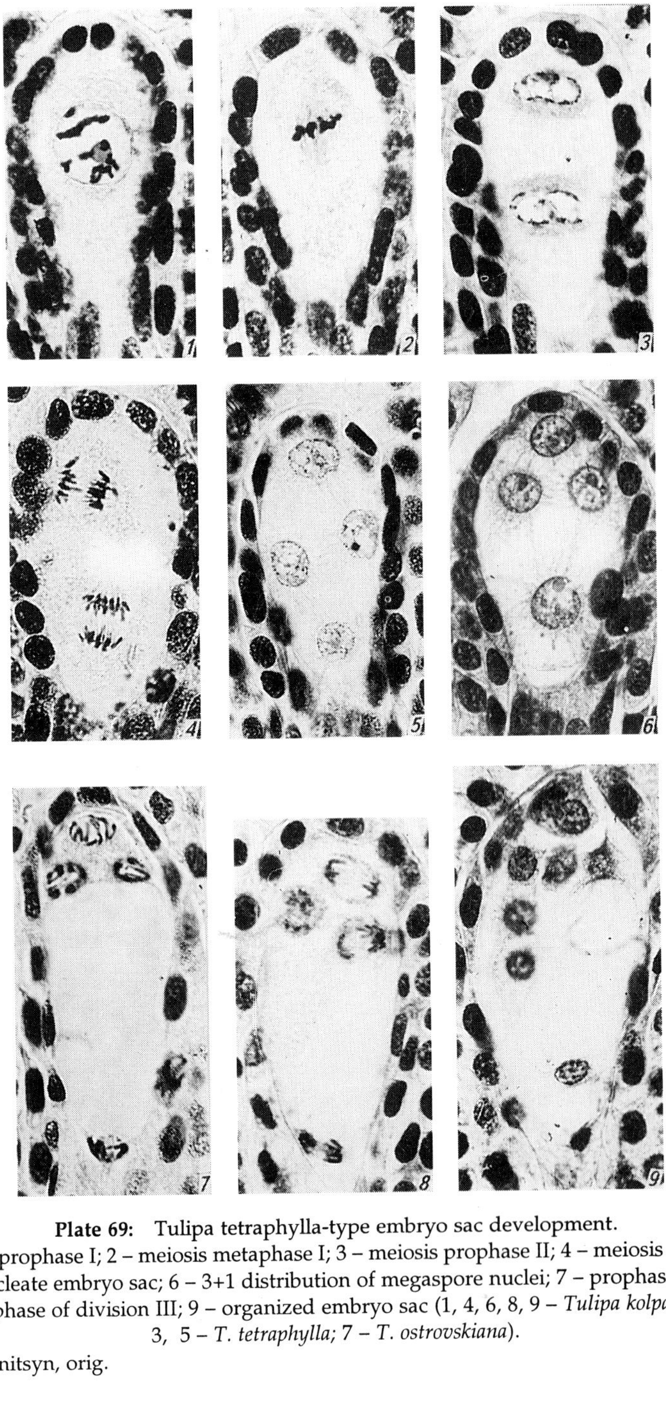

Plate 69: Tulipa tetraphylla-type embryo sac development.

1 – meiosis prophase I; 2 – meiosis metaphase I; 3 – meiosis prophase II; 4 – meiosis anaphase II; 5 – four-nucleate embryo sac; 6 – 3+1 distribution of megaspore nuclei; 7 – prophase of division III; 8 – anaphase of division III; 9 – organized embryo sac (1, 4, 6, 8, 9 – *Tulipa kolpakovskiana*; 2, 3, 5 – *T. tetraphylla*; 7 – *T. ostrovskiana*).

1-9 – Pechenitsyn, orig.

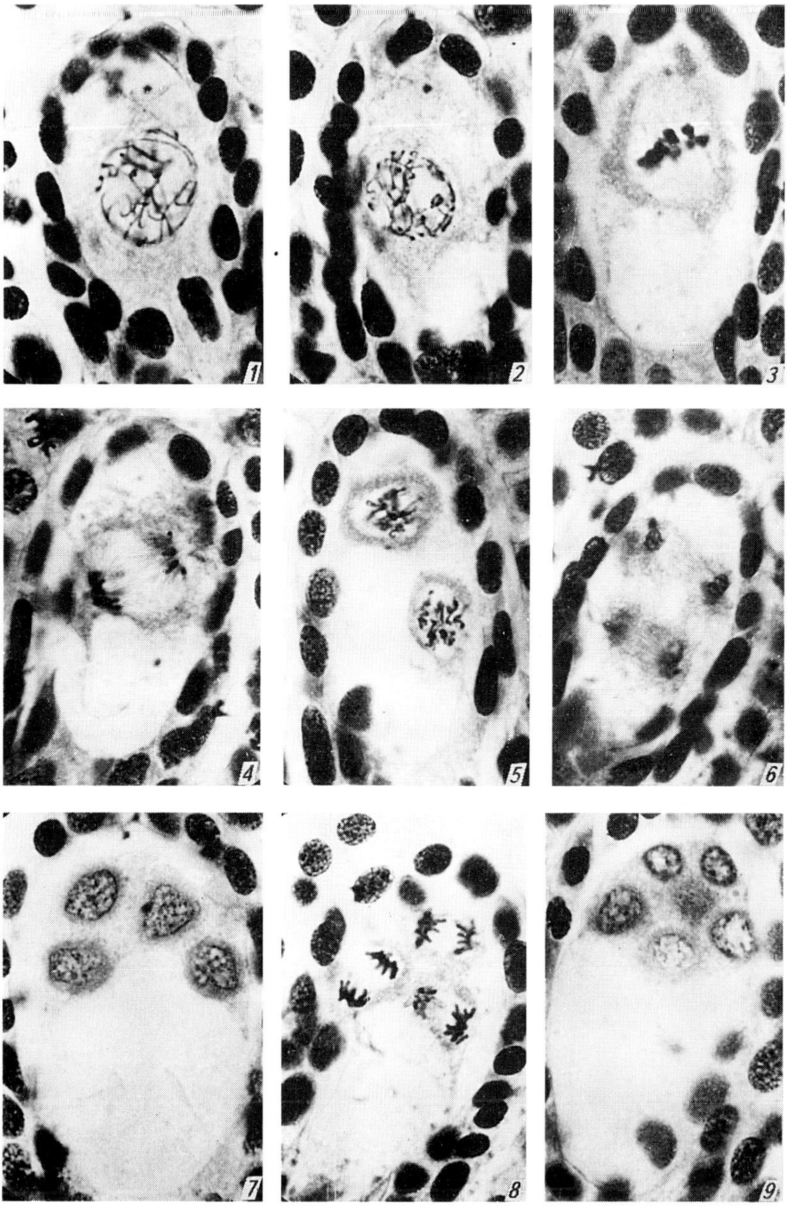

Plate 70: Eriostemones-type embryo sac development.

1-9 – development of embryo sac in *Tulipa neustruevae*: 1 – meiosis prophase I, beginning of cytoplasm vacuolization; 2 – meiosis prophase I, large vacuole seen in chalazal part; 3 – meiosis metaphase I; 4 – meiosis anaphase I; 5 – meiosis metaphase II; 6 – meiosis telophase II; 7 – four-nucleate embryo sac; 8 – anaphase of division III; 9 – just arisen eight-nucleate embryo sac.

1-9 – Pechenitsyn, orig.

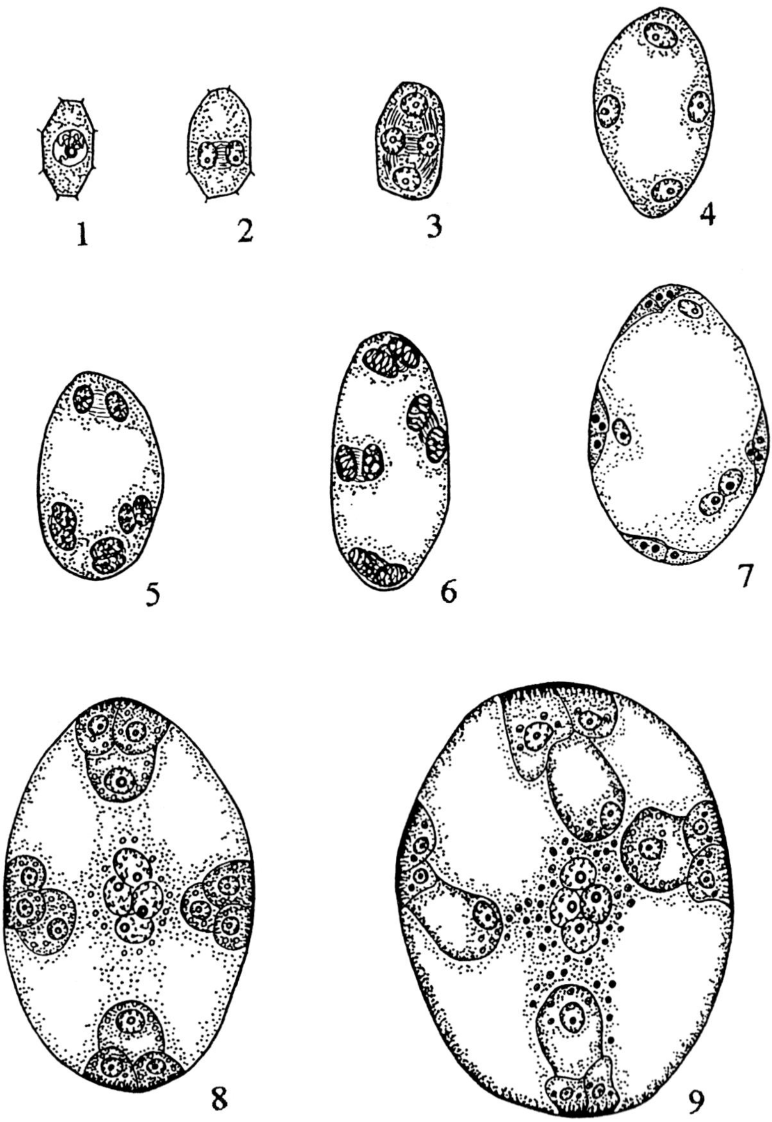

Plate 71: Penaea-type embryo sac development.

1-9 – development of embryo sac in *Biebersteinia multifida*: 1 – megasporocyte; 2 – meiosis I; 3 – meiosis II; 4 – four-nucleate coenocyte; 5,6 – first and second mitotic divisions of nuclei; 7 – organized embryo sac; 8 – maturating embryo sac, 9 – mature embryo sac.

1-9 – after Kamelina and Konnova, orig.

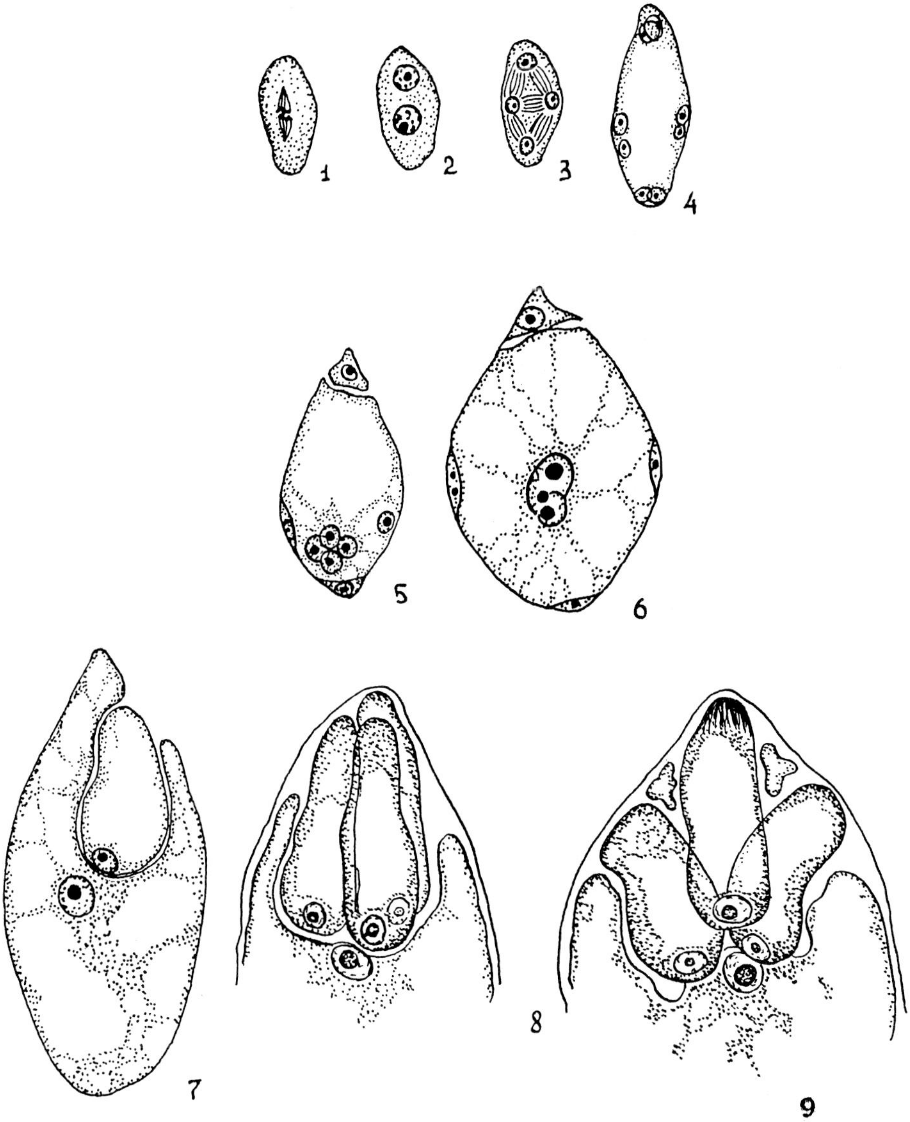

Plate 72: Plumbago-type embryo sac development.
1-7 – subsequential stages of embryo sac development in *Plumbago coccinea*; 8, 9 – *P. europaeum*: monopolar four-celled embryo sac, ready for fertilization (8), non-typical (dispersed) position of egg apparatus cells (9).

1-7 – after Boyes and Battaglia, 1951; 8,9 – after Korobova, 1983.

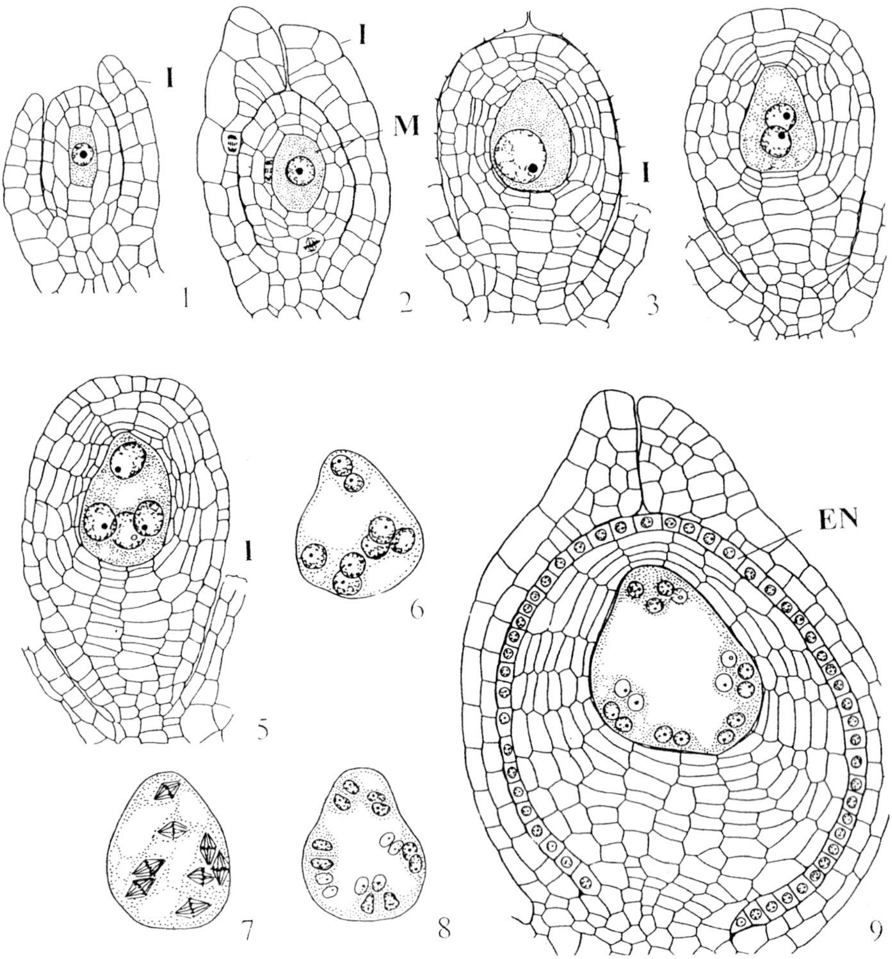

Plate 73: Peperomia-type embryo sac development.
1-5 – megasporogenesis in *Peperomia* species; 6-9 – eight- and 16-nucleate stages of embryo sac development; 10, 11 – embryo sac at moment of fertilization, note contact of male and female nuclei in an egg cell, two synergids (destroyed and intact), and cells bordered with an embryo sac wall (antipodals); in central cell group of polar nuclei and a sperm undergo fusion; 12-14 – embryo sac development; 15-16 – organized embryo sac with two- and three-celled egg apparatus; 17 – non-differentiated egg apparatus of *Peperomia blanda* (× 1500) (1, 9 – *P. obtusifolia*; 2-8, 10, 11, 17 – *P. blanda*; 12 – *P. verticillata*; 13 – *P. magnoliaefolia*; 14 – *P. rubrella*; 15 – *P.* sp.; 16 – *P. maculosa*); E – egg cell, I – integument, M – megasporocyte, EN – epidermis of nucellus, S – synergid, WC – cells bordered with embryo sac wall.

1-11 – after Nikiticheva *et al.*, 1981; 12-16 – after Smirnov, 1982; 17 – after Plyushch, 1992.

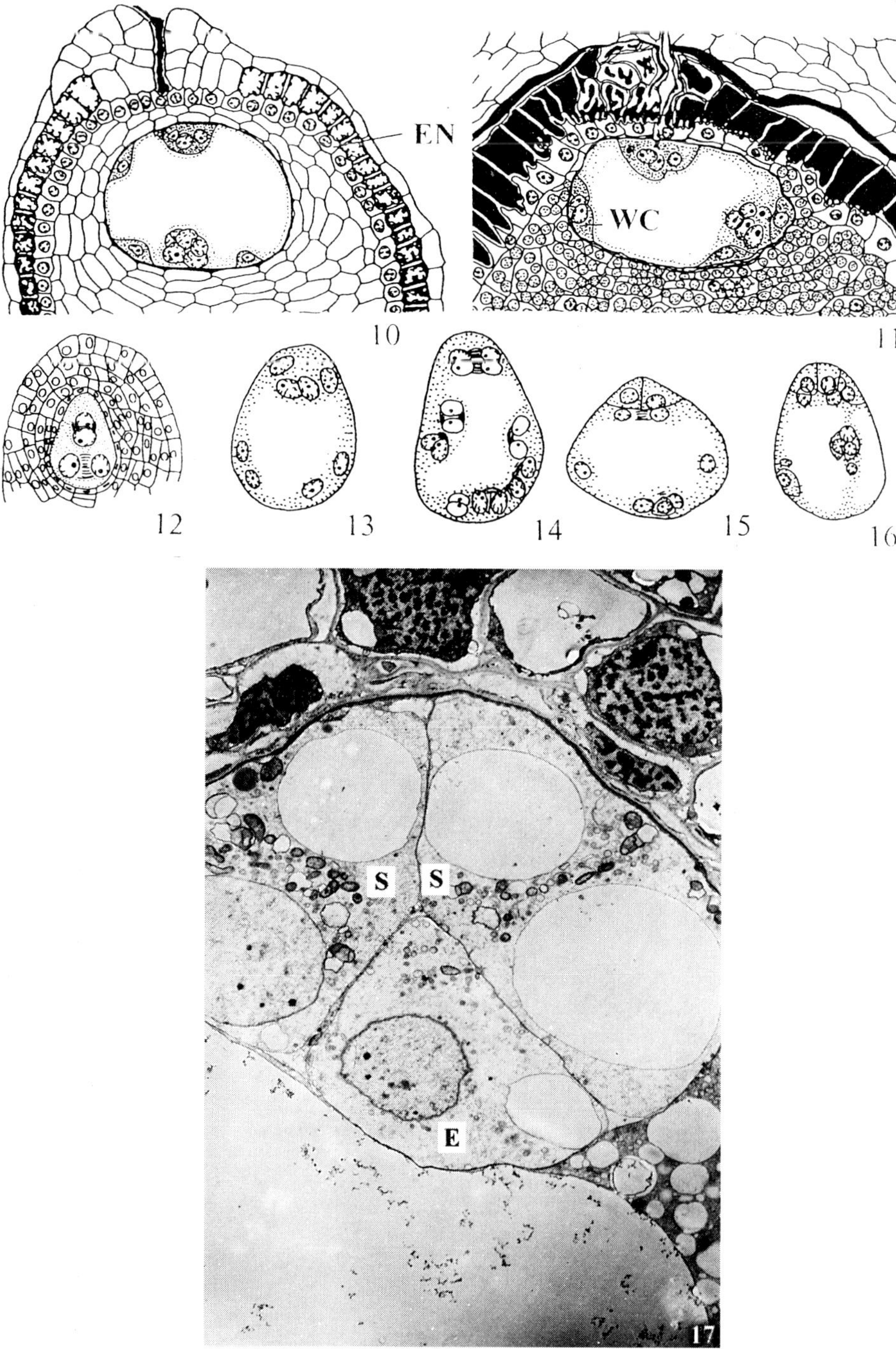

Plate 73 (*Contd.*)

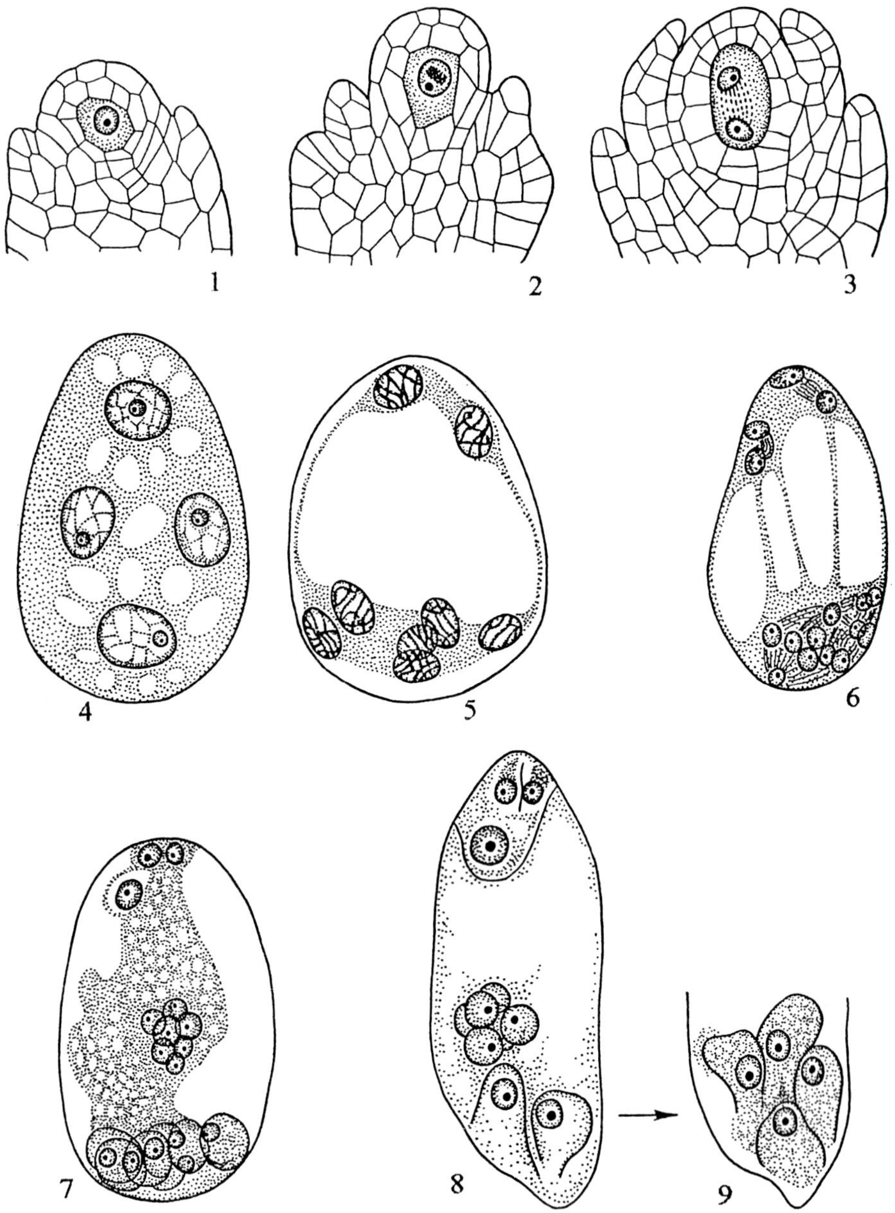

Plate 74: Gunnera-type embryo sac development.
1-7 – megasporogenesis (1-4) and embryo sac development (5-7) in *Gunnera macrophylla* ; 8, 9 – mature embryo sac of *G. chilensis,* containing three-celled egg apparatus, six-celled antipodal complex and central cell with seven nuclei.

1-7 – after Samuels, 1912; 8, 9 – after Modilewski, 1908.

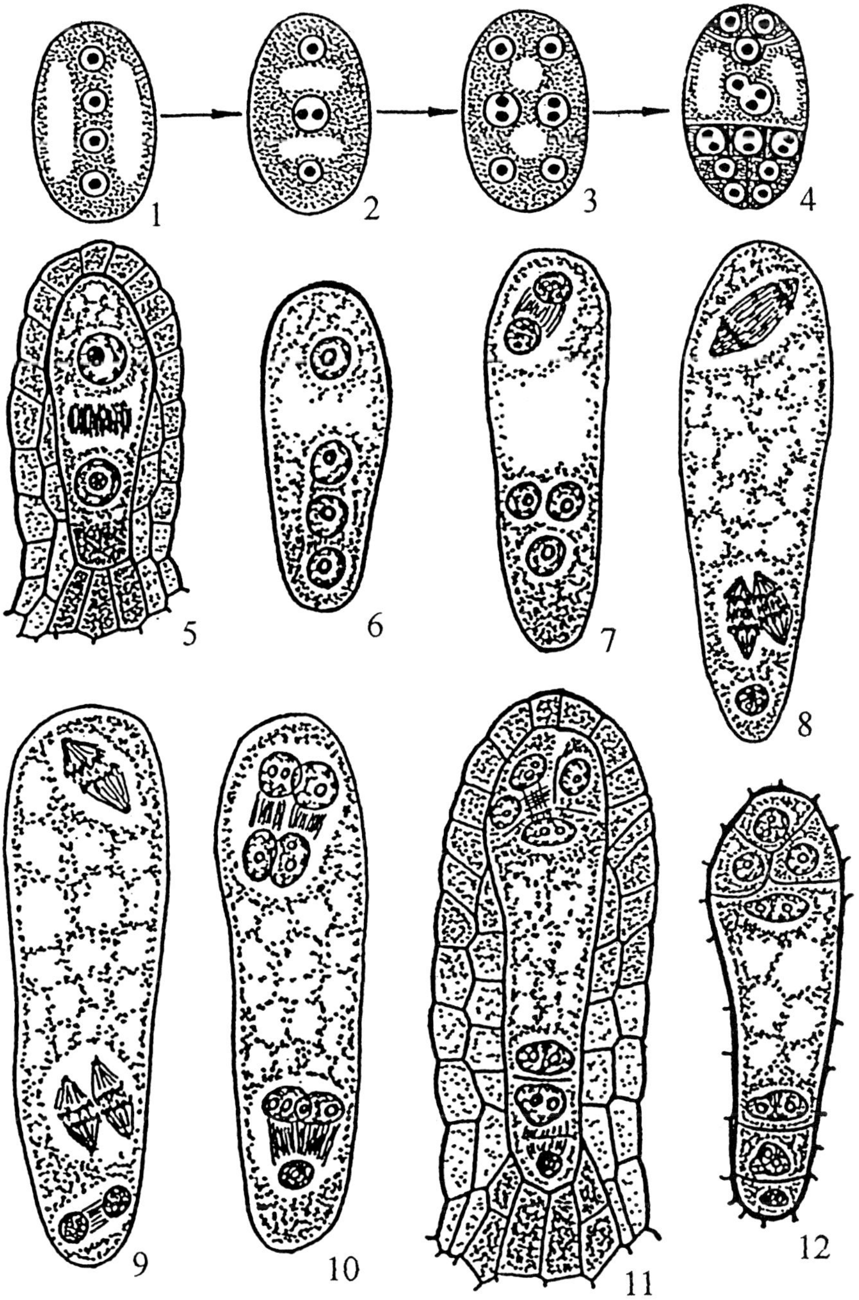

Plate 75: Chrysanthemum-type embryo sac development.
Embryo sac development in *Chrysanthemum cinerariaefolium* (1-4) and *Gagea stipitata* (5-12).
1-4 – after Martinoli, 1939; 5-12 – after Shamrov, 1990c.

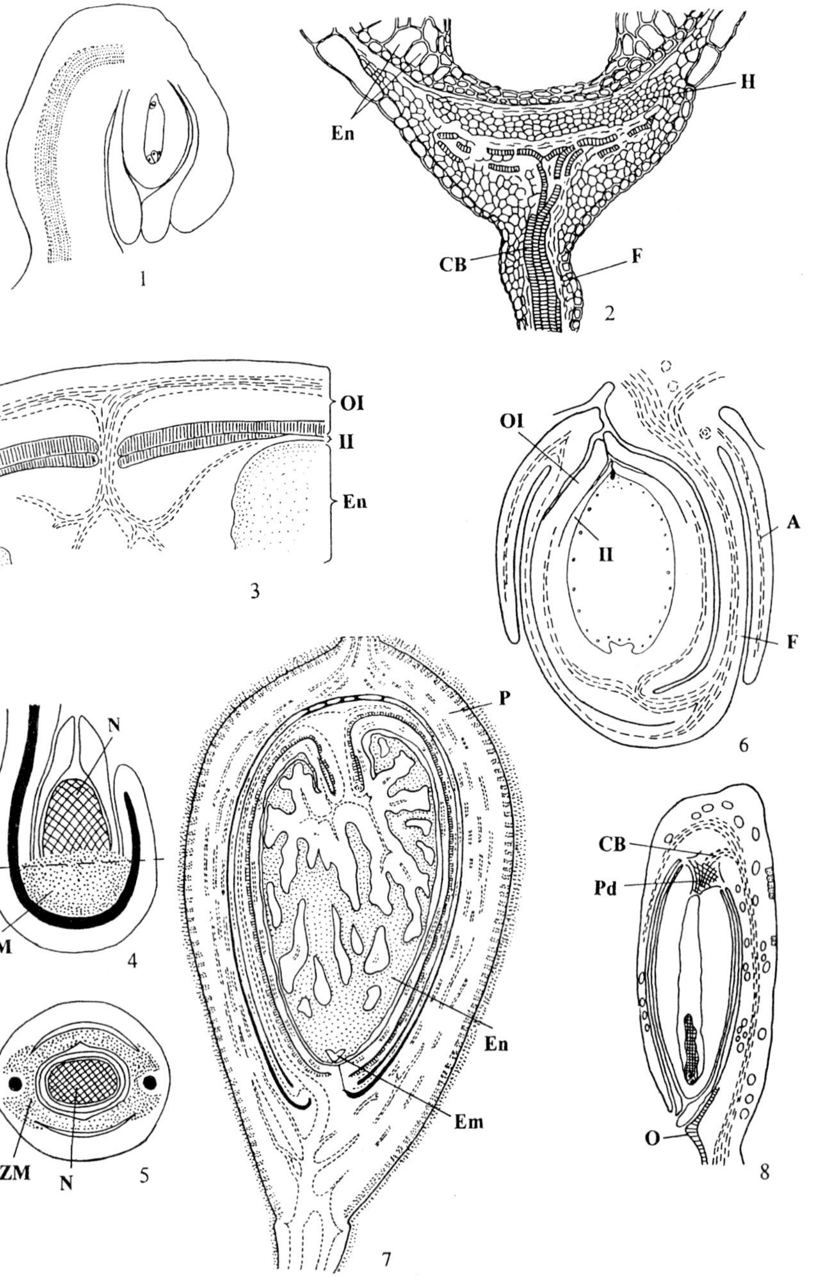

Plate 76: Ovule conductive system.

1-8 – allocation of conductive bundles in ovules of *Scheuchzeria palustris* (1), *Rumex* sp. (2), *Horsfieldia subglobosa* var. *brachiata* (3), *Cananga odorata* (4 – longitudinal section, 5 – cross section), *Myristica fragrans* (6), *Knema laurina* (7), *Magnolia stellata* (8); A – aril, CB – conductive bundle, Em – embryo, En – endosperm, F – funiculus, H – hypostase, II – inner integument, N – nucellus, O – obturator, OI – outer integument, P – pericarp, Pd – podium, ZM – zone of meristematic activity.

1 – after Stenar, 1935; 2 – after Netolitzky, 1926; 3, 7 – after Corner, 1976; 4, 5 – after Periasamy and Swamy, 1961; 6 – after Mametyeva and Kamelina, 1981; 8 – after Kapil and Bhandari, 1964.

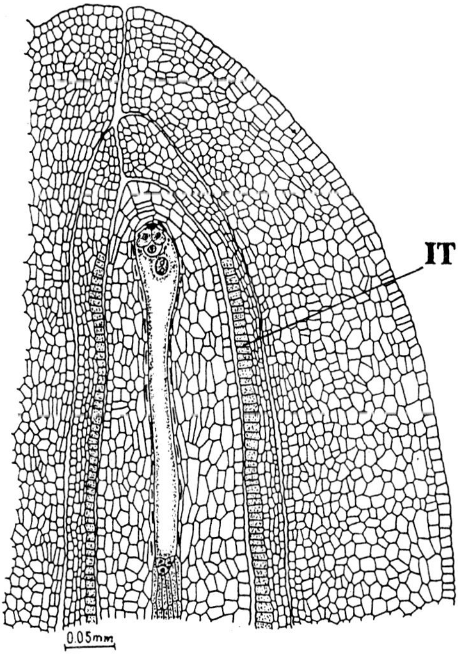

4

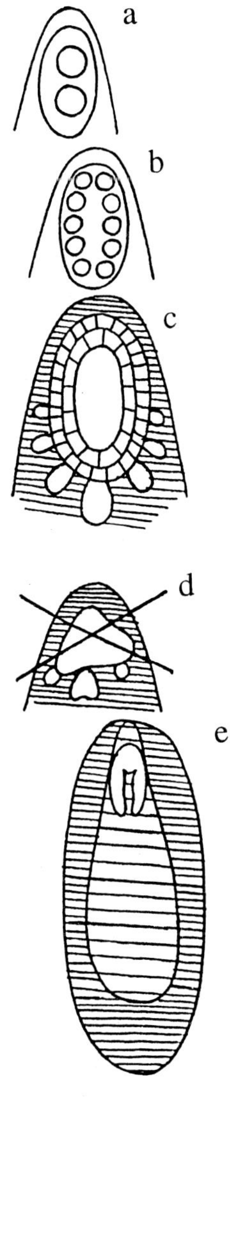

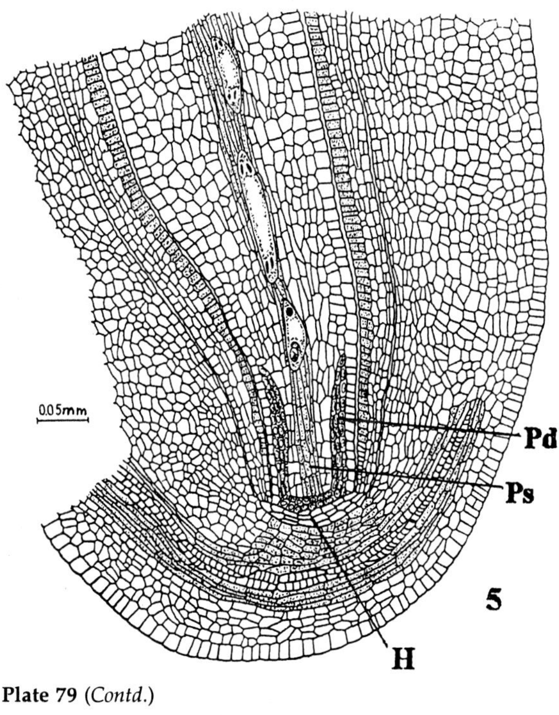

5

6

Plate 79 (*Contd.*)

INDEX